Electromagnetic Technologies in Food Science

Electromagnetic Technologies in Food Science

Edited by

Vicente M. Gómez-López

Universidad Católica San Antonio de Murcia (UCAM)
Murcia, Spain

Rajeev Bhat

Estonian University of Life Sciences (EMÜ)
Tartu, Estonia

WILEY

This edition first published 2022

Registered Offices
John Wiley & Sons, Inc., 111 River Street, Hoboken, NJ 07030, USA
John Wiley & Sons Ltd, The Atrium, Southern Gate, Chichester, West Sussex, PO19 8SQ, UK

Editorial Office
The Atrium, Southern Gate, Chichester, West Sussex, PO19 8SQ, UK

For details of our global editorial offices, customer services, and more information about Wiley products visit us at www.wiley.com.

Wiley also publishes its books in a variety of electronic formats and by print-on-demand. Some content that appears in standard print versions of this book may not be available in other formats.

Library of Congress Cataloging-in-Publication data applied for

Hardback: 9781119759515

Cover Design: Wiley
Cover Images: © AlexVector/Shutterstock, margouillat photo/Shutterstock, Dolgikh Pavel/Shutterstock, Phonkrit Ninchak/Shutterstock, ggw/Shutterstock

Set in 9.5/12.5pt STIXTwoText by Straive, Pondicherry, India
Printed and bound by CPI Group (UK) Ltd, Croydon, CR0 4YY

C9781119759515_221121

Contents

List of Contributors

Maimunah Mohd Ali
Department of Biological and Agricultural Engineering, Faculty of Engineering Universiti Putra Malaysia, Serdang, Selangor, Malaysia

Antonella Andreone
Universidad de Buenos Aires, Facultad de Ciencias Exactas y Naturales, Departamento de Industrias, Buenos Aires, Argentina.

CONICET - Universidad de Buenos Aires, Instituto de Tecnología de Alimentos y Procesos Químicos (ITAPROQ). Pabellón de Industrias, Buenos Aires, Argentina

Rajeev Bhat
ERA-Chair for Food By-products Valorization Technologies (VALORTECH), Estonian University of Life Sciences (EMÜ), Tartu, Estonia, EU

Mysore. L. Bhavya
Food Engineering Department, CSIR-Central Food Technological Research Institute, Mysuru, India

Camelia Berghian-Grosan
National Institute for Research and Development of Isotopic and Molecular Technologies, Cluj-Napoca, Romania

Amalia Conte
Department of Agricultural Sciences, Food and Environment, University of Foggia, Foggia, Italy

Daniel Cozzolino
Centre for Nutrition and Food Sciences, Queensland Alliance for Agriculture and Food Innovation, The University of Queensland, St. Lucia, Brisbane, Queensland, Australia

Matteo A. Del Nobile
Department of Agricultural Sciences, Food and Environment, University of Foggia, Foggia, Italy

Madhuresh Dwivedi
Department of Food Process Engineering, National Institute of Technology Rourkela, Rourkela, Odisha, India

Xuetong Fan
U. S. Department of Agriculture, Agricultural Research Service, Eastern Regional Research Center, Wyndmoor, PA, USA

Antoni Femenias
Applied Mycology Unit, Food Technology Department, University of Lleida, Agrotecnio Center, Lleida, Spain

Daniela Fenoglio
Universidad de Buenos Aires, Facultad de Ciencias Exactas y Naturales, Departamento de Industrias, Buenos Aires, Argentina.

CONICET - Universidad de Buenos Aires, Instituto de Tecnología de Alimentos y Procesos Químicos (ITAPROQ). Pabellón de Industrias, Buenos Aires, Argentina

Mariana Ferrario
Universidad de Buenos Aires, Facultad de Ciencias Exactas y Naturales, Departamento de Industrias, Buenos Aires, Argentina.

CONICET - Universidad de Buenos Aires, Instituto de Tecnología de Alimentos y Procesos Químicos (ITAPROQ). Pabellón de Industrias, Buenos Aires, Argentina

Benny P. George
Electron Beam Processing Section (EBPS), Isotope and Radiation Application Division (IRAD), Bhabha Atomic Research Centre (BARC), Trombay, Mumbai, India

Sunil K. Ghosh
Homi Bhabha National Institute, Anushaktinagar, Mumbai, India

Vicente M. Gómez-López
Cátedra Alimentos para la Salud, UCAM Universidad Católica San Antonio de Murcia, Murcia, Spain

Sandra N. Guerrero
Universidad de Buenos Aires, Facultad de Ciencias Exactas y Naturales, Departamento de Industrias, Buenos Aires. Argentina.

CONICET - Universidad de Buenos Aires, Instituto de Tecnología de Alimentos y Procesos Químicos (ITAPROQ). Pabellón de Industrias, Buenos Aires, Argentina

Norhashila Hashim
Department of Biological and Agricultural Engineering, Faculty of Engineering, Universiti Putra Malaysia, Serdang, Selangor, Malaysia.

SMART Farming Technology Research Centre, Faculty of Engineering, Universiti Putra Malaysia, Serdang, Selangor, Malaysia.

Laura M. Hinds
Food Chemistry and Technology Department, Teagasc Food Research Centre, Dublin, Ireland.

School of Biosystems and Food Engineering, University College Dublin, Dublin, Ireland

Rifna E. Jerome
Department of Food Process Engineering, National Institute of Technology Rourkela, Rourkela, Odisha, India

Shunshan Jiao
Department of Food Science and Technology, School of Agriculture and Biology, Shanghai Jiao Tong University, Shanghai, China

Yvan Llave
Department of Food Science and Technology, Tokyo University of Marine Science and Technology, Tokyo, Japan

Dana Alina Magdas
National Institute for Research and Development of Isotopic and Molecular Technologies, Cluj-Napoca, Romania

Sonia Marín
Applied Mycology Unit, Food Technology Department, University of Lleida, Agrotecnio Center, Lleida, Spain

Bibhuti B. Mishra
Homi Bhabha National Institute, Food Technology Division, Bhabha Atomic Research Centre, Mumbai, INDIA

Brendan A. Niemira
U. S. Department of Agriculture, Agricultural Research Service, Eastern Regional Research Center, Wyndmoor, PA, USA

Colm P. O'Donnell
School of Biosystems and Food Engineering, University College Dublin, Dublin, Ireland

José A. Pellicer
Molecular Recognition and Encapsulation Research Group (REM), Health Sciences Department, Universidad Católica de Murcia (UCAM), Campus de los Jerónimos, Guadalupe, Spain

Francesco E. Ricciardi
Department of Agricultural Sciences, Food and Environment, University of Foggia, Foggia, Italy

Noboru Sakai
Department of Food Science and Technology, Tokyo University of Marine Science and Technology, Tokyo, Japan

Bhaskar Sanyal
Food Technology Division, Bhabha Atomic Research Centre, Mumbai, India
Homi Bhabha National Institute Anushaktinagar, Mumbai, India

Eva Salazar
Departamento de Ciencia y Tecnologıa de Alimentos, UCAM Universidad Católica San Antonio de Murcia, Murcia, Spain

Marcela Schenk
Universidad de Buenos Aires, Facultad de Ciencias Exactas y Naturales, Departamento de Industrias, Buenos Aires, Argentina.

CONICET - Universidad de Buenos Aires, Instituto de Tecnología de Alimentos y Procesos Químicos (ITAPROQ). Pabellón de Industrias, Buenos Aires, Argentina

Brijesh K. Tiwari
Food Chemistry and Technology Department, Teagasc Food Research Centre, Dublin, Ireland.

School of Biosystems and Food Engineering, University College Dublin, Dublin, Ireland

Prasad S. Variyar
Homi Bhabha National Institute, Food Technology Division, Bhabha Atomic Research Centre, Mumbai, India

Michael Vollmer
Department of Engineering, University of Applied Sciences Brandenburg, Brandenburg an der Havel, Germany

Shaojin Wang
College of Mechanical and Electronic Engineering, Northwestern A&F University, Yangling, Shaanxi, China

Foreword

Scientific knowledge into the nature of electricity and magnetism was mainly developed throughout the eighteenth and nineteenth centuries through the work of world-famous scientists such as Coulomb, Ampère, Faraday, and Maxwell. Since then, the types of electromagnetic (EM) energy are broadly classified into the six main classes of electromagnetic radiation such as

1) Gamma radiation and X-ray radiation
2) Ultraviolet radiation
3) Visible light
4) Infrared radiation
5) Microwave radiation
6) Radio waves.

They are characterized by wavelength, frequency, and photon energy, which are distinctive for the type of radiation and result in different ways of interaction with the matter.

With the growth of scientific knowledge and understanding the specific features of interaction of EM energy with biological and organic matter, and the uniqueness of its effects in foods, these EM technologies found broad applications in food processing and food science. The first applications of radiation for food treatment were tested in the beginning of twentieth century, followed by microwave, infrared, and radiofrequency heating discovered in 1930–1950, and later at the end of twentieth century when light-based technologies found their own niche in food applications. Compared to traditional heat processing that was introduced in the beginning of nineteenth century, the electromagnetic technologies have much shorter history of exploration with limited commercialization and often are called emerging technologies.

The book "*Electromagnetic Technologies in Food Science*" explores two important aspects of the applications of EM technologies for foods. Firstly, the book offers the vision on how different properties of EM energies can benefit food processing and secondly how they can be used as specific tools to discover unique attributes of foods.

Additionally, the aim of this book is to provide a hybrid of advanced theoretical knowledge and practice of utilizing the EM spectrum in relatively new methods in food technology and food processing arena. To address this task, the book is focused on providing recent updates and technological challenges incurred in the agri-food technological arena. A range of topics have been covered, which are written by leading experts in the field. Some of the topics in this book cover: Physics of the electromagnetic spectrum, Dosimetry methods, Gamma irradiation, Electron beams, X-rays, UV-light, Visible light, Infrared radiation, Pulsed light, Radio waves, and Microwaves.

Though some of the topics addressed in this book can be found in other publications, the editors of this monography have envisioned that this book can serve as a single source of information on existing food technologies dealing with EM spectrum. A few chapters of the book cover a part of the concept that has been referred to as the "emerging processing technologies" or "non-thermal" or "mild" heating methods that are related to improving food safety, preservation, and improved food quality and healthier foods. Another concept of application of EM spectrum has been advantageous in developing quality control methods such as hyperspectral imaging that allows the control of continuous fast-speed processes within the food industry. Besides, few chapters also cover latest updates on Infrared spectroscopy, Visible light imaging, Hyperspectral imaging, Raman spectroscopy, and future challenges of employing electromagnetic spectrum in food technological applications. These methods are routinely being used in food industries that are driven by the market to produce safer food at a faster pace to fulfil the increasing demand of the world population.

The uniqueness of these two concepts of the book makes it difficult to trace back to similar publications. The perception of the book is to provide a single information source for readers interested in the use of these methods based on the EM spectrum as applied in food science and technology. Also, the objective of the authors and editors is that this book can be used as an important core text for graduate students, researchers, agri-food scientists, food industry personnel and other closely related professionals/researchers, policy makers, and to the Governmental and non-governmental organizations involved in establishing guidelines toward ensuring sustainable food production. In addition, it is expected that this book will become a valuable addition to all of the agri-food based scientific community.

It is foreseen that this book will become a valuable addition to all of the currently marketed books on food preservation/processing and can be an excellent resource for different categories of researchers in academia and industry. High level of appreciation goes to internationally renowned subject experts: Vicente Manuel Gómez-López and Rajeev Bhat who were able to bring out this type of useful book for the entire fraternity of personnel involved in the field of agri-food science and technology.

Tatiana Koutchma, PhD
Research Scientist, Agriculture and Agri-Food Canada/Government of Canada;
Past-Chair of Nonthermal Processing Division, Institute of Food Technologies (IFT);
Canada Ambassador, Global Harmonization Initiative and
Associate Editor [Critical Reviews in Food Science and Food Safety (IFT); Journal of Food Process Engineering (Wiley): Food Science and Technology International (SAGE)]

Preface

In the past few decades, the agri-food industry has witnessed evolution of a range of novel food processing techniques. This included both non-thermal and thermal methods that have wide industrial applications. Of late, technological advancements coupled with intense scientific integrities have led to a wide array of interesting scientific evidences being generated on the actual and potential applications, advantages and disadvantages of employing these methods. All of these methods have a fundamental feature in common, which includes the application of different portions of the electromagnetic spectrum. Furthermore, they also share this feature with many food analysis methods.

The efficiency of using these technologies depends totally on the clear understanding of the principles underlying on their reliability and operational principles. In this sense, this book aims to enhance the comprehension of various methods used in the field of food science and technology, which are based on the entire electromagnetic spectrum. The book contents are organized into four parts: the very first chapter provides an easy-to-grasp overview of the physics of the electromagnetic spectrum, which is designed to be understandable for general scientific audience and non-experts in physics. This is followed by the application of the electromagnetic spectrum to non-thermal food processing, which includes gamma irradiation, electron beams, X-rays, UV light, visible light, and pulsed light. Then, advanced heating methods based on the electromagnetic spectrum are covered, including infrared radiation, microwaves, and radiofrequency. These chapters discuss on the physics of working mechanism of these methods as well as case study examples of their current and potential applications. Further, the last part of the book deals with analytical techniques for quality control, and this is based on the electromagnetic spectrum like that of infrared spectroscopy, Raman spectroscopy, visible light imaging, and hyperspectral imaging.

As editors, we believe that the unique approach adopted in this book will help to view these methods as derived from a common root as well as to distinguish their identity, besides understanding the differences among them. In addition, all the book chapters open up space for more coherent studies aimed toward their optimum use in food processing and food analysis.

Written by a selected group of international experts, this book can be a key reference material for all the personnel who are dealing with food preservation and novel food processing techniques. As editors, we sincerely thank all the distinguished experts who have placed and shared their expertise view in this book. Further, we are highly indebted to

Miss. Kerry Powell, Miss. Rebecca Ralf, and the entire Wiley Blackwell team members involved in this book production. Besides, both of us are grateful to our respective University authorities for the continuous encouragement and support provided. Rajeev would like to specifically thank Prof. Ülle Jaakma (Vice-Rector for Research) and Prof. Toomas Tiirats (Director of Institute of Veterinary Medicine and Animal Sciences), Estonian University of Life Sciences, who have been supportive in all of his initiatives. Finally, we are thankful to our individual family members for their patience, and we dedicate this book to them with much love and affection. Gómez-López wants to make a special mention to the loving memory of his parents.

We hope this book will be a valuable resource material for all undergraduate and graduate students, researchers, academicians, industry personnel, and all other stake holders engaged in the field of food science and technology.

Vicente M. Gómez-López
Universidad Católica San Antonio de Murcia (UCAM)
Murcia
Spain

Rajeev Bhat
ERA-Chair in Valortech, Estonian University of Life Sciences
Tartu
Estonia
February 2021

1

Physics of the Electromagnetic Spectrum

Michael Vollmer

Department of Engineering, University of Applied Sciences Brandenburg, Brandenburg an der Havel, Germany

1 Introduction

From a physicist's point of view, electromagnetic (EM) radiation is used in three different ways in the food processing chain (besides the partially nontrivial use of light to advertise food products in super markets). First, food can be preserved by irradiation, second it may be heated, and third, its constituents may be characterized. In order to understand these quite differing uses, one must deal with the properties of electromagnetic radiation and respective possible descriptions as well as with its interaction with matter.

The most simple approach is to start with light which constitutes the visible part of electromagnetic radiation. With regard to food processing, light may, e.g. be focused onto food using appropriate mirrors, thus heating the food as in solar cookers. But what is light and how can light be described in physics terms? The first question is still a very difficult and rather a philosophic one, whereas the second can be readily answered.

Beginning at the onset of modern science in the seventeenth and eighteenth centuries, light has been described as either particle (e.g. Newton) or wave (e.g. Young). Both descriptions were based on interpretations of experiments. In modern language, light constitutes an example of the so-called wave–particle duality. Light may behave as wave in experiments which focus on wave phenomena (involving, e.g. interference), but light can also behave as particle in experiments which focus on respective particle properties (such as the photoelectric effect). As a consequence, the general description must be able to explain wave as well as particle properties of the entity which is called light. This may be condensed in the statement that light can be described in a complimentary way either as electromagnetic wave or as particles which are called photons.

While studying spectral properties of light, it became evident that there are similar entities with similar properties, but invisible to the human eye. In 1800, Herschel discovered infrared (IR) radiation beyond the red edge of a visible light spectrum which was able to heat thermometers, and soon after, ultraviolet (UV) radiation was discovered which

Electromagnetic Technologies in Food Science, First Edition. Edited by Vicente M. Gómez-López and Rajeev Bhat.

blackened photographic papers beyond the blue edge of visible light spectrum. It was already known around 1900 that UV radiation could kill germs and bacteria. An obvious question is whether there are more kinds of radiation which behave similar to light and may hence be described also in a very similar way? The answer is yes, and the general feature of all similar types of electromagnetic radiation is that they may behave as waves as well as as particles. In the following, we will summarize the basic features of electromagnetic waves and the complimentary particle description.

2 Description of Electromagnetic Waves

Electromagnetic phenomena are those which involve electric and magnetic forces between charges and electric currents. In a more generalized physics description, the effect of forces is substituted by the effects of electric and magnetic fields. In a very abstract way, all possible electromagnetic effects may be described by a set of just four differential (or integral) equations, the so-called Maxwell equations (see any standard physics, theoretical physics, or optics textbook, e.g. Berkeley 1965; Etkina et al. 2019; Feynman et al. 1964; Guenther 1990; Halliday et al. 2014; Hecht 2016; Jackson 1975; Saleh and Teich 2019; Tipler and Mosca 2007), and two material equations which relate the fields in vacuum to those in matter. Already in 1865, it was shown that one possible solution of these equations for the electric and magnetic fields are wave equations, i.e. differential equations of the type (shown for the electric field E which depends on space coordinates x, y, z, and time t)

$$\left(\frac{\partial^2}{\partial x^2}+\frac{\partial^2}{\partial y^2}+\frac{\partial^2}{\partial z^2}\right)\vec{E}(x,y,z,t)=\frac{1}{c^2}\frac{\partial^2\vec{E}(x,y,z,t)}{\partial t^2} \quad \text{with} \quad v=\frac{1}{\sqrt{\mu\mu_0 \in\in_0}} \tag{1}$$

v is the speed of propagation, $\varepsilon_0 = 8.8542\times10^{-12}$ AsVm^{-1}, and $\mu_0 = 1.2566\times10^{-6}$ VsAm^{-1} are well-defined constants; ε is the relative dielectric constant (permittivity) and μ is the relative permeability, both describing the electric or magnetic properties of matter. For isotropic materials, ε and μ are just dimensionless numbers. The speed of propagation is largest for the case of vacuum with $\varepsilon = 1$ and $\mu = 1$, which defines the speed of light in vacuum c

$$c=\frac{1}{\sqrt{\mu_0 \in_0}}=2.99792458\times10^8\,\text{ms}-1 \tag{2}$$

The existence of respective wave phenomena was first proven in 1887 by Hertz.

2.1 Properties of Waves

The simplest periodic processes are oscillations. Consider, e.g. a mass attached to a spring which is set into motion. It oscillates back and forth in the vertical direction as a function of time with a given time interval for one complete oscillation (e.g. from the highest position via the lowest position to again the highest position). This time interval T is called period and is related to the frequency ν of the oscillation according to $\nu = 1/T$.

Whereas oscillations are periodic processes as function of time alone, and the path in space is always the same, waves resemble disturbances which propagate in space. Most waves in physics are periodic processes in space and time. Think, e.g. of a stone which is thrown into a puddle or a lake with a flat surface of water. The stone initiates a deformation of the water surface with partially lower and partially higher water level. Denoting all regions with largest water level as peaks, the phenomenon can be described such that all water peaks propagate with circular geometry and a well-defined velocity in all directions outward. At given location, the water level rises and falls periodically which can be described by a transient period and the respective frequency ν. Besides, the distance between adjacent peaks is also a well-defined quantity. It resembles a spatial period called wavelength λ.

In surprisingly many cases, the geometric form of a snapshot of the elongation of a wave (here the difference of the water surface height with respect to the undisturbed water surface height) can be described by a so-called harmonic function (i.e. a sine or cosine function). Figure 1 depicts three snapshots of such a sine-like wave (think of elongation of water surface along an arbitrary horizontal direction x) recorded at three different times. Within one period T, the disturbance has propagated a distance λ, i.e. the speed of propagation is given by $c = \lambda/T$ which can also be written as

$$c = \nu\lambda \tag{3}$$

Obviously, the general description of a wave should depend on the spatial and transient periods. For mathematical convenience, they are included in the so-called spatial frequency $k = \frac{2\pi}{\lambda}$ and transient angular frequency $\omega = \frac{2\pi}{T}$. Usually, one deals with waves in three dimensions; in this case, the spatial frequency becomes the wave vector $\vec{k}$ which points into the direction of propagation of the wave.

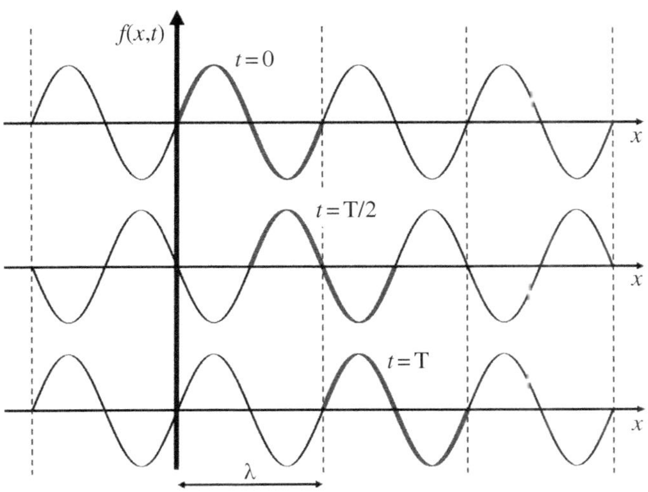

Figure 1 Example of elongation *f(x,t)* of a harmonic wave along a spatial coordinate *x* at three different times. (*See insert for color representation of this figure*).

In the water puddle example, the elongation of the water surface was perpendicular to the direction of propagation. Waves with this property are called transverse waves. Electromagnetic waves are transverse waves; in contrast, sound waves in gases are, e.g. longitudinal waves. The geometry of propagation also defines the type of waves. In the case of the water puddle wave (with two-dimensional water surface), the geometry was circular, so the water wave was a circular wave. A point light source in three-dimensional space, imagine, e.g. a distant star, constitutes an example of a spherical wave, as light can propagate in every direction with equal probability. Being very far away from such a point source (in a distance of many light years) means, however, that the radius of curvature is extremely large such that in practice, the wave front can be described as a plane (Figure 2). Whenever this is the case (optics has also other means to transfer a close-by point source into a plane wave), a wave is called a plane wave.

Most waves and therefore also electromagnetic waves are usually modeled as planar harmonic monochromatic waves (i.e. those having one well-defined frequency only). The reason behind is that once a problem is solved for such a plane harmonic monochromatic electromagnetic wave, it can easily be solved for any geometric form and any superposition of waves of different frequencies. The basis for the respective mathematical procedure of Fourier series and transforms is above the scope of this chapter. Whenever dealing with electromagnetic waves, we will therefore model the waves as plane harmonic waves (Figure 3). The electric field of such a wave is, e.g. given by

$$\vec{E}\left(\vec{r},t\right)=\vec{E}_0\sin\left(\vec{k}\cdot\vec{r}-\omega t+\varphi\right) \tag{4}$$

Each wave transports energy. A water wave close to the beach hitting someone can easily hurt. Similarly, being close to the sound waves of a loudspeaker can damage ears and intense direct exposure to sun light will cause sunburns. For electromagnetic waves, energy

Figure 2 Visualization of a plane wave originating from a point source in large distance.

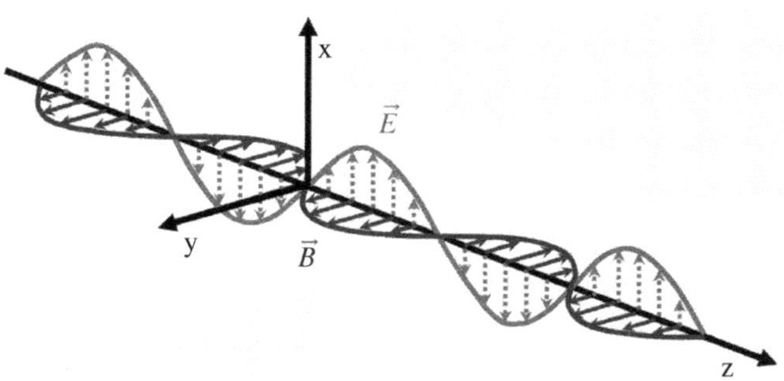

Figure 3 Snapshot of a plane harmonic three-dimensional EM wave according to Eq. (4). (*See insert for color representation of this figure*).

transport is described by the Poynting vector $\vec{S}(\vec{r},t)$, which describes the energy flux through a given area in J/(s m^2) = W m^{-2}.

The time averaged energy flux in non-ferromagnetic matter described by the index of refraction n ($n = \sqrt{\varepsilon}$) is given by

$$I = \langle |\vec{S}| \rangle = \frac{1}{2} c_0\, n\, \epsilon_0\, E_0^2 \tag{5}$$

2.2 Spectrum of Electromagnetic Waves

Electromagnetic waves are characterized by their frequencies ν and their wavelengths λ. White light usually consists of a distribution of different frequencies. This is easily investigated by passing light through a prism (Figure 4). Due to the index of refraction of glass, the deviation of the light depends on frequency (or wavelength) which leads to a spectrum behind the prism. Human eyes would see a colored streak of light with limiting borders of red and blue color. However, as mentioned before, experiments detected also radiation beyond the red (IR) and the blue edge (UV). The development of sensitive detectors in these regions allows to quantitatively investigate and display spectra of various sources.

Figure 5 shows an example of the optical spectrum of solar radiation outside of the atmosphere as well as at sea level at 45° latitude as a function of wavelength. It extends from the UV to the IR. In the nineteenth and twentieth centuries, many other sources of electromagnetic radiation have been developed and studied. Nowadays, the technically relevant spectrum of electromagnetic radiation used in food processing covers more than 15 orders of magnitude. Figure 6 depicts an overview.

The visible electromagnetic radiation is called light. Per definition it is given by the sensitivity of the standard human eye, which is tabulated for the wavelength range from 380 to 780 nm. The ranges for the adjacent spectral regions have initially been changed and/or renamed whenever new technologies allowed for new detectors and/or applications. Nowadays, they are mostly well defined, often spanning 3 or more orders of magnitude, allowing a variety of different applications. Therefore, there are, in addition, often subcategories, whose borders depend on the field of research. The visible spectral range (VIS) extends from 380 to 780 nm. Adjacent at longer wavelengths is the IR range, usually defined for wavelengths between 780 nm and 1 mm (typically used subcategories are, e.g. thermal

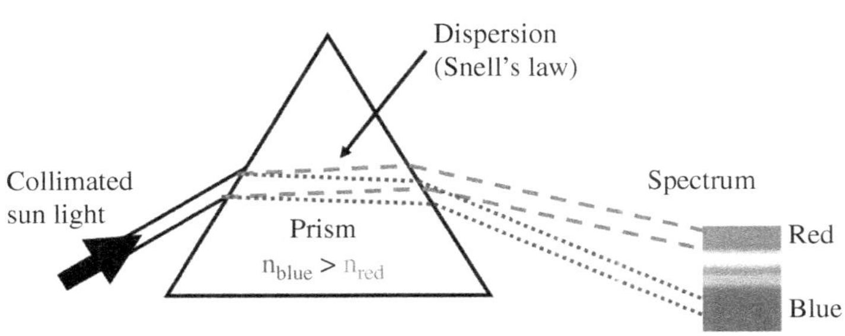

Figure 4 Light passing through a prism is deflected into a spectrum due to the wavelength (frequency)-dependent index of refraction. (*See insert for color representation of this figure*).

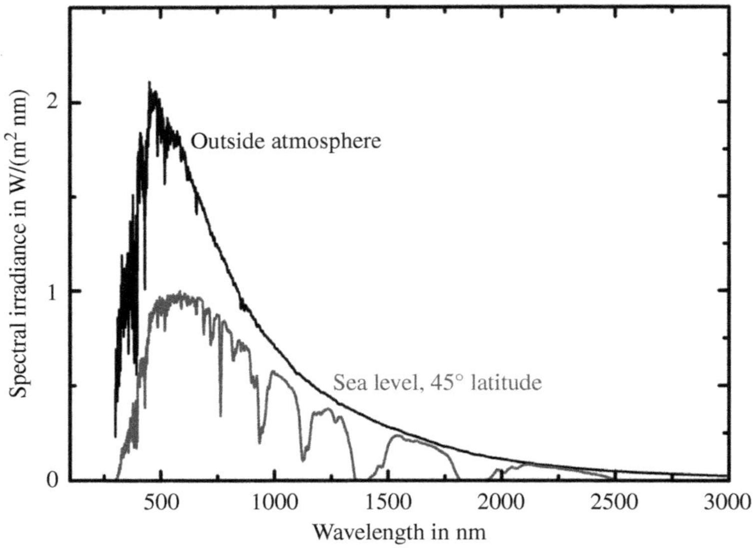

Figure 5 Spectral solar irradiance outside of the atmosphere and at sea level. Differences are due to interactions of the radiation with the constituents of the atmosphere. *Source:* Shaw and Vollmer (2017).

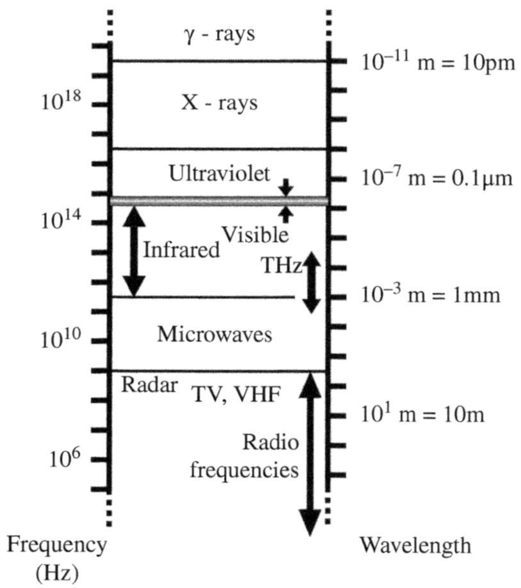

Figure 6 Various regions of the electromagnetic spectrum spanning more than 15 orders of magnitude in frequency or wavelength which are utilized in food science. The visible part of the spectrum is only a tiny part. The regions were mostly defined based on available sources and/or detectors. (*See insert for color representation of this figure*).

IR from 1 to 15 μm, used in thermal imaging or near [0.8–2.5 μm], mid [2.5–25 μm], and far IR [25 μm–1 mm] used in IR spectroscopy). It is followed by the microwave range between 1 mm and 1 m, beyond starts the regime of radiofrequency (RF) radiation which is utilized, e.g. for RADAR, radio, television, and all kinds of communication. Adjacent to the VIS range for shorter wavelengths is first the UV, extending from 380 nm down to 10 nm. Here, various subcategories are in use, depending on the field of research. Concerning solar radiation, the ability of the atmosphere, in particular the ozone layer, to absorb radiation leads to definition of soft UVA (above 315 nm), intermediate UVB (280–315 nm), and hard UVC (100–280 nm). Other definitions define extreme ultraviolet (EUV) from 10 to 121 nm (the 121 nm boundary being defined by the Lyman α line of the hydrogen atom), vacuum UV (10–200 nm, the name relating to the fact that air will effectively absorb any UV radiation below $\lambda = 200$ nm), as well as far (122–200 nm), middle (200–300 nm), and near UV (300–380 nm). Subsequently, X-rays range from $\lambda = 10$ nm to 10 pm. At even smaller wavelengths, the γ-ray region starts.

In the above description, all ranges were given as function of wavelengths; however, it is also quite common for people working in the field of some ranges to specify their borders with frequencies rather than wavelengths. The reason behind will become obvious when discussing the particle character of EM radiation (see Section 4).

One example, shown in Figure 6, indicates the typical problem in such rough categories defined by specific boundary wavelengths. Once THz detectors with new applications became available a few decades ago, the so-called THz range is used by the respective community as a range of its own, although, according to the above definition, it covers parts of the IR as well as the microwave region. Similarly, communication with radio frequencies of FM and AM radios was originally at rather low frequencies of around 1 MHz (AM) or 100 MHz (FM). However, the larger the frequency, the larger the usable bandwidth for communication. Therefore, modern communication frequencies of the 5G band extend well above the recently used ≤3 GHz regime to up to 40 GHz, which is – per the above definition – the microwave range. Obviously, the best way to characterize certain spectral regions is not names (which change with time) but wavelengths, frequencies, or energies related to the respective radiation.

In the introduction, it was stated that electromagnetic radiation is used in three different ways in food processing, first for preservation, second for heating, and third for characterization. Regarding the spectrum, one can roughly state that preservation is accomplished using short wavelengths extending from the UV to γ-rays, heating is mostly done using long wavelengths from the IR via the microwave to the RF range, whereas characterization is mostly done with VIS and near infrared (NIR) spectroscopy. The reason for the correlation of application with wavelength will become clear when dealing with the particle character of EM radiation.

3 Propagation of Electromagnetic Waves: Geometrical Versus Wave Optics

The general description of the propagation of EM waves can become quite complex. However, for certain conditions, the situation becomes much simpler as the propagation of light can show. When light is interacting with large obstacles, e.g. illuminating large objects,

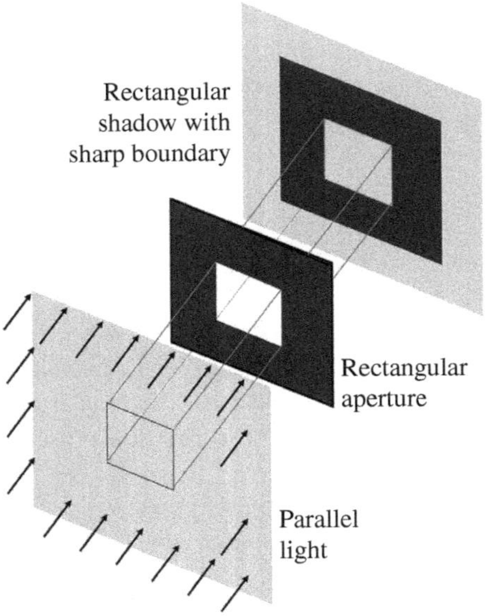

Figure 7 Rectilinear propagation of radiation can be demonstrated when irradiating opaque objects. If structures are large compared to the wavelength, they yield well-defined shadows with sharp boundaries.

some of them opaque, one can often observe very distinct shadows with sharp edges. This implies a rectilinear propagation of light (Figure 7). The respective field of optics is called geometrical optics: light is represented by rays which propagate rectilinearly within homogeneous matter (e.g. air). When light enters another homogeneous material, the propagation direction changes according to the law of refraction, but it then again propagates rectilinearly in the new material.

However, light can also propagate differently. Whenever light passes through small obstacles such as apertures or is reflected or transmitted from objects with small structures, one can observe deviations from rectilinear propagation, called diffraction. Figure 8 depicts an example where monochromatic light (light of a single frequency/wavelength) was passing through a small circular aperture. Behind the aperture the light hits a screen. Rather than just marking a circular spot on the screen, it also showed a number of ring-like structures around the center. This behavior can only be explained using wave optics and the principles of interference.

From these two ways of propagation, the geometrical optics one is much easier to treat. As a matter of fact, one can show that whether geometrical optics may be used depends on

Figure 8 Light after passing through an aperture: deviations from rectilinear propagation occur if the diameter of the aperture is comparable to the wavelength.

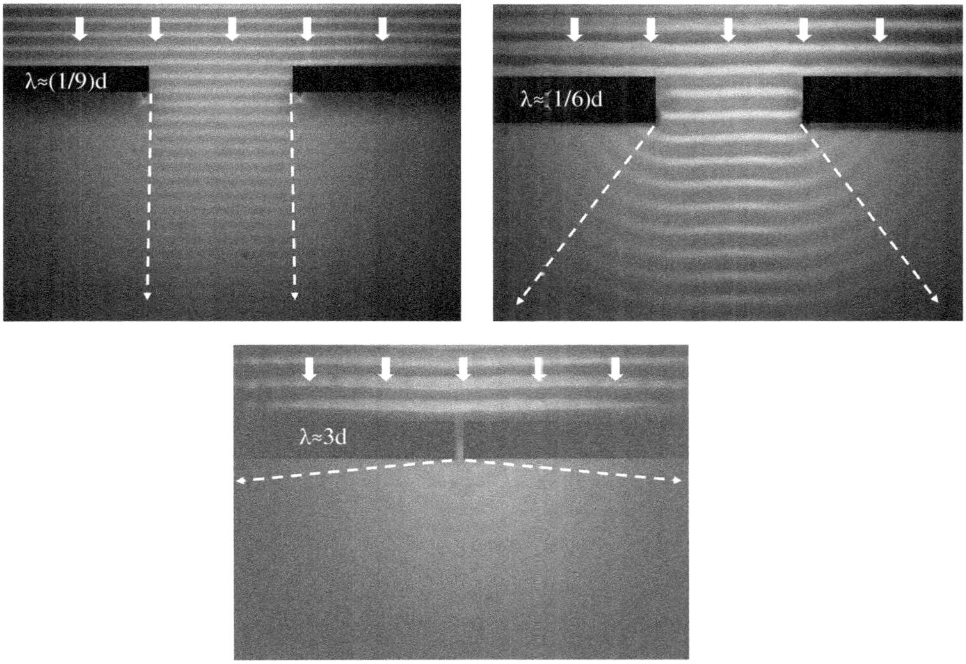

Figure 9 Transition between geometrical rectilinear and wave optics propagation for the example of water waves.

the ratio of wavelength λ of radiation and the typical dimension Ø of the obstacles in the path of the radiation.

If $\lambda \ll Ø \rightarrow$ description with geometrical optics is possible and a useful approximation

If $\lambda \approx Ø \rightarrow$ description with wave optics is needed, geometrical optics fails

As a consequence, geometrical optics can be derived as a limiting case of wave optics (Born 1972; Saleh and Teich 2019) with the direction of the light rays being given by the direction of the wave vector. Figure 9 illustrates the transition from the geometrical to the wave description for water waves.

A water wave of given wavelength (several mm) hits a slit-like aperture of variable width d. For λ small compared to the slit width (here $\lambda = d/9$), the wave travels with a well-defined shadow region behind the obstacle, whereas a much smaller width leads to a non-rectilinear circular diffracted wave spreading in all directions behind the slit ($\lambda = 3d$).

The same principle holds for any kind of wave as illustrated for a large object such as an advertising pillar of, say, 1 m diameter (Figure 10). An observer is on one side of the pillar, on the exact opposite side, a radio is placed with light emitting diode (LED) indicators and loudspeakers. The observer is not able to see the radio indicator lights as they are in the geometrical shadow of the pillar (the wavelength of the light is much smaller than the pillar diameter). On the other hand, the observer can hear the radio playing, i.e. the sound waves travel around the obstacle in a non-rectilinear way and reach the ear of the observer (wavelength of typical sound waves from voices or instruments are of the order of 1 m, i.e. they can readily bend around the 1 m pillar).

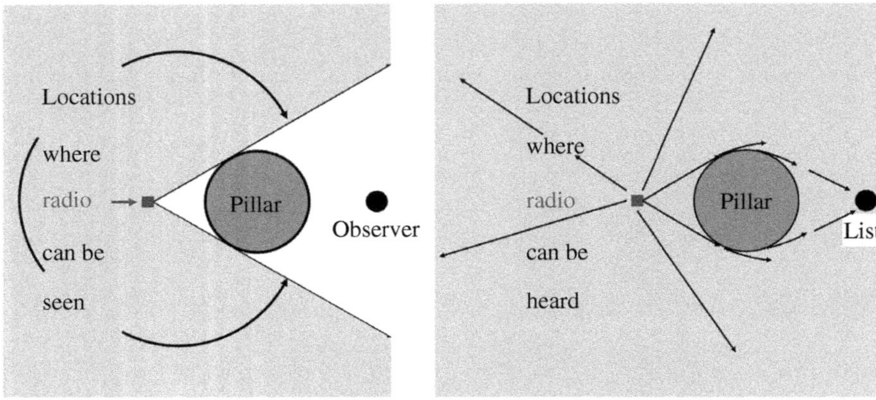

Figure 10 The transition between geometrical and wave optics propagation depends on the ratio of wavelength to object size. Light waves have a small wavelength compared to macroscopic objects of, e.g. 1 m size; therefore, light propagation can be described geometrically. Sound waves have, however, wavelengths of the same order of magnitude; they are diffracted around the obstacle.

The same principles of description also apply to all other kinds of EM radiation, i.e. one may describe the propagation of the radiation either using the rectilinear geometric model or the wave optics model depending on the ratio of λ to Ø.

4 Description of Particle Properties of Electromagnetic Radiation

In classical nonrelativistic mechanics, a particle is usually characterized by its kinetic energy $E = \frac{1}{2} m v^2$ and its momentum $p = mv$. Light is, however, moving at the speed of light and if treated as particle, called photon, must be described with the theory of relativity. A particle at rest is characterized by its rest mass and the mass increases for increasing velocity. As light particles travel at the speed of light, this is only possible if their rest mass is zero. Hence, it does not make sense to describe the momentum of light via the mass. In the complimentary concept of wave–particle duality in quantum mechanics, the momentum of the photons is related to the wave parameter wavelength λ and/or wave vector $\vec{k}$ via Eq. (6)

$$\vec{p} = \hbar \vec{k} \tag{6}$$

where $\hbar = h/2\pi$ with h being Planck's constant ($h = 6.626 \cdot 10^{-34}$ Js). Using $k = 2\pi/\lambda$, one finds that the momentum depends on the wavelength as $p = h/\lambda$.

Similarly, the energy of the photons is related to the wave parameter frequency ν and/or angular frequency ω via Eq. (7):

$$E = \hbar\omega = h\nu = \frac{hc}{\lambda}. \tag{7}$$

The SI unit of energy is Joule (J), which is a macroscopic quantity. However, when dealing with the interaction of radiation with matter, the typical energies involved relate to the microscopic systems atom or molecule or nucleus of atoms. There the relevant energies are, e.g. binding energies of electrons, vibrational or rotational energies of molecules, or the binding energies of the nucleons (protons and neutrons within the nuclei). In atomic, molecular, or nuclear physics, these energies are usually given in the units of electron volts (eV). The relation between eV and J follows from the work W done on a charge Q due to a potential difference U ($W = Q{\cdot}U$). Hence, 1 eV is the work done by 1 V potential difference on the electron charge $e = 1.6{\cdot}10^{-19}$ C, i.e. 1 eV=$1.5{\cdot}10^{-19}$ J.

The technologically used spectral range involves many orders of magnitude, hence the same applies to the photon energies. Usually – depending on the specific spectral range – they are given as meV = 10^{-3} eV, eV, keV = 10^3eV, MeV = 10^6eV, etc. If wavelength is given in nm or µm (for the UV, VIS, or IR range), the photon energies follow from Eq. (7) as

$$E(\text{eV}) \approx \frac{1240}{\lambda(\text{nm})} \approx \frac{1.24}{\lambda(\mu\text{m})}. \tag{8}$$

According to Eq. (8), the VIS extends from around 3 eV (blue) to 1.5 eV (red). For shorter wavelengths, the UV extends from 3 eV to around 100 eV, X-rays to around 100 keV, and γ-rays at still higher energies, typically several MeV. At longer wavelengths than the VIS, the IR extends from 1.5 eV to around 1 meV, the microwave region to 1 µeV, whereas RF frequency waves have still lower energy photons.

Knowing the energy of a single photon E as well as the radiative power P of the electromagnetic waves allows to estimate the respective number of photons N_{ph} according to

$$\frac{N_{\text{ph}}\,E}{t} = P, \text{d.h.} \quad N = \frac{P\,t\,\lambda}{hc} \tag{9}$$

For example, 1 W red (λ = 620 nm) or blue light (λ = 413 nm) source corresponds to around $3.1{\cdot}10^{18}$ photons s^{-1} or $2.1{\cdot}10^{18}$ photons s^{-1}.

5 Exponential Attenuation of Electromagnetic Radiation in Matter

Any application of EM waves to food with the purpose of preserving, heating, or characterizing requires the interaction of the waves with matter. First, a macroscopic description shall be given (Section 5) followed by a microscopic model of matter (Section 6) and some details of the interaction processes of EM waves with matter (Section 7). The latter will help to understand why various spectral regions have different applications.

Figure 11 schematically depicts the attenuation of collimated EM radiation, while passing through homogeneous matter, being characterized by its index of refraction n. In microscopic models, n is a complex quantity $n = n_1 + in_2$, its imaginary part n_2 describing absorption.

Neglecting scattering (for the moment), attenuation of EM waves in matter is solely due to absorption. After a distance z, the remaining radiation is given by the law of Lambert, Beer, and Bouguer (for monochromatic radiation of vacuum wavelength λ)

$$I(z) = I_0 e^{-\frac{z}{z_0}} \quad \text{with } z_0 = \frac{\lambda}{4\pi n_2(\lambda)} \tag{10}$$

where I_0 denotes any energy-related quantity of the wave, e.g. the incident energy flux.

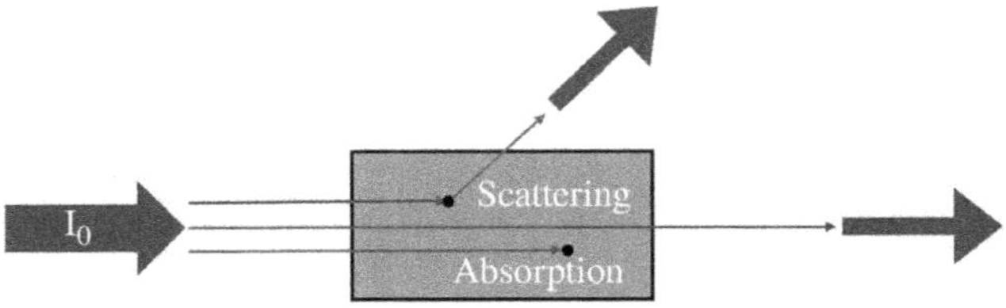

Figure 11 EM radiation passing through matter can be either absorbed or scattered. In the latter case, the light changes its direction, i.e. is missing in the forward direction. The scattered photons may have the same (elastic) or a different energy (inelastic).

A very similar exponential law also holds if the attenuation is due to singe scattering events (multiple scattering excluded). In general, both absorption and scattering contribute and

$$I(z) = I_0 e^{-\gamma x} \tag{11}$$

where $\gamma = \gamma_{abs} + \gamma_{scatt}$ is the total extinction coefficient. $z_0 = 1/\gamma$ describes the distance after which the energy flux has decreased to $1/e$ of its initial value. After a distance of 4.6 z_0, the wave has been attenuated by a factor of 100, i.e. $I(4.6z_0) = 0.01\ I_0$.

For pure absorption processes, z_0 can be estimated from tabulated data of refractive index (e.g. Palik 1985, 1991, 1998). If scattering is involved, more elaborate theoretical models are used. Vice versa, measuring the attenuation of EM radiation allows to experimentally determine the attenuation coefficient γ.

As an example, Figure 12 shows refractive index data for pure water (a very prominent and important ingredient in food) from radio frequencies to the EUV as a function of frequency. The VIS spectral region is indicated by the broken (red) and (blue) lines.

Assuming absorption processes only (which lead to heating of the water), Eq. (10) and the data from Figure 12 can be used to estimate the distance in water after which 99% of the radiation is absorbed, i.e. only 1% is still transmitted. Table 1 gives selected data from the

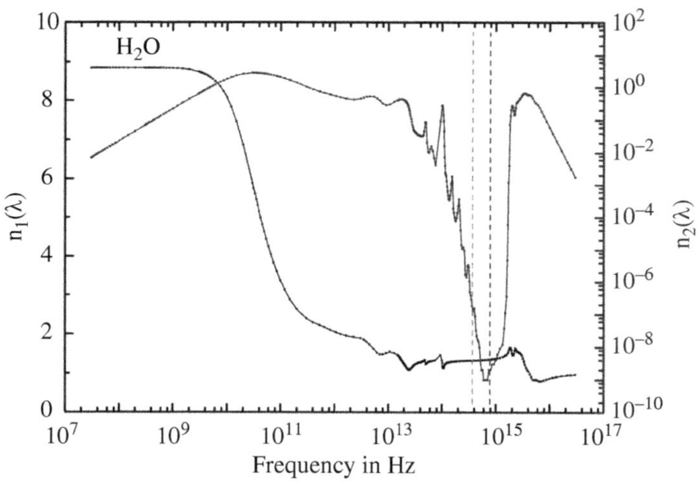

Figure 12 Index of refraction of pure water as a function of frequency. (*See insert for color representation of this figure*). *Source:* Palik (1991).

Table 1 Attenuation depths due to absorption only for pure water from the UV to the RF range according to Eq. (10) with refractive index data.

Frequency (Hz)	Wavelength	n_2	4.6 z_0 (drop to 0.01)
$3{\cdot}10^{16}$ (UV)	10 nm	0.001745	2.1 μm
$3.49{\cdot}10^{15}$ (UV)	86 nm	0.6573	48 nm
$7.5{\cdot}10^{14}$ (VIS blue)	400 nm	$2{\cdot}10^{-9}$	73 m
$3.75{\cdot}10^{14}$ (VIS red)	800 nm	$1.25{\cdot}10^{-7}$	2.3 m
$1.0{\cdot}10^{14}$ (IR)	3 μm	0.272	4.0 μm
$3{\cdot}10^{10}$ (30 GHz)	1 cm	2.83	1.3 mm
$2.4{\cdot}10^{9}$ (2.4 GHz)	12.5 cm	≈0.5	9.1 cm
$3{\cdot}10^{8}$ (300 MHz)	1 m	0.06727	5.4 m

Source: Palik (1985, 1991, 1998).

UV to the RF range. Obviously, the n_2 data have their absolute minima within the VIS with maximum distances above 70 m. On the one hand, this means that the deep sea is totally dark, as after a few hundred meters of water, no sunlight is left (of course, sea water will have salts and additional constituents, leading to even larger attenuation). On the other hand, the shallow ocean receives a lot of solar radiation, i.e. radiant energy, which is the reason why life evolved in the ocean and also why the perception of humans and many animals have their peak sensitivity in this spectral range.

Toward the IR, absorption increases dramatically (reason, see below) such that IR radiation is not even able to penetrate a layer of water with 1 mm thickness. Figure 13 visualizes the transmission of IR radiation through thin slabs of water of given thickness. This drastically different behavior with respect to the VIS should be a warning: the physical reaction of matter in any given spectral range can differ appreciably from that of an adjacent

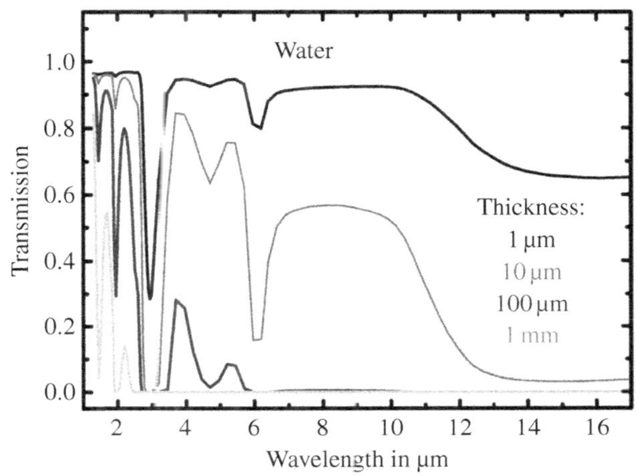

Figure 13 Transmission spectra of pure water in the near-infrared spectra range. Water very effectively absorbs IR radiation already within distances of less than 1 mm. (*See insert for color representation of this figure*).

spectral range, i.e. guessing material reaction to an EM wave by intuition from the knowledge of what typically happens in close-by spectral ranges may not work at all.

The attenuation distance in water (Table 1) increases slightly toward longer wavelengths, however, at the maximum n_2-value in the microwave band at around 30 GHz, the penetration depth for 99% absorption is still just 1.3 mm. Imagine a microwave oven which would operate at this frequency. Any food placed in the oven would absorb the radiation within an about 1-mm-thick surface layer, i.e. any food with larger dimensions than a few mm would be cold inside surrounded by a carbonized surface layer. This is the reason why commercial microwave ovens for the kitchen operate at a much lower frequency of around 2.4 GHz (Vollmer 2004). This allows heating of food with dimensions of several centimeters without problems. For still longer wavelengths toward the RF, the distance increases. Therefore, heating may still be accomplished in much larger samples.

Moving to smaller wavelengths, i.e. higher frequencies from the VIS range toward the UV will also dramatically increase the absorption, i.e. reduce attenuation distances.

A word of caution: whenever the incident radiation is not collimated, the theoretical description of how it propagates in matter due to absorption and scattering processes becomes much more complex, in particular if multiple scattering occurs. A typical everyday example is how the mixture of direct and diffusely scattered solar radiation can penetrate the water of oceans or lakes. The modeling of the respective forward and backward (in this case better called upward and downward) propagating radiation is way beyond the scope of this introductory chapter.

6 Microscopic Structure of Matter and Origin of EM Radiation

The generation of EM radiation is intimately related to the structure of matter. Atomic structure, in particular of the outer electrons in atoms, explains phenomena in the UV and VIS range, molecular structure is mostly related to features in the IR, inner electrons in heavy atoms are responsible for some X-ray features, and the structure of nuclei explains the origin of γ-radiation. Besides, the quantum nature of matter leads to emission of thermal radiation, and finally, basic classical electrodynamic concepts explain the generation of radio frequency radiation. For detailed descriptions, see standard physics textbooks, e.g. Berkeley 1965; Alonso and Finn 1968; Eisberg and Resnick 1985; Halliday et al. 2014, and Tipler and Mosca 2007.

6.1 UV–VIS and Atomic Spectra

In classical physics, particles are characterized by their kinetic energies and momentum. Consider, e.g. a billiard ball which was hit by a billiard cue. The possible energies of the ball are continuous, i.e. every imaginable energy value may be realized. This behavior dramatically changes when switching to microscopic particles such as electrons or protons in atoms. About 100 years ago, the then new *modern physics* called quantum mechanics was developed. As a result of the wave–particle dualism, electrons in an atom can no longer have any value for the energy but only certain well-defined discrete values. Their values depend on the charge in the nucleus (i.e. the chemical element, defined by the number of protons) and on the interaction of all electrons present in the atomic shells.

The allowed discrete states of an electron in an atom are described by four quantum numbers. For any given atom with a certain number of electrons (e.g. a neutral sodium atom has 11 protons in the nucleus and hence 11 electrons in atomic shells around the nucleus), the allowed discrete electronic states are filled up from lowest energy to highest energy following certain rules. This defines the so-called ground state of an atom. Of course, there are also many further allowed discrete states with higher energies which are so-called excited states. For example, the energies of allowed discrete electronic bound states (E_n) of hydrogen-like atoms with Z positive charges in the nucleus (an H atom has $Z = 1$) can be roughly described by a single quantum number n as

$$E_n(\text{eV}) = Z^2 \frac{13.6 \text{ eV}}{n^2} \tag{12}$$

This is interpreted such that the ground state ($n = 1$) of hydrogen ($Z = 1$) has a binding energy of 13.6 eV, the first ($n = 2$) and second ($n = 3$) excited states have energies of 13.6 eV/4 ≈ 3.4 eV and 13.6 eV/9 ≈ 1.5 eV, etc. Obviously, the energy differences between electronic states are of the order of eV, which justifies our earlier definition of the atomic energy scale of eV (Eq. (8)).

In order to simplify the discussion, one often draws a schematic energy-level diagram of the discrete electronic states within an atom with just a few states, usually the ground state (denoted, e.g. as E_1) and one higher lying excited state (denoted, e.g. as E_2).

Such a scheme can visualize what happens microscopically when a photon is incident on matter of the given kind of atoms (e.g. in a gas). Figure 14 depicts three different situations.

If the energy of an incident photon matches the energy difference between the electronic ground state and an excited state, it can transfer its energy to the electron, which thereby ends up in the excited state (Figure 14, left). As the photon is absorbed, it is now missing in the forward direction, i.e. the incident radiation is attenuated as was described by Eq. (10).

After a while, the electron returns to its ground state. Whether this is done by collisional energy transfer or via the reverse process to absorption depends on the preparation of the sample under study. The de-excitation by emission of a photon is called spontaneous emission (Figure 14, center). This emission is isotropic, i.e. it is possible in all directions. Last not least, there is a third possible process for already excited atoms (Figure 14, right): if a suitable photon is incident, it may stimulate the electron transfer back to the ground state.

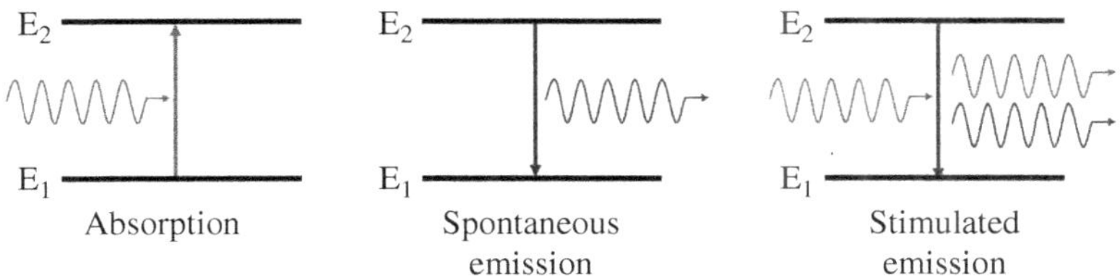

Figure 14 The three elementary microscopic interaction processes of EM radiation with atoms are absorption, spontaneous emission, and stimulated emission. (*See insert for color representation of this figure*).

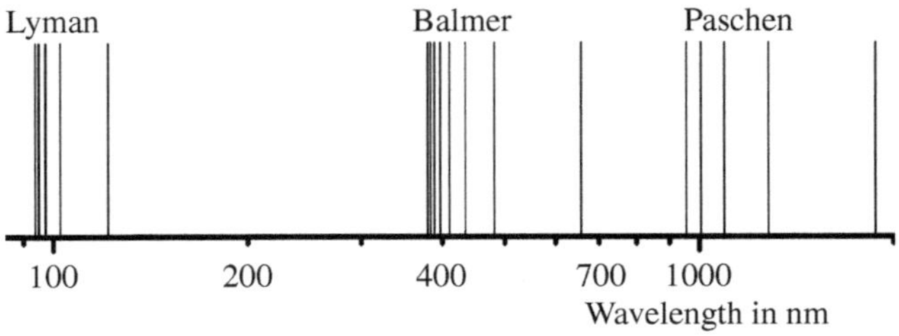

Figure 15 Position of absorption/emission features for hydrogen atoms. The indicated series refer to transitions starting/ending in $n = 1$ (Lyman), $n = 2$ (Balmer), and $n = 3$ (Paschen). Please note the logarithmic scale.

In this case, the emitted photon will have exactly the same energy and direction as the incident photon. This stimulated emission process is the basis for laser radiation.

Obviously, photon absorption or emission is related to the discrete energy levels of electrons in atoms. Therefore, atoms may absorb or emit a set of discrete energies of EM radiation. Figure 15 depicts the position of the respective wavelengths for hydrogen atoms. Other atoms have similarly well-defined spectra, all of them resembling atomic fingerprints which can be used for quantitative analysis. As can be seen from Figure 15, typical atomic spectra range in the UV, VIS, and NIR range.

From Eq. (12), the ground state of the electron in a hydrogen atom is 13.6 eV. If more energy is transferred to the electron (either via a collision or via absorption of a higher energy photon), the electron can be set free and the atom will be ionized. Therefore, the binding energy also corresponds to the ionization energy. For hydrogen, the ionization threshold of 13.6 eV corresponds to $\lambda \approx 91$ nm in the UV spectral range.

Most atoms have ionization thresholds for their least-bound outer electrons in the UV range. Binding energies of inner electrons can, however, well reach many keV (see Section 6.3). As discussed above, UV radiation extends from the blue edge of the VIS range to 10 nm wavelengths, i.e. it covers a photon energy range from around 3.25 eV ($\lambda = 380$ nm) via 10.2 eV ($\lambda = 121.4$ nm: Lyman α) and 13.6 eV ($\lambda = 91.2$ nm: ionization threshold H-atom) to 124 eV ($\lambda = 10$ nm). Therefore, EUV radiation can indeed be considered as ionizing radiation for the outer valence electrons of atoms in matter. However, the term ionizing radiation – which is not really exactly defined – is usually reserved for either charged particles such as electrons or EM waves such as X-rays and γ-rays of much larger energy (typically with initial particle energies above keV). In comparison, UV radiation may just be sufficient for one or a few ionization events, whereas the later considered keV to MeV ionizing radiation (see Section 7.4) leads to multiple ionization processes and a large number of further initiated reactions.

6.2 IR and Molecular Spectra

When two atoms come close together, it may happen that the overlay of the respective electron clouds leads to stable configurations. In this case, the repulsive Coulomb forces between the positive nuclei and between all electrons are overcompensated by the attractive Coulomb forces between the electrons and the positive nuclei. The electronic states in such a diatomic molecule (but also in complex molecules made of more than two atoms) will be characterized by discrete energies and quantum numbers; however, the situation is more complex than in atoms.

First, the molecules can also vibrate. Imagine a diatomic molecule with a given binding distance, typically in the range of 10^{-10} m. Imagine one atom is slightly displaced outward: it feels a restoring force and moves inward toward the other atom. It overshoots its equilibrium position and soon starts to feel repulsive Coulomb forces. As a consequence, it oscillates back and forth. In every oscillation, there is kinetic and potential energy, here of the oscillating atoms. The effect of the attractive and repulsive forces is usually visualized in terms of potential energy diagrams for the electrons. The equilibrium positions are at the minima of these curves. Excited electrons will usually have a different electron distribution and therefore in general a slightly different equilibrium position. The energy differences between such electronic states in molecules are similar to those of atoms, i.e. in the eV range.

Any potential energy curve with a minimum can be approximated to first order with a parabola. Theoretically describing the oscillations of the molecules around their equilibrium positions in such parabolic shapes leads to the so-called quantum mechanical harmonic oscillator. One ends up with another set of discrete energy levels for allowed vibrational energies of the molecule. These are first order equidistant to each other, however with much smaller energy spacing (compared to those between electronic levels) of the order to 0.1 eV.

Second, molecules can also rotate around two or three axis in space. Similar to classical rotating objects which have rotational energies depending on angular momentum and moment of inertia, microscopic rotating molecules have rotational energies, which have again, however, only a discrete set of allowed values. Their energy spacing is much smaller (than the one of vibrations) and of the order of 10^{-3} eV or even 10^{-4} eV. Hence, molecules are characterized by three quite different kinds of discrete energy levels with spacings of the order of eV for electronic levels, of 0.1 eV for vibrations, and of meV or below for rotational excitations. Figure 16 shows a schematic overview energy-level diagram of a

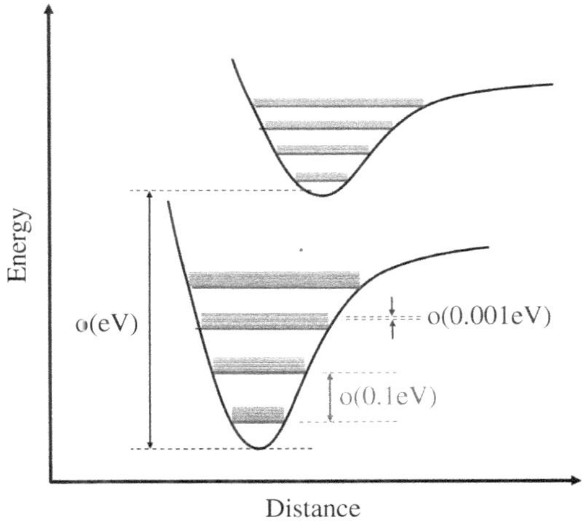

Figure 16 Schematic overview of energy levels of diatomic molecules: for each of the two drawn electronic levels with minima there are many discrete vibrational levels (with typical distances of 0.1 eV) and on top of each vibrational level, there are many discrete rotational energy levels (with typical distances of 0.001 eV). (*See insert for color representation of this figure*).

diatomic molecule. The spacing of the order of eV between electronic levels corresponds to the UV and VIS range, the one between vibrational levels of 0.1 eV is in the near IR range, and the one between pure rotational levels of meV or below is in the mid or far IR range.

Similar to atoms, photons can be absorbed or emitted by these molecular energy levels. However, due to the much larger variety of levels and level spacing, the respective line spectra are much more complex than those of atoms. A further complication arises from the fact that for any system with given energy levels, not all potential transitions between a given set of levels are possible.

In quantum mechanics, selection rules are formulated which define allowed optical transitions. One rather strict rule is that during an oscillation, there must be a change of the electric dipole moment if a photon is to be absorbed or emitted. Therefore, homo-nuclear diatomic molecules such as O_2 and N_2 do not have allowed optical transitions between vibrational levels of a given electronic state. In contrast, hetero-nuclear diatomic molecules (e.g. CO) or larger molecules such as CO_2 which already possess a dipole moment without any vibration can have radiative transitions between vibrational levels in the near IR range. This explains why trace gases as CO_2 contribute to the anthropogenic greenhouse effect, whereas the main constituents of air, O_2 and N_2, do not. We mention that a certain type of molecular spectroscopy, Raman spectroscopy, is complimentary to regular absorption spectroscopy in the sense that it can also detect *forbidden* direct optical transitions. The underlying Raman effect represents an inelastic scattering process of (mostly VIS) radiation from molecules. It allows to extract information about molecular vibrations if the polarizability changes during a vibration. This is a microscopic quantity which describes the extent and displacement of the electric charges within the molecule. For example, the polarizability of CO_2 changes during a symmetric vibration of the molecule, whereas the dipole moment does not. In contrast, the asymmetric vibration leads to a change of the dipole moment, whereas the polarizability stays constant. As a result, Raman spectroscopy is indeed complimentary to absorption spectroscopy: the latter allows to detect asymmetric vibrations, whereas the former can detect the symmetric vibrational modes. Therefore, both techniques are used for characterization of molecules (Günzler and Gremlich 2002).

6.3 X-Rays and Excitations of Inner Electrons in Atoms

X-rays extend the UV spectrum toward higher frequencies. Part of them, the so-called characteristic X-rays, are related to atomic structure. The energy levels and electronic transitions of the hydrogen atom were discussed in Section 6.1. If heavier atoms with Z protons and therefore also Z electrons for neutral atoms are treated, the energy of the lowest lying electrons (closest to the nucleus) increases with approximately Z^2 (Eq. (12)). Hence, atoms with, e.g. $Z = 42$ (Mo) would roughly have a binding energy of the lowest state ($n = 1$) of around 24 keV. In the electronic ground state of the neutral atom, all 42 electrons will fill up the electronic energy levels, starting at the lowest levels with $n = 1$. Due to the energetic order of the various levels (calculated from quantum mechanics), the configuration will be 2 electrons in state $n = 1$, 8 electrons in state $n = 2$, 18 electrons in state $n = 3$, 15 electrons in state $n = 4$, and finally one least-bound electron in the highest occupied state $n = 5$. The electrons with low n will have a high probability of being close to the nucleus; therefore, in a simplified model, they are visualized as occupying circular orbits close to the nucleus.

Figure 17 depicts such a schematic drawing for the 10 innermost electrons on levels $n = 1$ and $n = 2$ of a heavy atom (to be specific, we will discuss Mo). Now imagine that an inner electron (e.g. for $n = 1$) is missing. In practice, this is realized by bombarding Mo metal with high energy electrons, accelerated to an energy of, e.g. 35 keV. Some of these high-energy electrons hit some of the inner bound electrons and kick them out of the atom such that there is a vacancy, i.e. a missing electron.

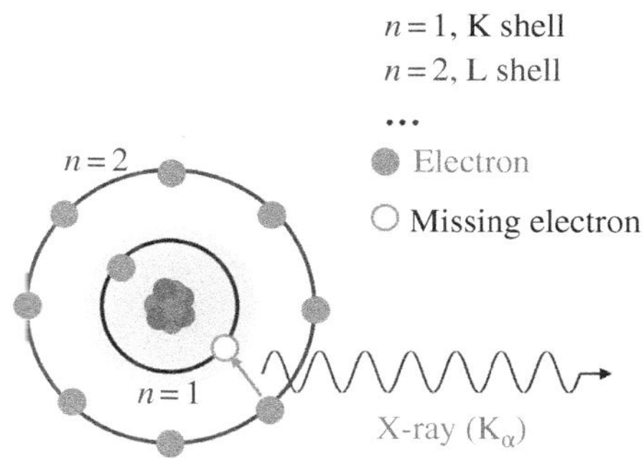

Figure 17 Scheme for generating characteristic X-rays: a vacancy in an inner shell is filled by electrons from other shells; the respective energy difference is emitted as a photon.

In Figure 17, the effect is illustrated for a missing electron in the innermost state $n = 1$. As the system moves back to its lowest energy state, outer electrons fill the vacancy. This resembles the same situation as that of an excited electron relaxing into the ground state via emission of a photon. The same happens here, however, as the energy scale is in the keV range, the respective photon energies are in the X-ray region.

For historic reasons, the respective transitions are denoted different than in outer atomic shells. In X-ray physics, the electronic energy levels with $n = 1, 2, 3$, etc. are denoted as K-shell, L-shell, M-shell, etc. Transitions ending in the K-shell are denoted as K_α, K_β, K_γ radiation, etc., depending on where the electron, which fills the vacancy, comes from ($n = 2$, $n = 3$, $n = 4$, etc.). As an example, the Mo K_α line has $\lambda \approx 70$ pm, corresponding to a photon energy of $\approx$18 keV.

Besides these characteristic X-rays, the electrons hitting the Mo metal also generate the so-called Bremsstrahlung. It is due to the fact that the incident electrons are slightly deflected (one speaks of scattering) while passing through the material. And deflection, i.e. change of initial direction of the electrons means that the electron is accelerated in the perpendicular direction and decelerated in the forward direction.

From a general law of classical electrodynamics, any acceleration/or deceleration of a charge must be accompanied by emission of radiation (see also Section 6.6). This results in a continuous spectrum with a maximum energy given by the electron initial kinetic energy. For example, a 35 keV electron beam would give a minimum wavelength (Eq. (8)) of the Bremsstrahlung of $\lambda \approx 1240/35\,000$ (nm) $\approx$35 pm. Both the Bremsstrahlung and the characteristic spectrum are superimposed. As a result, an X-ray source offers typically EM radiation with a wide range of energies in the X-ray region.

6.4 γ-Rays and Nuclear Spectra

Atoms consist of nuclei and electrons surrounding the nuclei, and each nucleus consists of a well-defined number of protons and neutrons. It is plausible to assume that similar to the discrete energies of electrons in an atom, there should also be only discrete allowed energies of protons and neutrons in a nucleus. There is a main difference in the description of the problem: electron energies are determined by the attractive Coulomb forces, i.e.

electromagnetic interaction between electron and positive nucleus. In a nucleus, the Coulomb forces between protons are, however, repulsive and the nucleus is only stable because there are the much stronger nuclear forces (described as strong interaction) which overcompensate the Coulomb forces. These stronger forces shift the value of the typical energy scale from eV into the MeV range, i.e. by about a factor of 10^6. As a consequence, if Figure 14, left would be interpreted as resembling the ground and excited state of a proton in a nucleus, the energy difference is now in the MeV range. From Eq. (8), a photon of 1.24 MeV energy corresponds to $\lambda = 1\,\text{pm}$ and $\nu = 3\cdot10^{20}\,\text{Hz}$, i.e. the γ-ray region of the electromagnetic spectrum.

For food processing, it is also quite common to use sources which emit γ-rays. Such sources are either natural or technically prepared radioactive samples. Stable nuclei cannot consist of an arbitrary number of protons and neutrons. The chemical identity of an atom is defined by the number of protons, i.e. the number of positive charges in the nucleus. However, the number of neutrons may vary, this means that there are a number of different nuclei (isotopes) possible for each atom. For example, chlorine atoms have always 17 protons, but two stable isotopes exist in nature, one with 18 neutrons and the other with 20 neutrons. Similarly, heavy elements can have many stable isotopes, but there are also quite a few natural or artificially produced nuclei, which are unstable. One may regard them as not being in their ground but an excited state. All systems in nature tend to propagate toward a state of lowest energy. For excited nuclei, there are mostly three pathways of how to reach such a state. These are α-decay (emission of He^{++} particles), β-decay (emission of electrons and neutrinos), and γ-decay (emission of photons). The latter process is the only one involving EM radiation.

6.5 Blackbody Radiation: Generating UV, VIS, and IR Radiation from Hot Objects

At the end of the nineteenth century, many physicists worldwide were studying so-called thermal radiation also called cavity radiation, blackbody radiation, or incandescent radiation. Think of the sun which has a surface temperature of around 5800 K and shines brightly in the VIS spectral range or an extremely hot plate of an oven which first starts to shine deep red and then changes toward orange when getting hotter and hotter. Scientists tried to understand how much energy a hot body could emit as function of frequency (or wavelength) and realized that theories based on classical electrodynamics and thermodynamics could not properly reproduce the accurate experimental results. The first one who could quantitatively describe these spectra was Max Planck. While developing the respective theoretical explanation he had to introduce a new concept, quantized energy states. Hence, blackbody radiation marks the birth of modern quantum physics.

In short: every blackbody object with a temperature above absolute zero (0 K = −273.15 °C) emits EM radiation, the amount of which can be described by Planck's law which only depends on the temperature T of the object under study (Vollmer and Möllmann 2018). In terms of spectral exitance M_ν as function of frequency ν, i.e. emitted radiant power within frequency interval $d\nu$ per unit area of a source in the hemisphere, it is written as (Eq. (13))

$$M_\nu(T)d\nu = \frac{2\pi h\nu^3}{c^2}\frac{1}{e^{\frac{h\nu}{kT}}-1}d\nu \tag{13}$$

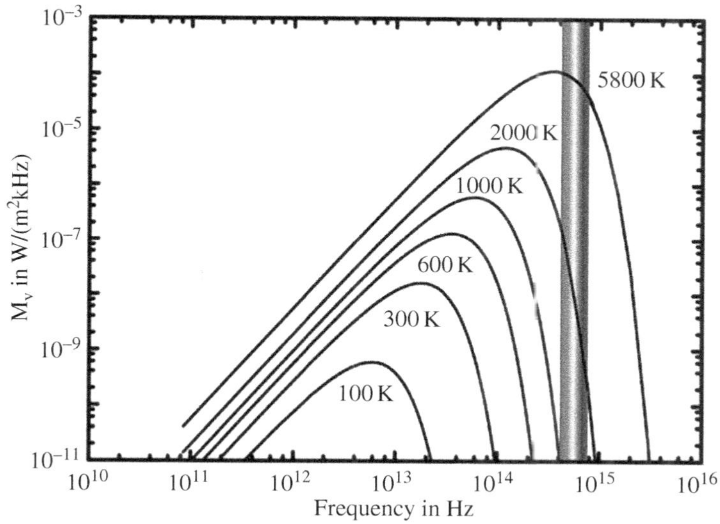

Figure 18 Blackbody radiation excitance spectra for object temperatures varying between 100 and 5800 K. (*See insert for color representation of this figure*).

where c is the speed of light in vacuum, h Planck's constant (see Eq. (6)), and k Boltzmann's constant ($k = 1.38065 \cdot 10^{-23}$ JK^{-1}). Real objects emit a bit less, described by the so-called emissivity (for many materials around or above 0.9); therefore, Eq. (13) describes the maximum emission of thermal EM radiation from an object of given temperature.

Figure 18 depicts blackbody radiation spectra for temperatures between 100 K and 5800 K. The VIS range is indicated by the colored band. Please note the double logarithmic scales. The excitance scale covers eight orders of magnitude, the frequency ranges from microwaves up to the UV. Solar emission is very large in the VIS and peaks in the near IR.

The lowest shown values of around 10^{-11} W/(m^{2}kHz) are indeed very difficult to detect. The 1000 K object, i.e. around 700 °C object, is just hardly detected with the human eye (which is a very sensitive light detector) as very deep red color. Similarly, typical commercial IR cameras are used in the near IR range operating around 10 μm, corresponding to $3 \cdot 10^{13}$ Hz. At this frequency, the lower limit corresponds to slightly above 100 K. Indeed, such low temperatures cannot be detected with commercial IR cameras. Even the sun as hottest object only emits a very small amount of far IR and microwave radiation.

In conclusion, blackbody radiation offers rather bright radiation sources in the UV, VIS, and very near IR range, but much less so for longer IR wavelengths. It definitely is not a bright source of EM radiation for even lower frequencies, i.e. within the RF range.

6.6 Generation of Microwave and RF EM Waves

All processes discussed so far can only explain the generation of tiny amounts of RF radiation. However, RF EM waves have many important applications, those in communication dominating our daily life but also in food processing. Their generation and detection can be easily explained by classical electrodynamics. As mentioned above, while discussing

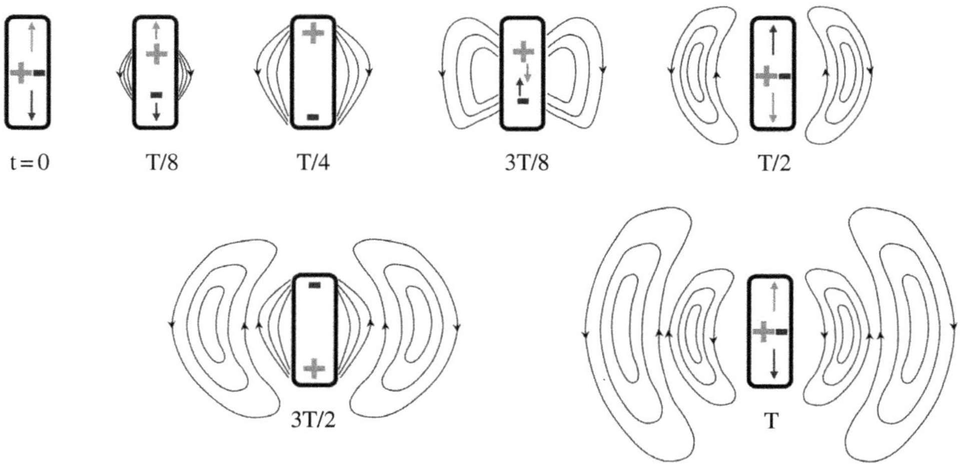

Figure 19 Schematic visualization of generation of RF EM waves by a dipole antenna. The magnetic fields are omitted here. (*See insert for color representation of this figure*).

Bremsstrahlung, any accelerated charge emits radiation. To optimize the emission of radiation of well-defined RF frequency, antennas are used. The basic principle how antennas emit radiation is schematically depicted in Figure 19.

Imagine an LC resonating circuit consisting of a parallel plate capacitor with capacitance C and a solenoid with inductance L resulting in a resonance frequency $\omega = \frac{1}{\sqrt{LC}}$. Imagine that you first reduce the number of turns of the solenoid to 1, second the area of the capacitor to the cross section of the wire. Both actions reduce L and C and hence increase the resonance frequency. If, third, the single loop is bend into a straight line, one ends up with a so-called dipole antenna. If one excites this (modified) resonant circuit, e.g. inductively at its center, currents will periodically flow within the antenna. Figure 19 depicts seven snapshots during one period. During the oscillation, the top and bottom of the wire are periodically charged. According to electrodynamics, electric field lines either start or end in charges. At $t = T/4$, the electric field is the one of the well-known electric dipole. While the oscillating positive and negative charges on the antenna approach the center and cancel each other at $t = T/2$, the previously present electric field lines can no longer start or end at the antenna surface. They detach themselves from the antenna and form closed-loop electric fields, which travel outwards, while the next new electric field lines build up when new surface charges develop on the antenna during the oscillation (each time-varying electric field is also associated with a time-varying magnetic field – according to Maxwell's equations – the corresponding magnetic field lines are just not shown here for clarity of the figure).

As a consequence, correlated electric and magnetic fields are propagating away from the antenna with the speed of light. Far away from it, they can be described by transverse electromagnetic waves. Of course, the frequency, and very importantly, the efficiency with which the EM waves detach from the wire, i.e. are transferred from the antenna to the surrounding medium (air or vacuum), depend on the construction of the antenna. Optimized efficient sources are mostly based on transmitting tubes. Here it suffices to state that one may produce a wide range of RF EM waves with high efficiency. For food processing, mostly RF frequencies in the range between 10 and 50 MHz are used for heating purposes. Microwave heating operates at higher frequencies, e.g. 2.45 GHz, efficient generation of microwaves is possible, e.g. with so-called magnetrons or klystrons. RF and/or microwave radiation with low power for devices such as mobile phones are usually generated differently with solid-state devices.

7 Interaction of EM Radiation with Food

The microscopic structure of matter revealed that atoms are mostly able to absorb EM radiation in the UV and VIS range; molecules have, in addition, characteristic features in the near IR range and some additional ones in the mid/far IR. Nuclei, on the other hand, have excitation energies in the γ-range. In this section, a brief summary is given how matter (such as food) responds to incident EM radiation of various frequencies from radio frequencies to γ-rays and why heating, characterization, and preservation happen in different spectral ranges. Focus is on interaction with atoms and molecules and the most important liquid in food: water. All three major applications in food processing, preservation, characterization, as well as heating, require inelastic interaction processes, i.e. those where the photon energy changes upon interaction. Therefore elastic scattering processes, the most well-known being Rayleigh scattering of EM radiation from the electrons of matter, are not treated here.

7.1 Low Frequencies: RF and Microwaves

At the low energy end of the used part of the electromagnetic spectrum in food processing, frequencies in the range from 10 to 50 MHz (RF) up to 2.5 GHz (microwaves [MW]) are used. They mostly act on the water molecules within food, i.e. they are used either for drying or for heating food products. The advantage to conventional heating (via convective heat transfer from outside to the product and then conductive heat transfer within the product) is that EM waves penetrate matter and this allows for volumetric heating. According to Eq. (10) and Table 1, the typical penetration depth z_0 of EM waves in water (i.e. attenuation to $1/e$) strongly depends on frequency. At RF frequencies, it can be above 1 m, but drops to around 2 cm at 2.45 GHz. This means that microwave heating works well for small dimensions but only heats the surfaces of large products. In contrast, RF heating with longer penetration depths can more gently heat within the whole volume. But what causes the water within the food to absorb energy?

Water is a polar molecule which can be represented as an electric dipole. Whenever a dipole is within an electric field, the electric forces generate a torque which rotates the dipole until it is parallel to the electric field (Figure 20). For static fields, this is equivalent to a macroscopic polarization. If the field orientation is changed by 180°, the dipole will accordingly rotate by 180° until it is again parallel to the field lines. If the electric field oscillates with a given RF frequency, so will the electric dipole. This means that the external electric field does work on the water molecules by inducing a rotation of the electric dipoles. During this induced rotation, the water molecules also collide with their

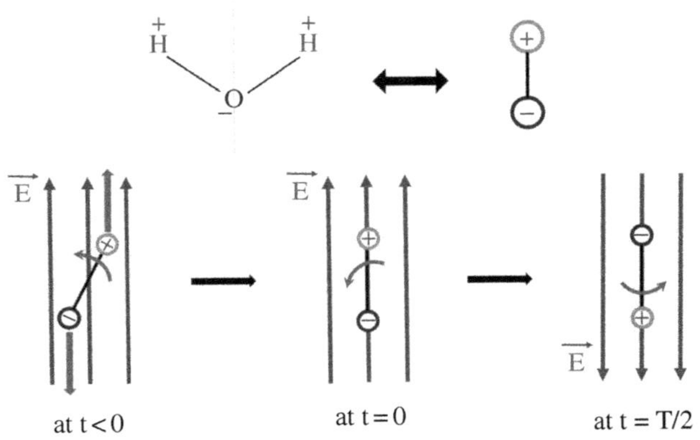

Figure 20 Water molecules are electric dipoles (top). When being in an electric field (bottom), they orient themselves parallel to the field. (*See insert for color representation of this figure*).

neighbors and hence efficiently transfer energy to their surroundings, i.e. heating up the surrounding matter.

There is, however, a constraint: if the mass of the dipole is very large and the frequency of oscillation is very high, the dipole may no longer be physically able to rotate by 180° in the same time the field changes by 180° due to its large inertia. As a consequence, the efficiency of the process will decrease for high frequencies. This explains the large drop of n_2 in the microwave region (Figure 12).

Additional microscopic excitation processes are not possible, as the photon energies associated with RF and microwave radiation are much smaller (for $\nu = 2.4\,\text{GHz}$, $E \approx 10^{-5}\,\text{eV}$) than those required for rotational or vibrational excitations and the available power is usually way below the levels required for multiphoton excitations.

Chapters 10 and 11 explain extensively the use of MW and RF, respectively, in food processing.

7.2 IR Radiation

Food systems are complex mixtures of different biochemical molecules, polymers, inorganic salts, and water (e.g. Krishnamurthy et al. 2008). As described in Section 6.2, molecules in the gas phase exhibit many rotational and vibrational energy levels. IR radiation has energies which correspond to respective energy-level differences. Hence, a major effect of incident IR radiation on food, containing molecules, is absorption due to excitations of rotations and/or vibrations.

Figure 21 schematically depicts the transmission spectrum through a gas with heteronuclear diatomic molecules. Transitions between rotational levels constitute the pure rotational bands at mid or far IR wavelengths. Transitions between a combination of rotational and vibrational levels lead to the so-called rotational–vibrational bands in the near IR range. These have very characteristic positions and usually consist of dozens of individual lines, thus constituting spectroscopic fingerprints of the molecule under study. It is mostly this part of the spectrum between around 1 and 20 μm which is used for spectroscopic characterization of molecules. Finally, if also a change of the electronic levels is taking place, one ends up with very complex band spectra in the UV and VIS range which consist again

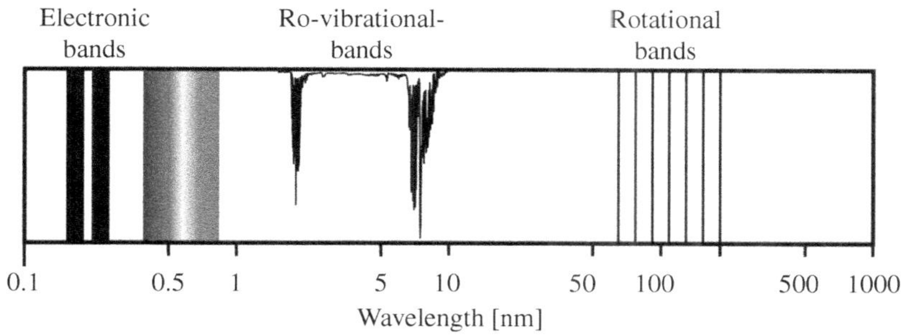

Figure 21 Schematic absorption spectrum of a hetero-nuclear diatomic molecule. (*See insert for color representation of this figure*).

of a large number of individual spectral lines in close proximity to each other. The absorption and re-emission spectra in the VIS range are often responsible for the perceived color. Besides, broad absorption bands may also serve heating purposes.

The energy-level diagrams of more complex molecules change of course; however, they all are characterized by a similar level structure, in particular ro-vibrational energy levels, which all constitute spectroscopic fingerprints for the given molecules allowing quantitative identification of constituents and, hence, characterization of a sample. Spectra of more complex molecules can show additional features such that, e.g. absorption and emission spectrum are shifted with respect to each other. Due to interaction with the surroundings, similar though spectrally less pronounced, i.e. broader, features result in liquids which contain a mixture of many different molecules.

In conclusion, there are two major application of IR radiation for food. First, one may use IR spectroscopy to analyze the constituents of food. Second, the absorption can be used for heating or drying purposes. This is possible as after absorption of IR photons, many large molecules in excited states have a high probability to return to the ground state not via IR radiation of the same energy (spontaneous emission in Figure 14) but rather via radiationless de-excitation processes (induced by interactions with neighbor molecules).

Unless extreme high power sources such as IR lasers are used, the IR radiation power will not suffice to initiate multiphoton processes, which could chemically alter the molecules.

Chapter 9 explains extensively the use of IR in food processing and Chapter 12 its use in food analysis. The use of Raman spectroscopy in food analysis is explained in Chapter 13 and the use of hyperspectral imaging, which can use IR, is explained in Chapter 15.

7.3 Visible and UV Radiation

Visible and UV radiation have much larger photon energies in the range of several to many electron volts. Concerning the microscopic structure of atoms and molecules, they can, therefore, not only lead to electronic excitations of atoms and molecules (Section 6), but also may have sufficient energy to ionize atoms or molecules or dissociate molecules. These three potential processes are schematically depicted in Figure 22 for diatomic molecules.

Once molecules are electronically excited (Figure 22a), they usually dissipate part of the energy radiationless via collisional energy transfer (or transfer from the electronic singlet

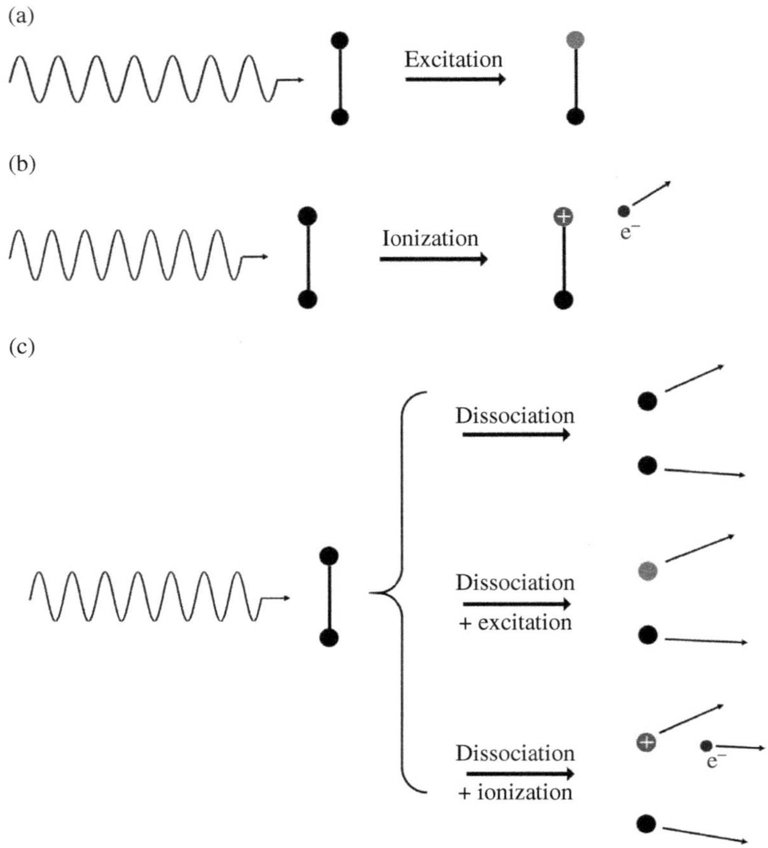

Figure 22 Three possible excitation processes of molecules by VIS or UV radiation. (*See insert for color representation of this figure*).

to the triplet system) but also reradiate a large part when returning to the electronic ground state. If, however, the photon energy is sufficiently large, the molecule may become ionized (Figure 22b). As a result, a positively charged molecule and an electron are separated.

Similarly, if the photon energy is sufficient to break the chemical bond between the atoms of the molecule (Figure 22c), it dissociates into its constituents, which may be either in the ground or in an excited state and in addition – depending on the available energy – they may even be ionized.

Ionized atoms or molecules in matter can also be radicals, i.e. chemically very reactive. In this case, the subsequent chemical reactions may lead to chemical changes. This means that UV excitations can severely change/damage the molecules and also affect their neighborhood. If this happens in organic living matter, the damage may destroy functionality of cells and cause diseases. A typical well-known example is skin cancer after prolonged intense UV irradiation.

As a result, VIS radiation may still be used for absorption and hence heating processes, whereas the higher energy UV radiation can already cause changes on the molecular level via ionization and dissociation.

Chapters 6 and 8 explain the use of UV light and pulsed light, respectively, in food processing, while Chapter 7 the application of visible light in food processing. Visible light imaging as a tool for food analysis is described in Chapter 14, and the use of hyperspectral imaging, which can use visible light, is explained in Chapter 15.

7.4 X-Rays and γ-Radiation

X-rays and γ-rays have still larger photon energies varying between 10 eV and 10 keV for X-rays and above 10 keV for γ-rays. Such photon energies are much larger than binding energies of the outer valence electrons, responsible for chemical bonds, and often still much larger than binding energies of inner electrons of heavy elements. Therefore, inelastic interaction processes usually involve ionization (see, e.g. Andreo et al. 2017; Eisberg and Resnick 1985). Three processes contribute to absorption of high-energy photons: the atomic photo effect, Compton scattering, and pair generation.

7.4.1 Atomic Photo Effect

Consider a heavy atom such as lead. The innermost electrons feel the attractive Coulomb force of a nucleus containing $Z = 82$ positive charges. The binding energy is quite similar to the one of a hydrogen atom (Eq. (12)), the main difference being that now $Z = 82$ and that there is some influence of the other 81 electrons. As a consequence, an innermost electron with quantum number $n = 1$ has a binding energy (i.e. an ionization energy) of around 88 keV. The electrons in the level with $n = 2$ are less tightly bound; their ionization energies are between 13 and 16 keV. Similarly, lighter atoms such as iron or aluminum have smaller positive charges in the nucleus and hence, respectively, lower ionization energies. For $n = 1$, they are around 7.1 keV for iron and about 1.56 keV for aluminum (for other values of all atoms, see LBL 2021).

Figure 23 depicts the absorption of a high-energy photon by a heavy atom. If the energy is large enough to excite an innermost electron into the continuum, the atom is ionized. As the electron is ejected (similar to the classical photoelectric effect), this process is called the

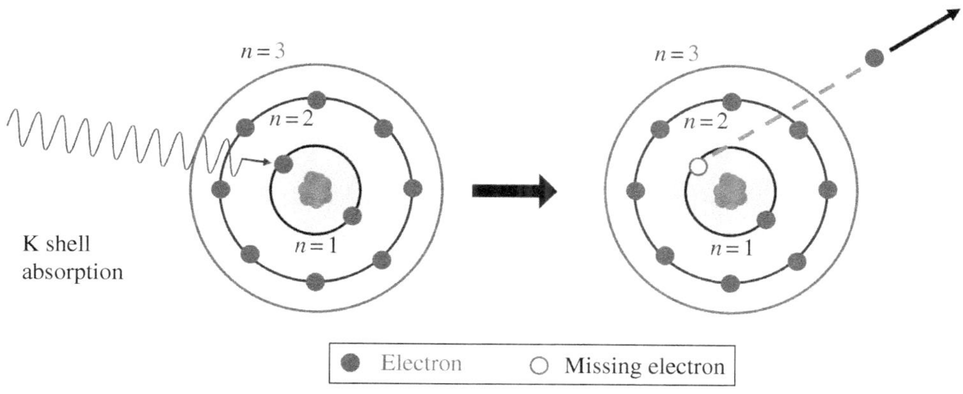

Figure 23 Atomic photo effect: a high-energy photon ionizes an atom by exiting inner bound electrons according to (atom) + hν (γ) → (atom)$^+$ + e$^-$.

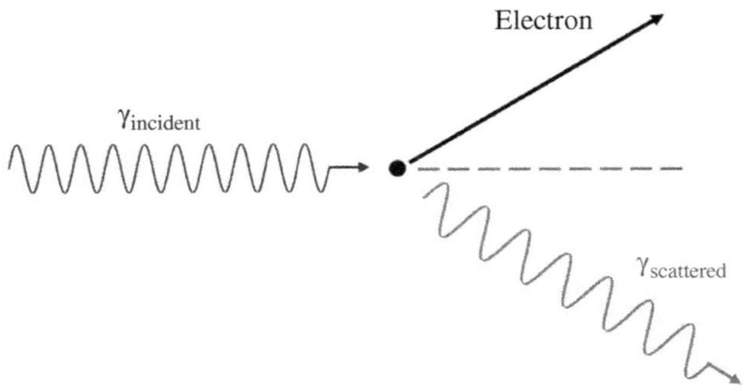

Figure 24 Compton effect: a high-energy photon is scattered from a nearly free electron according to (atom) + $h\nu_{inc}$ (γ) → (atom)$^+$ + $h\nu_{sca}$ (γ)→ e^-.

atomic photoelectric effect. Shortly after excitation, outer electrons may fill the vacancy either by emission of X-rays or transferring the energy difference to other outer electrons (Auger effect).

7.4.2 Compton Effect

About 100 years ago in 1922, Compton discovered that high-energy photons can be scattered from electrons in atoms, thereby losing part of their energy which is transferred to the electrons. Such a scattering process could only be understood in the particle description of the photon, i.e. while carrying energy as well as momentum. Figure 24 depicts the process schematically. The scattered photons may be scattered repeatedly or also be absorbed due to the photo effect.

7.4.3 Pair Generation Effect

Once the photon energy exceeds a certain threshold value, a third absorption process of high-energy photons becomes possible. According to Einstein's theory of relativity, each particle is associated with a rest energy $E_0 = m_0c^2$, where m_0 is the mass of the particle at rest. The elementary particles with the smallest mass are electrons with $m_0(e^-)$=9.109·10^{-31} kg. The corresponding rest energy of a single electron is m_0c^2 = 511 keV.

Another well-known feature of the modern theories of physics is that each elementary particle in our material world has a counterpart, known as anti-particle. It always has the same mass, but all other properties such as charge are the opposite of the real matter particle. In the case of electrons (e^-), the anti-particles are called positrons (e^+). Having the same mass, their equivalent rest energy is also 511 keV, as for the electrons. As a consequence, an electron–positron pair has a total rest energy of 1.022 MeV.

Whenever an anti-particle by chance meets a real article (i.e. when anti-matter encounters matter), both particles recombine, thereby creating energy in the form of high-energy photons, i.e. γ-rays. Of course, the reverse process is also possible, i.e. a high-energy photon may create an electron–positron pair (Figure 25) once the γ-energy is above the threshold value of 1.022 MeV. In order to satisfy energy and momentum conservation, a nucleus of an atom must be nearby; therefore, this process depends on the matter, traversed by the photons. Shortly afterward, the positron will annihilate again with the same or another

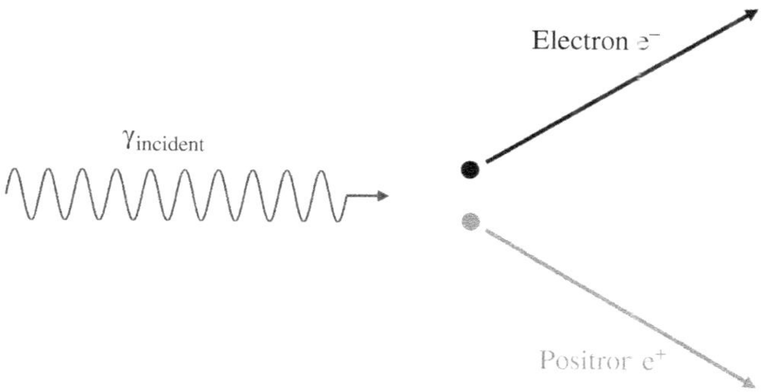

Figure 25 Pair generation: a high-energy photon with energy above 1.022 MeV can generate an electron–positron particle pair in the vicinity of a nucleus according to $h\nu\ (\gamma)$ + nucleus → $e^+ + e^-$ + nucleus.

electron. Due to energy and momentum conservation, two γ-ray photons with an energy of 511 keV each will be emitted in opposite directions.

7.4.4 Probabilities for Absorbing High-Energy Radiation

Figure 26 illustrates schematically the probabilities for absorption of high-energy photons due to the three processes discussed above for a material such as lead. Obviously, the photo effect will dominate at small energies. At intermediate energies, the Compton effect takes over, while at very large energies well above 1 MeV, pair generation will be most effective.

Plotting the absorption for lighter elements will first decrease the probabilities at given energy and second shift the photo effect structures (the K-shell, L-shell, etc., absorption features) to smaller energies.

Typically, the highest energy γ-rays used for food irradiation are from radioactive Co^{60} and Cs^{137} γ sources. Co^{60} nuclei emit γ photons with energies of 1.173 and 1.332 MeV, whereas Cs^{137} emits γ-rays of a lower energy of 661.7 keV. In practice (see Figure 26), this means that only the atomic photo effect and the Compton effect contribute to absorption in matter.

7.4.5 Consequence of Absorption of High-Energy Photons by Matter

γ-ray radiation is usually described in the context of radioactivity as ionizing radiation, similar to highly energetic α-particles (He^{++}-nuclei) and β-particles (electrons). Whereas He^{++} nuclei and electrons are themselves electrically charged and do therefore very effectively react with matter via electrodynamic interaction, γ-radiation does so indirectly. At first, the photons are absorbed via one of the three processes described above, in food processing mostly the photo and the Compton effects. Thereby, however, electrons are generated in matter, which still have quite high energies (though less than the initial γ-rays). It is these electrons (and additional ones created by further de-excitation effects and/or additional scattering) which – via their secondary effects – affect the molecules and matter.

Whenever free (i.e. not bound in atoms) electrons travel within matter, they suffer a large number of inelastic collisions due to electromagnetic interactions with the positive charges of the nuclei and the negative charges of the electron shells of the respective atoms.

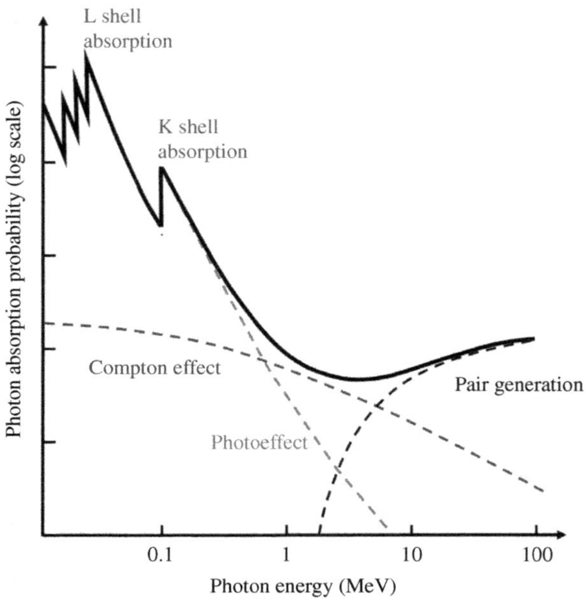

Figure 26 Schematic representation of the probability of absorption of high-energy photons due to the three processes: photo effect, Compton effect, and pair generation. The example is meant schematic, although its energy scale corresponds to about lead atoms. Changing the scale to smaller energies will explain the absorption for lighter atoms. However, the relative contributions between photo effect, Compton effect, and pair generation will change.

Focusing only on inelastic collisions, electrons transfer around between 30 and 100 eV during each collision. This energy is used up for excitation of atoms and molecules, ionization of atoms and molecules, as well as dissociation of molecules, i.e. breaking of chemical bonds. Thereby, also highly reactive molecular species, called radicals, are formed and subsequent chemical reactions take place. As a result of all of these processes, organic matter molecules frequently lose their biological function due to radiation-induced DNA changes, finally leading to cell death.

This is exactly the desired effect in irradiating food products with high-energy radiation (FDA 2021). It may, e.g. be used to eliminate illness-causing bacteria such as *Salmonella* or *Escherichia coli* or microorganisms that cause spoilage.

It is worth to mention that in the food industry, not only X-rays and γ-rays but also highly energetic electrons are used for food irradiation. Whereas X-rays and γ-rays generate secondary electrons due to the abovementioned processes within the irradiated product, one may also directly irradiate it from outside with electrons. The maximum allowed energy for electron beams in food processing is 10 MeV and therefore about a factor of 8 larger than the typically used γ-rays.

As a consequence, there are at least three major differences between electron and γ-ray irradiation based on the underlying physics.

First, it is easier to generate intense electron beams with higher energy per electron. Therefore, the energy which may be applied to a given food product in the same time interval is usually larger for electron beams than for γ-rays. The latter depend on radioactive sources, and for security reasons, their activity and hence the potential maximum number of photons from a γ-source per second are limited.

Second, an electron source may be just turned on and off, whereas a radioactive source will change its activity slowly according to the laws of radioactive decays. Therefore, safety issues while not irradiating are easier to comply with for the electron beams.

Third, however, and this is the only, but important disadvantage of electron beams and advantage of γ-rays, one must consider the respective penetration depth within the irradiated sample.

Electrons lose their energy fast: typical β-radiation can already be easily shielded by a few mm of ordinary matter; hence, electrons cannot penetrate food products very deeply. In water-based products or tissue, electrons of 10 MeV (or 1 MeV) energy have penetration depths of around 5 cm (or 5 mm). Therefore, it does not make much sense to irradiate much thicker samples. In contrast, γ-radiation has a much larger range, i.e. the secondary electrons which are causing the desired effects can be generated deeply within the volume of the product; hence, thicker products may be irradiated.

Chapter 2 gives details on radiation dosimetry, whereas Chapters 3 and 5 give a deep overview of the application of X-rays and γ-radiation in food processing. Chapter 4 does so for electron beam irradiation.

8 Outlook

The use of electromagnetic radiation in the food industry has changed appreciably in the past whenever either the range was extended or a previously unused intermediate range was becoming accessible due to the introduction of new technologies. Nowadays, it seems unlikely that the range will be further extended beyond the presently used high energetic γ-rays or the low-energy radio waves. However, it may well be – probably it should be even expected – that the development of new radiation sources and detectors in some subranges will lead to innovations, which will also find their way to food processing and characterization.

References

Alonso, M. and Finn, E.J. (1968). *Fundamental University Physics, Vol. III: Quantum and Statistical Physics*. Addison Wesley.

Andreo, P., Burns, D.T., Nahum, A.E., and Seuntjens, J. (2017). *Fundamentals of Ionizing Radiation Dosimetry*. Wiley.

Berkeley, *Berkeley Physics Course*, Vol. 1–5, McGraw Hill (1965–1967), various authors for the volumes, see in particular Vol. 2: *Electricity and Magnetism*; Vol. 3: *Waves*; and Vol. 4: *Quantum Physics*.

Born, M. (1972). *Optik*, original ed. 1932, reprint, 3e. Springer.

Eisberg, R. and Resnick, R. (1985). *Quantum Physics of Atoms, Molecules, Solids, Nuclei and Particles*. Wiley.

Etkina, E., Planinsic, G., and Van Heuvelen, A. (2019). *College Physics, Explore and Apply*, 2e. Pearson.

FDA (2021). Food Irradiation: What You Need to Know. https://www.fda.gov/food/buy-store-serve-safe-food/food-irradiation-what-you-need-know (accessed 23 June 2021).

Feynman, R.P., Leighton, R.B., and Sands, M. (1964). *The Feynman Lectures on Physics*. Addison Wesley, see also https://www.feynmanlectures.caltech.edu/info/.

Guenther, R. (1990). *Modern Optics*. Wiley.

Günzler, H. and Gremlich, H.-U. (2002). *IR Spectroscopy*. Wiley.

Halliday, D., Resnick, R., and Walker, J. (2014). *Fundamentals of Physics*, 10e. Wiley.

Hecht, E. (2016). *Optics*, 5e. Pearson.

Jackson, J. (1975). *Classical Electrodynamics*, 2e. Wiley.

Krishnamurthy, K., Khurana, H.K., Soojin, J. et al. (2008). Infrared heating in food processing: an overview. *Comprehensive Reviews in Food Science and Food Safety* 7: 2–13.

LBL (2021). Electron binding energies, in electron volts, for the elements in their natural forms. https://xdb.lbl.gov/Section1/Table_1-1.pdf (accessed 23 June 2021), data from the Lawrence Berkeley National Laboratory.

Palik, E.P. (ed.) (1985). *Handbook of Optical Constants of Solids*, vol. 1. Boston, MA: Academic Press.

Palik, E.P. (ed.) (1991). *Handbook of Optical Constants of Solids*, vol. 2. Boston, MA: Academic Press.

Palik, E.P. (ed.) (1998). *Handbook of Optical Constants of Solids*, vol. 3. Boston, MA: Academic Press.

Saleh, B.E.A. and Teich, M.C. (2019). *Fundamentals of Photonics*, 3e. Wiley.

Shaw, J.A. and Vollmer, M. (2017). Atmospheric optics in the near infrared. *Applied Optics* 56 (19): G145.

Tipler, P.A. and Mosca, G.P. (2007). *Physics for Scientists and Engineers*, 6e. W.H. Freeman.

Vollmer, M. (2004). Physics of the microwave oven. *Physics Education* 39 (1): 74–81.

Vollmer, M. and Möllmann, K.-P. (2018). *Infrared Thermal Imaging: Fundamentals, Research and Applications*, 2e. Wiley.

2

Dosimetry in Food Irradiation

Bhaskar Sanyal[1,2] *and Sunil K. Ghosh*[2]

[1] *Food Technology Division, Bhabha Atomic Research Centre, Mumbai, India*
[2] *Homi Bhabha National Institute, Anushaktinagar, Mumbai, India*

1 Introduction

Food is one of the basic necessities in life. The world population is increasing with much faster rate than the rate of food production. Therefore, to provide food to the growing population, production of food must be enhanced and at the same time postharvest food losses must be significantly reduced. Radiation processing of food with its unique advantages is one of the important approaches to mitigate the postharvest food losses (Ashraf et al. 2019). Soon after the discovery of X-rays in 1895, and γ-rays in 1896, it was discovered that germicidal properties of ionizing radiation can help preserve food. In early twentieth century, patents were filed for using ionizing radiation for food preservation. Food irradiation is an effective technique for value addition to food and agricultural commodities. This value addition includes prevention of losses caused by sprouting of bulbs and tubers, insect infestation of grains and pulses and their products during storage, ripening and senescence of fruits and vegetables. The technology is also used to pasteurize and even sterilize food and allied products for enhancing their shelf life and microbiological safety. It is also increasingly gaining market access for agricultural and horticultural produce to meet sanitary and phytosanitary requirements of international trade (Sharma and Madhusoodanan 2012).

Dosimetry is the measurement of the absorbed dose and a method to determine the dose which has been eventually absorbed by the food commodities during radiation processing. Radiation absorbed dose is the key quantity that governs the process. Proper delivery of dose to the food commodities directly controls the effectiveness of radiation processing of food (Codex Standard 106-1983, Rev.1-2003 2003). The dose ranges normally used in food irradiation to achieve various objectives can be classified into three groups, such as (i) applications of low-dose range (20 Gy–1 kGy), where sprouting of various agricultural produces such as potatoes, onions, garlic, shallots, etc., can be inhibited using ionizing radiation in the dose range of 20–150 Gy; physiological processes like ripening of fruits can be delayed in the dose range of 0.2–1 kGy. Insect disinfestation can be achieved by radiation in the dose range of

Electromagnetic Technologies in Food Science, First Edition. Edited by Vicente M. Gómez-López and Rajeev Bhat.

0.2–1 kGy to prevent losses caused by insect pests in stored grains, pulses, cereals, flour, spices, nuts, dried fishery products, etc. (ii) Applications at medium-dose levels (1–10 kGy) are associated with the enhancement of the keeping quality of certain foods through a substantial reduction in the number of spoilage causing microorganisms. A product-dependent dose in the range of 1–10 kGy can be delivered to fresh meat and seafood, as well as vegetables and fruits to achieve the objective. (iii) Applications at high-dose levels (10–70 kGy), where 10–30 kGy radiation dose is an effective alternative to the chemical fumigant ethylene oxide for microbial decontamination of dried spices, herbs, and other dried vegetables. A significant reduction in the total microbial load present in such products including pathogenic organisms is achieved. Radiation sterilization in the dose range of 25–70 kGy extends the shelf life for many food commodities almost indefinitely. This treatment is delivered to the enzymatically inactivated food or precooked food where steaming is normally employed for enzyme inactivation and kept in hermetically sealed containers. Therefore, it is obvious that the quality of the irradiated food product and success of this technology are governed by the controlled delivery of the radiation dose to the food commodities and their accurate measurements. In addition, adequate dosimetry practices are essential to gain confidence of the consumers of the irradiated foods and to comply with the regulatory requirements.

Industrial radiation processing such as irradiation of foodstuffs is highly regulated, about dose. Especially for food destined for international trade, accurate dosimetry techniques are of paramount importance for dose determination and the internationally accepted procedures must be in place to monitor the process. The process reliability is evaluated by the accurate dosimetry which should be traceable to national or international standards ensuring independent control of the process. The activities of principal concern in any food irradiation facility are process validation and process control. In addition, dosimetry is essential to scale up from the research-level food irradiation project to an economically viable industrial level.

The objective of this article is focused to give relevant information on general concept of dosimetry and its application in food. In this chapter, the fundamentals of radiation dosimetry associated with all the three types of approved ionizing radiations namely gamma (γ), electron beam, and X-rays are discussed in view of the physical aspects of radiation absorption. Dosimetry is a field of applied physics, and basic understanding of the interaction of ionizing radiation with matter is essential to employ the dosimetric tools in practical domain. A special emphasis has been given on various dosimetry techniques commonly used in commercial radiation processing facilities. The importance of dosimetry toward successful implementation of radiation technology for food is elaborated with various specified details. This chapter will provide information on food irradiation dosimetry to all the users engaged in academics as well as in industries associated with commercial radiation processing of food. Finally, the emerging areas in the field of food irradiation dosimetry has also been discussed to draw the attention of researchers to come up with new ideas and help address the needs of the industries.

2 Fundamentals of Dosimetry

In radiation processing of food, radiation sources with high intensity are used. These include electron accelerators, X-ray machines, and radionuclide-based irradiators containing either cobalt-60 or cesium-137 sources. The type and geometry of the source control the method of generating ionizing radiation. In case of monodirectional and scanned beams, electrons and X-rays are utilized. On the other hand, isotropic emission of γ radiation is achieved from the rectangular plaque or cylindrical sources available in radionuclide-based irradiation facilities. The limits of radiation energy for the machine sources suitable for food irradiation are approximately 7.5 MeV for X-rays and 10 MeV for electron beam. The approximate absorbed dose used in food processing is in the range of 0.02–25 kGy. It is essential that the personnel responsible for the operation of food irradiation facilities should possess basic understanding of radiation physics or engineering and dosimetry (Food and Agriculture Organization, World Health Organization, Codex General Standard 1984; McLaughlin et al. 1989; IAEA-ICGFI Document 1992; IAEA Technical Reports Series No. 409 2002). In addition, it is indispensable that the involved agencies such as the supplier or producer of the food product interested to use this technology and the facility management should have adequate understanding of the food engineering problems. This helps them to appreciate the importance of the radiation dosimetry in process control.

2.1 What is Dosimetry

Dosimetry is a field of science where radiation dose is determined by experimental measurement, theoretical calculation, or a combination of both. To get an accurate and trustworthy information on the relevant effects of ionizing radiation, the quantity "absorbed dose" is of immense importance in radiation processing of foodstuffs. Dosimetry is the only approach by which the regulatory authority responsible for the acceptance of the foodstuffs gets information that the dose uniformity inside the process load is achieved and the treatment is carried out within the acceptable dose range.

2.2 Absorbed Dose

The absorbed dose also referred as "dose," *D*, is the amount of energy absorbed per unit mass of irradiated material at a point in the region of interest. It is defined as the mean energy, $d\varepsilon$, imparted by ionizing radiation to the matter in a volume element divided by the mass, dm, of that volume element:

$$D = dD/dm \tag{1}$$

The SI derived unit of absorbed dose is the gray (Gy), which replaced the earlier unit of absorbed dose, the rad:

$$1\ \text{Gy} = 1\ \text{J kg}^{-1} = 100\ \text{rad}$$

The absorbed dose rate, *D*, is defined as the rate of change of the absorbed dose with time:

$$D = dD/dt \tag{2}$$

In practical situations, *D* and *D*˙ are measurable only as average values in a larger volume than is specified in the definitions, since it is generally not possible to measure these quantities precisely in a very small volume in the material.

It is essential to clearly mention the absorbed dose in the material of interest because for any given irradiation conditions different materials have different radiation absorption behaviors. Dosimeters measure the average dose in the material with which it is composed of over the volume of the dosimeter. It can be converted into absorbed dose in other materials using following relation:

$$D_M = D_D \times \left(\mu_{en}/\rho\right)_M / \left(\mu_{en}/\rho\right)_D \tag{3}$$

where, D_M is the absorbed dose to the material, D_D is the absorbed dose to the dosimeter, $(\mu_{en}/\rho)_M$ is the radiation energy absorption coefficient of the material, and $(\mu_{en}/\rho)_D$ is the radiation energy absorption coefficient of the dosimeter.

Regarding radiation interaction properties, most foods behave essentially as water because water is the major constituent of food. Therefore, the material of interest is generally specified as water in case of radiation processing of food.

2.3 Physical Aspects of Radiation Absorption

Multiple interactions take place that give rise to secondary particles when ionizing radiations such as X-rays, γ-rays, or electrons are incident on a medium. The interactions predominantly include ionization and excitation resulting production of secondary electrons and photons with lower energies (Attix 1986). These particles undergo further interactions producing the so-called cascading effect.

In the case of photons namely γ-rays from cesium-137 and cobalt-60, and X-rays approved for food irradiation, the main interactions are due to (i) photoelectric effect, (ii) Compton scattering, and (iii) pair production.

2.3.1 Photoelectric Effect

In this interaction, the photon imparts its complete energy to an orbital electron of an atom in the medium and the ejected electron is known as photoelectron.

A photon of energy E will release an electron with kinetic energy $E_e = E - \Phi$, where Φ is the binding energy of the electron in that orbit. Some salient features of photoelectric process are (i) process involves bound electrons, (ii) the probability of ejection of an electron is maximum when the photon energy higher than the binding energy of the electron, and (iii) the photoelectric absorption coefficient varies with energy.

2.3.2 Compton Scattering

In this process, a γ-ray interacts with a free or weakly bound electron and transfers part of its energy to the electron. This interaction involves the outer, least tightly bound electrons in the atom of the medium. In the process, the incoming photon is scattered through an angle with respect to its original direction.

2.3.3 Pair Production

A photon with energy greater than twice the rest mass of electron (1.022 MeV), under the influence of the strong electromagnetic field near a nucleus, can create an electron and positron pair. In this interaction, the photon is absorbed and the nucleus of the absorber material receives negligibly small amount of recoil energy to conserve momentum. Some salient features of pair production are (i) pair production involves an interaction between a photon and the nuclear field, (ii) threshold energy is 1.022 MeV, and (iii) probability of pair production increases with the square of the atomic number Z^2 of the medium and with increase in energy of the photon.

2.3.4 Interaction of Charged Particles

Charged particles lose energy by colliding with the electrons and nuclei of atoms in the medium. If the energy transferred to an electron in the medium is sufficient to remove it completely out of the atom, the process is referred to as ionization. If the electron is just raised to a higher energy state, the process is called excitation. Thus, ionization and excitation occur simultaneously along the path of charged particle in a medium. The number of ion pairs produced per unit path length of charged particle is called specific ionization. The specific ionization is directly proportional to the mass and charge of a particle and inversely proportional to its velocity. Energy absorbed in the medium per unit path length of the particle is called its linear energy transfer (LET). It is usually expressed in keV μm^{-1}. Biological effects depend on the rate of energy absorption in the medium. Therefore, the concept of LET is important to understand the biological end points. Electrons during their interactions with the material give rise to secondary particles with lower energies. These secondary particles mostly move forward at various angles to the primary beam direction.

In case of the monoenergetic incident radiation, initially there is an increase (buildup) of energy deposition (dose) near the incident surface of the interacting material if the angle of incidence at the surface of irradiation is approximately perpendicular and monodirectional. This region is known as buildup zone and extends up to the depth corresponding to the average range of the first interaction secondary electrons. The energy deposition is then followed by an exponential decay following the radiation attenuation behavior to greater depths as shown in Figure 1 (Rizzo 1971; IAEA Technical Reports Series No. 409 2002).

3 Dosimetry Systems for Food Irradiation Application

The requirement of accurate dose measurement techniques must not be underestimated because absorbed dose is the quantity which relates directly to the desired effect in a specific food. This is realized by analyzing the consequences of using inadequate techniques leading to under or overexposure of the product and consequently failure to administer an effective treatment. The consequences to the processor can be both legal and economic. In case of consumers, they may not only suffer an economic loss by having to discard an inadequately processed product but also lose confidence in the food irradiation technology. Dosimetry system consists of dosimeter with the readout tool. Appropriate dosimetry system is used to measure absorbed dose and to ensure that the food product receives the desired treatment. An appropriate dosimetry system is selected based on specific application and matching performance with the criteria. ASTM Standard Guide ASTM E1261 (2000) describes this selection process. There are several dosimetry systems available for

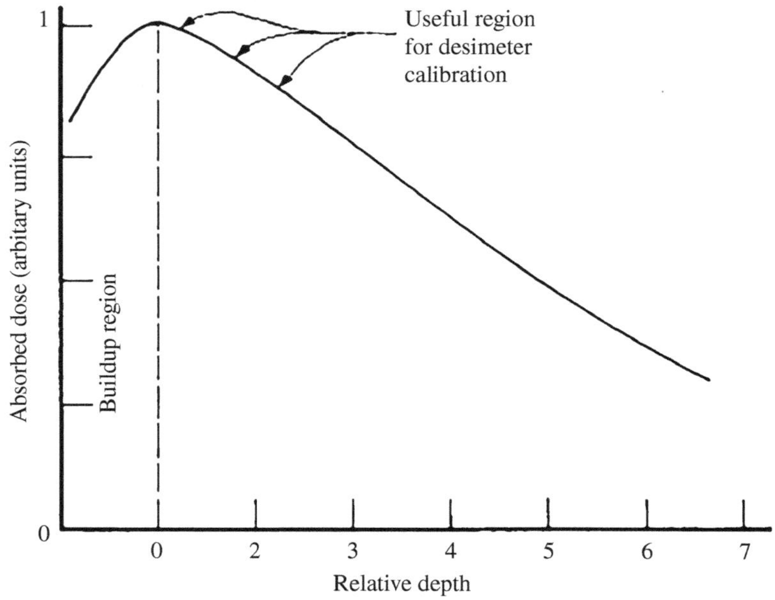

Figure 1 Radiation energy deposition as a function of thickness (depth) in an absorber showing the buildup region, the thickness required to achieve electron equilibrium (the dashed vertical line), and the useful region for dosimeter calibration. *Source:* IAEA Technical Reports Series No. 409 (2002).

different applications. Dosimetry systems may be categorized into the following four classes in accordance with their relative quality and areas of application:

- *Primary standard dosimetry systems*: Primary standard dosimeters are established and maintained by national standards laboratories for calibration of radiation fields. The two most commonly used primary standard dosimeters are ionization chambers and calorimeters.
- *Reference standard dosimetry systems*: Reference standard dosimeters are used to calibrate routine dosimeters and radiation fields. They may also be used as routine dosimeters. For high-dose applications such as for radiation processing of food, the acceptable reference standard dosimeters are ferrous sulphate (Fricke), dichromate, ceric-cerous sulphate, ethanol-chlorobenzene, and the alanine-EPR (EPR: electron paramagnetic resonance for readout).
- *Transfer standard dosimetry systems*: Transfer standard dosimeters are specially engaged in establishing traceability. The transfer dosimeters are deployed in transferring dose information from an accredited calibration laboratory or a national standards laboratory to an irradiation facility to carry out dose intercomparison studies. These dosimeters should be stable, sufficiently precise, and rugged during transport to establish the appropriate traceability for the processing facility. Examples of transfer standard dosimeters are alanine, dichromate solution, ceric-cerous sulphate solution, and ethanol-chlorobenzene solution.

- *Routine dosimetry systems*: The main functions of routine dosimeters are different from those of primary and reference standards dosimeters. Routine dosimeters are deployed for process control during product irradiation. They are mainly used as simple system to measure absorbed dose distributions in process loads (Humphreys et al. 1973; McLaughlin et al. 1982; Ehlermann 1996; Walker et al. 1997) and to monitor variations in radiation fields in regular basis. The routine dosimeters may not be as accurate or reproducible in their radiation response as reference dosimeters but they are generally more user-friendly and versatile for regular use. In addition, the routine systems must be relatively inexpensive and must have a response that can easily be measured since they are used in much greater number and frequency. Routine dosimeters may not be sufficiently stable and independent from environmental or radiation field conditions. They may also show intra- and inter-batch variations. In view of these, a frequent calibration of the routine dosimeter against reference or transfer standards is an essential need to achieve dose measurements with desired accuracy. Several dosimetry systems such as plastic plates, pellets or films, chemical solutions, dye systems, phosphors, or glasses are used for routine food irradiation dosimetry.

Food irradiation applications exhibit some very specific constraints which lead to limited choice of dosimetry systems and require adequate calibration procedures. In addition, there could be a significant variation in food product densities and radiation processing at sub-ambient temperatures (chilled and frozen) for the perishable food commodities. The dose range cannot be covered by a single dosimetry system and more than one system may be needed at food irradiation facilities (Table 1).

3.1 Characterization of Dosimetry Systems

The reliability of a dosimetry system increases with increasing understanding of its behavior. A prior characterization of the selected dosimetry system is essential before using for dose measurement in any food irradiation facility. The important steps of characterization are elaborated below.

3.1.1 Calibrating the Dosimetry System

Calibration is the relationship between the absorbed dose and the radiation-induced effects in dosimeters evaluated using the integrated measurement instrument. The major steps of calibration procedure are (i) irradiation of dosimeters to several known absorbed doses over the useful dose range of the system, (ii) response evaluation of the irradiated dosimeters using the calibrated measurement instrument, and (iii) construction of a calibration curve. Calibration must be carried out on each new batch of dosimeters along with the calibration of the measurement instrument.

3.1.2 Establishing Traceability

A mechanism should exist within each country to ensure that all measurements can be related to the national standard through an unbroken chain and may be termed as traceability chain. The food irradiation facility should be able to demonstrate by means of traceability chain of comparisons that a measurement agrees within the acceptable limits of uncertainty with the recognized standards.

Table 1 Useful dose ranges for various routine and reference standard dosimeters.

Dosimeter	Mode	Readout system	Usable absorbed dose range in Gy	ASTM No.
Radiochromic film	Electron/X-ray	Spectrophotometer	$1–1.5\times10^5$	51275
Polymethyl methacrylate	X-ray	Spectrophotometer	$10^2–1.5\times10^5$	51276
Radiochromic optical wave guide	X-ray	Spectrophotometer	$1–10^5$	51310
Dichromate	X-ray	Spectrophotometer	$2\times10^3–5\times10^4$	51401
Ethanol-chlorobenzene solution	X-ray	Spectrophotometer, color titration, high-frequency conductivity	$10–2\times10^5$	51538
Alanine	Electron/X-ray	Electron paramagnetic resonance spectrometer	$1–10^5$	51607
Calorimetric	Electron	Calorimeter	$10^2–5\times10^4$	51631
Cellulose tri acetate	Electron/X-ray	Spectrophotometer	$5\times10^3–3\times10^5$	51650
Fricke solution	X-ray	UV spectrophotometer	$20–4\times10^2$	51026
Ceric-cerous sulfate	X-ray	Potentiometry	$5\times10^2–5\times10^4$	51208

3.1.3 Determining Batch Homogeneity

Evaluation of the extent of the dose response variability of a given batch of dosimeters is a prerequisite before the deployment of the dosimeters. The coefficient of variation, CV (%) = 100×(standard deviation/mean), should be less than 2% for routine dosimeters (IAEA Technical Reports Series No. 409 2002).

3.1.4 Determining Uncertainty in the Measured Dose Value

The uncertainty in dose measurements is a parameter that characterizes the distribution of the values that could reasonably be attributed to the dose. The measured or estimated dose should be accompanied by the uncertainty statement for a meaningful measurement. The uncertainty provides the degree of accuracy of the measured dose.

3.1.5 Understanding and Quantifying Effects of the Influencing Quantities

There are several quantities which can influence the dose measurements in varying degrees. Some of the common quantities are temperature, humidity, oxygen content of the dosimeter, dose rate, radiation type (γ-rays or electrons), and energy of radiation. It is important that these effects should be carefully understood and their influences should be taken into consideration to reduce the uncertainty in the dose measurement.

3.2 Specific Dosimetry Systems for Food Irradiation Applications

Food irradiation specifications usually include an upper and lower limit of absorbed dose and minimum dose to ensure the intended beneficial effect and a maximum dose to avoid product degradation. One or both values may be prescribed by regulations that have been established based on available scientific data. There are several potential dosimeters available to cover the entire dose range of food irradiation and few are discussed below.

3.2.1 Chemical Dosimeter (Fricke and Ceric-cerous Sulphate)

Chemical dosimeter is a system where absorbed dose is derived from concentration changes in reaction products or in the solute when it is exposed to radiation. In all chemical dosimeters, radiation-induced chemical reactions produce at least one initially absent species which is long-lived to determine its quantity or the change in properties of the initial system. The concentration change is proportional to the absorbed dose. In chemical dosimeter, the knowledge of a quantity known as radiation chemical yield (G) is required to calculate the absorbed dose. There are different types of chemical systems such as liquid, solid, and gaseous. In liquid systems, the aqueous dosimeters such as Fricke and Ceric-cerous sulphate dosimeters are of paramount importance in food irradiation dose measurements.

a) *Fricke dosimeter:* It is the best known and most broadly applied system for low-dose measurements in food irradiation in the range of 20–400 Gy. The Fricke dosimeter is widely accepted as a standard in radiation chemistry and because of its accuracy (1–2%) and reliability, it is often used to calibrate other dosimeters (ASTM Standard, E1026 2004). The standard dosimetric solution consists of air-saturated solution of ferrous sulphate or ferrous ammonium sulphate (1 mol m^{-3}) in sulphuric acid (400 mol m^{-3}) and sodium chloride (1 mol m^{-3}). The ferrous ions in Fricke solution are oxidized to Ferric ions upon exposure to ionizing radiation. The concentration of ferric ions gives the measure of absorbed dose. The concentration is measured spectrophotometrically. The major advantage of this system is that trained dosimetrist can prepare this dosimeter in laboratory. Intercomparinson studies with national laboratory should be carried out before use. The limitation of Fricke system is its narrow usable dose range. However, for low-dose requiring agricultural commodities, this dosimeter is most preferred.
b) *Ceric-cerous sulphate dosimeter:* It is a chemical system based on radiation-induced reduction of ceric ions to cerous ions in acidic aqueous solution. This dosimeter is normally standardized against Fricke system and provides a reliable means for measuring absorbed dose to water. As a reference dosimeter, spectrophotometry is used to evaluate absorbed dose by a change (decrease) in optical absorbance at specified wavelength in the ultra-violet region. However, for routine dosimetry, where large numbers of dosimeters are to be evaluated, potentiometry is used. The potentiometric reader directly reads the absorbed dose when each of its two electrodes is dipped in unirradiated and irradiated solutions, respectively. The recommended concentrations for the ceric-cerous dosimeter to measure absorbed doses from 5 to 50 kGy (high-range dosimeter) are 15-mM ceric sulfate and 15-mM cerous sulfate. On the other hand, for the measurement of absorbed doses in the range of 0.5–10 kGy (low-range dosimeter), the recommended concentrations are 3 mM of both ceric and cerous sulphate (ISO/ASTM 51205:2002(E) 2002). Ceric-cerous sulphate system is extremely useful for its wide dose

ranges based on the concentrations. The preparation of this system requires trained chemist and sterilized glass wares because of its sensitivity toward organic impurities. Both the chemical systems can be used during irradiation at ambient temperature with appropriate temperature correction factors. In food irradiation, several perishable foodstuffs are irradiated at sub-ambient temperatures such as chilled (0–4 °C) and frozen (−20 °C) state where these chemical systems cannot be used. However, the dosimeters can be placed at reference position where the temperature is at ambient and required doses can be calculated with the established mathematical relations.

3.2.2 Alanine Dosimeter

The alanine, one amino acid, is a well-accepted dosimeter for various applications in radiation dosimetry. Alanine-EPR dosimetry systems are used as reference or transfer standard. This system can also be employed for routine dosimetry in food irradiation application. The dosimeter contains crystalline alanine and registers the absorbed dose by the formation of alanine-derived free radicals. Identification and measurement of alanine-derived free radicals are performed by EPR spectroscopy. With careful adjustment of the parameters of EPR spectrometer, the dosimeters can be employed to measure the dose in a wide range of 1 Gy to 100 kGy. The noteworthy advantage of this system is that the measurement of free radicals by EPR spectroscopy is nondestructive and can be read out repeatedly resulting archival usage. The stability of the signal intensity in the dosimeter is dose-dependent and the fading is less than 3% in a year for doses below10 kGy (Sleptchonok et al. 2000). The EPR signal amplitudes of irradiated alanine dosimeters have been shown to be equivalent for photon and electron-absorbed doses (Onori et al. 1990). The stability and ruggedness of the alanine dosimeter have made this system as the first choice for the dose intercomparison studies. The limitation of this dosimetry system is its cost.

3.2.3 Radiochromic Dosimeter

The leuko dyes are colorless dyes which change color when exposed to ionizing radiation. Radiation-induced darkening of leuko dyes has been used as the basis of various dosimeters for the measurement of ionizing radiation doses suitable for food irradiation applications. There are two types of radiochromic systems, one is film and the other is radiochromic liquid dosimeter. In case of radiochromic film, chemical reactions take place creating or enhancing, or both, optical absorption bands. Absorbance is determined within these radiation-induced absorption bands using a spectrophotometer or photometer (ISO/ASTM 51275:2013(E) 2013). The film dosimeters are most suitable for the dosimetry in electron beam accelerator-based food irradiation facilities. The useful dose range is wide starting from 1 Gy to 100 kGy. In case of radiochromic liquid dosimeter, ionizing radiations influence chemical reactions in the radiochromic solution modifying the amplitudes of optical absorption bands. Absorbance values are measured at the selected wavelength(s) within these affected absorption bands (ISO/ASTM 51540:2004(E) 2004). A dose range of 0.5 Gy to 40 kGy can be measured. Because of this wide dose range, radiocromic dosimeters can cover entire dose range suitable for food irradiation. This system is used as routine dosimeter, especially for the phytosanitary application of fruits using radiation to overcome quarantine barrier for international trade.

3.3 Role of Product Density in the Absorbed Dose

Ionizing radiation can modify physical, chemical, and biological properties of the irradiated food within the desirable limits if the product is subjected to adequate dose. Usually, each food product has a pair of dose limits ($Dose_{max}$ and $Dose_{min}$) defining the acceptable dose window in such a way that every part of the product receives dose within the stipulated dose range. The ratio of the upper dose limit to the lower dose limit usually referred to as dose uniformity ratio (DUR). In a commercial food irradiator, the dimension of the product container and the source geometry are fixed. Therefore, the non-uniformity in absorbed dose would mainly be influenced by the variation in the bulk density of the process load. The behavior of DUR and absorbed doses (D_{max} and D_{min}) with increasing bulk density of the process load was studied in a category IV irradiation facility having both side irradiation geometry (Sanyal et al. 2017). It was observed that the $Dose_{min}$ which is the technologically required dose to achieve the objective of the food irradiation should be measured carefully with bulk density of the actual process load. On the other hand, DUR showed a monotonous increase with increasing bulk density of the process load. The results suggested that the shape and size of the individual food products and the reference materials used to simulate the actual process load would influence the final absorbed dose in a commercial food irradiation facility.

4 Dosimetry in Food Irradiation Facility

Absorbed dose to the food products is affected by the design and construction of the food irradiation facility. The irradiator design is therefore of paramount importance to ensure adequate dose delivery to the food commodities. The major focus in the design of a food irradiation facility is associated with the dose uniformity in the irradiated product, efficient utilization of the radiation energy, and cost-effectiveness. The sources of ionizing radiation approved for food irradiation are (i) radionuclide (cobalt-60 and cesium-137) and (ii) electron beam and X-ray from linear accelerator (LINAC). Radionuclide sources emit radiation in all directions. Thus, the core of the irradiation chamber is designed in such a way that the process loads are distributed around the source leaving reasonably small gaps among product containers to facilitate effective absorption of the emitted radiation energy. On the other hand, in case of the machine sources (LINAC), the emitted electrons are unidirectional beam and even when converted into bremsstrahlung X-ray the forward direction is predominant. Consequently, for effective utilization of the emitted beam energy, the process loads to be irradiated must be brought close to the beam exit window of the accelerator. Electron beam is narrow inside the accelerating tube and are broadened by scanning or scattering to obtain a homogeneous dose distribution over process loads. Radiation processing facilities may further be categorized by (i) mode of operation such as batch or continuous type, (ii) conveyor system, for example continuous or shuffle-dwell, and (iii) product handling type such as container or bulk flow. The process loads in food irradiation facilities move around the source and pass through the locations with different dose rates. Appropriate shuffling, linear transfer, and identical motion of the process load around the source assembly are therefore adequately designed to achieve uniform dose distribution.

4.1 Dosimetry in Radionuclide-Based Irradiation Facility

In a commercial food irradiation facility equipped with radionuclide source, the dosimetry is employed to characterize the radiation facility in operational qualification (OQ), dose mappings in irradiated products during performance qualification (PQ) programs, and routine dose measurements during product processing to monitor the irradiation process.

During OQ, the purpose of dosimetry is to demonstrate that the irradiator, as installed, is capable of operating and delivering appropriate doses within defined acceptance criteria. OQ also helps to verify the capabilities of the irradiator to establish the role of the key operating parameters. It also helps assessing the effect of absorbed dose to the product during variability of the key operating parameters. One of the important operational parameters is conveyor speed or dwell time and control the dose to the product. Dose also depends on the bulk density of the process load. Delivery time of the same dose to a product increases with the increase of the bulk density. These relationships should be experimentally evaluated and established during PQ. The understanding of all the controlling parameters is of practical help during the operation of the facility. In case of PQ, the dosimetry must bring the proof that the minimum required dose necessary for the technological objective is achieved and that the maximum acceptable dose is not exceeded. Dose mapping is performed on either real products or dummy products to determine the locations of the minimum and maximum dose zones, their values and their relationships to monitor dose measurements during routine product processing.

4.1.1 Dose Mapping Experiment

The dose mapping exercise is designed to find out the minimum and the maximum dose positions inside the process load containing actual food or dummy products. The bulk density of the dummy products should be chosen close to the mean bulk density of the actual food products intended to be irradiated at the irradiation facility. The product box is subdivided into small segments by making a grid of dosimeters with the help of several vertical planes created by inserting cardboard sheets or any other sheets made with the material having similar density as that of the food. A pair of dosimeters are placed at the cross section of the column and rows. The choice of number of column and rows on the carboard sheet depends how closely one wants to monitor the dose pattern. Normally, the dosimeters are placed in locations where the minimum and maximum doses are expected. The product box with dosimeters are then exposed to radiation setting the appropriate conveyor speed of the product handling system. Finally, the dosimeters are analyzed to obtain the relationships between the variables followed by the evaluation of the absorbed dose. Radionuclide-based irradiators having plaque source (γ) is designed to deliver radiation dose to both the sides of the rectangular process loads. The maximum dose (D_{max}) is observed somewhere in the outside or surface planes parallel to the source plaque and the dose minimum (D_{min}) is somewhere in the midplane, which is also parallel to the source plaque. This is illustrated in Figure 2a. For product overlap geometry, the positions of D_{max} are found along a center line of the outside planes of the process load parallel to the direction of motion, and the positions of D_{min} are found on the edges of the midplane. However, for the facility with source overlapping irradiation geometry, the position of D_{min} moves from the edges of the midplane into the inside of the process load but still on this plane. The depth–dose

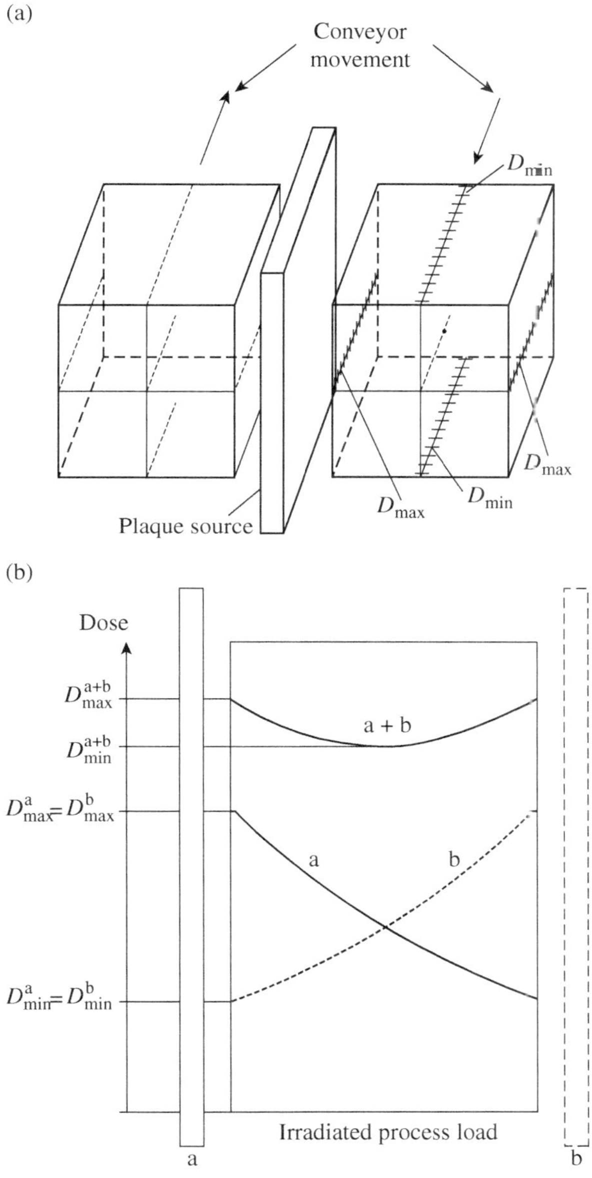

Figure 2 (a) Regions of D_{min} and D_{max} (indicated by hatching) for a rectangular process load after two passes, one on each side of a stationary γ-ray plaque source, and (b) depth–dose curves in a process load irradiated from two sides with a radionuclide plaque source. During the first pass, the source is on the side "a" relative to the process load, while during the second pass the source is on the side "b" relative to the process load. The curves a and b represent the dose contributions for the first and second passes separately, and the curve a + b represents the accumulated dose from both the passes. *Source:* IAEA Technical Reports Series No. 409 (2002).

distribution in the food products usually resembles an approximately exponential curve. Figure 2b exhibits the dose distribution inside a product box for both sided irradiation geometry. Finally, the DUR defined as the ratio of the maximum dose to the minimum dose is determined.

In case of special type of food irradiation such as phytosanitary application and processing of perishable food products, the dose minimum position inside the process load is not accessible. Alternative positions may be used for absorbed dose monitoring during routine product processing. The relationships between the absorbed doses at these alternative reference positions and the absorbed doses shall be established during process qualification with adequate reproducibility and documentation.

4.1.2 Routine Processing of Food Product

During routine processing, the facility must demonstrate that the radiation process is completely under control. It requires attention to all process parameters that can affect absorbed dose including the use of dosimetry measurements. During routine dosimetry, the dosimeter sets are placed on the middle plane inside the process load and on the surface planes where minimum and maximum doses are determined, respectively, during process qualification. The sets of dosimeters may also be placed at the reference locations whenever necessary in accordance with the established relations to find out the regulatory dose values. After the process, the dosimeters are retrieved from the process loads and the dose values are evaluated to ensure appropriate dose delivery to the food products.

4.2 Dosimetry in Linear Accelerator (LINAC) Facility

The installation qualification (IQ), OQ, PQ, and routine dosimetry are also essential for the characterization of an accelerator irradiator deployed for radiation processing of food. Two features distinguish electron beam dosimetry from that of photon (γ and X-ray) dosimetry. First, the fact that the particles are charged, the information about the beam can be obtained by measuring the charge that they carry. The second characteristic arises from the greater rate of energy loss by electrons leading to large variation in absorbed doses over a comparatively small distance. The beam characterization involves measurement of the mean energy of the electron beam and scan width. These parameters help to ensure that the dose is uniformly delivered on the surface of a process load. The penetration of the electrons depends on the beam energy. It is measured by determining the depth–dose distribution along the beam axis in a reference material, usually water or polystyrene. Figure 3a gives a typical depth–dose distribution which is measured by exposing several thin film dosimeters at different depths in the reference material. The thickness designated as r_{50} (half-value depth in water) can be used to estimate the mean electron beam energy based on the following relationship (ASTM Standard Practice ASTM E1649 2000):

$$E_{\text{mean}} = 2.33 r_{50} \tag{4}$$

Like the characterization in radionuclide-based irradiation facility, the relationships of conveyor speed with bulk density and dose should be established. In case of electron facilities, parameter like pulse frequency, i.e. pulse repetition rate (number of pulses per second)

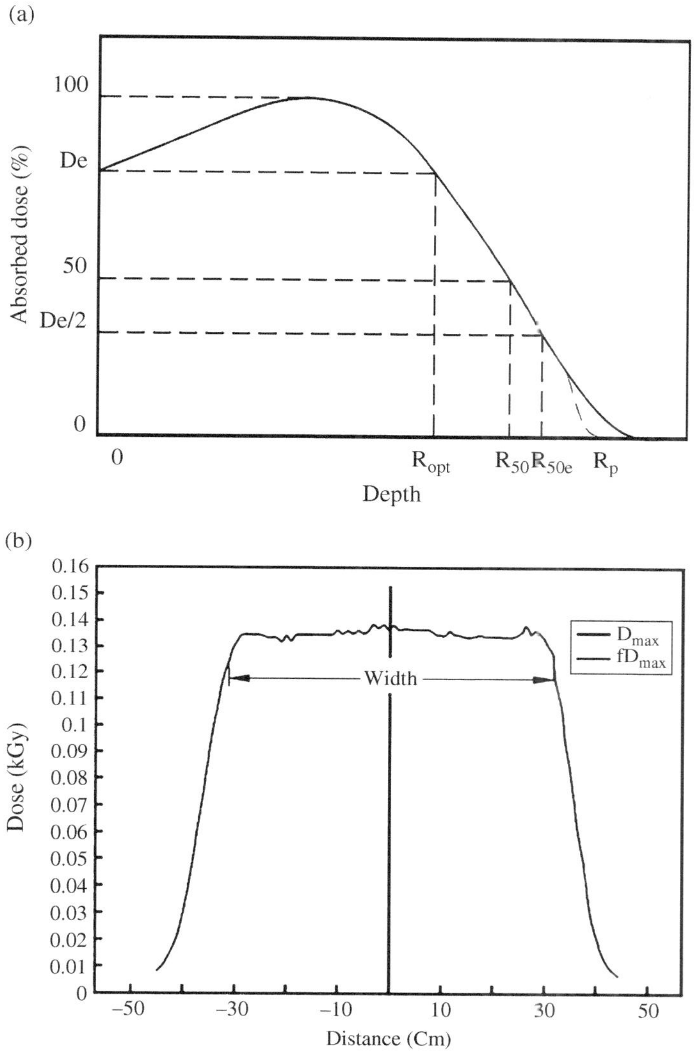

Figure 3 (a) Depth–dose curve for 10-MeV electrons in water, where the entrance (surface) dose is marked as D_e. The various ranges are identified as: R_{opt} is the depth at which the dose equals the entrance dose, R_{50} is the depth at which the dose equals half of the maximum dose, R_{50e} is the depth at which dose has decreased to 50% of D_e, and R_p is the depth at which extrapolated straight line of descending curve meets depth axis, and (b) example of electron beam dose distribution along the scan direction, where the beam width is specified at a defined fractional level f of the average maximum dose D_{max}. *Source:* ISO ASTM 51649:2015(E) (2015).

can suitably be controlled to vary the absorbed doses to the products instead of changing the conveyor speed. This gives a larger flexibility to such irradiators. It is important to realize that in case of electron beam, a dose buildup region within the product is always seen (ASTM Standard Practice ASTM E1649 2000). Therefore, the maximum dose will always occur under the surface, i.e. inside the product. Penetration of the electron beam in a homogeneous material is nearly proportional to their initial energy. This relationship between the energy and penetration is used to determine the energy of the electron beam. The measurement of the most probable beam energy can be performed in accordance with ISO/ASTM 51649:2015(E), using aluminum stack/wedge and radiochromic films. The most probable energy (E_p, in MeV) of the incident electron beam shall be determined from the practical range (R_p, in cm) derived from the depth–dose profile (Figure 3a) using following empirical relation. The experiment shall be repeated at least three times to obtain statistically stable value for energy (ISO/ASTM 51649:2015(E) 2015):

$$E_p = 0.423 + 4.69R_p + 0.0532 \times R_p^{\ 2} \tag{5}$$

The information on the scan width of the beam is necessary to ensure that the radiation zone covers the lateral size of the process load expected to be irradiated. Dose distribution along the scan direction is measured by placing strips of dosimeter film or arrays of single dosimeters in the scan direction. More dosimeters should be placed in zones of expected high-dose gradients such as at the extremes of scan, and less where dose distribution is expected to be uniform. The dosimeter array or long strips may be mounted on a fixture with homogeneous backing material. The dosimeter fixture is irradiated by passing it through the electron beam using a known set of operating parameters. The center line of the dosimeter array shall correspond to the expected center line of the beam width. The overall width of the dosimeter array shall be large enough to compensate for any possible differences in centering. Dose values shall be plotted as a function of measurement location. Beam width and the variation of the measured dose along the scan direction (scan uniformity) are then determined as depicted in Figure 3b.

Similar to the radionuclide-based irradiation, the distribution of dose inside the process load is accomplished through dose mapping experiments in accelerator-based irradiator. In case of one-sided irradiation and beam is coming from the top, the maximum dose (D_{max}) is likely to lie along a line parallel to the direction of motion of the product, through its center and closer to the surface (Figure 4a). D_{max} may extend over a midplane through this zone to the edges of the slab. On the other hand, the minimum doses (D_{min}) are normally identified along the lines parallel to the direction of the motion of the product running through the side edges at the bottom of the process load. In case of both sided (two pass, with turning of the process load) irradiation geometry and beam is coming from the top (Figure 4b), the position of the maximum dose (D_{max}) will be observed somewhere midway between the two outside planes, along two lines parallel to the motion of the product and about halfway between the top and bottom surfaces. However, the minimum dose (D_{min}) will be detected along lines parallel to the direction of motion of the product either through the side edges at the top and bottom of the process load or in the midplane at the side edges.

The same LINAC can also be utilized as a source of X-ray by inserting a converter plate made up of high Z materials such as tungsten or molybdenum. The irradiation of food and allied products using X-ray can be carried out in accordance with the regulatory bodies when X-rays are generated from a machine operated at or below 7.5 MeV. The dosimetry procedures for X-ray mode of the LINAC is like the electron beam mode. However, once X-rays are generated, the penetration of radiation beam enhances and other than film dosimeters can also be used for product dosimetry.

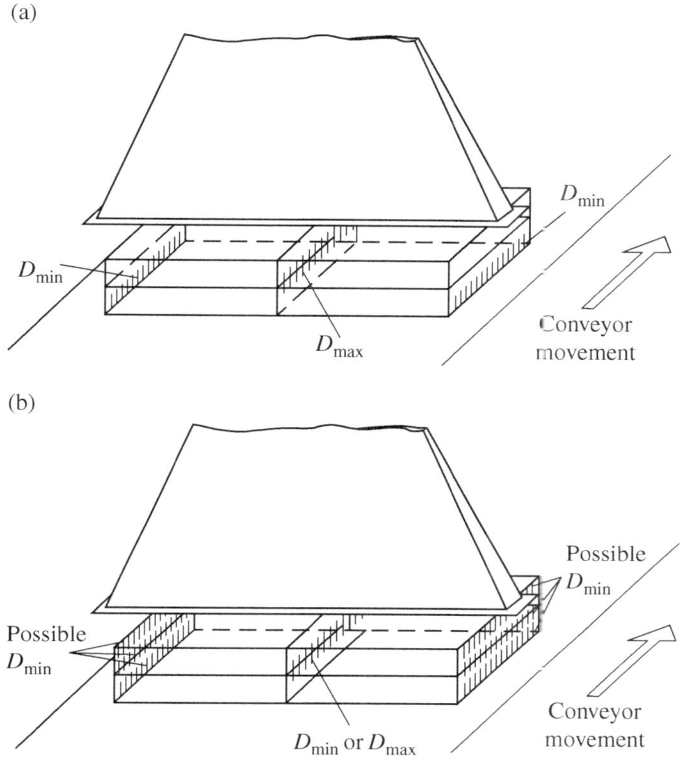

Figure 4 (a) Regions of D_{min} and D_{max} (indicated by hatching) for a rectangular process load for single-pass electron beam irradiation, and (b) regions of D_{min} and D_{max} (indicated by hatching) for a rectangular process load for two pass, two-sided electron beam irradiation. (Depending on the process load thickness with respect to the electron energy/range, either D_{min} or D_{max} may occur in the center.) *Source:* IAEA Technical Reports Series No. 409 (2002).

5 Emerging Field of Dosimetry in Low-Energy Accelerator Irradiator for Surface Treatment of Food

There is a growing pressure on food businesses to reduce their environmental impact namely, decrease emissions of carbon dioxide, reduced physical size (land usage), use water more efficiently, avoid chemicals, and reduce energy demands. This has led to the beginning of a new approach of radiation processing of food and agricultural products using low-energy beams. In many instances, foodborne organisms reside at or on the surface of foods and certain segments of the food industry are primarily concerned with in-house or in-line decontamination of the surfaces of commodities such as spices, grains, seeds, nuts, and similar dry ingredients. Radiation surface treatment of bulk food products such as spices, herbs, cereals, or seeds, with low-energy electron beam (LEEB) is being studied for many years (Jahn et al. 2005). A deeper understanding of the science, technology, and the processes associated with LEEB and low-energy X-ray (LEEX) sources and associated equipment can facilitate their use in the food industry for in-house solutions for food safety, phytosanitary, and quality challenges.

The efficiency of LEEB irradiation depends on the dose delivered and penetration ability of the beam, which must be controlled using appropriate dosimetric system. However, this application raises new questions and adds additional specific constrains when it comes to dosimetry. In the case of LEEB irradiation, the measurement of absorbed dose is a challenging task. Dosimetric approach used for high-energy electrons is not suitable when using LEEB, due to significant thickness of dosimeters in comparison to penetration ability of LEEB. There is limited number of dosimetric system which can be used for dose measurements when irradiating with LEEB. Currently, the most suitable is thin dosimetric foil. However, the challenge is to develop dosimetric system dedicated to LEEB which would be able to measure the dose on the surface, to monitor dose distribution on the surface and to control depth at which electrons can penetrate food products. Matsui et al. (2016) showed that in case of evaluation of on-site dosimetry and process design in industrial use of ultra-low-energy electron beam (ULEB) processes, the energy deposition can be estimated using a thin radiochromic film and Monte Carlo simulation.

6 Conclusion and Future Outlook

The main characteristics of an ideal dosimeters includes (i) a linear dose response over a wide dose range, (ii) good reproducibility of 1–5%, (iii) good pre- and post-irradiation stability, (iv) independency on energy of the radiation and dose rate of the radiation generating machine, (v) systems should be independent of small changes in composition, and (vi) easy, simple to use, and cost-effective. Therefore, it is obvious that a dedicated research and development process is essential to achieve the desired characteristics of a potential dosimeter in general. In addition, dosimetry in food has several specific requirements and thorough research is necessary to address those needs. For example, with increasing need of greater throughput and irradiation of perishable food commodities at sub-ambient temperatures, LINACs are emerging as suitable option. Aqueous chemical systems are not suitable to measure absorbed dose from accelerated electrons. Hence, synthesis of novel solid state systems and evaluation of their dosimetric characteristics are gaining interest in dosimetry research for food irradiation application. Luminescence technique could be an interesting area of research to measure higher doses incurred during various radiation processing applications, such as food irradiation, radiotherapy, and medical product sterilization. Dosimetry systems having polycrystals as host matrices showed saturation in dose response in higher dose range. Development of amorphous systems and their thermo- and photoluminescence characteristics could be another potential area of dosimetry research for higher dose measurements. In addition, computational dosimetry and thin film-based systems are also emerging as suitable approaches to measure the dose for the surface treatment of food products using low-energy accelerators. Dosimetry in a food irradiation facility is part of the quality assurance plan. It is of paramount importance that the personnel involved in dosimetry in an irradiation facility should have sufficient understanding of the dose measurements with adequate training. The owner of the facility should extend financial support in the various dosimetry-related activities to explore the commercial scope of the facility.

References

Ashraf, S., Sood, M., Bandral, J.D. et al. (2019). Food irradiation: a review. *International Journal of Chemical Studies* 7 (2): 131–136.

ASTM Standard, E1026 (2004). *Standard practice for using Fricke reference standard dosimetry system*. West Conshohocken, United States: ASTM International.

ASTM Standard Guide ASTM E1261 (2000). *Selection and calibration of dosimetry systems for radiation processing*. Annual Book of ASTM Standards. Vol. 12.02, Philadelphia, PA: ASTM.

ASTM Standard Practice ASTM E1649 (2000). *Dosimetry in an electron beam facility for radiation processing at energies between 300 keV and 25 MeV*. ASTM Standards, 12.02, Philadelphia, PA.

Attix, F.H. (1986). *Introduction to Radiological Physics and Radiation Dosimetry*. New York: Wiley.

Codex Standard 106-1983, Rev.1-2003 (2003). Revised codex general standard for irradiated foods. pp. 1–10.

Ehlermann, D.A.E. (1996). Automated dosimetry using radiation sensitive films and validation of routine dosimetry data. *Acta Alimentaria* 25: 363–366.

Food and Agriculture Organization, World Health Organization, Codex General Standard (1984). *Irradiated foods and recommended international code of practice for the operation of radiation facilities used for the treatment of food*, Codex Alimentarius, Vol. 15, Rome: FAO/WHO.

Humphreys, J.C., Chappell, S.E., McLaughlin, W.L., and Jarrett, R.D. (1973). Measurement of depth–dose distributions of 10 MeV incident electrons, *Rep. 73–413*, Washington, DC: National Bureau of Standards.

IAEA Technical Reports Series No. 409 (2002). Dosimetry for Food Irradiation. Vienna: International Atomic Energy Agency.

IAEA-ICGFI Document (1992). IAEA Training Manual. Operation of Food Irradiation Facilities, International Consultative Group on Food Irradiation. *ICGFI Document No. 14*, Vienna: ICGFI, IAEA.

ISO/ASTM 51205:2002(E) (2002). *Practice for use of a ceric-cerous sulfate dosimetry system*, ISO/ASTM International.

ISO/ASTM 51275:2013(E) (2013). *Practice for use of a radiochromic film dosimetry system*, ISO/ASTM International.

ISO/ASTM 51540:2004(E) (2004). *Practice for use of a radiochromic liquid dosimetry system*, ISO/ASTM International.

ISO/ASTM 51649:2015(E) (2015). *Standard practice for dosimetry in an electron beam facility for radiation processing at energies between 300 keV and 25 MeV*, ISO/ASTM International.

Jahn, M., Röder, O., and Tigges, J. (2005). Electron treatment of cereal crop seeds – overview and appraisal of field trials. *Mitteilungen aus der Biologischen Bundesanstalt für Land- und Forstwirtschaft* 399: 66–128.

Matsui, S., Mori, Y., Nonaka, T. et al. (2016). Energy deposition evaluation for ultra-low energy electron beam irradiation systems using calibrated thin radiochromic film and Monte Carlo simulations. *Review of Scientific Instruments* 87: 053309.

McLaughlin, W.L., Jarrett, R.D., and Olejnik, T.A. (1982). Dosimetry. In: *Preservation of Food by Ionizing Radiation*, vol. 1 (eds. E.S. Josephson and M.S. Peterson), 189–245. Boca Raton, FL Chapter 8: CRC Press.

McLaughlin, W.L., Boyd, A.W., Chadwick, K.H. et al. (1989). *Dosimetry for Radiation Processing*. London and New York: Taylor and Francis.

Onori, S., Bartolotta, A., Caccia, B. et al. (1990). Dosimetric characteristics of alanine-based ESR detectors in electron beams used in radiotherapy. *Radiation Protection Dosimetry* 34: 287.

Rizzo, F.X. (1971). Irradiator design calculational techniques based on centre line depth dose distributions. *International Journal of Radiation Engineering* 1: 549–584.

Sanyal, B., Prakasan, V., Chawla, S.P., and Ghosh, S.K. (2017). A study to assess the role of bulk density of process load in Co60 based food irradiation facility. *BARC News Letter*: 17–19.

Sharma, A. and Madhusoodanan, P. (2012). Techno-commercial aspects of food irradiation in India. *Radiation Physics and Chemistry* 81: 1208–1210.

Sleptchonok, O.F., Nagy, V., and Desrosiers, M.F. (2000). Advancements in accuracy of the alanine dosimetry system. Part 1. The effects of environmental humidity. *Radiation Physics and Chemistry* 57: 115–133.

Walker, M.L., McLaughlin, W.L., Puhl, J.M., and Gomes, P. (1997). Radiation-field mapping of insect irradiation canisters. *Applied Radiation and Isotopes* 48: 117–125.

3

Gamma Irradiation

Xuetong Fan and Brendan A. Niemira

U. S. Department of Agriculture, Agricultural Research Service, Eastern Regional Research Center, Wyndmoor, PA, USA

1 Introduction

γ-rays were discovered by French scientist Paul Villard in 1900 when he was studying radiation from radium (Gerward 1999). γ-rays are different from α and beta (β^-) rays, which were discovered a few years earlier. α and β-rays are particles with mass and electric charged, while γ-rays have no mass or charge. Like visible light and radio waves, γ-rays are photons comprised of waves of electromagnetic radiation with zero rest mass. Different from non-ionizing radiation, such as visible light and radio waves, γ-rays have a much higher frequency, shorter wavelength, and higher energy. Non-ionizing radiation can deposit energy in the materials through which it passes but does not have sufficient energy to directly break molecular bonds. In contrast, ionizing radiations like γ and X-rays have enough energy to strip electrons from atoms and break molecular bonds, resulting in the creation of electrically charged particles (ions).

Food waste and loss are major problems facing both developing and developed countries. In the United States, over one-third of all available food is lost (Buzby et al. 2014). Food loss can be the result of food spoilage (quality deterioration to a point that is not acceptable by consumers), development of undesirable characters such as sprouting of tubers and bulbs, and contamination with human pathogens or harmful chemicals. Ionizing radiation is capable of reducing food losses by destroying microorganisms of both spoilage and human pathogens, slowing down senescence of fresh produce, and inhibiting sprouting. There are a number of recent books, book chapters, and reviews on food irradiation which do not necessarily deal with γ-rays (Ferreira et al. 2017; Pillai and Shayanfar 2017; Fan and Niemira 2020; Ahn et al. 2017). The current chapter will focus on γ-rays, including their generation from radioactive isotopes, interaction with matter, mechanisms, benefits, recent developments regarding application, and factors affecting the efficacy of γ-rays.

Electromagnetic Technologies in Food Science, First Edition. Edited by Vicente M. Gómez-López and Rajeev Bhat.

2 Characteristics and Generation of γ-rays

Ionizing radiation takes a number of forms: α-rays, β-rays (electron), neutrons, γ-rays, and X-rays. There are several differences among various types of ionizing radiation (Table 1). γ, β, and neutron are particles, which have electric charge and mass, while γ and X-rays have no charge or mass and are known as electromagnetic radiation. γ and X-rays travel at the speed of light, while the three particles travel at speeds lower than that of light. The penetration ability of these varying types of radiation ranges from low penetrability for α-rays to great penetrability of neutrons (Figure 1). There are three common types of ionizing irradiation that are used for food preservation: γ-rays, X-rays, and electrons. Each has its advantages and disadvantages. Compared to electron beams, γ-rays have a much higher penetration ability. The penetration depth for alpha particles is too limited for food

Table 1 Comparison of α, β, γ, and X-rays and neutron.

Type of radiation	Symbol	Description	Mass (amu)	Charge	Speed
α	${}^{4}_{2}He$ or ${}^{4}_{2}\alpha$	Helium nucleus comprised of 2 protons and 2 neutrons	4	+2	1/10 of speed of light (<speed of light)
β	${}^{0}_{-1}e$ or ${}^{0}_{-1}\beta$	High-energy elections	1/1823 (0.0005486)	−1	90% of speed of light (270 000 $km\,s^{-1}$)
γ/X-rays	${}^{0}_{0}\gamma$	High-energy, high-frequency electromagnetic radiation	0	0	Speed of light (299 792 $km\,s^{-1}$)
Neutron	${}^{1}_{0}n$	Very high frequency	1	0	7.6×10^{-6} of speed of light (2.19 $km\,s^{-1}$)

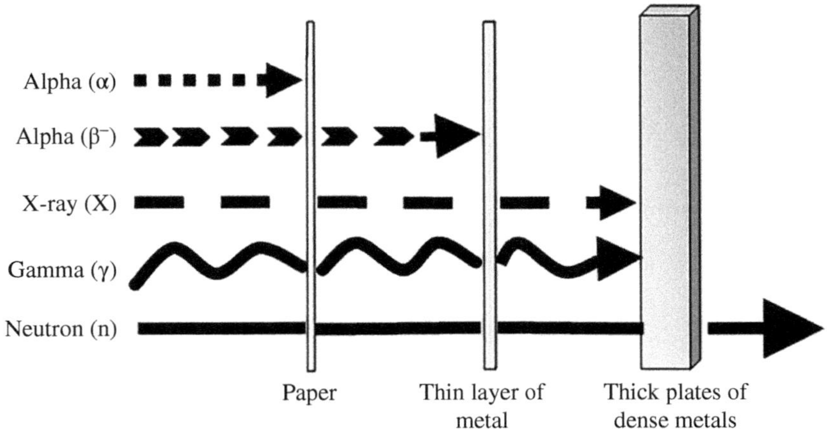

Figure 1 Penetration ability of various types of ionizing radiation. *Source:* Modified from Nuclear Regulatory Commission (https://www.nrc.gov/about-nrc/radiation/health-effects/radiation-basics.html).

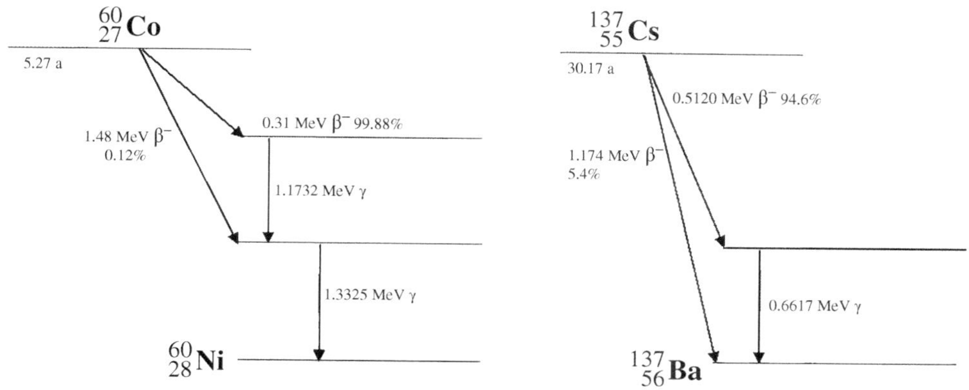

Figure 2 Decay of Co-60 and Cs-137.

irradiation purposes, and X-ray radiation is considered to be less practical due to its low generation efficiency from electrons as well as its associated higher cost than γ-rays. While neutrons have good penetration and are quite effective in destroying bacteria, they are not considered practical for food application because of the potential for making food radioactive.

γ-rays are produced from radioactive isotopes that are one or more forms of the same element having the same atomic number but different mass numbers. Cobalt-60 (Co-60) and cesium-137 (Cs-137) are the most widely used and permitted sources of γ-rays for food processing. As unstable radioactive materials, Co-60 or Cs-137 decay to more stable elements over time (Figure 2). During the decay process, γ-rays and β-rays are emitted. However, only a small percentage of radiation is emitted as β-rays, which have much lower penetrability. A layer of plastic film can block β-rays. Most of the radiation from the decay of Co-60 and Cs-137 is in the form of γ emissions. Co-60 has a half-life of 5.27 years and produces γ-rays with energies of 1.173 and 1.332 MeV. Cs-137, however, has a half-life of 30.17 years and produces γ-rays with an energy of 0.662 MeV, thus, γ-rays produced from Co-60 have higher energy than those from Cs-137. Therefore, the γ-rays from Co-60 have slightly higher penetration ability and are the main source of γ-rays for commercial applications. For both isotopes, the γ-rays have enough energy to dislodge electrons from their orbits, but do not have energy high enough to dislodge neutrons from the nucleus, and therefore cannot induce radioactivity in the irradiated products. Cs-137 is much longer-lived than Co-60, with more than five times the half-life. Cs-137 systems, therefore, require far less frequent replenishment, since the source does not diminish in strength as quickly. However, Cs-137 is not commonly used because it is soluble in water, increasing the challenges in containment in case of Cs-137 being released. In addition, its availability has been strictly regulated. Cs-137 is often used in small self-contained, dry storage irradiators primarily for research and other uses such as insect sterilization. Due to its shorter half-life, Co-60 needs to be replenished much more frequently or in higher amounts than Cs-137 to maintain the similar dose rate, normally 10% per year. Co-60 and Cs-137 are often associated with the nuclear industry as they are produced in nuclear reactors. The concerns over environmental waste, safe, and secure handling and transportation of radioactive materials have a negative impact on the wide acceptance of irradiated foods by the general public. Development of new strategies for consumer education and communication of γ irradiation may be employed to facilitate the acceptance of the technology (Bearth and Siegrist 2019).

3 Compton Effect

The Compton effect or Compton scattering is one of the principle forms of γ-ray (photon) interaction with matter. It occurs as a result of the interaction between the photons and loosely bound valence shell (outer shell) electrons of an atom (Figure 3). In this process, absorbed energy by an outer shell electron from photons leads to ejection of the electron from the atom. After interaction, the incident photon continues traveling through the absorber (food), but in a changed direction, and with less energy and frequency than the original photon. The incident photon takes part in further Compton events to free an orbital electron from the atom (Figure 4). The ejected fast electrons (primary electrons) from Compton events interact with matter (food) resulting in multiple excitation and ionization events. Depending on the atom, the energy (ionization potential) required to free electrons from the atom is between 4 and 20 eV, while γ-rays have energy much greater than the required 4–20 eV (Urbain 1986). Therefore, it is expected that primary γ-rays and scattering photons of the Compton effect will eject multiple electrons. Each Compton event is accomplished by tens of thousands of excitation and ionization events initiated by fast electrons. When the amount of energy derived from the radiation is less than that needed for ionization, most of the excitation energy in molecules is converted to heat.

Another type of photon–matter interaction is the photoelectric process in which photons eject electrons from the bound shells (k shell) of the atom resulting in complete disappearance of the photons. The photoelectric process occurs with relative low-energy photons such as visible light and in materials with high atomic numbers. The photelectric process in food is negligible for high-energy photons such as γ-rays.

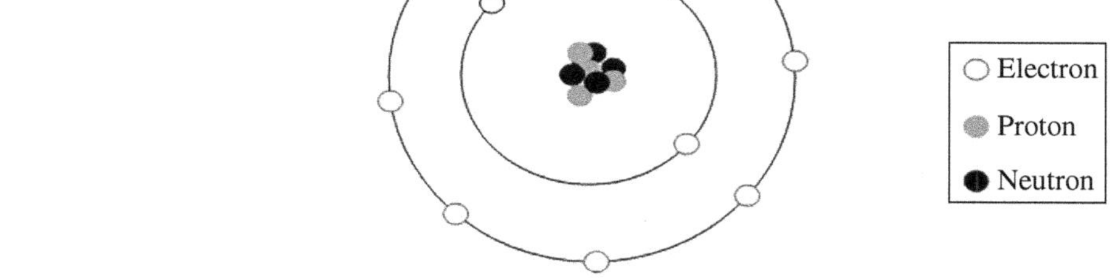

Figure 3 Compton scattering resulting from interaction of photons and electrons.

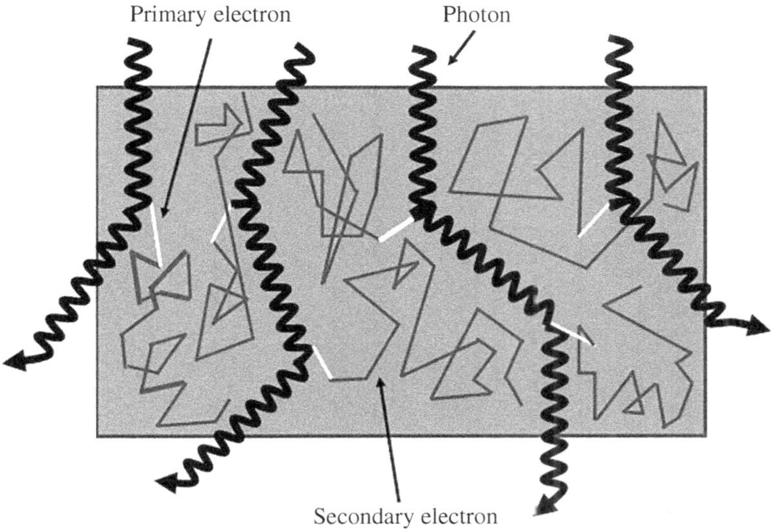

Figure 4 Productions of primary and secondary electrons from interaction of photons and electrons.

4 Basic Effects on Food: Interaction of γ-rays with Matter

γ-rays influence their effects on foods via two mechanisms: direct and indirect effects (Figure 5). For the direct effect, interaction of γ-rays and electrons from Compton scattering in matter causes formation of electronically excited molecules by the ejection of electrons from atoms, which destabilizes molecules and breaking bonds between the atoms. For example, γ-rays can result in the cleavage of phosphodiester bonds of DNA, which renders bacterial cells losing their ability of replication and eventfully leads to the death of bacterial cells. A dose of 0.1 kGy γ-rays can cause changes in 2.8% DNA, 0.14% enzymes, and 0.005% amino acids in bacterial cells (Diehl 1995).

The indirect effect of γ irradiation involves radiolysis of water, which is a major component of most food and biological systems (Figure 5). When water is irradiated, hydrated electrons (e^-_{aq}), hydroxyl radicals ($^{\cdot}OH$), hydrogen peroxide (H_2O_2), hydrogen ion (H_3O^+), dihydrogen (H_2), and hydrogen atom ($^{\cdot}H$) are generated (Simic 1983):

$$H_2O \rightarrow e^-_{aq} + {}^{\cdot}OH + {}^{\cdot}H + H_3O^+ + H_2 + H_2O_2$$

In addition, secondary reactive oxygen species such as superoxide radicals ($O_2^{\cdot-}$) are formed, probably within one picosecond (10^{-12} seconds) (Singh and Singh 1982). These chemical species can be strong oxidants ($^{\cdot}OH$, $O_2^{\cdot-}$, H_2O_2) or reductants (e^-_{aq}, $^{\cdot}H$). The H_3O^+ ion is considered to be the same as the H^+ ion as it is the H^+ ion joined to a water molecule, which makes water slightly acidic. Nevertheless, the action of $^{\cdot}H$ and H_2 are poorly understood in food and biological systems.

The chemical species from the radiolysis of water formed in the proximity of biomolecules directly attack DNA, cell membranes, proteins, and other components of

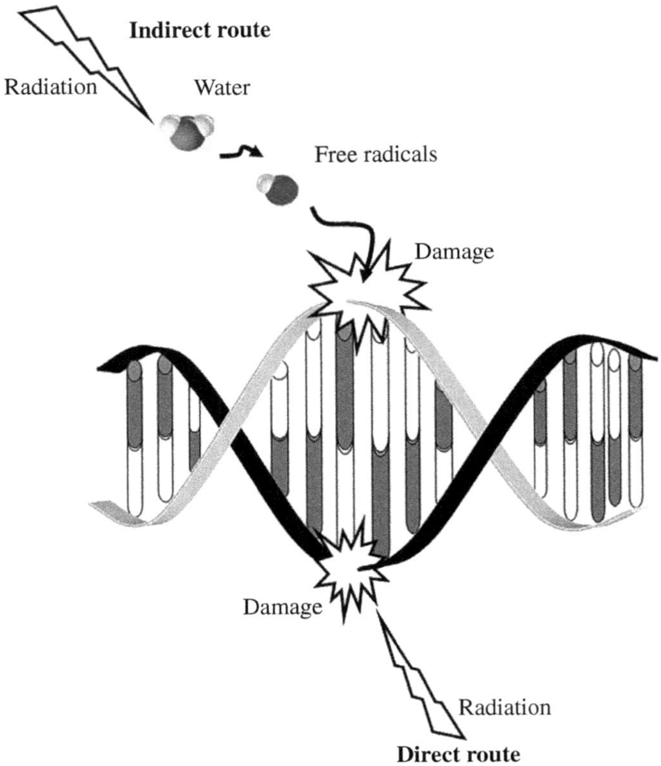

Figure 5 Direct and indirect effect of γ-rays on DNA.

microorganisms (Figure 6). Among the water radiolysis products, ˙OH is particularly destructive to cellular and food components (Reisz et al. 2014). Being a powerful oxidizing agent, hydroxyl radicals (˙OH), generated in the hydration layer of DNA, react with the base and sugar moieties of DNA, resulting in alteration of bases and sugars, formation of DNA–DNA, and DNA–protein cross-links, breakage of sugar-phosphate bonds and single and double strands, and DNA clustering, which lead to the loss of cell replication (Ohshima et al. 1996). Hydroxyl radical (˙OH) and other water radiolysis oxidative species also attack lipids, particularly unsaturated double bonds of polyunsaturated fatty acids leading to increased membrane permeability, lipid oxidation, disruption of ion gradients and other transmembrane processes, and altered activity of membrane-associated receptors and proteins (Reisz et al. 2014). Hydroxy radicals (˙OH) attack proteins by reacting with conjugated and heterocyclic rings, and side chains of amino acids, leading to carbonylation of proteins and cleavage of the protein backbones (Menon et al. 2016; Simic 1983). The reductants (e^-_{aq} and ˙H) from water radiolysis are also capable of denaturing proteins and enzymes by reacting with amino acids such as methionine and cysteine residues of proteins, leading to protein desulfurization and loss of functionality (Reisz et al. 2014).

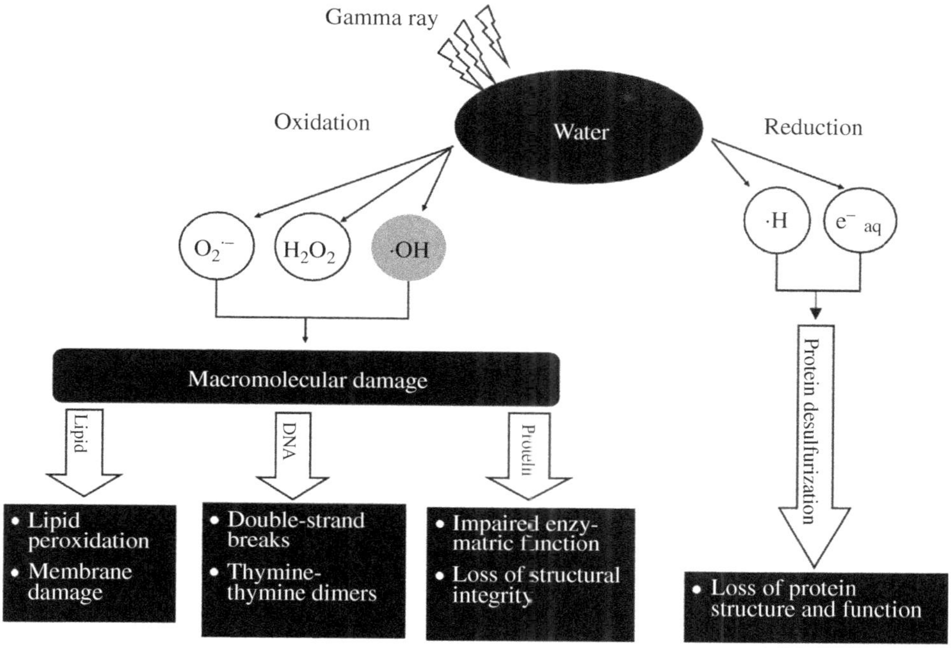

Figure 6 Damages of macromolecular by reactive species from radiolysis of water. *Source:* Modified from Menon et al. (2016).

In most foods, such as fresh produce, meat and meat products, where water is the major component, and biomolecules are often surrounded by water, thus indirect effects play the dominant role in the chemical changes and microbial inactivation caused by irradiation in those foods.

5 Dose Unit, Dose Rate, and Dose Distribution

To study the impact of γ-rays on biological and chemical aspects of foods and microorganisms, we need to know the absorbed dose, which is the energy that is deposited into foods which are exposed to γ-rays. The international (SI) unit for the absorbed dose is the gray (Gy), which is defined as 1 joule of energy deposited in 1 kilogram of water. One joule equals 0.24 calories, therefore, temperature of water would increase only by 0.24 °C after 1 kGy γ-ray irradiation. Therefore, ionizing irradiation is a truly nonthermal processing technology. The older unit of radiation dose is the rad, which stands for "radiation absorbed dose." For means of comparison, 1 Gy = 100 rad.

Dosimetry is the measurement and calculation of absorbed doses that foods receive from ionizing radiation (Kuntz and Strasser 2016). The dosimetry system should be calibrated in accordance with appropriate international or national standards to ensure the reliability and traceability of the measurement. It is important to produce 3-D dose map distributions using proper dosimetry to achieve acceptable dose uniformity throughout the product. Factors that commonly affect dose mapping include density and composition of foods, homogeneity of foods, shape and size, and variations in orientation of the product, stacking, volume, and packaging. Major types of dosimeters for food irradiation include those measuring change in optical density such as radiochromic film and those that measure radiation-induced free radicals with an electron paramagnetic resonance (EPR) spectrometer (such as alanine) (Hansen 2001; Miller et al. 2005).

The primary objective of applying γ-rays to food is to deliver specified minimum doses to all parts of the products to meet the requirements of achieving desired benefits. In addition, it is necessary to estimate maximum doses absorbed by the products to ensure that products at all parts can tolerate the doses, without significant deterioration in food quality or doses not exceeding the maximum allowable dose set by regulatory agencies. A dose map, or dose distribution study, needs to be established using numerous dosimeters throughout the product load to establish the minimum and maximum doses within the load.

6 γ-ray Facility

Approximately 180 γ-ray commercial facilities are used for various applications in the world of which approximately 50 facilities are in the United States, mainly for sterilizing medical devices and for food irradiation (GIPA 2014). As the dominant source of γ-rays, Co-60 is produced by placing metallic slugs of stable Co-59 in a nuclear power reactor in which Co-59 absorbs neutrons. The activated metal (Co-60) as slugs is doubly encapsulated in stainless steel tubes before being released to radiation facilities. Co-60 sources are often in the form of pencils that are 45.2 cm in length of 1.1 cm in diameter. The pencils are loaded into flat, vertical racks as shown in Figure 7. The Co-60 sources in racks are stored in water-filled pools where water serves as a shield when not in use. To irradiate products, the racks are raised above the pool water where products are exposed to γ-rays. The treatment room is often surrounded by thick concrete walls, which protects operating workers from the γ radiation when the source rack is in the raised position (Cleland 2013). The absorbed doses depend on the amount of time that the irradiated products are in the irradiation field as well as the distance of the product from the source rack. To capture as many γ-rays as possible from the Co-60 source, products are positioned in close proximity to the source.

7 Applications of γ-ray Radiation in Foods

γ-rays have been investigated for food applications for over a century with extensive efforts occurring after the "Atoms for Peace Program" was established in the 1950s (Farkas and Mohácsi-Farkas 2011). γ-rays have a number of applications, including improvement of microbial food safety by inactivating foodborne pathogens, phytosanitary treatment of fresh produce, preservation of various foods by reducing populations

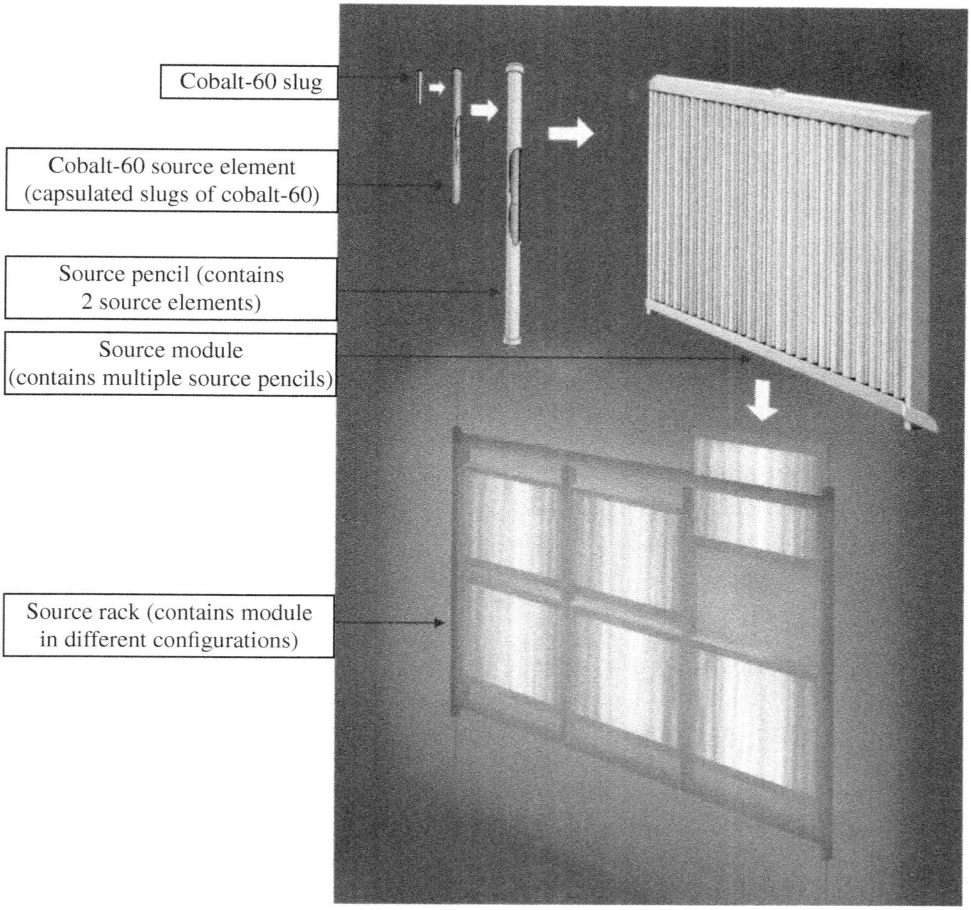

Figure 7 Illustration of a typical cobalt source rack built from slugs, pencils, and modules. (*See insert for color representation of this figure*). *Source:* Courtesy of MDS Nordion.

of spoilage microorganisms, and food sterilization. Other potential applications such as reductions of mycotoxin and allergens have also been explored. In general, low doses of radiation (<1.0 kGy) are often used for insect disinfestation of fruits and vegetables; Moderate doses (1–5 kGy) may be applied to inactivate vegetative bacteria for the purpose of microbial food safety; while higher doses (>10 kGy) are used for sterilizations of foods.

7.1 Improving Microbial Safety

It has long been known that irradiation is capable of inactivating bacteria. In fact, the first patent application for food preservation by irradiation was filed in 1905 (Farkas and Mohácsi-Farkas 2011). In recent years, irradiation has been investigated to inactivate microorganisms that cause foodborne illness, such as *Salmonella* spp., pathogenic *Escherichia coli*, and *Listeria monocytogenes*. The radiation resistance of a pathogen is often represented by D values, which are the amounts of radiation energy required to inactivate 90% of a given pathogen population. The *D* values can be calculated from the linear responses of pathogen reduction to irradiation doses, even though there may be some tailing effects of irradiation on microorganisms at the low-dose range. Differences in radio resistance have been reported for the same pathogens among studies reported in the literature. For example, it appears that *D* values for many human pathogens are higher on meats than those on fresh produce (Fan 2012). The differences may be caused by varying strains/isolates of pathogens used in disparate studies, composition of the food, dose rates, and irradiation temperature.

In general, radiation sensitivity of a microorganism is inversely related to the size and complexity of the organism (Fan and Niemira 2020). A small sized microorganism is more resistant to ionizing radiation compared to a large one. Viruses are small in size and have very tiny genomes (compared to bacteria and fungi); therefore, being more resistant to irradiation. Furthermore, bacterial spores are much more resistant to ionizing irradiation than vegetative bacteria and require higher doses of irradiation (van Gerwen et al. 1999). After analysis of *D* values reported in the literature, the average *D* values for spores and vegetative bacteria cells were estimated to be 2.11 and 0.420 kGy, respectively (van Gerwen et al. 1999).

Viruses are highly resistance to γ-rays. High doses of radiation (up to 20 kGy) were necessary to achieve a significant reduction of up to 1.26 log norovirus (Molina-Chavarria et al. 2020). The main viral inactivation mechanism is thought to be the destruction of replication-competent nucleic acid, either by direct radiolytic cleavage or cross-linking of genetic material or by the action of free radicals on nucleic acids, and to a lesser degree, proteins (Hume et al. 2016). Compared with arenavirus, bunyavirus, filovirus, flavivirus, orthomyxovirus, and paramyxovirus, SARS-CoV (severe acute respiratory syndrome coronavirus), which harbors the largest genome of all studied viruses was completely inactivated by a dose of 1 Mrad (10 kGy)(Feldmann et al. 2019). By contrast, a dose of 4 Mrads was needed to inactivate other viruses, confirming that radiation resistance is inversely correlated with genome size (Feldmann et al. 2019).

The widespread presence of antibiotic resistance as a result of misused or overused antibiotics has raised concerns over public health. It has been shown that antibiotic-resistant *Salmonella* spp. had similar or higher sensitivity to γ-rays than antibiotic-sensitive *Salmonella* (Niemira and Lonczynski 2006; Niemira et al. 2006). However, another study indicated that the antibiotic-resistant *L. monocytogenes* strains were generally more resistant to γ-rays than were antibiotic-sensitive *L. monocytogenes* (Skowron et al. 2018). How antibiotic resistance affects radiosensitivity of bacteria needs further investigation.

Parasites such as *Toxoplasma gondii*, *Cryptosporidium parvum*, *Trichinella spiralis*, and *Cyclospora* can be inactivated by γ-rays. Doses less than 0.5 kGy may be enough for decontaminating *T. gondii* oocysts on fruits and vegetables. Raspberries inoculated with sporulated *T. gondii* oocysts were rendered innocuous after irradiation at 0.4 kGy (Dubey et al. 1998). Similarly, irradiation treatments at 0.2 kGy effectively inactivated *T. gondii* oocysts on blueberries (Lacombe et al. 2017). It is unclear why *T. gondii* oocysts are sensitive to γ-rays, perhaps due to their size and composition.

7.2 Preservation of Food

Shelf life extension can be extended by controlling spoilage, retarding physiological processes associated with ripening and senescence of fresh fruits and vegetables, and inhibiting the sprouting of vegetables such as potato tubers, onion, and garlic bulbs. γ-rays can inactivate spoilage microorganisms on foods, such as poultry, meat, and seafood which are prone to spoilage caused by bacteria, resulting in reduced development of microbial-induced discoloration, off-odor, and slime (Thayer 1993; Arvanitoyannis et al. 2008; Rodrigues et al. 2020).

γ-rays also extend the shelf life of fresh fruits and vegetables by inhibiting ripening and senescence. Irradiation suppressed ethylene production by apples consistent with lower activity of 1-aminocyclopropane-1-carboxylate oxidase, a key enzyme in the synthesis of ethylene (Kheshti et al. 2019). The sensory shelf life of celery treated with 1 kGy of γ radiation was 29 days compared to 22 days for the control (chlorinated samples) (Prakash et al. 2000). A 5-kGy radiation dose and 10 °C storage temperature were found to keep peeled ginger samples microbe free and acceptable for 70 days, whereas non-irradiated (control) peeled ginger spoiled within 40 days under similar storage conditions (Mishra et al. 2004). Irradiation (2 kGy; 10 °C) of gourd (*Benincasa hispida*) cubes resulted in 7 more days of shelf life compared to the non-irradiated controls (Tripathi et al. 2013). Irradiation has also been shown to extend the shelf life of less common fresh produce and edible fungi. For example, irradiation (3 kGy) preserved the quality of bamboo shoots (Wang et al. 2019). γ irradiation (2.5 kGy) extended the shelf life of summer truffles packaged under modified atmosphere without significant effects on bioactive compounds or sensory parameters during 42 days of storage at 4 °C compared to non-irradiated controls (Tejedor-Calvo et al. 2020). The impact of irradiation on shelf life and quality of fresh produce depends on many factors such as cultivar, maturity, and pre- and postharvest handling.

When combined with other treatments such as modified atmosphere packaging (MAP) and natural antimicrobials, irradiation may be more effective in extending food shelf life (Jeong and Jeong 2018). For example, γ irradiation in combination with an essential oil coating reduced aerobic plate counts and populations of *Pseudomonas putida*, achieving at least a 12-day extension of shelf life of cooked shrimp (Ouattara et al. 2001). While irradiation (4 kGy) had a greater effect in extending the shelf life of chicken as compared to MAP (70% CO_2/30% N_2), the combination of irradiation and MAP resulted in the highest shelf-life extension by 12 days, compared to the air-packaged samples (Chouliara et al. 2008).

Sprouting of tuber and bulbs is an undesirable phenomenon which results in loss and reduction of marketability. γ radiation is an effective technique to prevent sprouting of potatoes when applied at proper doses and times (Sarkar and Mahato 2020). Irradiation (100 Gy) treatments on the 10 and 30 days following harvest completely prevented sprouting of two varieties of potatoes (Rezaee et al. 2013). The effectiveness of irradiation in inhibiting sprouting depends on irradiation doses and variety and developmental stages of tubers/bulbs. Irradiation decreases sprouting when applied in stages when tubers are in an active metabolic state which are more sensitive to irradiation. Irradiation disrupts nucleic acid, nucleotide, and hormonal syntheses of potatoes (Rezaee et al. 2013). The irradiation inhibition of sprouting results in changes in several chemical components of sprouts, such as growth regulators, DNA, RNA, proteins, soluble carbohydrates, and lipids (Pérez et al. 2007). It is also shown that γ-rays at 120 Gy minimize sprouting of onions and garlic while maintaining best quality for 3 and 4 months, respective of storage at ambient temperature (Sharma et al. 2020a,b).

7.3 Phytosanitary Treatment

As a result of globalization, there is an increase in trade and transportation of agricultural commodities, which may facilitate the introduction and spread of invasive pests to new areas, if not properly managed. Fresh fruits and vegetables can harbor many pests and potentially increase crop damage and the cost to control the pests when introduced into new production areas. A phytosanitary treatment is required to prevent the spread of regulated pests from infested to non-infested areas (Hallman and Loaharanu 2016).

The commonly used phytosanitary treatments include exposing commodities to elevated or reduced temperatures with or without changing atmospheres; fumigation with various chemicals such as methyl bromide, and ionizing radiation (Hallman et al. 2016). Phytosanitary irradiation differs from other phytosanitary treatments in that irradiation does not induce immediate deaths of pests, rather, prevention of further development or reproduction of pests (Follett and Griffin 2013). Therefore, the regulated pests may be present and alive when entering the importing area. The United States Department of Agriculture (USDA) has played an important role in facilitating the commercial application of phytosanitary irradiation and growth in the market access of irradiated fresh produce by publishing the rules on irradiation phytosanitary treatments, including the approved generic dose of 150 Gy for tephritid fruit flies and 400 Gy for all insects except the pupae and adults of Lepidoptera (Hallman and Loaharanu 2016). Similarly, International Plant Protection Convention, the international body recognized as the authority on plant health and phytosanitary measures, declares 150 Gy to be the dose to ensure the non-emergence of adults of all tephritid fruit flies (Follett and Neven 2020). In 2015, 25 000 tons of fresh produce were irradiated worldwide for the phytosanitary purposes (Hallman and Loaharanu 2016). γ-ray facilities are predominantly used for phytosanitary irradiation of fresh produce. Due to their high penetrability, γ-rays can be used to treat full pallet loads of fresh produce by delivering irradiation doses throughout the palleted loads to meet minimum doses to achieve the phytosanitary standards. At the doses used for phytosanitary application (150–400 Gy), most fresh fruits and vegetables may tolerate radiation without significant changes in quality (Hallman 2011; Jain et al. 2017).

7.4 Applications on Low-Moisture Foods

Spices are mainly crops (or parts of crops) grown in the field, which, like other agricultural commodities, are known to be contaminated with high levels of spoilage organisms such as bacteria, molds, and yeasts (Sjöberg et al. 1991). The majority of fungal and bacterial populations present in spices are spores. For example, *Bacillus* spp., as one of the major species of total microbial populations in spices, occurs as spores. Molds are present in spices usually at levels of 10^3 per gram. Spices are commonly treated with ethylene oxide or propylene oxide and heat treatment to inactivate microbes. However, concerns on chemical residues after chemical fumigation have been raised, while some microbes in spore forms are highly heat-resistant. In addition, high-temperature treatments may change the flavor or taste of spices which alters their main purpose in foods. Therefore, γ irradiation has been extensively studied as a means of reducing the microbial contamination of spices. Experiments indicate that spices, with water contents of 4.5–12%, are very resistant to physical or chemical interventions. A dose of 4.0–7.5 kGy is often needed to reduce the microbial population of spices to below the acceptable levels (10^4 CFU per gram). In addition, irradiation may be combined with other treatments to achieve synergistic reductions of microorganisms as surviving microflora after irradiation become more sensitive to subsequent food processing treatments (Kiss and Farkas 1988). When γ irradiation is used to decontaminate spices for shelf life extension, it also enhances microbial safety as human pathogens are inactivated.

Spices and other low-moisture foods have been increasingly found to be associated with foodborne pathogens such as *Salmonella* spp. and *L. monocytogenes*. Foods with water activity <0.85 are generally regarded as low-moisture foods and include cereals, flours, seeds, nuts, and jerky, in addition to spices (Niemira 2014; Anderson 2019). Pathogens in low-moisture foods are much more radio-resistant than in high-moisture foods such as fresh produce and meats. For example, the *D*-values for *Bacillus* in mesquite flour with a *aw* of 0.31 was 1.27 kGy (Fan et al. 2015). *D*-values of *L. monocytogenes*, *E. coli* O157:H7 and *S.* Typhimurium on pistachios were 1.02, 0.85, and 0.86 kGy, respectively (Song et al. 2019), while the same pathogens often had D-values of 0.1–0.2 kGy in fresh produce (Niemira et al. 2005; Niemira 2008). There are also studies on irradiation sterilization of low-moisture, high-value foods such as food for immunocompromised patients, infant formula, etc., which generally requires high doses (10 kGy) (Feliciano 2018).

7.5 Potential Uses of γ Irradiation for Degradation of Mycotoxin and Allergen

There are many studies demonstrating that γ-rays reduce, destroy, or degrade mycotoxins and allergens. γ-rays can inactivate mycotoxin-producing fungi; therefore, reducing the chance of producing mycotoxins from the fungi. There is also the possibility that irradiated fungus can produce more mycotoxins than original strains (Calado et al. 2014). Therefore, after foods are irradiated to decontaminate fungi and other microorganisms, proper storage conditions need to be established to minimize recontamination and regrowth of remaining fungi. Mycotoxins in low-moisture foods are extremely radio-resistant, even though mycotoxins are sensitive to irradiation in aqueous solutions. As a result, γ radiation, in general, is not a feasible technology for the direct detoxification of mycotoxin in foods as high doses of irradiation are needed (Calado et al. 2018).

Studies have also been conducted regarding the effects of γ irradiation on the allergenicity of peanut extract and found that the allergenicity of peanut extracts decreased with increasing radiation dose. Furthermore, it has been shown that irradiation induced significant changes in the secondary and tertiary structures of Ara h 6, a major peanut allergen, correlating with the loss in α-helix and IgG binding (Oh et al. 2009). In addition, the amount of other allergens was reduced by γ irradiation, depending on the dose (Byun et al. 2002). Therefore, irradiation may be used to reduce the allergenicity of foods. However, the feasibility of using irradiation for this application remains unclear as high doses are normally required to achieve significant reduction of allergens in real foods (Luo et al. 2013).

8 Factors Impacting the Efficacy of γ-rays

The efficacy of irradiation in achieving the above-mentioned benefits is influenced by many factors, which may be related to treated objects (food, microorganisms, and insects) and conditions where irradiation is conducted. Examples of the factors include pH, water activity, and physical state of foods, maturity of fresh produce, temperature, atmosphere, growth state of microorganisms and insects, dose, and dose rate. Furthermore, differences in radiation sensitivity of living organisms may be related to the ability to repair nucleic acid damage rather than an inherent irradiation resistance of the nucleic acid (Hansen 2001).

8.1 Temperature

The temperature of the product during irradiation affects the intensity of irradiation-induced chemical changes and the efficacy of irradiation against microorganisms (Farkas 1998). The lower the temperature, the slower the chemical reaction and higher the radio resistance of microbes will be. Free radicals generated from radiolysis of water cannot migrate freely when food is frozen. As a result, much less damage occurs to food quality, and much slower rates of bacteria inactivation are observed in the frozen state. *D*-values of *L. monocytogenes* in frozen vegetables increased significantly with decreasing temperature from −5 to −20 °C (Niemira et al. 2002). At an irradiation temperature of −5 °C, *D*-values ranged from 0.505 to 0.613 kGy while at −20 °C, *D* values were 0.767–0.916 kGy. It also demonstrated (Sommers 2012) that *D*-values of *L. monocytogenes* in meat irradiated at refrigerated and frozen temperatures were 0.40–0.47 and 1.21 kGy, respectively. Similarly, higher *D*-values (0.84 kGy) was observed for *L. monocytogenes* in the frozen salmon than at 25 °C (0.64 kGy) (Skowron et al. 2018).

8.2 Atmosphere

The composition (O_2 and CO_2) of the atmosphere in which foods and microorganisms are treated with irradiation has an influence on chemical and quality changes of foods and radiation resistance of microorganisms. Oxygen impacts the course of water radiolysis and the subsequent reactions of the primary radicals from water. It oxidizes free radicals and leads to formation of hydrogen peroxide, peroxides, and hydroperoxides (Simic 1983). In oxygen-free MAP, γ-rays exhibited much less impact on the quality of black pepper and cumin as compared to non-MAP (i.e. air) samples (Kirkin and Gunes 2018). Similarly, sensory evaluations demonstrated that the combination of irradiation at 4 kGy and MAP (70% CO_2/30% N_2) resulted in the highest shelf life extension of chicken breast by 12 days compared to the air-packaged samples (Chouliara et al. 2008). Reducing oxygen levels in MAP can decrease lipid oxidation of meat and poultry during and after irradiation (Ahn and Lee 2006). While low-oxygen levels may be favored for the preservation of food quality, high oxygen is generally preferred for increased radiosensitivity of microorganisms. For example, increased concentrations of oxygen in packages significantly increased the radiation sensitivity of *Salmonella* spp. and *Listeria* spp., leading to 7–25% reduction in D-values (Gomes et al. 2011). When grated carrots were irradiated at doses ⩾0.3 kGy, no *E. coli* was detected during 7 days of storage at 4 °C in samples treated under MAP (60% O_2, 30% CO_2, and 10% N_2), while levels of 1–2 log CFU g^{-1} of *E coli* were detected in irradiated samples treated under ambient air (Lacroix and Lafortune 2004). The increased radiosensitivity of microorganisms due to the presence of oxygen could be related to the production of ozone as a result of air (oxygen) irradiation; as ozone, a strong oxidant, is formed from oxygen in air during irradiation. During post-irradiation storage, the interactions of oxygen and CO_2 with microorganisms and with irradiation-generated free radicals are important factors influencing the survival of microorganisms and chemical/quality changes (Gazsó 2005). CO_2 has a bacteriostatic function against microorganisms probably due to enzymatic decarboxylation and change in permeability of bacterial membranes (Arvanitoyannis and Stratakos 2012).

8.3 Water Activity

As mentioned earlier, in high-moisture foods, γ-rays exert their impact via radiolysis of water. In dry foods, chemical changes and pathogen reductions are mostly results of direct effect, since water availability and diffusion of free radicals from water radiolysis are limited. Therefore, much higher doses are required to either achieve the same population reductions or to exert the same impact on food quality or chemical changes in low-moisture foods. For example, the *D*-values for *B. cereus* were 0.431 on chicken meat (Thayer and Boyd 1994), while in mesquite flour with *aw* of 0.31, the *D*-values were 1.27–1.55 kGy (Fan et al. 2015).

8.4 Composition of Foods (Antioxidants)

Antioxidants which are radical scavengers may protect microorganisms from irradiation injury or quality changes. Antioxidants terminate the reactions of free radicals and reactive spices generated from radiation of water. It has been shown that irradiation resistance of *L. monocytogenes* was higher in a solution of calcium ascorbate (an antioxidant) than in a buffer (Fan et al. 2005). Similarly, the *D*-values of *L. monocytogenes* increased when suspended in 0.1% sodium erythorbate solution (Sommers et al. 2002). However, no differences in radiation resistance (ranging from 0.67 to 0.70 kGy) were observed when *L. monocytogenes* was inoculated onto cooked meat products containing 0.05% sodium erythorbate, suggesting that the level of the antioxidants in the cured meats was not sufficient to protect the pathogen against the lethal effects of ionizing radiation.

9 Conclusion

γ-rays are high-energy photons emitted from the isotopes Co-60 and Cs-137 as permitted sources of γ-rays for food irradiation. As high-energy photons, γ-rays interact with matter through Compton scattering in which photons eject electrons from the orbit of the atom. There are two types of effects γ-rays have on foods: (i) direct disruption of biomolecules such as DNA or other components in food and microbes, and (ii) radicals and ions from water radiolysis reacting with molecules (indirect effect). Direct effect plays a major role in food in the dry and frozen state where water availability or diffusion of free radicals is limited, while the indirect effect dominates in food with free water as the major component. γ-rays have been applied or investigated for phytosanitary treatments of fresh fruits, microbial food safety enhancement, extension of foods by reducing spoilage, retarding ripening and senescence and inhibiting sprouting, and degradation of mycotoxin and food allergens. In spite of many decades of research and commercialization, widespread applications of γ-rays and other ionizing radiation still face challenges, including undesirable changes in some quality attributes, reluctance in consumer acceptance, and increased cost. To apply γ-rays and other irradiation technologies, factors such as nature and type of food, irradiation conditions (atmosphere, temperature, dose, dose uniformity, etc.), and growth stage and strain of organisms need to be considered and optimized for their intended purposes.

Acknowledgments

The authors thank Jessica Baik for drawing most of the figures in the book chapter, and Dr. Joshua Gurtler for reviewing the manuscript. Mention of trade names or commercial products in this publication is solely for the purpose of providing specific information and does not imply recommendation or endorsement by the US Department of Agriculture. USDA is an equal opportunity provider and employer.

References

Ahn, D.U. and Lee, E.J. (2006). Mechanisms and prevention of quality changes in meat by irradiation. In: *Food Irradiation Research and Technology* (eds. X. Fan and C.H. Sommers), 127–142. Ames, IA: Blackwell Publishing.

Ahn, D.U., Lee, E.J., and Mendonca, A. (2017). Meat decontamination by irradiation. In: *Advanced Technologies for Meat Processing*, 2e (eds. F. Toldrá and L.M.L. Nollet), 197–226. Boca Raton, FL: CRC Press.

Anderson, N.M. (2019). Recent advances in low moisture food pasteurization. *Current Opinion in Food Science* 29: 109–115.

Arvanitoyannis, I.S. and Stratakos, A.C. (2012). Application of modified atmosphere packaging and active/smart technologies to red meat and poultry: a review. *Food and Bioprocess Technology* 5 (5): 1423–1446.

Arvanitoyannis, I.S., Stratakos, A., and Mente, E. (2008). Impact of irradiation on fish and seafood shelf life: a comprehensive review of applications and irradiation detection. *Critical Reviews in Food Science and Nutrition* 49 (1): 68–112.

Bearth, A. and Siegrist, M. (2019). "As long as it is not irradiated"–influencing factors of US consumers' acceptance of food irradiation. *Food Quality and Preference* 71: 141–148.

Buzby, J.C., Farah-Wells, H., and Hyman, J. (2014). The estimated amount, value, and calories of postharvest food losses at the retail and consumer levels in the United States. USDA-ERS Economic Information Bulletin [online]. https://www.ers.usda.gov/publications/pub-details/?pubid=43836 (accessed 8 January 2021).

Byun, M.W., Lee, J.W., Yook, H.S. et al. (2002). Application of gamma irradiation for inhibition of food allergy. *Radiation Physics and Chemistry* 63 (3–6): 369–370.

Calado, T., Venâncio, A., and Abrunhosa, L. (2014). Irradiation for mold and mycotoxin control: a review. *Comprehensive Reviews in Food Science and Food Safety* 13 (5): 1049–1061.

Calado, T., Fernández-Cruz, M.L., Verde, S.C. et al. (2018). Gamma irradiation effects on ochratoxin A: degradation, cytotoxicity and application in food. *Food Chemistry* 240: 463–471.

Chouliara, E., Badeka, A., Savvaidis, I., and Kontominas, M.G. (2008). Combined effect of irradiation and modified atmosphere packaging on shelf-life extension of chicken breast meat: microbiological, chemical and sensory changes. *European Food Research and Technology* 226 (4): 877–888.

Cleland, M.R. (2013). Advances in electron beam and X-ray technologies for food irradiation. In: *Food Irradiation Research and Technology* (eds. X. Fan and C.H. Sommers), 9–27. Ames, IA: IFT/Blackwell.

Diehl, J.F. (1995). Potential and current applications of food irradiation. In: *Safety of Irradiated Foods*, 2e (ed. J.F. Diehl), 291–338. New York, USA: Marcel Dekker Inc.

Dubey, J.P., Thayer, D.W., Speer, C.A., and Shen, S.K. (1998). Effect of gamma irradiation on unsporulated and sporulated *Toxoplasma gondii* oocysts. *International Journal for Parasitology* 28 (3): 369–375.

Fan, X. (2012). Ionizing radiation. In: *Decontamination of Fresh and Minimally Processed Produce* (ed. V.M. Gómez-López), 379–405. Somerset, NJ: Wiley.

Fan, X. and Niemira, B.A. (2020). Gamma ray, electron beam, and X-ray irradiation. *Food Safety Engineering*: 471–492. Springer, Cham [online]. Available at: https://link.springer.com/chapter/10.1007/978-3-030-42660-6_18 (accessed 8 January 2021).

Fan, X., Niemera, B.A., Mattheis, J.E. et al. (2005). Quality of fresh-cut apple slices as affected by low-dose ionizing radiation and calcium ascorbate treatment. *Journal of Food Science* 70 (2): S143–S148.

Fan, X., Felker, P., and Sokorai, K.J. (2015). Decontamination of mesquite pod flour naturally contaminated with *Bacillus cereus* and formation of furan by ionizing irradiation. *Journal of Food Protection* 78 (5): 954–962.

Farkas, J. (1998). Irradiation as a method for decontaminating food: a review. *International Journal of Food Microbiology* 44 (3): 189–204.

Farkas, J. and Mohácsi-Farkas, C. (2011). History and future of food irradiation. *Trends in Food Science & Technology* 22 (2–3): 121–126.

Feldmann, F., Shupert, W.L., Haddock, E. et al. (2019). Gamma irradiation as an effective method for inactivation of emerging viral pathogens. *The American Journal of Tropical Medicine and Hygiene* 100 (5): 1275–1277.

Feliciano, C.P. (2018). High-dose irradiated food: current progress, applications, and prospects. *Radiation Physics and Chemistry* 144: 34–36.

Ferreira, I.C.F.R., Antonio, A.L., and Verde, S.C. (2017). *Food Irradiation Technologies Concepts, Applications and Outcomes*, 434. Cambridge, UK: Royal Society of Chemistry.

Follett, P.A. and Griffin, R. (2013). Phytosanitary irradiation for fresh horticultural commodities: research and regulations. In: *Food Irradiation Research and Technology*, 2e (eds. X. Fan and C.H. Sommers), 227–254. Ames, IA: Blackwell Publishing.

Follett, P.A. and Neven, L.G. (2020). Phytosanitary irradiation: does modified atmosphere packaging or controlled atmosphere storage creating a low oxygen environment threaten treatment efficacy? *Radiation Physics and Chemistry*: 108874.

Gazsó, L.G. (2005). Physical, chemical and biological dose modifying factors. In: *Radiation Inactivation of Bioterrorism Agents*, vol. 365 (eds. L.G. Gazsó and C.C. Ponta), 59. Amsterdam, Netherlands: IOS Pres.

Gerward, L. (1999). Paul Villard and his discovery of gamma rays. *Physics in Perspective* 1 (4): 367–383.

van Gerwen, S.J., Rombouts, F.M., Riet, K.V.T., and Zwietering, M.H. (1999). A data analysis of the irradiation parameter D 10 for bacteria and spores under various conditions. *Journal of Food Protection* 62 (9): 1024–1032.

GIPA (2014). Comparison of cobalt-60 gamma and x-ray technologies. http://gipalliance.net/wp-content/uploads/2013/01/GIPA-Whitepaper-Comparison-of-Cobalt-60-Gamma-X-ray-Technologies.pdf (accessed 8 January 2021).

Gomes, C., Moreira, R.G., and Castell-Perez, E. (2011). Radiosensitization of *Salmonella* spp. and *Listeria* spp. in ready-to-eat baby spinach leaves. *Journal of Food Science* 76 (1): E141–E148.

Hallman, G.J. (2011). Phytosanitary applications of irradiation. *Comprehensive Reviews in Food Science and Food Safety* 10 (2): 143–151.

Hallman, G.J. and Loaharanu, P. (2016). Phytosanitary irradiation–development and application. *Radiation Physics and Chemistry* 129: 39–45.

Hallman, G.J., Hénon, Y.M., Parker, A.G., and Blackburn, C.M. (2016). Phytosanitary irradiation: an overview. *Florida Entomologist* 99: 1–13.

Hansen, S.L. (2001). Content of free amino acids in onion (*Allium cepa* L.) as influenced by the stage of development at harvest and long-term storage. *Acta Agriculturae Scandinavica, Section B - Soil and Plant Science* 51 (2): 77–83.

Hume, A.J., Ames, J., Rennick, L.J. et al. (2016). Inactivation of RNA viruses by gamma irradiation: a study on mitigating factors. *Viruses* 8 (7): 204.

Jain, A., Ornelas-Paz, J.J., Obenland, D. et al. (2017). Effect of phytosanitary irradiation on the quality of two varieties of pummelos (*Citrus maxima* (Burm.) Merr.). *Scientia Horticulturae* 217: 36–47.

Jeong, M.A. and Jeong, R.D. (2018). Applications of ionizing radiation for the control of postharvest diseases in fresh produce: recent advances. *Plant Pathology* 67 (1): 18–29.

Kheshti, N., Melo, A.A.M., Cedeno, A.B. et al. (2019). Physiological response of 'Fuji'apples to irradiation and the effect on quality. *Radiation Physics and Chemistry* 165: 108389.

Kirkin, C. and Gunes, G. (2018). Modified atmosphere packaging and gamma-irradiation of some herbs and spices: effects on antioxidant and antimicrobial properties. *Journal of Food Processing and Preservation* 42 (8): e13678.

Kiss, I. and Farkas, J. (1988). Irradiation as a method for decontamination of spices. *Food Reviews International* 4 (1): 77–92.

Kuntz, F. and Strasser, A. (2016). The specifics of dosimetry for food irradiation applications. *Radiation Physics and Chemistry* 129: 46–49.

Lacombe, A., Breard, A., Hwang, C.A. et al. (2017). Inactivation of *Toxoplasma gondii* on blueberries using low dose irradiation without affecting quality. *Food Control* 73: 981–985.

Lacroix, M. and Lafortune, R. (2004). Combined effects of gamma irradiation and modified atmosphere packaging on bacterial resistance in grated carrots (*Daucus carota*). *Radiation Physics and Chemistry* 71 (1–2): 79–82.

Luo, C., Hu, C., Gao, J. et al. (2013). A potential practical approach to reduce Ara h 6 allergenicity by gamma irradiation. *Food Chemistry* 136 (3–4): 1141–1147.

Menon, S.S., Uppal, M., Randhawa, S. et al. (2016). Radiation metabolomics: current status and future directions. *Frontiers in Oncology* 6: 20.

Miller, G.Y., Liu, X., McNamara, P.E., and Barber, D.A. (2005). Influence of *Salmonella* in pigs preharvest and during pork processing on human health costs and risks from pork. *Journal of Food Protection* 68 (9): 1788–1798.

Mishra, B.B., Gautam, S., and Sharma, A. (2004). Shelf-life extension of fresh ginger (*Zingiber officinale*) by gamma irradiation. *Journal of Food Science* 69 (9): M274–M279.

Molina-chavarria, A., Félix-Valenzuela, L., Silva-Campa, E., and Mata-Haro, V. (2020). Evaluation of gamma irradiation for human norovirus inactivation and its effect on strawberry cells. *International Journal of Food Microbiology* 330: 108695.

Niemira, B.A. (2008). Irradiation compared with chlorination for elimination of *Escherichia coli* O157: H7 internalized in lettuce leaves: influence of lettuce variety. *Journal of Food Science* 73 (5): M208–M213.

Niemira, B.A. (2014). Irradiation, microwave, and alternative energy-based treatments for low-water activity foods. In: *The Microbiological Safety of Low Water Activity Foods and Spices* (eds. J.B. Gurtler, M.P. Doyle and J.L. Kornacki), 389–401. New York, NY: Springer.

Niemira, B.A. and Lonczynski, K.A. (2006). Nalidixic acid resistance influences sensitivity to ionizing radiation among *Salmonella* isolates. *Journal of Food Protection* 69 (7): 1587–1593.

Niemira, B.A., Fan, X., and Sommers, C.H. (2002). Irradiation temperature influences product quality factors of frozen vegetables and radiation sensitivity of inoculated *Listeria monocytogenes*. *Journal of Food Protection* 65 (9): 1406–1410.

Niemira, B.A., Fan, X., and Sokorai, K.J. (2005). Irradiation and modified atmosphere packaging of endive influences survival and regrowth of *Listeria monocytogenes* and product sensory qualities. *Radiation Physics and Chemistry* 72 (1): 41–48.

Niemira, B.A., Longczynski, K.A., and Sommers, C.H. (2006). Radiation sensitivity of *Salmonella* isolates relative to resistance to ampicillin, chloramphenicol or gentamicin. *Radiation Physics and Chemistry* 75 (9): 1080–1086.

Oh, S., Jang, D.I., Lee, J.W. et al. (2009). Evaluation of reduced allergenicity of irradiated peanut extract using splenocytes from peanut-sensitized mice. *Radiation Physics and Chemistry* 78 (7–8): 615–617.

Ohshima, H., Iida, Y., Matsuda, A., and Kuwabara, M. (1996). Damage induced by hydroxyl radicals generated in the hydration layer of γ-irradiated frozen aqueous solution of DNA. *Journal of Radiation Research* 37 (3): 199–207.

Ouattara, B., Sabato, S.F., and Lacroix, M. (2001). Combined effect of antimicrobial coating and gamma irradiation on shelf life extension of pre-cooked shrimp (*Penaeus* spp.). *International Journal of Food Microbiology* 68 (1–2): 1–9.

Pérez, M.B., Aveldaño, M.I., and Croci, C.A. (2007). Growth inhibition by gamma rays affects lipids and fatty acids in garlic sprouts during storage. *Postharvest Biology and Technology* 44 (2): 122–130.

Pillai, S.D. and Shayanfar, S. (2017). Electron beam technology and other irradiation technology applications in the food industry. In: *Applications of Radiation Chemistry in the Fields of Industry, Biotechnology and Environment* (eds. M. Venturi and M. D'Angelantonio), 249–268. New York, NY: Springer, Cham.

Prakash, A., Inthajak, P., Huibregtse, H. et al. (2000). Effects of low-dose gamma irradiation and conventional treatments on shelf life and quality characteristics of diced celery. *Journal of Food Science* 65 (6): 1070–1075.

Reisz, J.A., Bansal, N., Qian, J. et al. (2014). Effects of ionizing radiation on biological molecules – mechanisms of damage and emerging methods of detection. *Antioxidants & Redox Signaling* 21 (2): 260–292.

Rezaee, M., Almassi, M., Minaei, S., and Paknejad, F. (2013). Impact of post-harvest radiation treatment timing on shelf life and quality characteristics of potatoes. *Journal of Food Science and Technology* 50 (2): 339–345.

Rodrigues, I., Baldini, A., Pires, M. et al. (2020). Gamma ray irradiation: a new strategy to increase the shelf life of salt-reduced hot dog wieners. *LWT Food Science and Technology* 135: 110265.

Sarkar, P. and Mahato, S.K. (2020). Effect of gamma irradiation on sprout inhibition and physical properties of Kufri Jyoti variety of potato. *International Journal of Current Microbiology and Applied Sciences* 9 (7): 1066–1079.

Sharma, P., Sharma, S.R., Dhall, R.K., and Mittal, T.C. (2020a). Effect of γ-radiation on post-harvest storage life and quality of onion bulb under ambient condition. *Journal of Food Science and Technology* 57 (7): 2534–2544.

Sharma, P., Sharma, S.R., Dhall, R.K. et al. (2020b). Physio-chemical behavior of γ-irradiated garlic bulbs under ambient storage conditions. *Journal of Stored Products Research* 87: 101629.

Simic, M.G. (1983). Radiation chemistry of water-soluble food components. In: *Preservation of Food by Ionizing Radiation*, vol. 2 (eds. E. Josephon and M. Peterson), 1–73. Boca Raton, FL: CRC Press.

Singh, A. and Singh, H. (1982). Time-scale and nature of radiation-biological damage: approaches to radiation protection and post-irradiation therapy. *Progress in Biophysics and Molecular Biology* 39: 69–107.

Sjöberg, A.M., Manninen, M., Pinnioja, S. et al. (1991). Irradiation of pices and its detection. *Food Reviews International* 7 (2): 233–253.

Skowron, K., Grudlewska, K., Gryń, G. et al. (2018). Effect of electron beam and gamma radiation on drug-susceptible and drug-resistant *Listeria monocytogenes* strains in salmon under different temperature. *Journal of Applied Microbiology* 125 (3): 828–842.

Sommers, C.H. (2012). Microbial decontamination of food by irradiation. In: *Microbial Decontamination in the Food Industry* (eds. A. Demirci and M.O. Ngadi), 322–343. Cambridge, UK: Woodhead Publishing.

Sommers, C.H., Handel, A.P., and Niemira, B.A. (2002). Radiation resistance of *Listeria monocytogenes* in the presence or absence of sodium erythorbate. *Journal of Food Science* 67 (6): 2266–2270.

Song, W.J., Kim, Y.H., and Kang, D.H. (2019). Effect of gamma irradiation on inactivation of *Escherichia coli* O157: H7, *Salmonella* Typhimurium and *Listeria monocytogenes* on pistachios. *Letters in Applied Microbiology* 68 (1): 96–102.

Tejedor-Calvo, E., Morales, D., García-Barreda, S. et al. (2020). Effects of gamma irradiation on the shelf-life and bioactive compounds of *Tuber aestivum* truffles packaged in passive modified atmosphere. *International Journal of Food Microbiology* 332: 108774.

Thayer, D.W. (1993). Extending shelf life of poultry and red meat by irradiation processing. *Journal of Food Protection* 56 (10): 831–833.

Thayer, D.W. and Boyd, G. (1994). Control of enterotoxic *Bacillus cereus* on poultry or red meats and in beef gravy by gamma irradiation. *Journal of Food Protection* 57 (9): 758–764.

Tripathi, J., Chatterjee, S., Vaishnav, J. et al. (2013). Gamma irradiation increases storability and shelf life of minimally processed ready-to-cook (RTC) ash gourd (*Benincasa hispida*) cubes. *Postharvest Biology and Technology* 76: 17–25.

Urbain, W.M. (1986). Food Irradiation. In: *Elsevier*. Cambridge: MA.

Wang, J., Jiang, J., Wang, J. et al. (2019). The influence of gamma irradiation on the storage quality of bamboo shoots. *Radiation Physics and Chemistry* 159: 124–130.

4

Electron Beams

Rajeev Bhat[1], *Benny P. George*[2], *and Vicente M. Gómez-López*[3]

[1] *ERA-Chair for Food By-products Valorization Technologies (VALORTECH), Estonian University of Life Sciences (EMÜ), Tartu, Estonia*
[2] *Electron Beam Processing Section (EBPS), Isotope and Radiation Application Division (IRAD), Bhabha Atomic Research Centre (BARC), Trombay, Mumbai, India*
[3] *Cátedra Alimentos para la Salud, UCAM Universidad Católica San Antonio de Murcia, Murcia, Spain*

1 Introduction

Currently, world over, a wide array of sustainability challenges is being witnessed along the entire food supply chain. The major emphasis laid by the concerned personnel engaged in the field of agri-food technology (e.g. academicians, researchers, policymakers, industrialist, stakeholders, etc.) is to establish an efficient sustainable food production system with the main focus intended on ensuring food safety and overcoming certain food security challenges. Of late, evolution of circular economy concepts, minimizing industrial wastes/by-products generation, developing novel low-cost food processing technologies, designing new adoptable business models, setting up innovative plans to overcome unexpected pandemic situations (such as that of COVID-19), revolutionizing food policy and regulations are some of the themes of deliberation for the engaged scientific community.

Today's society is a reflection of smart consumers who have access to updated information's on safety, quality, and wholesomeness of a processed food product. Accordingly, the present day research focus is mainly driven around meeting consumers' needs and other quests such as identifying food authenticity, food traceability, adulteration, mislabeling of a product, etc. Nevertheless, majority of the consumers' world over are demanding foods with minimal processing and additives, and those which retains original flavor even after processing. In this regard, currently, a wide range of novel sustainable technologies are being designed, the focus of which is ensuring quality and safety of a processed food commodity. Till date, one of the most widely employed, popular, and effective modes of food processing/preservation had been the heat/thermal treatments. However, this method always had its own limitations wherein quality deterioration of food product occurred. Among a wide array of technologies available in the present day market (e.g. microwave,

Electromagnetic Technologies in Food Science, First Edition. Edited by Vicente M. Gómez-López and Rajeev Bhat.

pulse electric field, pulse light, microwave, ultrasound, radio frequency [RF]), food irradiation (use of ionizing radiations) has gained high recognition.

Food irradiation is considered as a non-thermal (cold pasteurization) mode of food processing technology, which is proved to be highly efficient in accomplishing disinfestation, decontamination, and retaining the wholesomeness of treated agri-food-based commodities. This technology produces minimal/negligible heat during treatment or on exposing a food product for irradiation treatments. Food irradiation involves exposure of agri-foods and other consumable products to γ rays (generated via cobalt-60 and cesium-137) and/or machine generated electron beams (EBs) or X-rays. This technology is considered as the best alternative for the routinely employed conventional thermal or chemical-based food disinfection processes.

During initial days of proposing radiation treatments for food application (during early 1950s), there were controversies on the safety in food products. Nevertheless, this was overcome based on a series of scientific evidences provided by the research community. Following the first meeting in 1964 organized by the '*Joint Expert Committee on Food Irradiation of the United Nations*', radiation treatments were duly established as a safer mode of processing. Further, in 1980, it was concluded by that food irradiation (up to a dose of 10 kGy) is safe and does not introduce any complications (nutritionally or microbiologically) and is safe for human consumption. After several decades of R & D activities, commercialization of this technology gained much success in food treatments (IAEA 1995; WHO 1999). Based on the requirements, radiation treatments can be achieved at low dose (<2 kGy, for insects larva disinfestation, induce delays in seed sprouting), medium dose (up 10 kGy for microbial decontamination), and high dose (>10 kGy for sterilization purposes).

Radiation treatment is measured as dose delivered and is expressed in Grays (Gy) or in kilo Grays (kGy) ($1\,\text{kGy} = 1000\,\text{kJ}\,\text{kg}^{-1}$). However, delivery of radiation dose depends entirely on the types of raw material or the food commodity as well as on the reasons for which radiation treatment is required. As per the International Atomic Energy Agency (IAEA) and the Food and Agriculture Organization of the United Nations (FAO), nuclear technologies offer competitive and distinctive solutions to overcome certain food security challenges such as fight against hunger and malnutrition and enhance environmental sustainability and certifies that foods are safe (https://www.iaea.org/topics/food-and-agriculture, access date: 12 November 2020).

Initially, food irradiation was performed by using γ rays, and this had gained much attention and was scientifically proved to be a success. However, negative perceptions among consumers were coupled with reluctance from the scientific communities who had opposed the use of γ rays in treating foods (raising concerns on safety due to the generation of radiation degradation products (RDP's) and being non-environmental-friendly), for many decades, this non-thermal technique remained quite unpopular compared to other conventional modes of food processing. However, later scientific advancements on use of EBs for food irradiation purpose were much appreciated. Historical, EB source was identified by Van der Graaff (in 1930), which was later followed by the development of EB irradiation and commercialization by Ethicon (Johnson & Johnson in 1950). Further in 1952, researcher Dr. Arthur Charles identified the effect of EB on cross-linking of polyethylene. This was followed by the production of polyethylene wires in Japan (1961) and setting up of a pilot-scale industrial radiation processing plant using EB (1970). After this, future decades witnessed tremendous amount of scientific research works being undertaken on the potential applications of EB irradiation in various applications.

EB accelerators are widely used for various applications in radiation processing applications (Bhat et al. 2008; Chmielewski et al. 1995; Cleland and Parks 2003; IAEA 2002, 2008; Zimek et al. 1993; Sarma et al. 2011). Using EB irradiation in food treatment is considered to be an innovative non-thermal (cold sterilization), highly efficient, reliable, and an environmental-friendly technology. EB irradiation finds wide applications in food industries where they are being employed for disinfestation and decontamination of processed food products as well as raw materials (e.g. spices, coffee beans, pulses, etc.), sprout inhibition in seeds, bulbs, and tubers and inducing delays in fruit ripening. In the chemical industries, EB irradiation is reported to be of use in cross-linking and polymerization of biopolymers, polymer modifications; in curing and grafting, vulcanization, gemstones coloring; in medical field as a source of radiotherapy; and for sterilization of medical disposable materials. Besides, EB treatments are also being employed in environmental remediation such as in the disinfection of sewage sludge, flue gas cleaning, wastewater treatments, etc. Of late, various types of EB accelerators have been developed with varying beam power (10–200 kW; e.g. Cyclotron, Dynamitron, Microtron, Rhodotron, Linac).

In the present chapter, deliberations are made on use of EB irradiation technology, types, and classification of industrial EB accelerators as a source of ionizing radiation, dosimetry, measurement traceability, and operational qualifications (OQs) with some classical case studies on food applications. This is a brief discussion on legislations imposed on using EBs for food applications. Future gaps of research and overcoming certain sustainability changes are also being highlighted.

2 Accelerator as a Source of Ionizing Radiation

Electrons are negatively charged particles, and it can be produced either through thermionic emission or photoemission from appropriate materials. Electrons can be energized by moving them in the high electric field. Accelerators are devices that produce beams of energetic charged particles. These particles can eject electrons from atoms when they interact with material and subsequently deposit energy in the material. Therefore, accelerated electrons are ionizing radiation, and they are suitable energy sources for the radiation processing applications. Other energy sources suitable for radiation processing applications are γ rays produced by radioactive nuclides and X-rays (bremsstrahlung) emitted by energetic electrons when it interacts with high atomic number material. The gamma and X-rays are electromagnetic radiation, while electrons are particles. Penetration of the electron in the medium is limited by its energy and density of the medium. The EB has limited penetration in the medium compared to that of electromagnetic radiation of same energy. EB is characterized by energy, current, and power. The electron energy is expressed in the electron volt or their multiples in kilo electron volt (keV) or mega electron volt (MeV). Energy of the beam determines thickness of the product that can be treated. Beam current defines number of electrons incident on the surface of the product per second, and it is expressed in milli ampere (mA). Power and beam current determine the throughput of the processing.

EB accelerators have many advantages over γ ray sources. EB accelerator is a machine source of ionizing radiation, and hence, there is no need of storage, transport, and disposal of radioactive material. This is an alternate technology to high activity radioactive source, and it has more acceptances in public for this technology. EB accelerators are electric machines, and it can be switched "on or off" as and when required, while radioactive sources always emit radiation irrespective of its use. Radioactive sources emit radiation in all direction, while EB is mono-directional and therefore efficient utilization of radiation can be achieved. EB accelerators provide higher dose rates resulting in short irradiation times and high throughputs. Additionally, EB machines can be hooked up with the industry for online processing of finished goods.

Compared to γ rays, EB has a shallow penetration in the material due to the greater rate of energy loss by electrons. The EB would be completely stopped by the thickness of absorber that would barely attenuate high-energy X-ray and γ rays. EB treatment of homogenous material produces absorbed dose distribution that varies with depth within the material. Further, EB and X-rays being machine-generated, a constant supply of electricity remains as a necessity. The γ rays and X-ray have long penetration compared to EB and exponentially attenuated.

To meet diverse applications in industry for various products, a range of low-, medium-, and high-energy electron accelerators are commercially available. These wide ranges of accelerators are suitable for applications in food irradiation, medical products sterilization, time control in semi-conductor power devices, flue gas treatment, as well as waste water treatment and polymer modification. Of late, the number of EB accelerator facilities for processing products for various purposes is increasing worldwide. In Sections 7–10 of this chapter, we have aimed to provide information regarding dosimetric procedures to be followed for process validation of an industrial EB accelerator.

3 Working Principle of EB Accelerator

The working principle of EB accelerator is based on the action of the electric field to increase the energy of beam of electrons, and further this is scanned by using the magnetic field. The electric field comes directly from high-voltage electrodes or indirectly via electromagnetic field. The major components of an accelerator to produce EB include: (i) the cathode, from which electrons are produced; (ii) electric field to accelerate electrons; and (iii) vacuum condition to prevent scattering of electrons by gas molecules. Due to energy range and beam power required for different radiation processing applications, different types of accelerators are designed.

4 Types of Industrial Electron Accelerators

Different types of EB accelerators are employed for radiation processing applications. The type of accelerator suitable for radiation processing depends on electron energy and beam power.

- *Direct current* (DC) *type*: These accelerators use a static potential difference between two electrodes. A cathode which emits electrons on heating is connected to the high voltage source and anode is maintained at ground potential. Electrons pass through the voltage and gain kinetic energy. DC accelerators produce continuous beams of electrons. If V_0 is the voltage and q is the charge of electron, then the kinetic energy acquired by the electron is qV_0 Joules. Cockcroft–Walton generators and Van de Graf generators are DC type accelerators. There is a limitation of potential difference that can be achieved by these accelerators. This limitation can be avoided by using Microwave and RF accelerators.

- *Microwave linear accelerators*: The particles are accelerated using many small resonant cavities energized by klystron and are designed to operate in either standing wave or traveling wave mode. The electron gains energy from electromagnetic waves.
- Radiofrequency (RF) type: Electrons are accelerated with a high-frequency electromagnetic field generated by RF auto generator in a single toroidal resonator cavity. Electrons gain energy in the single pass along the axis of the cavity. ILU accelerator produced by the Budker Institute of Nuclear Physics (Russia) and Rhodotron produced by Ion Beam Applications (IBA) (Belgium) are examples of RF type accelerators.

Maximum energy of the EB is limited up to 10 MeV for the irradiation of food to avoid any induction of radioactivity in irradiated food through photonuclear reactions.

5 Classification of Industrial Electron Beam (EB) Accelerators

Accelerators used for radiation processing are classified into the following categories based on the energy.

- *Low energy*: In this category, accelerators operate in the range of 70–300 keV energy and the beam currents range between 25 and 250 mA. The main use of these accelerators is in the field of the surface curing of thin films, antistatic and antifogging films, wood surface coatings, etc.
- *Medium energy*: Accelerators in the energy range of 300 keV–5 MeV are medium-energy categories. Cross-linking of wire and cables insulations, heat shrinkable plastic tubes, medical sterilization, and food irradiation are some of the major applications using this category of accelerators.
- *High energy*: The accelerators whose energy range is 5–10 MeV belong to this category, and these accelerators are suitable for bulk product irradiation.

6 Absorbed Dose

The absorbed dose is referred to as the quantity of ionizing radiation dose delivered and is the amount of energy absorbed per unit mass of an irradiated product. SI unit of absorbed dose is Gray (Gy), and it is equivalent to one Joule per kg. For simplicity, the absorbed dose is commonly specified to as "dose."

When EB treatments are used for enhancing physico-chemical properties of a material or to reduce the microbial load in a product, the changes induced are not easy to measure. While these changes are related to absorbed dose, it can be easily measurable for quantifying these changes. Radiation dosimetry is used for measuring and estimating the dose absorbed in the product.

7 Radiation Dosimetry

7.1 Theoretical Aspect of EB Dosimetry

When high-energy electrons pass through a material, it loses energy continuously through Coulomb interactions with valence electrons and nuclei of the atoms of that material (collision and radiative losses). The energy lost per unit path length is given by a quantity known as the linear stopping power of the material; this quantity is often divided by density of a material to give the mass stopping power. Units of the linear and mass stopping powers are $\text{MeV}\,\text{cm}^{-1}$ and $\text{MeV}\,\text{cm}^{2}\,\text{g}^{-1}$, respectively. Energy lost by fast electrons can either be absorbed by the medium or radiated as bremsstrahlung radiation. The stopping power thus consists of two components, a collision component, which describes interaction leading to absorption of energy close to the electron track and a radiative component, which describes bremsstrahlung production. Bremsstrahlung production is important for high energy beam and high atomic number material. For radiation processing of food, the bremsstrahlung is essentially negligible as food is low atomic number material. Particle fluence, ϕ, is the field quantity characterizing the EB, and it is the number of electrons cross per unit area. The unit of particle fluence is cm^{-2}.

Absorbed dose in a dosimeter or at a reference point in a given medium can be determined as the product of the electron fluence ($\phi(E)$) at that point and the mass collision stopping power $\left(\frac{S_{\text{col}}(E)}{\rho}\right)$ (ICRU/Report 35 n.d.; ICRU/Report 60 n.d.).

$$D = \int_{0}^{E_{\max}} \varphi(E)\frac{S_{\text{col}}(E)}{\rho}dE$$

For very small dosimeters, and well-matched systems where the fluence in the dosimeter placed in medium and fluence in the medium without dosimeters are the same, the dose in the medium, D_{m}, can be determined using,

$$D_{\text{m}} = \frac{(S/\rho)_{\text{m}}}{(S/\rho)_{\text{d}}} D_{\text{d}}$$

where $(S/\rho)_m$ and $(S/\rho)_d$ are mass collision stopping power for the medium and dosimeter, respectively. D_d is the absorbed dose in the dosimeter. Collision mass stopping power $\left(\frac{S_{col}(E)}{\rho}\right)$ depends on electron energy and the nature of product. Different products irradiated under the same beam will have different absorbed doses. Therefore, it is required to specify an absorbed dose in a reference material. For food irradiation purpose, the reference material is water as radiation interaction properties of most of the food is the same as water.

7.2 Practical Aspect of EB Dosimetry

In an industrial EB accelerator, radiation dosimetry revolves around practical measurement of absorbed dose and dose distribution in the irradiated product via use of dosimeters. Dosimeters are the devices used for the measurement of the dose. For the applicability of EB irradiation, a precise estimate of dose absorbed by a commodity is vital, as the absorbed dose and its distribution in the product need to be precisely monitored in the process. Dosimetry is carried out in setting process parameters of the facility, dose mapping in products to ensure that the dose is delivered to the product in acceptable dose limits, and for routine monitoring of processed products at the facility (IAEA 2013; ISO/ASTM 51649 2002). The absorbed dose in a commodity needs to be adequate to achieve the desired effects, but this should not exceed the levels required to gain the desired effect.

In EB processing technology, dose ranges from few Grays (e.g. sprout inhibition, insect disinfestation) up to few kilo Grays (for sterilization of ready-to-eat food items). A single dosimetry system cannot be used for this dose span. There are many different dosimetry systems available to cover a wide range of doses. Accuracy of the measurements can be achieved by establishing traceability of measurements to recognize and well-accepted national/international measurement standards. The dosimetry data are used for regulatory purpose for verification and control of licensing. The routine dosimetry measurement at the facility is used for quality control of the processed product, and the testing of individual processed products is not required routinely. Also, routine dosimetry is an independent method to verify that the operating parameters are under control to deliver the desired dose.

7.3 Dosimetry Systems

Dosimetry systems are used to measure the absorbed dose and involve dosimeters and other measurement instruments. Dosimeters are classified based on the accuracy of dosimetry systems and the field of application (primary, reference, transfer standards and the routine dosimeters). Primary standard dosimeters are maintained at national or international laboratories. Transfer standard dosimeters are employed to transfer dose information from calibration laboratories to the EB processing facilities in order to maintain the accuracy of measurement. Reference standard dosimeters are generally employed at a facility to calibrate routine dosimeters. Routine dosimeters are used for quality control of the product and process monitoring. Accuracy of routine dosimeters is much lesser when compared to reference standard dosimeters. However, these are of low cost and easy to use. Some of the commercially available dosimeters used for EB dosimetry are listed in Table 1.

Table 1 Dosimeters for electron beam accelerator facility.

Dosimeter system	Method of analysis	Useful dose range (Gy)	References
Alanine	EPR spectrometry	$1–10^5$	ISO/ASTM 51607 (2002)
FWT film	VIS-spectro-photometry	$10^3–10^5$	ISO/ASTM 51275 (2004)
B3 film	VIS-spectro-photometry	$10^3–10^5$	ISO/ASTM 51275 (2004)
CTA film	UV-spectro-photometry	$10^4–10^6$	ISO/ASTM 51650 (2002)
Calorimeter	Resistance/temperature	$1.5\times10^3–5\times10^4$	ISO/ASTM 51631 (2002)

The readout technique for each dosimeter, useful dose range, and ISO/ASTM standard practice to use them is also highlighted in this table.

The dosimeters used for EB processing need to thin in size. This is due to high dose gradient in distribution of the dose in a product subjected to EB irradiation. Radio chromic film dosimeters are convenient dosimeters useful to carry out dose mapping of a product. Optical density of these films changes upon irradiation. Change in optical density is measured using a UV/visible spectrophotometer. These film dosimeters can cover a wide dose range. Some of the film dosimeters have a dose rate-dependent response. Calibration of these dose rate-dependent dosimeters should be carried out in same irradiation conditions during normal operation of the facility.

7.4 Calibration of Dosimetry Systems

Dosimetry system is calibrated to establish the relationship between response of a dosimeter and absorbed dose. Calibration of dosimetry system involves calibration of dosimeters, readout system, and any other equipment used for the measurements. Calibration of routine dosimeters can be carried out either in the plant in which the dosimeters are routinely used or at a recognized calibration laboratory. Calibration at the irradiation facility is the most preferred option as it accounts for the same irradiation conditions during routine use (ISO/ASTM 51261 2002; ISO/ASTM 51431 2002). Data and information on radiation dosimetry as applied to food irradiation are detailed in the IAEA technical report (IAEA 2002).

7.4.1 Performance Check of Measuring Instruments

Before the dosimeter calibration, measuring instruments should be checked for their performance. For radio chromic film dosimeter measurement, the UV/visible spectrophotometer should be checked by using the standard absorbance and wavelength standards.

7.4.2 Calibration of Routine Dosimeters

Routine dosimeters can be calibrated either at the calibration laboratory or at irradiation facility. If the dosimeter is calibrated at the calibration facility, irradiation conditions may not be the same as it encounters during actual irradiation at the facility. Calibration at the facility is preferred as same irradiation conditions exist during plant operation. For calibration of films in EB accelerators, alanine dosimeters are used as reference dosimeters. Initially, depth dose distribution is measured in a stack of unit density material like plastic sheets by placing film dosimeters on top and intervening sheets and irradiated under EB. Total thickness of the stack should be maintained more than one and half time the total range of electrons in the material. From the depth dose curve, position of minimum dose gradient should be determined. After finding minimum dose gradient in the stack, four holes are drilled on the sheet where minimum dose gradient was found to accommodate alanine pellets. Film dosimeters are then irradiated together with alanine dosimeters at the minimum gradient position in the stack for different doses in the required range of doses. At each dose point, four dosimeters of each type should be used for taking average.

7.4.3 Establishing Measurement Traceability to National/International Standards

It is mandatory that all the measurements should be traceable and up to the national/international standards. For dose inter-comparison, routine dosimeter values are compared with the dose measured by transfer standard dosimeter. The routine dosimeters and transfer standard dosimeters are placed side by side during irradiation at the facility. Thus, accuracy of dose measurement is maintained by calibration and establishing traceability to national/international standards.

8 Scanning Characteristics of the Electron Beam Accelerator

Unlike gamma sources, EB from the accelerator is mono-directional. It is scanned using magnet for delivering the uniform dose to the product during irradiation. Beam width refers to the dimension of beam sweep, while beam length refers to dimension of the irradiation zone upright to the width of beam (ISO/ASTM 51649 2002). Beam width should be adequate to cover the irradiation zone for processing products. Beam length is very important for processes where product is stationary under the beam and for limiting speed of the conveyor.

9 Depth Dose Profile of Electron Beam

When EB passes through products, the electrons undergo multiple scattering due to Coulomb interactions between the incident electrons and the nuclei of the medium. Due to this multiple scattering, build-up of the dose occurs from the entrance surface to a depth of about half the maximum electron range and then decreases from there to the end of the range. An experimentally measured depth dose curve for 5 MeV EB in aluminum is shown in Figure 1. It is to be noted that the initial part of measured depth dose curve is higher than that of the calculated values for the mono-energetic beam reported in literature (ISO/ASTM 51649 2002). This is due to the incident EB may consist of low-energy electrons at different angles in actual measurement, whereas, in the Monte Carlo calculations, the parallel beam of mono-energy is considered (Benny et al. 2013; Vandana et al. 2018).

The parameters such as optimum thickness (R_{opt}), practical range (R_p), and half-value depth (R_{50}) can be obtained from the depth dose curve. Optimum thickness (R_{opt}) is considered as the depth where dose equals the entrance dose, half value depth (R_{50}) is the depth at which the dose equals half of maximal dose and practical range (R_p) is the depth from the surface where EB enters to the point where tangent at the steepest point on the straight descending depth dose curve meets depth axis. The depth dose profile depends on type of material and energy of the beam. Half-value depth and practical range are used to estimate average and most probable energy (ICRU/Report 35 n.d.; ISO/ASTM 51649 2002).

Absorbed dose profile in unit density product irradiated from one side under 4.5 MeV beam is shown in Figure 2. The dose in the product initially increases from the entrance surface to a depth of about half the maximum electron range in the material and decreases from there to the end of the range. Process thickness is the thickness where the decreasing dose equals the entrance dose for one-sided irradiation. Process thickness is important for

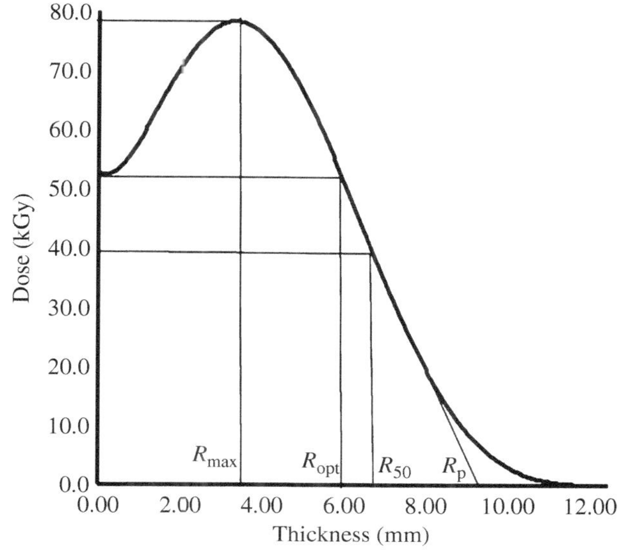

Figure 1 Measured depth–dose curve in Aluminum block under 5 MeV electron beam.

optimizing thickness of the product for uniform irradiation of the product. The ratio of maximum dose to entrance dose provides dose uniformity ratio. It should be close to unity for uniform dose distribution in the product.

For industrial applications of bulk products such as a food product, two-sided irradiations are employed (IAEA 2002). Dose distribution in unit density material for two-sided irradiations under 4.5 MeV beam is shown in Figure 2. The thickness for treatment from opposite sides is more than that of one-sided irradiation. Therefore, two-sided irradiations of products are economically viable. Measurement of dose distribution inside a product to be subjected for irradiation is vital to confirm the treatment is adequately uniform.

10 Process Validation of Industrial EB Accelerator

Process validation concerns the establishment (qualifying) of irradiation conditions that will yield products with desired quality. For this, an EB facility must be competent enough for its ability to deliver required controllable doses in a reproducible manner. After qualifying the facility, processing of the products within the dose limit must be qualified. There are many process parameters that are essentially need to be controlled for proper delivery of the dose. Process parameters consist of product characteristics (such as size, bulk density, and heterogeneity), irradiation conditions (for example, multi-sided irradiation, and number of passes under the beam), and operating parameters. Operating parameters include machine operating parameters (energy, average beam current, and pulse rate), parameters associated with scanning of the beam to cover the size of the product (beam width and scan frequency), and conveyor speed controls. All these parameters need to be optimized for processing of products to achieve desired effect. Therefore, when setting up

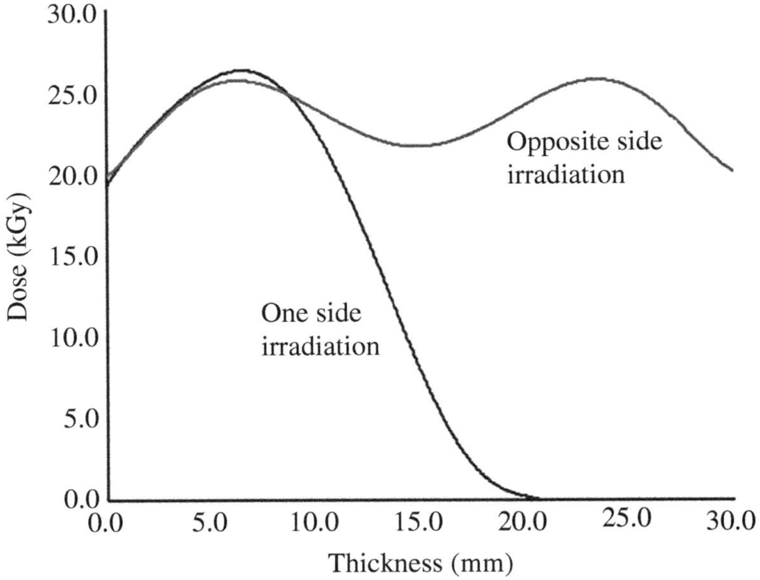

Figure 2 Absorbed dose profile in unit density product for one-side and two-side irradiation under 4.5 MeV electron beam.

an industrial EB accelerator facility, the three important activities required to be carried out with documentation for process validation of the facility includes:

(i) Installation qualification (IQ), (ii) operational qualification (OQ), and (iii) performance qualification (PQ) (ISO/ASTM 51649 2002).

These are pursued by using calibrated dosimeters. The IQ and OQ are used for the qualification of operation of an irradiator, while PQ is used for qualification of processing of the product.

10.1 Installation Qualification (IQ)

In installation qualification, the beam characteristics such as the electron energy, the beam width, homogeneity of the scanned beam, and the beam spot are measured.

- **Determination of electron beam energy**
  Penetration of EB in the material is determined by its energy and density of the material. Therefore, beam energy is the one of the important characteristics. Depth dose curve in a suitable reference material (e.g. Aluminum, graphite, water) is measured using dosimeters. Practical range (R_p) and half-value depth (R_{50}) are measured from the depth dose curve (Figure 1). The most probable and mean electron energy is estimated from these range values by using appropriate equations given in ISO/ASTM 51649.
- **Beam width**
  Beam width and dose uniformity are measured by irradiating strips of dosimeter films parallel to the direction of scan and measure absorbance along the dosimeter strips. The

length of dosimeter strip should be more than the dimension of the beam extraction window. Strips of long dosimeters mounted on Perspex plate are used for irradiation for this purpose. It should be ensured that beam width cover the irradiation zone adequately for processing industrial products. The dose should be uniform within the acceptable limit.

- **Beam spot**
  Measurement of beam spot is carried out by exposing strips of dosimeter films transverse to the scan direction in a static mode of operation and measuring absorbance along the dosimeter strips. These distributions look like Gaussian. Full width at half maximum (FWHM) is measured for beam spot. Beam spot is critical for processes where the product is stationary under the beam. Also, the speed of the conveyor is limited to ensure that the subsequent pulses overlap during irradiation to deliver uniform dose to a product.

10.2 Operational Qualification (OQ)

Following installation qualification of the facility, operational qualification has to be carried out using the suitable dosimetry system. There are mainly three aims for carrying out this activity. The first aim is to establish a relation between absorbed dose in a product and the various operating parameters over its expected range of operation. For EB irradiation, processing of a product is ensured by controlling as well as by monitoring operating parameters. The operating parameters consist of beam current, pulse frequency, scanning, and conveyor speed. These parameters are controlled to obtain reproducibility in the irradiation process, and the products receive the same dose as long as these parameters remain unchanged. The second aim is to establish capability of the irradiator to deliver the dose in an acceptable range by measuring minimum and maximum dose in a reference process load. A three-dimensional dose mapping study is carried out in a product container filled with a reference material by using film dosimeters. This study can help to identify the location of minimum and maximum dose. The third aim is to assess the dose variation when operating parameters fluctuate statistically during normal operation.

10.3 Performance Qualification (PQ)

The aim of PQ is the measurement of dose distribution in the actual product (i.e. dose mapping). Dose mapping is the measurement of the absorbed-dose distribution in an irradiation unit through the use of dosimeters placed at specified locations within the irradiation unit. Placement patterns may be selected which can most probably identify locations of absorbed dose maxima and minima. As a consequence of absorbed dose mapping, a positive link can be recognized between absorbed doses at a reference position relative to a maximum or minimum absorbed-dose position. This relationship is most useful for dosimeter placement during routine processing when the minimum or maximum absorbed-dose position is inaccessible. In the EB processing, thin-film dosimeters are used for dose mapping. Selection of proper product dimension (along with operational parameters for EB processing) is very important for the uniform dose distribution in the product.

10.4 Routine Monitoring

Dosimetry is used for routine monitoring of the radiation processing facility. Dose required in food irradiation spans from low to high dose. Hence, different dosimetry systems are required for routine dose monitoring. Dosimeters should be placed at appropriate locations on the products at regular intervals to monitor the dose delivered to the product. Routine dosimetry is an independent method to monitor the processing of the facility. Calibration of the routine dosimeters should be carried out periodically, and it should be traceable to the national or international laboratory.

11 EB Irradiation in Food Applications

Consumers' perception on employing ionizing radiations in food applications has been well comprehended (Bearth and Siegrist 2019; Bruhn 1998; Galati et al. 2019; Maherani et al. 2016; Roberts 2014). Application of radiation treatments in food processing/preservation has considerably contributed toward fulfilling vital food safety criteria and contributing toward minimizing economic loss of a region (Arvanitoyannis 2010; Loaharanu 1994; Bruhn 1998; Huang et al. 2020). As indicated earlier, microbial decontamination, insect disinfestation, and shelf-life extension remain as one of the top priorities to employ EB irradiation of agri-food commodities. However, enhancement in bioactive compounds in treated vegetal samples and cross-linking effects induced in polymers remains as an added advantage (Reyes and Cisneros-Zevallos 2007; Huang et al. 2020). Effects of both high energy and low energy EB treatments have been studied in various food matrices. Generally, treatment with EB irradiation at low energy (up to 1 MeV) is used for surface sterilization and for aseptic food packaging, while medium energy (between 1 and 5 MeV) is employed for pasteurization and decontamination of foods in packaging, whereas high energy (between 5 and 10 MeV) are employed for sterilization and pasteurization purposes, all depending on the raw material (Bhat et al. 2008; Gryczka et al. 2018, 2020; Smith et al. 2020). In Table 2, details on some interesting studies undertaken on application of EB irradiation in food are provided.

In the international trading arena, radiation processing is mainly recommended to overcome fundamental trade barriers to ensure consumer safety. EB treatments have been undertaken successfully on various types of fruits, vegetables, grains, cereals, in poultry, and in the seafood's to achieve various technical feasibilities. Surface decontamination of grains, cereals, and legumes has been achieved by application of soft electrons (energy level being 300 kV or lower) (Röder et al. 2009; Todoriki and Hayashi 2000; Todoriki et al. 2002; Urgiles et al. 2007). Low-energy EB treatment has low penetration and is generally more effective on the surface of a contaminated food product (Urgiles et al. 2007). Besides, low-energy electrons have been described to produce much more secondary electrons when compared with high-energy electrons (Nikjoo and Lindborg 2010).

Effects of EB irradiation on the quality characteristics of fruits such as strawberries, mango, oranges, watermelon, goji-berry, cherry tomato, and others have been studied (Madureira et al. 2019; Nam et al. 2019; Reyes and Cisneros-Zevallos 2007; Rodrigues et al. 2020; Smith et al. 2017; Yu et al. 1995). Decontamination of vegetables such as

Table 2 Some interesting studies undertaken on application of electron beam (EB) irradiation in food products.

Products	Purpose of irradiation	Observations	References
Vacuum- or air-packaged minced chicken meat	Effect of irradiation with X-rays or electrons followed by storage temperature, and post-irradiation cooking on thiamin content was evaluated.	Samples irradiated with 3 kGy X rays (50 Gy/min) or electrons (5 kGy/min) had less thiamin than control specimens. Highest loss of thiamin (31 and 28%) was recorded for X-rays and EB, respectively, in vacuum-packaged specimens stored and irradiated at 5 °C.	Van Calenberg et al. (1999)
Gum arabic	Raw samples were initially contaminated with bacteria, fungi, and spore-forming bacteria (e.g. *Enterococcus faecalis*, *Bacillus cereus*, and *Clostridium perfringens*).	The samples were completely decontaminated by irradiation at 10 kGy with either γ-rays or electrons.	Zaied et al. (2007)
Underutilized legumes (*Mucuna* sp.)	Proximate and nutritional composition, as well as functional properties of raw and EB irradiated samples (0–30 kGy dose range) were investigated.	EB irradiation increased protein, carbohydrates, and in vitro protein digestibility (dose-dependent). Improvement in functional properties was observed with cooking time of being significantly reduced on irradiation.	Bhat et al. (2008)
Lotus seeds	Nutritional, anti-nutritional, and functional properties of EB irradiated (doses: 0, 2.5, 5, 7.5, 10, 15 and 30 kGy) seeds were studied.	EB irradiation showed significant dose-dependent increase in total phenolics and tannins, with phytic acid eliminated at 5 kGy. Functional properties were improved in irradiated samples.	Bhat and Sridhar (2008)
Lotus seeds	Microbial decontamination (EB and gamma irradiation were compared at 0–30 kGy).	Significant dose-dependent decrease in the fungal and yeast contaminants occurred. A dose range between 10 and 15 kGy is recommended for complete decontamination.	Bhat et al. (2010)
Peanut butter	Microbial decontamination.	Reduction in *Salmonella* Tennessee and *S.* Typhimurium by 1-log after exposure to <1 kGy. However, these pathogens survived in peanut butter when exposed to non-lethal doses of EB.	Matak et al. (2010)

(*Continued*)

Table 2 (Continued)

Products	Purpose of irradiation	Observations	References
Maize seeds	Inactivation effects of EB irradiation (accelerated electrons, 0.1–6.2 kGy) on fungal load of naturally contaminated seeds were studied.	Fungal contamination significantly decreased corresponding to increase in irradiation dose. Sensitivity of fungi to EB treatment followed the order *Penicillium* spp. > *Fusarium* spp. > *Aspergillus* spp. (total inactivation achieved at doses of 1.7, 2.5, and 4.8 kGy, respectively).	Nemţanu et al. (2014)
Fresh produce (strawberries and lettuce)	Viral decontamination: Tulane virus (TV) and murine norovirus (MNV-1).	TV was reduced from 7 log10 units to undetectable levels (doses of 16 kGy or higher). MNV-1 was much more resistant to EB treatments when compared with TV.	Predmore et al. (2015)
Plasticized fish gelatin film	Effect of EB accelerator doses on properties of plasticized fish gelatin film.	EB irradiation increased thermal stability of the film by increasing the glass transition temperature and degradation. Besides, tensile strength was significantly increased by 30% (at 60 kGy). Also, surface tension, its polar component increased significantly with increase of wettability.	Benbettaïeb et al. (2016)
Almonds	Decontamination of *Salmonella* strains.	Non-significant differences between irradiation and thermal treatment were observed. D10-values for *Salmonella* Enteritidis PT30 and *Salmonella* Senftenberg 775W were 0.90 and 0.72 kGy.	Cuervo et al. (2016)
Sea buckthorn oil	Effects of accelerated EB irradiation (up to 8 kGy) on overall quality characteristics (color, polyphenolic content, carotenoids, antioxidant activity, and oxidative status) were investigated.	Minimal undesirable changes in oil quality were observed up to 3 kGy dose.	Nemţanu and Braşoveanu (2016)
Beef loin	Vacuum-packaged, EB irradiated samples (0 or 2 kGy) and stored for 10 days (aging temperatures of 2, 14, or 25°C) were evaluated for microbial and physiochemical qualities.	Inactivation of food-borne pathogens (*Listeria monocytogenes* and *Escherichia coli* O157:H7) was observed with reduced total aerobic bacterial counts. Shear force value decreased and lipid oxidation was reduced at short aging time and at low aging temperature. The color values in irradiated samples were lower than non-irradiated samples throughout the aging time.	Yim et al. (2016)

Table 2 (Continued)

Products	Purpose of irradiation	Observations	References
Heat-induced gels prepared with pork muscles	Investigated the physicochemical and textural properties.	Irradiation treatment (γ rays, EB, and X-rays at 5 kGy) showed lower water-holding capacity, protein solubility, viscosity, hardness, springiness, cohesiveness, gumminess, and chewiness when compared with control gel prepared with non-irradiated pork muscle, thus significantly affecting the gelling properties.	Kim et al. (2017)
Corn flour	Effects of EB irradiation on gelatinization and physicochemical properties of corn flour irradiated at 0–5.40 kGy were studied.	EB irradiation resulted in significant reduction of total starch and crude fiber with a corresponding increase in moisture and reducing sugar. Inhibition in the gelatinization of corn flour was recorded which was expected to be of use while designing new food formulations.	Xue et al. (2017)
Quail meat	Effects of EB treatment on microbial load and on shelf life were studied (doses of 0.5, 1, and 3 kGy and stored at 4 ± 1 °C for 15 days).	EB irradiation at doses of 1.5 and 3 kGy extended the shelf life and preserved the quality of meat. Even though thiobarbituric acid increased, irradiation did not have any significant effects on sensory properties.	Derakhshan et al. (2018)
Shrimp (*Solenocera melantho*)	Effect of EB irradiation on IgG binding capacity and conformation of tropomyosin in shrimp meat was studied (doses of 0, 1, 3, 5, 7, 9 kGy).	Significant decrease in IgG binding capacity of tropomyosin from irradiated shrimp meat was recorded (significant at 7 kGy). Results showed irradiation to reduce immune-reactivity of shrimp tropomyosin by altering the structure.	Guan et al. (2018)
Tegillarca granosa (blood cockle)	Effect of EB irradiation on biochemical properties and structure of myofibrillar protein were studied.	Irradiation dose of 3–5 kGy was effective toward preserving meat quality.	Lv et al. (2018)
Raw salmon fillets	Effect of gamma radiation and high-energy EB doses on inactivation of antibiotic-susceptible and antibiotic-resistant *Listeria monocytogenes* strains.	The study showed gamma radiation to be more effective in elimination of microbes than high-energy EB. Antibiotic-resistant *L. monocytogenes* strain was more resistant to both of the radiation methods.	Skowron et al. (2018)

(*Continued*)

Table 2 (Continued)

Products	Purpose of irradiation	Observations	References
Medicinal herbs (*Cnidii rhizoma* and *Alismatis rhizoma*)	Reduce microbial loads (doses ≤10 kGy) Bioactive compounds like chlorogenic acid, z-ligustilide, senkyunolide A, and ferulic acid were evaluated.	No significant changes were recorded for bioactive components or in anti-inflammatory activity.	Baek et al. (2019)
Red pepper powder and gochujang (red pepper paste)	Effects on *A. flavus* and on physicochemical and sensory quality were evaluated.	Gamma rays and EB irradiation (at 3.5 kGy dose) were effective in eliminating *A. flavus* (>4 log), present in red pepper powder and gochujang, but X-ray irradiation was not effective. No deterioration in other quality were noticed.	Byun et al. (2019)
Fresh mushrooms (*Agaricus bisporus* Portobello)	Effects of gamma and electron beam radiation on chemical and nutritional composition (irradiated at 1, 2 or 5 kGy) were analysed at different times (0, 4 and 8 days).	A dose of 5 kGy, independent of irradiation source, showed higher protein level.	Cardoso et al. (2019)
White rice and starch	Effects on physical properties (swelling power and major pasting viscosities including peak, breakdown, and setback viscosities) were studied.	EB irradiation at Low dose (<4 kGy) as a pre-treatment improved the quality of white rice during storage.	Du et al. (2019)
Edible lavers (*Porphyra umbilicalis*)	Effects of EB irradiation on inactivation of microbial populations (aerobic bacteria plate count, fungi, and coliforms) at 4 and 7 kGy were studied. In addition color values, soluble pigments, chlorophylls, and carotenoid content immediately and after storage were studied up to 12 weeks of storage.	Irradiation at 7 kGy maintained microbial safety and overall quality attributes during storage (up to 12 weeks).	Lee et al. (2019)
Mandarin oranges (*Citrus unshiu* (Swingle) Marcov)	Effect of EB irradiation on changes in microbial and physicochemical characteristics during storage at 4 °C for 15 days was studied.	Irradiated mandarins revealed dose-dependent inhibition of microbes (up to 15 days). Irradiation at 0.4 and 1 kGy doses did not prevent changes in stored mandarins. Dose of 0.4 kGy did not affect the constituents and physical quality of mandarins.	Nam et al. (2019)

Table 2 (Continued)

Products	Purpose of irradiation	Observations	References
Fresh noodles	Decontamination and inactivation of inoculated *Listeria innocua* pathogen.	EB treatment improved microbiological safety and shelf life. At 3 kGy, significant reduction was recorded in bacterial and fungal counts (mould and yeasts), and at 4 or 5 kGy, *L. innocua* population was inhibited to undetectable levels.	Shi et al. (2020)
Onion	Enhancing shelf life of onions, using energy of 1.25 MeV for both gamma irradiation and EB.	The prepared onion samples exposed to EB and gamma irradiations (Doses of 200 and 500 Gy at 25 °C) revealed minimal damages at 200 Gy doses with nutritional properties preserved, up to 30 days. This dose also eliminated pathogenic microorganisms.	Firouzi et al. (2020)
Red paprika	Microbiological quality and color properties.	Moulds, yeasts, and sulfite-reducing clostridia were the most resistant species, although a 10-kGy dose of irradiation leads to optimum sanitation. No significant differences were recorded in color properties of irradiated and non-irradiated samples.	Nieto-Sandoval et al. (2000)
Cereal and cereal-based products	Reduction in mycotoxin and microbial decontaminations.	EB irradiation is described to significantly reduce microbial load and some of the mycotoxins and is recommended to be effective when combined with fermentation techniques.	Khaneghah et al. (2020)
Pumpkin seeds (*Cucurbita pepo* var. styriaca) and golden flax seeds (*Linum usitatissimum*)	Decontamination of *Salmonella* cocktail with five serovars or a two-strain *Escherichia coli* cocktail comprising pathogenic strains, including *E. coli* O157:H7.	High-energy EB treatment (5 MeV) revealed (on both seeds), *Salmonella* cocktail to exhibit higher tolerance (at doses of 4 kGy) than the *E. coli* cocktail.	Henz et al. (2020)
Rice proteins (RPs).	Effects of EB irradiation on hydrolysis and antioxidant activity of RPs were studied.	Pre-treatment with EB significantly improved degree of hydrolysis and increased the DH value (15%) at a dose of 30 kGy. Besides, pre-treatments before enzymatic hydrolysis showed highest potential to obtain hydrolysates with higher bioactivity.	Zhang et al. (2020)

(*Continued*)

Table 2 (Continued)

Products	Purpose of irradiation	Observations	References
Probiotic *Lactobacilli* (*Lactobacillus acidophilus* DDS®-1, from Lacto-G (a marketed synbiotic formulation) and probiotic, *L. rhamnosus* Vahe)	EB irradiation for the improvement of biofilm formation by probiotic *Lactobacilli*.	EB irradiation (50–100 Gy) treatment was recommended to be considered for product sterilization, quality improvement and for packaging practices.	Pepoyan et al. (2020)

cabbage, cantaloupe, crushed garlic, lettuce seeds, spinach, and white button mushrooms is reported on EB irradiation (Espinosa et al. 2012; Grasso et al. 2011; Kim et al. 2014; Trinetta et al. 2011; Yurttas et al. 2014). Further, reports are available wherein high-energy EB radiation treatments at different dose ranges have been applied for various types of dry products, cereals, and pulses to achieve decontamination/sterilization, disinfestation, sprout inhibition, and for knocking off selected anti-nutrient compounds (Bhat et al. 2008, 2010; Hayashi and Todoriki 2001; Imamura et al. 2004a,b; Kottapalli et al. 2006; Nemţanu et al. 2014; Schoeller et al. 2002; Shawrang et al. 2011).

With regard to poultry industry, use of linear EB accelerator (Linacs delivering 2.5 kGy dose) resulted in non-significant changes in the overall qualities of egg (yolk or in protein) (Huang et al. 1997). Further, use of "Linac" in decontamination of poultry meat from pathogenic *Salmonella* sp. is reported by Sadat and Volle (2000). Decontamination (from bacteria) was achieved in boneless and skinless chicken breasts at EB irradiation dose of 1.0 and 1.8 kGy (Lewis et al. 2002). Nanosecond EB treatment to achieve complete sterilization in industrial poultry farming is reported by Kotov et al. (2003). Reduction in viral load (avian influenza virus) in egg and poultry meat on using high-energy (10 MeV) EB irradiation is illustrated by Brahmakshatriya et al. (2009).

In reference to seafood (e.g. salted, seasoned, fermented oyster), EB treatment (0, 0.5, 1, 2, and 5 kGy) has been successful against managing some of the food-borne pathogens such as *Staphylococcus aureus*, *Listeria monocytogenes*, and *Vibrio parahaemolyticus* (Song et al. 2009). Smoked salmon exposed to 2 kGy of EB and stored (for 35 days at 5 °C) was successful in controlling the microbial loads along with retention of odor (Medina et al. 2009). Further, the effectiveness of EB treatment on reduction of pathogenic fungal load and degradation of mycotoxins is reported. The mechanism involves breakage in the organic molecules and destruction of basic structure of the mycotoxin (Kottapalli et al. 2006; Luo et al. 2017; Wang et al. 2015). Besides, this can occur due to the mediation of free radicals produced by radiolysis of water molecules (water present in a food matrix). In addition, studies on combination treatment of EB irradiation with other processing technologies (e.g. natural botanicals, modified atmospheric packaging, combination with probiotics) for various food products have been a success (Balayan et al. 2019; Boynton et al. 2006; Smith et al. 2017; Yang et al. 2020).

11.1 Mechanism

EB technology generates high-energy electrons, while other ionizing radiation source such as gamma and X-rays uses energetic to achieve microbial inactivation. The mechanism of action to achieve decontamination and disinfestation relies on the established fact that irradiation treatment inactivates microbes as well as insects (larvae) by directly destructing the cell membrane and cell components. Besides, low amounts of free radicals produced after irradiation can also disintegrate the cell of an organism. EB irradiation effects are considered to be much higher and targeted directly onto a contaminated food product when compared with gamma wherein radiation emitted remains scattered (Diehl 1990). Tauxe (2001) has established EB radiation to directly damage the DNA of organisms by generation of stream of high-energy electrons which by introducing cross-linkages leads to incapability to proliferate and/or reproduce. Low-energy EB were identified to cause ionization of water molecules which lead to the generation of free radical-mediated reactive species (e.g. hydroxyl and/or superoxide radicals), which ultimately affected cell membrane and DNA (Chalise et al. 2004, 2007; Tahergorabi et al. 2012). In Figure 3a,b, direct and indirect effects of microbial inactivation by EB are depicted. Accordingly, in the direct effect, microbial inactivation by e-beam targets the genetic material (DNA and RNA) and breaks the base pairs G–C (guanine–cytosine) and T–A (thymine–adenine) which results in reproductive death of microorganisms, whereas in indirect effect, this is owed to free radicals generated from water, which undergoes radiolysis when subjected to e-beam yielding species with unpaired electrons thus affecting the membrane.

12 Legislations on Electron Beams Application

Research and the developmental activities undertaken after rigorous scientific assessments and based on domestic and international standards, today nearly 60 countries have set regulations to allow use of irradiation in food products, and thus opening up new market opportunities. Cross border international trade of irradiated fruits and vegetables has been a success mainly in Asia-Pacific and America. IAEA along with FAO and the IPPC (International Plant Protection Convention) has set certain guidelines on using irradiation as a phyto-sanitary measure and treatments for regulated/invasive pests (https://www.iaea.org/topics/food-irradiation, assessed on 12 November 2020).

In the European Union (EU), consumers' acceptance has been very poor on the use of ionizing radiations for food applications and this is mainly linked with the possible fear of inducing carcinogenicity in foods. Further link with term "radioactivity" (Frewer et al. 2011; Schweiggert et al. 2007) is also constraint for not accepting this technology. With regard to food applications, standard references have been put-forth for irradiated foods by Codex Alimentarius International Organization for Standardization and dosimetry in irradiation facilities (CAC-Codex Alimentarius 2003a,b; ISO 2005, 2012; IAEA 2015).

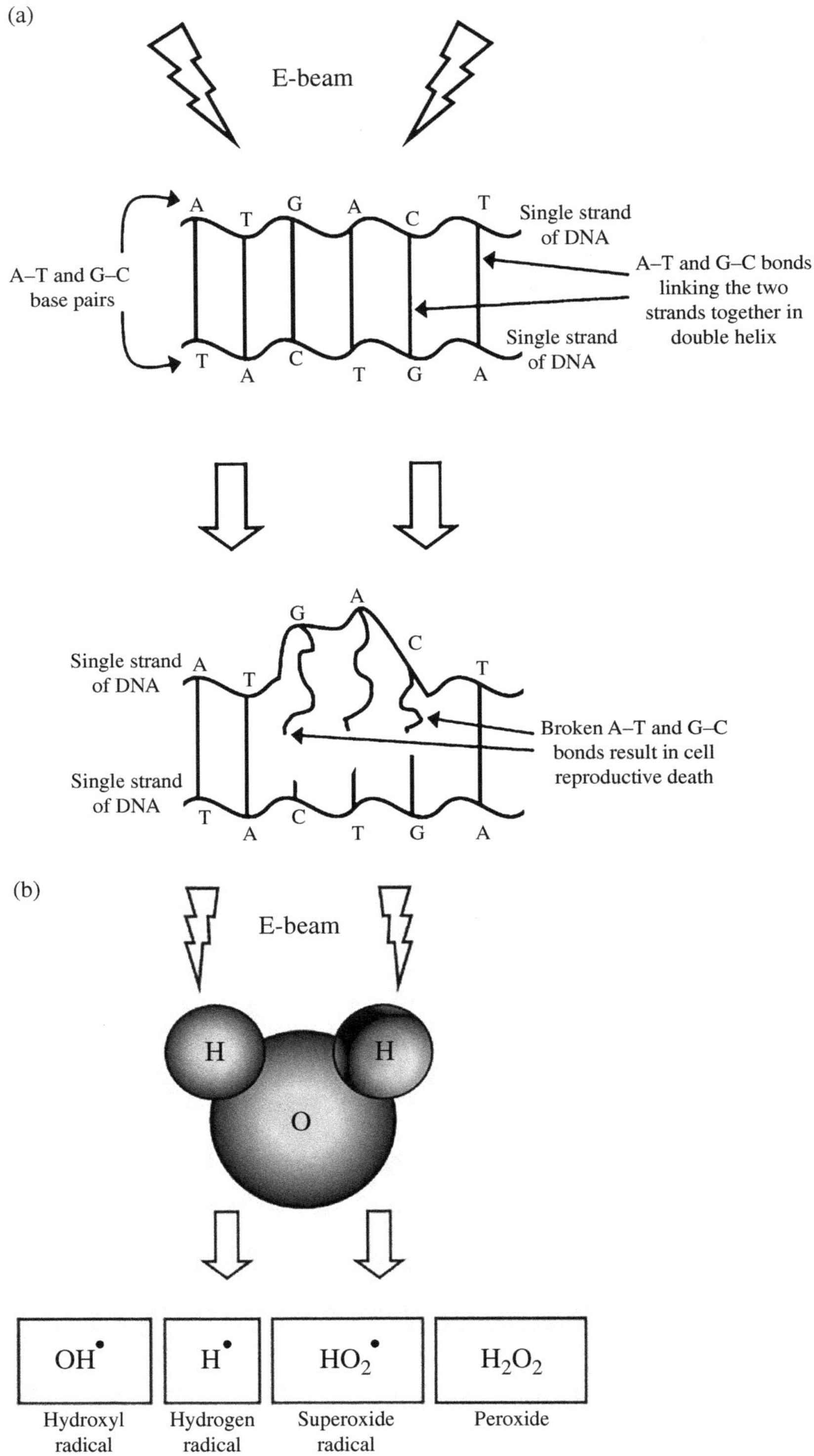

Figure 3 Direct (a) and indirect effects (b) of microbial inactivation by e-beams. *Source:* reproduced with kind permission from Elsevier: License no. 4946561351698, dated 12 November 2020; Tahergorabi et al. (2012).

Irradiated foodstuffs are subjected to special regulations in addition to all regulations applicable to other foodstuffs.

Irradiation of foods and food packaging materials is regulated in the USA by the Code of Regulations of the Food and Drug Administration in part 179 (21CFR179) sections 179.25 and 179.26 (FDA 2020). This regulation sets the maximum irradiation doses allowed, depending on the purpose of irradiation treatment (e.g. control of *Trichinella spiralis* in pork, control of food-borne pathogens in poultry), physical state of a food product (non-frozen or frozen, wet or dry), etc. A series of protocols have been put-forth by the USDA-APHIS (The United States Department of Agriculture and Animal and Plant Health Inspection Service) for employing ionizing radiation such as gamma, X-ray or EBs to treat selected agri-food-based commodities.

Application of ionizing radiation in food applications in EU is regulated by the Directive 1999/2/EC of the European Parliament and of the Council of 22 February 1999 (EC 1999a). The directive sets the conditions that must be fulfilled for achieving the approval of this treatment for specific uses and the protocol to calculate overall average absorbed dose. Furthermore, Directive 1999/3/EC authorizes the only kind of irradiated food allowed in the EU, specifically, dried aromatic herbs and spices, and vegetable seasonings treated at a maximum overall average absorbed radiation dose of 10 kGy (EC 1999b). EU countries like Belgium, Czech Republic, France, Italy, The Netherlands, Poland, Romania, and the United Kingdom can opt to irradiate specific foods and food ingredients at maximum allowed doses according to a list of authorization published by the EU (EC 2009). These foodstuffs include: fruits and vegetables (including root vegetables), cereals, cereal flakes, condiments, casein, caseinates, egg white fresh meat and poultry, frog leg, gum arabic, rice flour, spices, fish, shellfish, raw milk, blood products, etc. In 2011, European Food Safety Authority has provided a scientific opinion to consider food irradiation to be a safe, reliable, and processing method to effectively reduce food borne pathogens and can be explored to be coalesced as a multi-hurdle strategy (EFSA 2011).

The old restrictions imposed by the European Union to food irradiation have been opined by the International Irradiation Association to be responsible for impairing the development of this technology in Europe (IIA 2020). Furthermore, possibilities on the occurrence of carcinogenesis-promoting radiation-induced molecules (radiolytic products of fatty acid) like 2-dodecylcyclobutanone and 2-tetradecylcyclobutanone remain much disputed (Yamakage et al. 2014).

The diversity witnessed on the legal status of food irradiation across the world is a barrier to global trade of irradiated foodstuffs. The Global Harmonization Initiative (GHI) is an organization working toward harmonization of food safety legislation and regulations worldwide, which has recently revised on the legislation regarding food irradiation in different countries. This organization provides advice to the international regulatory bodies to recognize and accept scientific evidences on the safety of irradiated foodstuffs (GHI 2018). Emphasis is made in these regulations on labeling of irradiated foods and on placement of labeling using the "Radura" symbol as well as including statements such as: "treated with radiation" or "treated by irradiation" (FDA 2020).

13 Conclusions and Future Outlook

In the present chapter, attempts were made to emphasize on the importance of EB radiation technology for food industry applications to overcome food safety and food security issues. There are several research articles that have been published in this novel, nonthermal mode of food processing. However, in this chapter, our focus is laid mainly on covering technical aspects and providing valuable information's on advantages of using EB irradiation in food applications. Decades of research activities undertaken have now confirmed that treating foods at appropriate dose levels are beneficial and can offer economic gains. Safety aspects of employing food irradiation have been cleared and certified by world's leading bodies like IAEA, WHO, and FAO of the UN leading to approval in more than 60 countries. In the future, it is expected that more countries will embarrass this novel technology to ensure safety of a produce.

As in case of any other novel processing technologies, there are certain sustainability challenges to be overcome on using EB treatments for food applications on a pilot scale. EB technology can be considered as the best alternative over routinely employed conventional thermal or chemical-based treatments/processing. Operating of EB accelerators remains much simple (just like an on/off switch), and there is no radioactivity materials involved as in the case of γ rays. EB radiation technology being environmental-friendly also holds much advantage over gamma radiation treatments and ensures safety and wholesomeness of a food products. As of today, only selected agri-food produce has been studied for the effects on EB treatments. Hence, this needs to be extended for more number of commodities with expanded impacts studied (e.g. effects on healthy bioactive components such as antioxidant compounds, vitamins, effects on spoilage microorganism, food-borne virus, etc.). Extending beyond this, effects of EB treatment to valorize food industrial wastes and by-products (fruits and vegetables) can also be explored. Also, the impacts of EB treatment on possible generation of RDP's need to be carefully monitored in dry and wet foods.

One of the major disadvantages of using EB irradiation is the necessity of having a constant power supply throughout the controlled treatments. Besides, in certain cases, low penetrability of electrons has been reported, which can lead to inconsistency and inefficacy of the treatments. Low penetration capability can be a major issue when microbes or insects larva are found to be endophytic or present deep inside a product. Nevertheless, high penetration depth of high energy electrons can lead to non-uniform dose distribution inside a packaged food product. To ensure maximal effects, size, density, exposure time, accuracy in dose delivered, and packaging material need to be carefully monitored. Moreover, as witnessed with regard to other novel food processing technologies, EB treatments cannot assure reversing effects in the physico-chemical processes. Additionally, though scientific evidences have been provided, in majority of the countries, consumers still have negative perceptions on the use of terms such as "radiation", "radioactivity", "nuclear energy," "radioactive wastes," etc. Additionally, on certain instances, there is exclamation that setting of EB accelerates are expensive and cannot be afforded in low- or middle-income generating countries. However, it is envisaged that in the coming years, this will be overcome. Though, the need of the hour would be to educate the consumers and enhance their knowledge on the use of EB treatments (ionizing radiation) in food applications. Besides, EB is an environmental-friendly technique that supports global initiates of creating less stress on the environment.

Acknowledgements

The first author (RB) acknowledges ERA-Chair in VALORTECH project at the Estonian University of Life Sciences, which has received funding from the European Union's Horizon 2020 Research and Innovation Program under grant agreement No 810630.

Conflict of Interest Statement

The authors declare no conflicts of interest.

References

Arvanitoyannis, I.S. (2010). Consumer behavior toward irradiated food. In: *Irradiation of Food Commodities: Techniques, Applications, Detection, Legislation, Safety and Consumer Opinion*, 673–698. Cambridge, MA: Academic Press https://doi.org/10.1016/B978-0-12-374718-1.10017-3, ISBN: 978-0-12-374718-1.

Baek, M.E., Jo, Y., Chung, N. et al. (2019). Effect of E-beam irradiation on microbial load, stability of active components, and anti-inflammatory activity of *Cnidii rhizoma* and *Alismatis rhizoma*. *Journal of Medicinal Food* 22 (10): 1067–1077.

Balayan, M.H., Pepoyan, A.Z., Manvelyan, A.M. et al. (2019). Combined use of E-beam irradiation and the potential probiotic *Lactobacillus rhamnosus* Vahe for control of foodborne pathogen *Klebsiella pneumoniae*. *Annals of Microbiology* 69 (13): 1579–1582.

Bearth, A. and Siegrist, M. (2019). "As long as it is not irradiated"–influencing factors of US consumers' acceptance of food irradiation. *Food Quality and Preference* 71: 141–148.

Benbettaïeb, N., Karbowiak, T., Brachais, C.H., and Debeaufort, F. (2016). Impact of electron beam irradiation on fish gelatin film properties. *Food Chemistry* 195: 11–18.

Benny, P.G., Selvam, T.P., and Sarma, K.S.S. (2013). Comparison of graphite calorimeter dosimetry system with Monte Carlo simulation at an industrial electron beam accelerator. *Nuclear Instruments and Methods in Physics Research Section A: Accelerators, Spectrometers, Detectors and Associated Equipment* 703: 98–101.

Bhat, R. and Sridhar, K.R. (2008). Nutritional quality evaluation of electron beam-irradiated lotus (*Nelumbo nucifera*) seeds. *Food Chemistry* 107 (1): 174–184.

Bhat, R., Sridhar, K.R., Young, C.C. et al. (2008). Composition and functional properties of raw and electron beam-irradiated *Mucuna pruriens* seeds. *International Journal of Food Science & Technology* 43 (8): 1338–1351.

Bhat, R., Sridhar, K.R., and Karim, A.A. (2010). Microbial quality evaluation and effective decontamination of nutraceutically valued lotus seeds by electron beams and gamma irradiation. *Radiation Physics and Chemistry* 79 (9): 976–981.

Boynton, B.B., Welt, B.A., Sims, C.A. et al. (2006). Effects of low-dose electron beam irradiation on respiration, microbiology, texture, color, and sensory characteristics of fresh-cut cantaloupe stored in modified-atmosphere packages. *Journal of Food Science* 71 (2): S149–S155.

Brahmakshatriya, V., Lupiani, B., Brinlee, J.L. et al. (2009). Preliminary study for evaluation of avian influenza virus inactivation in contaminated poultry products using electron beam irradiation. *Avian Pathology* 38 (3): 245–250.

Bruhn, C.M. (1998). Consumer acceptance of irradiated food: theory and reality. *Radiation Physics and Chemistry* 52 (1–6): 129–133.

Byun, K.H., Cho, M.J., Park, S.Y. et al. (2019). Effects of gamma ray, electron beam, and X-ray on the reduction of *Aspergillus flavus* on red pepper powder (*Capsicum annuum* L.) and gochujang (red pepper paste). *Food Science and Technology International* 25 (8): 649–658.

CAC (Codex Alimentarius Commission), (2003a). Revised codex general standard for irradiated foods. (CODEX STAN 106-1983, REV.1-2003).

CAC (Codex Alimentarius Commission), (2003b). Recommended international code of practice for radiation processing of food. (CAC/RCP 19-1979, Rev. 2-2003).

Cardoso, R.V., Fernandes, Â., Barreira, J.C. et al. (2019). Effectiveness of gamma and electron beam irradiation as preserving technologies of fresh *Agaricus bisporus* Portobello: a comparative study. *Food Chemistry* 278: 760–766.

Chalise, P.R., Rahman, M.S., Ghomi, H. et al. (2004). Bacterial inactivation using low-energy pulsed-electron beam. *IEEE Transactions on Plasma Science* 32 (4): 1532–1539.

Chalise, P., Hotta, E., Matak, K., and Jaczynski, J. (2007). Inactivation kinetics of *Escherichia coli* by pulsed electron beam. *Journal of Food Science* 72 (7): M280–M285.

Chmielewski, A.G., Iller, E., Zimek, Z. et al. (1995). Industrial demonstration plant for electron beam flue gas treatment. *Radiation Physics and Chemistry* 46 (4–6): 1063–1066.

Cleland, M.R. and Parks, L.A. (2003). Medium and high-energy electron beam radiation processing equipment for commercial applications. *Nuclear Instruments and Methods in Physics Research Section B: Beam Interactions with Materials and Atoms* 208: 74–89.

Cuervo, M.P., Lucia, L.M., and Castillo, A. (2016). Efficacy of traditional almond decontamination treatments and electron beam irradiation against heat-resistant *Salmonella* strains. *Journal of Food Protection* 79 (3): 369–375.

Derakhshan, Z., Conti, G.O., Heydari, A. et al. (2018). Survey on the effects of electron beam irradiation on chemical quality and sensory properties on quail meat. *Food and Chemical Toxicology* 112: 416–420.

Diehl, J.F. (1990). *Biological Effects of Ionizing Radiation in Safety of Irradiated Foods.* New York, NY: Marcel Dekker Inc, ISBN 0-8247-8137-6.

Du, Z., Xing, J., Luo, X. et al. (2019). Characterization of the physical properties of electron-beam-irradiated white rice and starch during short-term storage. *PLoS One* 14 (12): e0226633.

EC, European Commission (1999a). Directive 1999/2/EC. https://eur-lex.europa.eu/legal-content/EN/ALL/?uri=CELEX%3A31999L0002 (accessed 4 November 2020).

EC, European Commission (1999b). Directive 1999/3/EC. https://eur-lex.europa.eu/legal-content/EN/ALL/?uri=CELEX%3A31999L0003 (accessed 4 November 2020).

EC, European Commission (2009). List of Member States' authorisations of food and food ingredients which may be treated with ionising radiation. https://eur-lex.europa.eu/legal-content/EN/ALL/?uri=CELEX:52009XC1124(02) (accessed 4 November 2020).

EFSA (2011). European Food Safety Authority: scientific opinion on the efficacy and microbiological safety of irradiation of food. *EFSA Journal* 9: 2103–2201.

Espinosa, A.C., Jesudhasan, P., Arredondo, R. et al. (2012). Quantifying the reduction in potential health risks by determining the sensitivity of poliovirus type 1 chat strain and rotavirus SA-11 to electron beam irradiation of iceberg lettuce and spinach. *Applied and Environmental Microbiology* 78 (4): 988–993.

FDA, Food and Drug Administration (2020). Code of Federal Regulations 21CFR179. https://www.ecfr.gov/ (accessed 4 November 2020).

Firouzi, S., Khorshidi, A., Soltani-Nabipour, J. et al. (2020). Evaluation of gamma and electron radiations impact on vitamins for onion preservation. *Applied Radiation and Isotopes* 167: 109442.

Frewer, L.J., Bergmann, K., Brennan, M. et al. (2011). Consumer response to novel agri-food technologies: implications for predicting consumer acceptance of emerging food technologies. *Trends in Food Science & Technology* 22 (8): 442–456.

Galati, A., Moavero, P., and Crescimanno, M. (2019). Consumer awareness and acceptance of irradiated foods: the case of Italian consumers. *British Food Journal* 121: 1398–1412.

GHI, Global Harmonization Initiative (2018). Consensus document on food irradiation. Discordant international regulations of food irradiation are a public health impediment and a barrier to global trade. https://www.globalharmonization.net/sites/default/files/pdf/GHI-Food-Irradiation_October-2018.pdf (accessed 4 November 2020).

Grasso, E.M., Uribe-Rendon, R.M., and Lee, K. (2011). Inactivation of *Escherichia coli* inoculated onto fresh-cut chopped cabbage using electron-beam processing. *Journal of Food Protection* 74 (1): 115–118.

Gryczka, U., Migdał, W., and Bułka, S. (2018). The effectiveness of the microbiological radiation decontamination process of agricultural products with the use of low energy electron beam. *Radiation Physics and Chemistry* 143: 59–62.

Gryczka, U., Kameya, H., Kimura, K. et al. (2020). Efficacy of low energy electron beam on microbial decontamination of spices. *Radiation Physics and Chemistry* 170: 108662.

Guan, A., Mei, K., Lv, M. et al. (2018). The effect of electron beam irradiation on igg binding capacity and conformation of tropomyosin in shrimp. *Food Chemistry* 264: 250–254.

Hayashi, T. and Todoriki, S. (2001). Low energy electron irradiation of food for microbial control. In: *Irradiation for Food Safety and Quality* (eds. P. Loaharanu and P. Thomas), 118–128. Lancaster: Technomic Publishing Co. Inc.

Henz, S., Nitzsche, R., Kießling, M. et al. (2020). Surrogate for electron beam inactivation of *Salmonella* on pumpkin seeds and flax seeds. *Journal of Food Protection* 83 (10): 1775–1781.

Huang, S., Herald, T.J., and Mueller, D.D. (1997). Effect of electron beam irradiation on physical, physiochemical, and functional properties of liquid egg yolk during frozen storage. *Poultry Science* 76 (11): 1607–1615.

Huang, T., Fang, Z., Zhao, H. et al. (2020). Physical properties and release kinetics of electron beam irradiated fish gelatin films with antioxidants of bamboo leaves. *Food Bioscience* 36: 100597. https://doi.org/10.1016/j.fbio.2020.100597.

IAEA (1995). Food irradiation newsletter, Supplement #19. Vienna, Austria: International Atomic Energy Agency.

IAEA (2015). *Manual of Good Practice in Food Irradiation: Sanitary, Phytosanitary and Other Applications*; Technical Reports Series. Vienna, Austria: International Atomic Energy Agency. ISBN: 978-92-0-105215-5.

ICRU/Report 35 (1984). International Commission on Radiation Units and Measurements, Radiation dosimetry: Electron beams with energies between 1 and 50 MeV, ICRU Report 35, ICRU, Bethesda, MD.

ICRU/Report 60 (1998). International Commission on Radiation Units and Measurements, Fundamental Quantities and Units for Ionizing Radiation, ICRU Report 60, ICRU, Bethesda, MD.

IIA, International Irradiation Association (2020). Food irradiation factsheet. https://iiaglobal.com/public-resources/ (accessed 4 November 2020).

Imamura, T., Todoriki, S., Sota, N. et al. (2004a). Effect of "soft-electron" (low-energy electron) treatment on three stored-product insect pests. *Journal of Stored Products Research* 40 (2): 169–177.

Imamura, T., Miyanoshita, A., Todoriki, S., and Hayashi, T. (2004b). Usability of a soft-electron (low-energy electron) machine for disinfestation of grains contaminated with insect pests. *Radiation Physics and Chemistry* 71 (1–2): 213–215.

International Atomic Energy Agency (IAEA) (2002). Dosimetry for Food Irradiation. *Tech. Rep. Ser. 409*. Vienna: IAEA.

International Atomic Energy Agency (IAEA) (2008), Trends in radiation sterilization of health care products. Vienna: IAEA.

International Atomic Energy Agency (IAEA) (2013). Guidelines for Development, Validation and Routine Control of Industrial Radiation Processes. *Radiation Tech. Ser. 4*. Vienna: IAEA.

ISO, International Organization for Standardization (2005). ISO/ASTM 51431:2005. Practice for dosimetry in electron beam and X-ray (bremsstrahlung) irradiation facilities for food processing. https://www.iso.org/standard/39913.html (accessed 4 November 2020).

ISO, International Organization for Standardization (2012). ISO 14470:2011. Food irradiation – requirements for the development, validation and routine control of the process of irradiation using ionizing radiation for the treatment of food. https://www.iso.org/standard/44074.html (accessed 4 November 2020).

ISO/ASTM 51261, (2002). Standard Guide for Selection and Calibration of Dosimetry Systems for Radiation Processing. ASTM International, 100 Barr Harbor Drive, PO Box C700, West Conshohocken, PA 19428-2959, United States.

ISO/ASTM 51275, (2004). Standard Practice for Use of a Radiochromic Film Dosimetry System. ASTM International, 100 Barr Harbor Drive, PO Box C700, West Conshohocken, PA 19428-2959, United States.

ISO/ASTM 51431, (2002). Standard Practice for Dosimetry in Electron and bremsstrahlung Irradiation Facilities for Food Processing. ASTM International, 100 Barr Harbor Drive, PO Box C700, West Conshohocken, PA 19428-2959, United States.

ISO/ASTM 51607, (2002). Standard Practice for Use of the Alanine-EPR Dosimetry System. ASTM International, 100 Barr Harbor Drive, PO Box C700, West Conshohocken, PA 19428-2959, United States.

ISO/ASTM 51631, (2002). Standard Practice for Use of Calorimetric Dosimetry Systems for Electron Beam Dose Measurements and Dosimeter Calibrations ASTM International, 100 Barr Harbor Drive, PO Box C700, West Conshohocken, PA 19428-2959, United States.

ISO/ASTM 51649, (2002). Standard Practice for Dosimetry in an Electron Beam Facility for Radiation. Processing at Energies Between 300 keV and 25 MeV. ASTM International, 100 Barr Harbor Drive, PO Box C700, West Conshohocken, PA 19428-2959, United States.

ISO/ASTM 51650, (2002). Standard Practice for Use of Cellulose Acetate Dosimetry Systems. ASTM International, 100 Barr Harbor Drive, PO Box C700, West Conshohocken, PA 19428-2959, United States.

Khaneghah, A.M., Moosavi, M.H., Oliveira, C.A. et al. (2020). Electron beam irradiation to reduce the mycotoxin and microbial contaminations of cereal-based products: an overview. *Food and Chemical Toxicology*: 111557. https://doi.org/10.1016/j.fct.2020.111557.

Kim, H.Y., Ahn, J.J., Shahbaz, H.M. et al. (2014). Physical-, chemical, and microbiological-based identification of electron beam-and γ-irradiated frozen crushed garlic. *Journal of Agricultural and Food Chemistry* 62 (31): 7920–7926.

Kim, H.W., Kim, Y.H.B., Hwang, K.E. et al. (2017). Effects of gamma-ray, electron-beam, and X-ray irradiation on physicochemical properties of heat-induced gel prepared with salt-soluble pork protein. *Food Science and Biotechnology* 26 (4): 955–958.

Kotov, Y.A., Sokovnin, S.Y., and Balezin, M.E. (2003). A review of possible applications of nanosecond electron beams for sterilization in industrial poultry farming. *Trends in Food Science & Technology* 14 (1–2): 4–8.

Kottapalli, B., Wolf-Hall, C.E., and Schwarz, P. (2006). Effect of electron-beam irradiation on the safety and quality of *Fusarium*-infected malting barley. *International Journal of Food Microbiology* 110 (3): 224–231.

Lee, E.J., Ramakrishnan, S.R., Kim, G.R., and Kwon, J.H. (2019). Storage stability of soluble pigments, chlorophylls, and carotenoids in electron-beam-irradiated edible lavers (*Porphyra umbilicalis*) with impact on microbial safety and sensory characteristics. *Journal of the Science of Food and Agriculture* 99 (8): 3860–3870.

Lewis, S.J., Velasquez, A., Cuppett, S.L., and McKee, S.R. (2002). Effect of electron beam irradiation on poultry meat safety and quality. *Poultry Science* 81 (6): 896–903.

Loaharanu, P. (1994). Food irradiation in developing countries: a practical alternative. *IAEA Bulletin* 36: 30–35.

Luo, X., Qi, L., Liu, Y. et al. (2017). Effects of electron beam irradiation on zearalenone and ochratoxin a in naturally contaminated corn and corn quality parameters. *Toxins* 9 (3): 84.

Lv, M., Mei, K., Zhang, H. et al. (2018). Effects of electron beam irradiation on the biochemical properties and structure of myofibrillar protein from *Tegillarca granosa* meat. *Food Chemistry* 254: 64–69.

Madureira, J., Severino, A., Cojocaru, M. et al. (2019). E-beam treatment to guarantee the safety and quality of cherry tomatoes. *Innovative Food Science & Emerging Technologies* 55: 57–65.

Maherani, B., Hossain, F., Criado, P. et al. (2016). World market development and consumer acceptance of irradiation technology. *Foods* 5 (4): 79.

Matak, K.E., Hvizdzak, A.L., Beamer, S., and Jaczynski, J. (2010). Recovery of *Salmonella enterica* serovars Typhimurium and Tennessee in peanut butter after electron beam exposure. *Journal of Food Science* 75 (7): M462–M467.

Medina, M., Cabeza, M.C., Bravo, D. et al. (2009). A comparison between E-beam irradiation and high pressure treatment for cold-smoked salmon sanitation: microbiological aspects. *Food Microbiology* 26 (2): 224–227.

Nam, H.A., Ramakrishnan, S.R., and Kwon, J.H. (2019). Effects of electron-beam irradiation on the quality characteristics of mandarin oranges (*Citrus unshiu* (Swingle) Marcov) during storage. *Food Chemistry* 286: 338–345.

Nemţanu, M.R. and Braşoveanu, M. (2016). Impact of electron beam irradiation on quality of sea buckthorn (*Hippophae rhamnoides* L.) oil. *Journal of the Science of Food and Agriculture* 96 (5): 1736–1744.

Nemţanu, M.R., Braşoveanu, M., Karaca, G., and Erper, I. (2014). Inactivation effect of electron beam irradiation on fungal load of naturally contaminated maize seeds. *Journal of the Science of Food and Agriculture* 94 (13): 2668–2673.

Nieto-Sandoval, J.M., Almela, L., Fernandez-Lopez, J.A., and Munoz, J.A. (2000). Effect of electron beam irradiation on color and microbial bioburden of red paprika. *Journal of Food Protection* 63 (5): 633–637.

Nikjoo, H. and Lindborg, L. (2010). RBE of low energy electrons and photons. *Physics in Medicine & Biology* 55 (10): R65–R109.

Pepoyan, A.Z., Manvelyan, A.M., Balayan, M.H. et al. (2020). Low-dose electron-beam irradiation for the improvement of biofilm formation by probiotic *Lactobacilli*. *Probiotics and Antimicrobial Proteins* 12 (2): 667–671.

Predmore, A., Sanglay, G.C., DiCaprio, E. et al. (2015). Electron beam inactivation of *tulane* virus on fresh produce, and mechanism of inactivation of human norovirus surrogates by electron beam irradiation. *International Journal of Food Microbiology* 198: 28–36.

Reyes, L.F. and Cisneros-Zevallos, L. (2007). Electron-beam ionizing radiation stress effects on mango fruit (*Mangifera indica* L.) antioxidant constituents before and during postharvest storage. *Journal of Agricultural and Food Chemistry* 55 (15): 6132–6139.

Roberts, P.B. (2014). Food irradiation is safe: half a century of studies. *Radiation Physics and Chemistry* 105: 78–82.

Röder, O., Jahn, M., Schröder, T. et al. (2009). Die eventus Technologie – eine Innovation zur nachhaltigen Reduktion von Pflanzenschutzmitteln it Empfehlung für Bio-Saatgut/E-ventus technology – an innovative treatment method for sustainable reduction in the use of pesticides with recommendation for organic seed. *Journal für Verbraucherschutz und Lebensmittelsicherheit* 4 (2): 107–117.

Rodrigues, F.T., Koike, A.C.R., da Silva, P.G. et al. (2020). Effects of electron beam irradiation on the bioactive components of goji-berry. *Radiation Physics and Chemistry* 179: 109144. https://doi.org/10.1016/j.radphyschem.2020.109144.

Sadat, T. and Volle, C. (2000). Integration of a linear accelerator into a production line of mechanically deboned separated poultry meat. *Radiation Physics and Chemistry* 57 (3–6): 613–617.

Sarma, K.S.S., Rawat, K.P., Benny, P.G., and Kader, S.A. (2011). Developments in Electron Beam Processing Technology. BARC News Letter, ISSN 0976-2108.

Schoeller, N.P., Ingham, S.C., and Ingham, B.H. (2002). Assessment of the potential for *Listeria monocytogenes* survival and growth during alfalfa sprout production and use of ionizing radiation as a potential intervention treatment. *Journal of Food Protection* 65 (8): 1259–1266.

Schweiggert, U., Carle, R., and Schieber, A. (2007). Conventional and alternative processes for spice production–a review. *Trends in Food Science & Technology* 18 (5): 260–268.

Shawrang, P., Sadeghi, A.A., Behgar, M. et al. (2011). Study of chemical compositions, anti-nutritional contents and digestibility of electron beam irradiated sorghum grains. *Food Chemistry* 125 (2): 376–379.

Shi, F., Zhao, H., Song, H. et al. (2020). Effects of electron-beam irradiation on inoculated *Listeria innocua*, microbiological and physicochemical quality of fresh noodles during refrigerated storage. *Food Science & Nutrition* 8 (1): 114–123.

Skowron, K., Grudlewska, K., Gryń, G. et al. (2018). Effect of electron beam and gamma radiation on drug-susceptible and drug-resistant *Listeria monocytogenes* strains in salmon under different temperature. *Journal of Applied Microbiology* 125 (3): 828–842.

Smith, B., Ortega, A., Shayanfar, S., and Pillai, S.D. (2017). Preserving quality of fresh cut watermelon cubes for vending distribution by low-dose electron beam processing. *Food Control* 72: 367–371.

Smith, B., Shayanfar, S., Walzem, R. et al. (2020). Preserving fresh fruit quality by low-dose electron beam processing for vending distribution channels. *Radiation Physics and Chemistry* 168: 108540.

Song, H.P., Kim, B., Jung, S. et al. (2009). Effect of gamma and electron beam irradiation on the survival of pathogens inoculated into salted, seasoned, and fermented oyster. *LWT – Food Science and Technology* 42 (8): 1320–1324.

Tahergorabi, R., Matak, K.E., and Jaczynski, J. (2012). Application of electron beam to inactivate *Salmonella* in food: recent developments. *Food Research International* 45 (2): 685–694.

Tauxe, R.V. (2001). Food safety and irradiation: protecting the public from foodborne infections. *Emerging Infectious Diseases* 7 (Suppl 3): 516–521.

Todoriki, S. and Hayashi, T. (2000). Disinfection of seeds and sprout inhibition of potatoes with low energy electrons. *Radiation Physics and Chemistry* 57 (3–6): 253–255.

Todoriki, S., Kikuchi, O.K., Nakaoka, M. et al. (2002). Soft electron (low energy electron) processing of foods for microbial control. *Radiation Physics and Chemistry* 63 (3–6): 349–351.

Trinetta, V., Vaidya, N., Linton, R., and Morgan, M. (2011). A comparative study on the effectiveness of chlorine dioxide gas, ozone gas and e-beam irradiation treatments for inactivation of pathogens inoculated onto tomato cantaloupe and lettuce seeds. *International Journal of Food Microbiology* 146 (2): 203–206.

Urgiles, E., Wilcox, J., Montes, O. et al. (2007). Electron beam irradiation for microbial reduction on spacecraft components. In: *2007 IEEE Aerospace Conference*, 1–15. IEEE.

Van Calenberg, S., Philips, B., Mondelaers, W. et al. (1999). Effect of irradiation, packaging, and post irradiation cooking on the thiamin content of chicken meat. *Journal of Food Protection* 62 (11): 1303–1307.

Vandana, S., Benny, P.G., and Selvam, T.P. (2018). Comparison of measured and Monte Carlo-calculated electron depth dose distributions in aluminium. *Indian Journal of Pure & Applied Physics* 56: 48–52.

Wang, R., Liu, R., Chang, M. et al. (2015). Ultra-performance liquid chromatography quadrupole time-of-flight ms for identification of electron beam from accelerator degradation products of aflatoxin B 1. *Applied Biochemistry and Biotechnology* 175 (3): 1548–1556.

WHO (1999). High-dose irradiation. Wholesomeness of food irradiated with doses above 10 kGy. Report of a Joint FAO/IAEA/WHO Study Group on High Dose Irradiation. *WHO Tech. Rep. Ser.* Geneva: World Health Organization; Food Irradiation Clearances.

Xue, P., Zhao, Y., Wen, C. et al. (2017). Effects of electron beam irradiation on physicochemical properties of corn flour and improvement of the gelatinization inhibition. *Food Chemistry* 233: 467–475.

Yamakage, K., Sui, H., Ohta, R. et al. (2014). Genotoxic potential and in vitro tumour-promoting potential of 2-dodecylcyclobutanone and 2-tetradecylcyclobutanone, two radiolytic products of fatty acids. *Mutation Research, Genetic Toxicology and Environmental Mutagenesis* 770: 95–104.

Yang, K., Li, K., Pan, L. et al. (2020). Effect of ozone and electron beam irradiation on degradation of zearalenone and ochratoxin A. *Toxins* 12 (2): 138.

Yim, D.G., Jo, C., Kim, H.C. et al. (2016). Application of electron-beam irradiation combined with aging for improvement of microbiological and physicochemical quality of beef loin. *Korean Journal for Food Science of Animal Resources* 36 (2): 215–222.

Yu, L., Reitmeier, C.A., Gleason, M.L. et al. (1995). Quality of electron beam irradiated strawberries. *Journal of Food Science* 60 (5): 1084–1087.

Yurttas, Z.S., Moreira, R.G., and Castell-Perez, E. (2014). Combined vacuum impregnation and electron-beam irradiation treatment to extend the storage life of sliced white button mushrooms (*Agaricus bisporus*). *Journal of Food Science* 79 (1): E39–E46.

Zaied, S.F., Youssef, B.M., Desouky, O., and El Dien, M.S. (2007). Decontamination of gum arabic with γ-rays or electron beams and effects of these treatments on the material. *Applied Radiation and Isotopes* 65 (1): 26–31.

Zhang, X., Wang, L., Chen, Z. et al. (2020). Effect of high energy electron beam on proteolysis and antioxidant activity of rice proteins. *Food & Function* 11 (1): 871–882.

Zimek, Z., Waliś, L., and Chmielewski, A.G. (1993). EB industrial facility for radiation sterilization of medical devices. *Radiation Physics and Chemistry* 42 (1–3): 571–572.

5

X-Rays

Francesco E. Ricciardi, Amalia Conte, and Matteo A. Del Nobile

Department of Agricultural Sciences, Food and Environment, University of Foggia, Foggia, Italy

1 Introduction

1.1 Thermal and Non-thermal Technologies

The heat treatment methods used for food sanitization offer a good yield in terms of shelf life, guaranteeing a microbial inactivation and safety of the product itself, but the drawback of these technologies is basically represented by the alteration of the organoleptic and/or nutritional properties of the treated food, as the heat treatments favor a decrease of certain nutrients and vitamins (water soluble) by changing the intrinsic characteristics of the product itself (Vicente and Castro 2007; Barba et al. 2016). Over the last few years, consumers prefer healthier and safer foods. This has pushed researchers toward the study of non-thermal technologies for food sanitation capable of guaranteeing consumer safety on the one hand and maintaining nutritional power and/or organoleptic value of the food on the other (Hernàndez-Hernàndez et al. 2019). These new technologies offer an interesting alternative to normal heat-based treatments because, in addition to food safety and quality, they offer the possibility of shortening treatment times and reducing both energy consumption and carbon dioxide emissions. The effectiveness of these technologies, in compliance with the quality of the treated product, depends on the type of food and treatment parameters (treatment time, treatment intensity, etc.). This has entailed a continuous and accurate study in the different areas of application of these methods (Jambrak et al. 2018).

In recent years, a wide range of papers have been officially published to certify the effectiveness of these emerging technologies on the main altering and pathogenic microorganisms that are harmful to food (Rifna et al. 2019). The term "emerging technologies or non-thermal technologies" can define all technologies that are currently being developed by researchers in the near future (after 5 or 10 years), or already developed and marketed (Misra et al. 2017). Emerging technologies such as high pressure processing (HPP) and pulsed electric fields (PEFs) have been the subject of many studies over time. The historically well-known technology is HPP, and it is currently used for both scientific and

Electromagnetic Technologies in Food Science, First Edition. Edited by Vicente M. Gómez-López and Rajeev Bhat.

industrial purposes, as it has proven its efficiency in terms of increasing shelf life in post-packaging applications not only on certain ready-to-eat foods (meat and salami) but also on matrices with a low pH value and fruit juices (Barbosa-Canovas and Bermúdez-Aguirre 2010). The treatment of food with high hydrostatic pressures (HHPs) is a technology in use since about 20 years, where the packaged solid foods are placed inside a high-pressure system, generally a chamber where the food is placed and then filled with liquid (usually water is used). The system, with the aid of pistons, exerts a pressure (from 100 to 800 MPa) on the packaged food matrix, for a variable time, from one millisecond to several minutes, in order to lead to the activation of microorganisms and enzymes (Khouryieh 2020).

A large number of foods and microorganisms have been tested with PEF technology, which has reported good results especially on liquid foods (Barbosa-Canovas and Bermúdez-Aguirre 2010).

In the PEF system, the food is subjected to an equipment with electrodes to allow the application of electric pulses at an intensity ranging from 10 to 80 $kV\,cm^{-1}$ for a time, variable from micro- to milliseconds, for microbial reduction, enzyme and vegetative cell inactivation, without clearly altering the nutritional properties of the product (Khouryieh 2020). Another non-thermal technology used for food sanitization is represented by UV light, which is distinguished by the wavelength in UV-A (315–480 nm), UV-B (280–315 nm), and UV-C (200–280 nm). In this case, microbial killing is caused by DNA damage (Zhang and Jiang 2019). Pulsed light (PL) is a very similar technology, but which unlike UV, the system consists of intermittent action lamps. PL systems can work continuously (Chaine et al. 2012), even if many laboratories use this emerging technology with machines that work in discontinuous mode (Ricciardi et al. 2020). In fact, the food, placed at a certain distance from the lamps, is subjected to intermittent pulsations of light. In this case, sanitization takes place thanks to the use of a broad-spectrum light in the wavelength range of 100–1100 nm. The most important factor that determines the effectiveness of microbial inactivation of PL is the dose of energy (or fluence) incident on the sample. The mechanism of PL inactivation, as well as the effectiveness of this technology in microbial reduction, is closely related to DNA damage (Kramer et al. 2015). Ultrasound treatment (US) causes a reduction of microbial load, thus following the occurrence of the phenomenon of cavitation, a process of growth and collapse of bubbles. This phenomenon locally generates pressure variations and cell rupture (Hernàndez-Hernàndez et al. 2019). Cold plasma treatment (CP) is a new non-thermal technology in which a partially ionized gas is used. The food is invested by this gas which contains a set of reactive species (photons, anions, cations, free radicals, etc.) which cause microbial inactivation (Barba et al. 2020). Among the emerging technologies, "dense carbon dioxide" can also be found. Carbon dioxide placed under supercritical conditions becomes a very soluble liquid that has properties similar to those of a gas, in addition to its characteristic extractive capacity. It is used to break down the microbial load in food by promoting the extraction of vital cell components and reducing endocellular acidity (Yuk et al. 2010). Irradiation can be defined as the non-thermal process in which foods subjected to the treatment record a clear reduction of the altering and pathogenic microflora due to the penetrating power of radiation, with a consequent increase in shelf life. Irradiation can be used for different purposes (Figure 1). The changes in food following irradiation, including nutritional effects, are generally acceptable

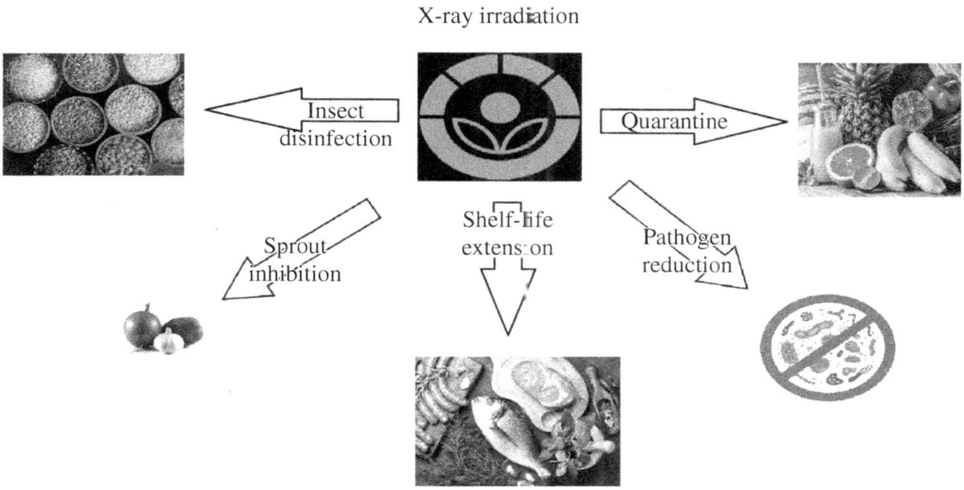

Figure 1 Applications of X-rays. (*See insert for color representation of this figure*).

(Ravindran and Jaiswal 2019). Among all these new technologies, ionizing radiations show a remarkable ability to reduce the microbial load in food (Lacivita et al. 2019). A focus on food irradiation has been provided in the subsequent paragraphs.

1.2 Irradiation Technology

Technologies based on ionizing radiation are basically represented by X-rays (≤7.5 MeV in some countries and ≤5 MeV in others), electron beams (≤10 MeV), and γ radiation (generated using radionuclides such as ^{60}Co or ^{137}Cs, but this latter isotope is not taken into consideration in the food sector because it is water soluble) (Lung et al. 2015). The advantages of this technology consist in the reduction of the processing time and the excellent penetration in food, even packaged products, thus avoiding recontamination. The use of ionizing radiation is allowed in some countries because it shows a reliable food quality (with an improvement of flavor in some foods), with few losses, suitable for large-scale production (especially dry food) and low energy cost (Khan et al. 2018; Pan et al. 2017). However, ionizing radiation has limitations, such as high capital costs, risks localized by radiation, poor consumer understanding, and unwanted changes due to high treatment intensity (Olatunde and Benjakul 2018; Pan et al. 2017; Shahbaz et al. 2016). Irradiation is a process that does not involve any increase in temperature and which unlike a normal heat treatment, generally does not significantly affect the nutritional value of the irradiated food, but this depends on factors such as: type of food, intensity of irradiation, process temperature, and packaging type (Ravindran and Jaiswal 2019). In some researches carried out on flour treated with γ rays, it has been recorded that the amylose content has increased by 25–36% with a reduction in viscosity. Irradiation leads to a mutation of the structure of the lipid fraction of the food due to the onset of specific reactions (oxidation, polymerization, decarboxylation, and dehydration) and, depending on the composition of the lipids, release of different compounds. However, these changes can be avoided if the food is packed (avoiding the effects of light and oxygen), and the irradiating treatment is carried out at low temperature (Bashir et al. 2017). Even foods that are important sources of fat-soluble vitamins, i.e. vitamins A and E, such as dairy products (which are generally not suitable for the sanitizing process through ionizing radiation), can be irradiated up to 40 kGy at a temperature of −78° C without alteration of their molecular structure (Hashisaka et al. 1990). A study supported by Hajare et al. (2006) on fruit and vegetable products treated with ionizing radiation at an intensity of 0.1 kGy demonstrated the efficacy of radiation, and consequently, products were stored for six months.

The reactive radicals that are generated after the irradiation process can lead to the formation of radiolytic products (formaldehyde and hydrocarbons), that can react with some food constituents. Interesting is the radiolysis of triglycerides that generates the 2-alkylcyclobutanones (2-ACB). These unique molecules in nature are used as markers to identify products subjected to the irradiation process. Recent studies have shown that the 2-ACBs have no toxicity effect if ingested at low concentrations (Song et al. 2014).

Irradiation shows positive effects on nitrosamines in seasoned meat products, additives such as nitrites and nitrates added to these foods give flavor on the one hand, but have carcinogenic effects on the consumer, on the other. Furthermore, it has been found that a high-dose treatment drastically reduces the formation of nitrosamines while maintaining the flavor of some cured meats unchanged. On bacon, it has been verified that an irradiation treatment of 30 kGy performed at a temperature of −40 °C, drastically reduces the nitrosamine content making the food similar to a product without the addition of nitrites and nitrates (Hui 2012). However, a study conducted on the damage to the genetic material of Indian children, following the consumption of stored and irradiated wheat, has led to negative results because there have been breaks along the DNA strands compared to children fed with non-irradiated wheat (Bhaskaram and Sadasivan 1975).

Irradiation was first used for commercial purposes in Stuttgart (Germany) in the late 1950s, when a manufacturer found an improvement in the sanitation quality of his spice-based products. In 1970, the International Field of Food Irradiation project (IFIP) was created by 19 countries. The goal was to carry out a research program on a global scale concerning the hygiene and health safety of irradiated foods. The number of nations grew to 24 with the help of Food and Agriculture Organization (FAO), International Atomic Energy Agency (IAEA), and OECD. Later, OMS also joined as an IFIP consultant. The research investigations were mainly aimed at evaluating the potential chemico-physical changes of the irradiated products (with a maximum dose of 10 kGy). In fact, tests were carried out on animals fed with irradiated food for a very long period of time and short-term screening tests (Diehl 1995). In 1980 after a careful evaluation of the results on the health of irradiated foods by the FAO, IAEA, and OMS expert committee, it was concluded that any food treated with a dose lower than 10 kGy does not present toxicological risks or particular microbiological and nutritional risks (WHO 1981). In 1982, it was established by IFIP that foods are made safe if irradiated up to 10 kGy; official research papers were drawn up (Elias and Cohen 1983). In 1983, the International Consultative Group on Food Irradiation (ICGFI) was created by 45 member countries and provides a wide range of publications on the safety of irradiated food and, in the same year, the Codex Alimentarius commission adopted a specific general standard for irradiated food products and a recommended international code of conduct for the operation of irradiation facilities (Diehl 2002).

In 1997, the study group set up by FAO, IAEA and OMS examined the safety research results carried out on foods that had undergone treatment at doses >10 kGy. An objective loss of sensory quality of treated food was found, from which it can be deduced that very few foods tolerate doses higher than 10 kGy (OMS 1999).

1.3 X-Rays

The X-ray radiation technique is a particular interesting irradiation method. X-rays are generated mechanically without the production of dangerous radioactive substances. Furthermore, given their high penetration capacity, this technology is used in post food packaging, considering that X-rays are able to cross materials up to 30–40 cm deep. In addition, unlike other radiation technologies, the intensity dose used for X-ray treatment of food matrices must not exceed the 10 kGy threshold (Roberts 2016). Therefore, given the proven effectiveness and efficiency of X-rays, the low environmental impact, and the potential for direct installation on an industrial scale, in recent years, this technology has attracted attention as an excellent strategic source for reduction of microbial load in foods (Moosekian et al. 2012). The energy spectrum of X-rays depends on three factors: angle of radiation, energy dependence on the absorption of X-rays, and the probabilistic scattering of the particles in the food matrix. The irradiation doses must be homogeneously distributed in the product to avoid an overdose. Therefore, models necessary to predict the distribution of the dosage in the food product (expressed in kGy) and to evaluate the efficiency of the irradiation intensity necessary for the reduction of the microbial load have been put in place (Miller 2003).

Due to the relevance of irradiation in food sector, this overview study aims to be a collection of various research on radiation reported in the literature. It represents the first attempt in the last five years to overview the topic and collect basic information on the X-rays technology, real applications to food sector, efficacy on food products, and effects on its quality and safety. Specifically, the chapter will focus mainly on: (i) the mechanisms of action, with an emphasis on some case studies of interest, (ii) the effect of radiation on packaging, and finally (iii) brief mention of regulation aspects related to irradiated foods. The proposed chapter might very well target a broad audience with an interest in food science, thus including academia (graduate level), food-related industries, and trade associations; in fact, topics are treated a degree of in-depth analysis suitable for science community (Master and PhD students) and industrial operators.

2 Mechanism of Action of X-Rays

X-rays are generated within a system when a high-energy electron beam is fired at a target (generally tungsten); once the electrons reach this metal structure, the atoms that constitute it are excited and give off an electron; it follows that an electron of the upper orbital of the atom immediately falls into the lowest energy level releasing energy (photon). There are two fundamental factors for the functioning mechanism of the X-rays on which penetration depends, they are energy and current. The current (expressed in mA) is closely related to the number of photons generated following the treatment, and the energy refers to that of the photon when it comes out of the tube (Haff and Toyofuku 2008). When a food is subjected to the irradiating action of X-rays, an atom ionization phase occurs in the food matrix when the latter reacts with photons. This ionization strictly depends on the chemico-physical properties of the irradiated food and on the irradiating intensity that is used for sanitizing the product. Two different energy transfer processes can occur: photoelectric absorption and compton dispersion. Photoelectric absorption occurs when low energy photons are used; in this case, the energy of the photon is completely absorbed by the target electron, and in turn, the latter can interact with the other atoms. Compton dispersion occurs when at high energy level electrons are targeted on the outer orbit; in this case, the same photon and the ions (anions and cations) generated due to the interactions (hence the term ionizing radiation) can continue to interact with other atoms that make up the food.

The activity of the water in the food is a fundamental element for obtaining a good efficiency of an X-ray irradiation treatment because its presence promotes the formation of highly reactive free radicals, such as hydrogen and hydroxide (devoid of the hydrogen bond). These water molecules in the excited state generate further reactions up to the production of stable products from the water radiolysis process, which are hydrogen and hydrogen peroxide (Moosekian and Ryser 2011). Therefore, based on the aforementioned principles, the sanitizing activity of this technology can be divided into two types of effects: primary and secondary effects. The primary effects occur when the electrons, that are released from the interaction with photons, stimulate other nearby atoms favoring ions formation. Secondary effects occur when these ions generated following primary reactions react with other molecules to form highly reactive components, such as free radicals, electrons and ions (Miller 2005). A general scheme of the mechanism of action is reported in Figure 2.

The effects of radiolysis depend on the dose of energy adsorbed per unit of mass. The unit of measurement that defines the adsorbed dose is the Gray (Gy), which corresponds to the unit value of $1\,J\,Kg^{-1}$, even if in literature many times it is reported the unit of measurement expressed in "radiation adsorbed dose" (rda): 1 rad = 0.01 Gy (Miller 2005).

In many cases, the shelf life of a food is closely related to microbial inactivation. X-rays cause damage to the cellular structure including DNA, killing the cell, and making it

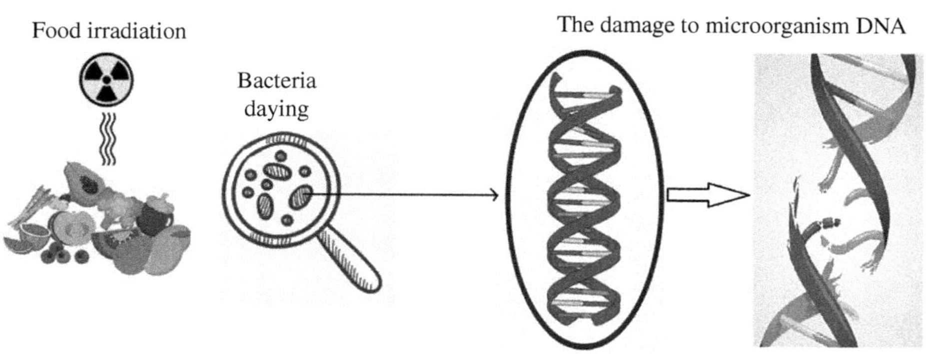

Figure 2 Mechanism of action of X-rays. (*See insert for color representation of this figure*).

unable to replicate and exercise its biological function in food (Jaiswal 2016). In percentage terms, 20% of the attack occurs on sugars that make up the genetic material and the remainder occurs on nitrogen bases. Of the latter, thymine has a higher sensitivity than cytosine, adenine, and guanine (Moseley et al. 1990). Due to the orientation of the DNA in the cell, damage to the double strand is more difficult (Jaiswal 2016). During an ionizing treatment, most of the DNA strand breaks are due to the cleavage of the C-3 phosphodiester bond, which produces 5-PO_4 and 3-PO_4 (3 : 1 ratio) (Johnston and Stevenson 1990). In addition to DNA, harmful effects can occur on proteins and membranes, compromising the life of the microorganism (Ravindran and Jaiswal 2019).

After an irradiation treatment, the potential survival of microorganisms depends on multiple factors, such as the chemical compositional structure of the food in which the microorganisms are present, the temperature, the pH, and the intensity of the treatment. The spores are generally more resistant than the vegetative cells, even if a greater resistance to a radiating treatment is shown by yeasts, molds, and viruses (Monk 1994). However, microbial species capable of resisting the action of treatments with ionizing radiation have been disclosed, an example is the *Deinococcus radiodurans*, which, through its proteomic system, is able to make repairs to DNA damaged following a radiating treatment. Specifically, this microorganism is able to resist up to a treatment intensity of 30 kGy (Ravindran and Jaiswal 2019). A study conducted by Krisko and Radman (2013) testifies that regardless of the bacterial strain of *D. radiodurans*, cell death is closely related to protein damage and that a protected proteome capable of resisting irradiating treatment ensures good repair to DNA and other cell constituents up to a dose of 15 000 Gy.

3 Case Study

Numerous X-ray applications in a wide range of food matrices are reported in the literature. The goal has always been to apply the treatment to control the proliferation of microbial load, both spoilage and pathogen microorganisms, and at the same time evaluate the effects of the treatment on chemical, physical, and sensory quality. The food categories to which the X-rays have been applied are different and include fish products, dairy products, fresh and dried fruits, and processed meat products. Below are some details of the individual experiments carried out, divided by food categories, that highlight the effects of X-rays according to the intensities used and the microbial target.

3.1 Seafood Products

In this section, different fish products treated by X-rays at different doses are reported and effects on fish quality are underlined. Table 1 summarizes the most recent applications of X-rays, highlighting the main effects on food microbiological quality. Specifically, X-ray irradiation tests were carried out on smoked red mullets vacuum packed up to the maximum dose of 2 kGy (Robertson et al. 2006). More specifically, the authors applied X-rays to mullet inoculated with a cocktail consisting of five different *L. monocytogenes* strains at a cellular concentration of 10^4 CFU g^{-1}. After contamination, the fish was vacuum-packed and exposed to different radiation intensities: 0.5, 1.0, 1.5, and 2.0 kGy. The samples were

Table 1 Effects of X-rays on different food categories.

Seafood products	Dose treatment X-rays (kGy)	Microorganism inoculation	Results	References
	0.5	—	Presence of *L. monocytogenes* equal to 27% of the total microbial load, compared to 40% of the untreated product	
Catfish fillets	1.0	—	Presence of *L. monocytogenes* equal to 0% of the total microbial load, compared to 40% of the untreated product	Moosekian et al. (2012)
	1.5	—	Presence of *L. monocytogenes* equal to 7% of the total microbial load, compared to 40% of the untreated product	
	0.5	—	Reduction of the microbial load of the inoculum equal to 1.1 log CFU g^{-1}	
Mullet	1.0	*L. monocytogenes* 10^4 (CFU ml^{-1})	Reduction of the microbial load of the inoculum equal to 1.6 log CFU g^{-1}	Robertson et al. (2006)
	1.5	—	Reduction of the microbial load of the inoculum equal to 2.1 log CFU g^{-1}	
	2.0	—	<Limit of detection (10^2 CFU g^{-1})	
Salmon	1.0	*L. monocytogenes* ~10^4 (CFU ml^{-1})	Microbial population below the limit of detection up to 30 days at the temperature of 5 °C	Mahumud (2012)
	2.0	*E. coli* O157:H7 10^8–10^9 (CFU ml^{-1})	<Limit of detection (10^2 CFU g^{-1})	
Shrimp	4.0	*S. enterica* 10^8–10^9 (CFU ml^{-1})	Reduction by 6 log CFU g^{-1}	Mahmoud (2009b)
	3.0	*S. flexneri* 10^8–10^9 (CFU ml^{-1})	Reduction by 6 log CFU g^{-1}	
	3.0	*V. parahaemolyticus* 10^8–10^9 (CFU ml^{-1})	Reduction by 6 log CFU g^{-1}	
	0.1		Reduction to 5.4 ± 0.1 (HSO) log CFU g^{-1} and to 6.1 ± 0.1 (WSO) log CFU g^{-1}	
	0.5		Reduction to 5.1 ± 0.2 (HSO) log CFU g^{-1} and to 6.0 ± 00 (WSO) log CFU g^{-1}	
	0.75		Reduction to 3.3 ± 0.1 (HSO) log CFU g^{-1} and to 5.8 ± 0.05 (WSO) log CFU g^{-1}	

Half shell oyster (HSO) and Whole shell oysters (WSO)	1.0	*V. parahaemolyticus* 10^7–10^8 (CFU ml^{-1})	Reduction to 3.0 ± 0.01 (HSO) log CFU g^{-1} and to 5.2 ± 0.2 (WSO) log CFU g^{-1}	Mahmoud and Burrage (2009)
	1.5		Reduction to 2.1 ± 0.02 (HSO) log CFU g^{-1} and to 5.0 ± 0.4 (WSO) log CFU g^{-1}	
	2.0		<Limit of detection (HSO) (10 CFU g^{-1}) and to 4.0 ± 0.1 (WSO) log CFU g^{-1}	
	3.0		<Limit of detection (HSO) (10 CFU g^{-1}) and to 3.5 ± 0.2 (WSO) log CFU g^{-1}	
	5.0		<Limit of detection (HSO) and (WSO) (10 CFU g^{-1})	
Fruit and vegetables	Dose treatment X-rays (kGy)	Microorganism inoculation	Results	References
Spinach	1.0	*E. coli* O157:H7; *L. monocytogenes; S. enterica; S. flexneri* 10^8–10^9 (CFU ml^{-1})	Reduction of bacterial load >5 log CFU g^{-1}	Mahmoud et al. (2010)
Iceberg lettuce	2.0		Reduction of bacterial load >5 log CFU g^{-1}	Mahmoud 2010a
Roma tomatoes	1.5		Reduction of bacterial load >5 log CFU g^{-1}	Mahmoud 2010b
Dried fruit products	Dose treatment X-rays (kGy)	Microorganism inoculation	Results	References
Almonds	0.226	*S.* Enteritidis *PT30* 10^8 (CFU ml^{-1})	Reached value D_{10}, in the specific case: 5 log CFU g^{-1}	Jeong et al. (2012)
	0.431	*S.* Tennessee 10^8 (CFU ml^{-1})	Reached value D_{10}, in the specific case: 5 log CFU g^{-1}	
Walnuts	0.474	*S.* Enteritidis *PT30* 10^8 (CFU ml^{-1})	Reached value D_{10}, in the specific case: 5 log CFU g^{-1}	
	0.930	*S.* Tennessee 10^8 (CFU ml^{-1})	Reached value D_{10}, in the specific case: 5 log CFU g^{-1}	

stored at 3 and 10 °C for 90 and 17 days, respectively. In all fish irradiated at 2.0 kGy, regardless of storage temperature, the concentration of *L. monocytogenes* dropped to undetectable levels. For treatments at 0.5, 1.0, and 1.5 kGy, reductions in microbial population occurred at 1.1, 1.6, and 2.1 log CFU g^{-1}, respectively. A sensorial test was also carried out with panelists, who found no significant differences between the treated fish and the control sample, therefore suggesting that X-rays treatment can be advantageously used to control the microbiological quality, without altering the food sensory properties.

A similar experiment was reported on catfish fillets in order to study the effect on mesophilic, psychrotrophic, coliforms, as well as on the pathogenic population of *L. monocytogenes* (Moosekian et al. 2012). The samples were irradiated at three different intensities (0.5, 1.0, and 1.5 kGy) and then stored at 3 °C for 17 days. The estimated *L. monocytogenes* percentage concentration in the products treated at 0.5, 1.0, and 1.5 kGy was, respectively, 27%, 0%, and 7% compared to 40% of the control sample. Sensory quality did not show significant differences in the treated samples, while in the untreated fish there was a decrease in aroma over time, probably due to the progressive and natural deterioration of non-irradiated product.

X-rays were also applied to salmon inoculated with a cocktail of three different strains of *L. monocytogenes* at concentrations of ~10^4 CFU g^{-1} (Mahumud 2012). The irradiation occurred at 0.1, 0.5, 1.0, and 2.0 kGy. Right after the treatment, the 1.0 kGy test significantly reduced the inoculated microbial population to not detectable levels; however, after 30–35 days of storage at 5 °C, the microorganisms developed to a detectable extent. The inoculum always remained below the detection threshold for the sample treated at 2.0 kGy.

In 2009, Mahmoud and Burrage experimented X-rays on live oysters to verify the inactivation efficacy of *Vibrio parahaemolyticus* (Mahmoud and Burrage 2009). To this end, the oysters were immersed in artificial seawater contained in a polyethylene tank where a cocktail of three different strains of *V. parahaemolyticus* was introduced at a load equal to 10^7–10^8 CFU ml^{-1}. The oysters stayed in this solution in contact with the microorganisms for a period of 12–18 hours before the irradiating treatment at a temperature of 20–22 °C. The microbial load of *V. parahaemolyticus* was evaluated before and after the irradiation treatments at intensities of 0.1, 0.5, 0.75, 1.0, 1.5, 2.0, 3.0, 5.0 kGy in pure culture, in half-shell and whole shell oyster. The pure culture before treatment showed an initial value of 7.1 log (CFU ml^{-1}), while in the treated samples, the microbial concentration was 4.4 and 3.1 log (CFU ml^{-1}) at 0.1 and 0.5 kGy, respectively; at 0.75 kGy, a value was recorded below the detection limit (<1.0 log CFU ml^{-1}). The half-shell oysters showed a starting value of 7.5 log (CFU g^{-1}), and in the treated products at 0.1, 0.5, 0.75, 1.0, and 1.5 kGy, the values were, respectively, 5.4, 5.1, 3.3, 3.0, and 2.1 log (CFU g^{-1}). At the intensity value equal to 2.0 kGy, a concentration below the detection limit of *V. parahaemolyticus* was found (<1.0 log CFU g^{-1}). Whole shell oysters, exposed to the radiating action of X-rays, have shown a higher microbial load than the previous ones. The initial concentration of *V. parahaemolyticus* was 7.2 log (CFU g^{-1}), while in the treated samples, the values gradually decreased with increasing the intensity from 0.1 to 3.0 kGy, respectively, 6.1, 6.0, 5.8, 5.2, 5.0, 4.0 and 3.5 log (CFU g^{-1}). At 5.0 kGy, *V. parahaemolyticus* dropped below the detectable limit (<1.0 log CFU g^{-1}).

In the same year, Mahmoud conducted a similar study on oysters, inoculating three different strains of *Vibrio vulnificus* (Mahmoud 2009a). The cocktail, having an initial concentration of 10^7–10^8 CFU ml^{-1}, was added to the water containing the oysters, and after several hours of contact, the samples were subjected to different radiation intensities ranging from 0.1 to 3.0 kGy. Again, the results showed a reduction in microbial load greater than 6 log CFU. The effect increased as the radiation intensity increased, in pure culture, half-shell, and full-shell oysters. It is also worth noting that the 3.0 kGy treatment did not adversely affect the survival of half-shell oysters.

X-ray treatments have also been adopted on crayfish inoculated with a starter consisting of three strains of four different pathogens having a cellular concentration of 10^8–10^9 CFU ml^{-1}. The samples were immersed in the solution containing the microorganisms for 15 minutes at 22 °C, and in order to allow the bacterial attack, they were left to dry in the open air in sterile glasses at the temperature of 22 °C for 30 minutes, after which they were packed in sterile bags and irradiated by X-rays at the following intensities: 0.1, 0.2, 0.3, 0.5, 0.75, 1.0, 2.0, 3.0, and 4.0 kGy. The experiment demonstrated a 6 log CFU g^{-1} reduction for *Escherichia coli* O157:H7, *Salmonella enterica*, *Shigella flexneri,* and *V. parahaemolyticus*, at the three highest intensities (Mahmoud 2009b).

The findings reported in the paragraph highlight the efficacy of X-ray treatment on fish products. In particular, this irradiation technique is effective on both spoilage microorganisms and pathogens; the effects are more pronounced by increasing the intensity of X-ray treatment. No particular defects are underlined after treatments on chemical and sensory quality of different fish-based food.

3.2 Fresh and Dried Fruit

In this paragraph the effects of X-rays on both fresh and dried fruit are reported. More concisely, Table 1 lists data on the effects of X-rays applied to spinach (Mahmoud et al. 2010), iceberg lettuce (Mahmoud 2010a), and Roma tomatoes (Mahmoud 2010b). The same author worked on the application of various irradiation intensities (1.0, 2.0, and 1.5 kGy) to different products, previously inoculated with four different microorganisms: *E. coli* 0157:H7, *L. monocytogenes, S. enterica,* and *S. flexneri*. The results demonstrated the effectiveness of X-rays in reducing the bacterial load by 5 log CFU at all the intensities tested on the different products.

The test carried out by Jeong et al. (2010) on iceberg lettuce inoculated with three different strains of *E. coli* O157:H7 has shown that a radiation intensity of 0.04 kGy is able to reduce the bacterial population by 90%, compared to 0.136 kGy of γ rays used for obtaining the same microbial killing (Niemira et al. 2002). For the test, 10 iceberg lettuce leaves were stacked, irradiated from both sides at an intensity of 1 kGy, the intensity gradually decreased as the photons penetrated to the heart of the product, reaching in the central area 0.2 kGy, and causing a microbial reduction of the *E. coli* O157:H7 inoculum equal to 5 log CFU g^{-1}.

An experimental plan with X-rays has also been conducted on the Clemenules mandarin variety, against *Ceratitis capitata*, the Mediterranean fruit fly (Alonso et al. 2007). To this aim, the effects of the treatment were assessed on fruit quality indices, such as rind color, firmness, juice yield, maturity index, internal birds, deterioration index, and sensory quality. The study did not report negative effects for radiation intensities of 0.195 and 0.395 kGy. In fact, at these values of irradiation, a light browning of the rind of the fruit was found. In the juice of these mandarins, no differences were found with respect to the juice of the control sample.

X-rays have also been tested on papaya, rambutan, and kau oranges at an intensity of 0.75 kGy (Boylston et al. 2002). After the treatment, the fruits were stored at 2 °C for nine days. Research results showed that the color was influenced by the treatment for rambutan and kau oranges, while the aroma and flavor were slightly stronger in all irradiated fruits. Regarding the parameters related to the content in vitamin C (ascorbic acid), pH, titratable acidity, total soluble solids, and content in carotenoids, the irradiation did not cause any significant change.

Even, matrices such as walnuts and almonds were inoculated with two different pathogenic species, *Salmonella* Enteritidis PT30 and *Salmonella* Tennessee, at four different water activities: 0.23, 0.45, 0.64, and 0.84. The samples inoculated at a microbial concentration of 10^8 CFU g^{-1} were irradiated until a microbial reduction of 90% was obtained. The experimental test has shown that regardless of the water activity applied, values of 0.226 and 0.431 kGy are sufficient for almonds, while for walnuts 0.474 and 0.930 kGy, respectively, for *S. enteridis* PT30 and *S.* Tennessee. A sensory quality analysis was also carried out. No significant difference was found between irradiated and non-irradiated almonds, while for walnuts, a difference in flavor was detected in the irradiated samples (Jeong et al. 2012). The findings of the literature about fresh and dried fruit demonstrate the efficacy of irradiation technique also on these food categories in terms of both microbial reduction and sensory acceptability. For dried samples like walnuts, some off-flavors are perceived after the irradiation, most probably due to the decomposition of some aromatic compounds.

3.3 Dairy Products

Milk and dairy products are highly regarded products in the food sector, bearing in mind that milk is the most complete food present in nature in terms of nutritional characteristics. This sub-paragraph shows results of a study performed on irradiated milk intended for direct consumption and the unique two recent applications of X-rays to dairy products, such as ricotta (artisanal and industrial) and mozzarella (fiordilatte).

In 2009, Mahmoud demonstrated the inactivation of *Enterobacter sakazakii* in milk (Mahmoud 2009c). For this purpose, a microbial starter containing six different strains of *Cronobacter (E. sakazakii)* having an initial microbial population of 10^8–10^9 CFU ml^{-1} was prepared. Several milk samples were inoculated based on their fat percentage: skim milk (0% fat), low-fat milk (1% fat), low-fat milk (2% fat), and whole milk (3.5% fat), and it was performed an X-ray treatment at the intensities of 0.1, 0.5, 0.75, 1.0, 2.0, 3.0, 4.0, 5.0, and 6.0 kGy. The tests performed showed a reduction in the microbial load of *Cronobacter*, greater than 7 log CFU ml^{-1} on skim milk at 5.0 kGy, recording a value below the detection limit (<1 log CFU ml^{-1}) on milk containing a fat concentration of 1%, 2%, and 3.5% at the intensity of 6.0 kGy. The radiation treatment was also carried out on the liquid medium used for the broth culture, and the same microbial reduction occurred at the intensity of 4.0 kGy.

Ricciardi et al. (2019) applied X-rays to artisanal and industrial ricotta cheese. The first type of ricotta was packaged in high barrier plastic bags and irradiated at 0.5, 2.0, and 3.0 kGy in order to evaluate the effects of X-rays on the microbial load of *Enterobacteriaceae*, yeasts, mesophiles, *Pseudomonas* spp., and *Bacillus cereus* and on sensory quality during refrigerated storage (4 °C). The results translated in terms of shelf life are particularly interesting since the control sample reached the limit of acceptability within four days, while the irradiated samples remained acceptable for 14 days if treated at 0.5 kGy and over 24 days if treated with intensities of 2.0 and 3.0 kGy. The same experiment, at the same intensities, was also conducted on industrial (pasteurized) ricotta. In this case, there was also a significant increase in the shelf life compared to the untreated sample, which decayed due to sensory defects after about 40 days, while all the other ricotta samples irradiated at 0.5, 2.0, and 3.0 kGy recorded a shelf life value greater than 84 days (Figure 3).

Lacivita et al. (2019) studied X-ray treatment on fiordilatte. Also in this case, the shelf life test carried out on both treated and untreated samples by X-rays at intensities of 0.5, 2.0, and 3.0 kGy showed the effects of radiation on the microbiological and sensory quality of

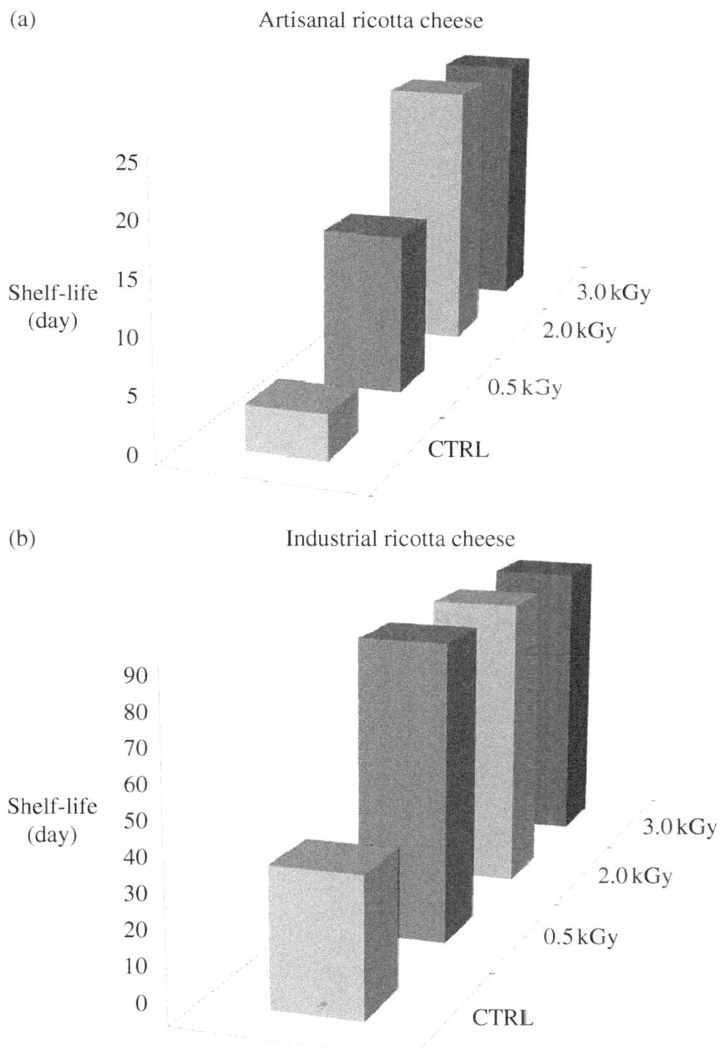

Figure 3 Shelf life of artisanal (a) and industrial (b) ricotta cheese.

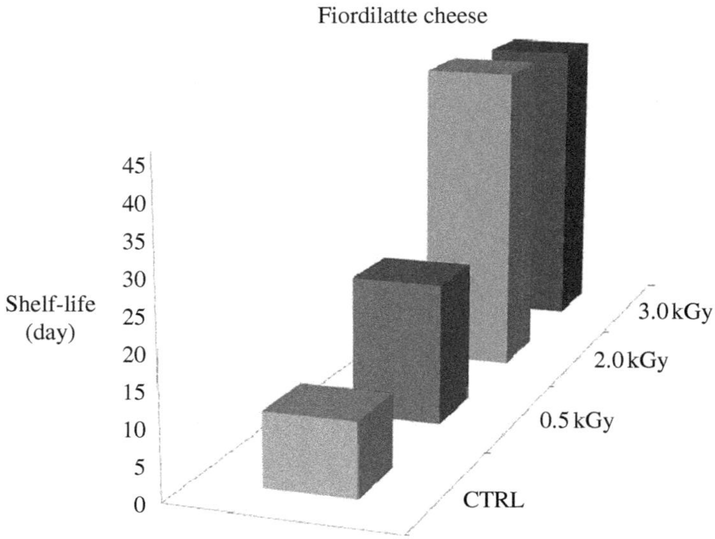

Figure 4 Shelf life of Fiordilatte cheese.

fresh cheese. The results showed that the control sample became unacceptable after 10 days of storage, the products treated at 0.5 kGy showed a double increase in shelf life compared to the control, while those treated at 2.0 and 3.0 kGy reached the limit of acceptability just over 40 days, without any sensory changes that compromise its acceptability (Figure 4).

Irradiated milk and dairy products indicated that X-rays can promote a significant shelf life prolongation. This is most probably due to the type of food matrix but in particular to the type of microbial contamination. No high intensities are necessary to exert antimicrobial effects on milk or fresh cheese, and in addition, if X-rays are combined with other sanitizing methods, as pasteurization, very low level of treatment is yet enough to record great results.

3.4 Meat-Based Foods

This brief sub-section is dedicated to highlight the unique recent application of X-rays to meat-based foods. The authors Cho and Ha (2019) studied the effects of X-rays on ham inoculated at a microbial concentration of 10^7–10^8 CFU g^{-1} with a starter comprising three different strains, *Salmonella* Typhimurium, *L. monocytogenes,* and *E. coli* O157:H7. After contamination, the samples were packed and treated by X-rays at the following intensities: 0.2, 0.4, 0.6, and 0.8 kGy. At the highest treatment intensity, the voluntarily added microbial population showed results equal to 5.7, 7.2, and 6.9 log CFU g^{-1}, respectively, for *S.* Typhimurium, *E. coli* O157:H7, and *L. monocytogenes*. After the irradiation treatment, qualitative aspects of the product, such as color and structure, have not undergone any detectable changes with respect to the untreated samples. In addition, the X-ray lethality mechanism determined by fluorescence was also studied and it emerged that the main microbial lethality factors due to the irradiating effect are represented, at the intracellular level, by enzymatic inactivation and damage to the genetic material.

The evidences recorded on X-ray-treated ham suggest that mild irradiation (below 1.0 kGy) was already enough to decrease the microbial population to acceptable level. The study also highlights that treated products were qualitatively similar to the untreated ones. Therefore, the interesting findings promote further studies in the sector to explore the effects of X-rays on different meat-based foods.

4 Effects of X-Rays on Packaging

The assessment of the suitability of packaging materials used for the treatment of food with X-rays, γ rays, or electron beam is an aspect of fundamental importance since radiation can induce changes to the plastic packaging material, thus generating unwanted compounds that could migrate into the food, compromising its safety (Komolprasert 2013). The intensity of irradiation can generate a chemical dissociation reaction of the macromolecules, the so-called radiolysis, from which decomposition products called radiolysis products (RPs) can be formed. These RPs in turn depend on various factors, such as the intensity of irradiation, the dose absorbed by the polymer, the presence of additives in the packaging, the temperature during treatment, and the chemical composition of the atmosphere (in particular the content of oxygen). The various factors can trigger two different types of reactions, which are random and can also coexist: cross-linking reactions and degradation reactions (Paquette 2004). When antioxidants and stabilizers are present in the plastic packaging, these protect the polymer. In the event that the irradiating process is performed in the presence of oxygen, these compounds degrade to a greater extent than the polymer, consequently leading to low levels of polymer RP and higher levels of RP of adjuvant compounds. The presence or absence of oxygen is important on the prevalence of the type of reaction that leads to the formation of RP. When a food is packaged in anaerobic conditions (vacuum or modified atmosphere) and is irradiated, the cross-linking reaction is the one that prevails, while in the presence of oxygen, the oxidative degradation reaction of the macromolecules dominates. In the latter case, RP compounds can be identified as oxidative radiolysis product (ORP). In the presence of oxygen, the ORP content (aldehydes, ketones, alcohols, carboxylic acids etc.) is directly proportional to the intensity of irradiation.

The raffinose family oligosaccharides (RFOs) adversely affect the organoleptic quality of the food matrix and consequently its shelf life, as well as all the safety aspects (Franz and Welle 2004). Therefore, when an already packaged food is subjected to an irradiating treatment, it is appropriate to analyze the following parameters: absorbed irradiating dose, presence/absence of oxygen during treatment, dosage speed, temperature, time after irradiation, and contact with food simulants. During an irradiating treatment, at a given allowed dose for a food, in the absence of oxygen, the RP concentrations increase linearly with the absorbed dose. On the other hand, in the presence of oxygen, for a given dose, the low rate of transmission of γ rays can lead to a higher concentration of RP compared to an irradiating treatment performed by X-rays or electron beam. With regard to temperature, it has no effect on the concentration of RP, unless the temperature during the irradiating phase is below the polymer glass transition temperature (Paquette 2004). After a food is irradiated, the time following the irradiation phase affects the migration of RPs from packaging to food.

The radiation also induces the formation of peroxide radicals that remain trapped in the packaging, thus helping to increase the concentration of RP.

In order to study the migration from the polymer to the food of RP coming exclusively from the packaging, it is assumed that the migration is total (100%) and recourse is made to the use of mathematical models describing the migration phenomenon. The results obtained can then be validated by experiments on plastic packaging containing food simulants (Komolprasert 2016). Polymers such as polypropylene (PP), polystyrene (PS), low density polyethylene (LDPE), ethylene vinyl acetate (EVA), polyethylene terephthalate (PET), Nylon 6 (PA), and polyvinyl chloride (PVC) irradiated in a range of intensity 10–50 kGy with γ or electron beam sources, in the presence of oxygen (atmospheric conditions) and at room temperature, were analyzed after 24 hours from irradiation (Paquette 2004). Different RPs of organic nature were identified and quantified, thus allowing to determine the exposure of the food matrix to the RP compounds. The study carried out was supported by mathematical modeling and led to the determination of the organic RP dietary concentrations (DCs) that the seven different polymer object of the study can transfer to food. Test results showed that most RPs had DC <0.5 ppb. Komolprasert (2014) developed an approach based on the identification and quantification of low-molecular-weight RP (volatiles) through the use of gas chromatography, combined with mass spectrometry. Once the qualitative and quantitative estimation of the RP has been performed, the concentrations of the compounds that can be transferred to the food were determined on the basis of mathematical migration approaches. The threshold of regulation (TOR) corresponds to the negligible limit of a substance, i.e. the concentration such as to make its carcinogenicity unlikely. On the basis of the TOR values, packaging materials that can be used to irradiate already packaged foods have been properly identified (Paquette 2004).

5 Regulation of X-Ray Irradiation

According to data of the World Health Organization (WHO)/IAEA/FAO expert committee, irradiated foods up to 10 kGy do not present toxicological, nutritional, and microbiological risks (JECFI 1981). The Food and Drugs Administration (FDA) has issued regulations concerning all sources authorized to carry out food inspections, general provisions for food irradiation, and rules for ascertaining the use of both ionizing and other radiation emerging technologies, such as radio frequency, ultraviolet radiation, pulsed light, and CO_2 laser (Komolprasert 2016). Australia and New Zealand have conducted studies on the safety of irradiated food (FSANZ 2012). Safety studies have also been conducted by specialized organizations, such as the International Committee for Microbiology and Food Hygiene (ICFMH 1982). Food irradiation has also been approved as a safe process by the American Medical Association and the American Dietetic Association (Achenson and Steele 2001; Wood and Bruhn 2000). Over 30 countries have granted national authorizations for the irradiation of one or more food products. For a wide range of foods deemed suitable for irradiation treatment, post-treatment detection methods have been developed and standardized (CEN, European Committee for Standardization). In the case of Europe, the European Parliament and the Council of the European Union have drawn up a much more concise legislative criterion than the references used by the specialized agencies of the

United Nations, namely Directive 1999/2/EC and Directive 1999/3/EC. With regard to the former, it arises from the need to standardize the legislative aspect of the member states of the European Union on the treatment of food and food ingredients to be subjected to the action of ionizing radiation. Directive 1999/3/EC, on the other hand, presents the list of irradiated foods and food ingredients (EP 1999). The national regulations of countries, such as France, Belgium, Spain, the Netherlands, Italy, and United Kingdom, established before the entry into force of the Directive 1999/3/EC, move according to the laws of their country (Anon 2003). According to the commission of the Codex Alimentarius, i.e. the body responsible for the standards relating to human health, irradiation must be carried out according to good management practices and in compliance with the Codex standard of food hygiene principles (CAC 2003a).

A food should be irradiated with an irradiation dose suitable for the specific technological purpose to be achieved, and the maximum absorption dose should be below the safety threshold for the consumer, below the limit that affects the sensory quality and the structural and functional integrity of the food matrix. As previously mentioned, the maximum allowable absorption dose must not exceed 10 kGy, except in the case in which particular technological aim is desired for a specific food (CAC 2003b).

Few countries have a regulation that allows a food to be irradiated in compliance with the Codex standard. For example, in Brazil, any food can be irradiated at a specific dose, provided that it is tolerable by the food itself, so that there is the achievement of a specific purpose. The labeling of irradiated foods must comply with the general rule of the code for the labeling of pre-packaged foods (CAC 2010). This rule requires that an irradiated food has a written declaration on the label of the treatment to which it has been subjected. When it comes to a food ingredient, the latter must be declared in the list of food ingredients (Roberts 2016). The European Union, Australia, and New Zealand require that all whole foods and all ingredients are subjected to mandatory labeling, regardless of the irradiation dose undergone, and that this labeling is also maintained within the restaurant. In Europe, the text on the label is normally indicated, while in Australia and New Zealand, the text, as well as the "Radura" logo (used to indicate irradiated foods), is optional (EU 1999; ANZFSC 2016). The "Radura" symbol is used by US regulations, accompanied by the wording "treated with radiation or irradiation," but if irradiated ingredients are added to a non-irradiated food, no special labeling is required on retail products (USFDA 2015). Canada and Malaysia have established that labeling is not required when the irradiated ingredient is less than 5% and 10%, respectively, for the two countries, of the total weight of the food (MOHM 2011; CFIS 2014). Of fundamental importance is the regulation of packaging for products to be irradiated. In this regard, the FDA has established a regulation of packaging that undergoes an irradiating treatment in post-packaging according to which the packaging must not adulterate the food in its close contact and must be subjected to regulations based on specific tests (Morehouse 2002).

In 1999, the irradiation of pre-packaged foods was regulated through a petition on food additives that led to the promulgation of a regulation, 21 CFR 179 (Chapters B and C) published in the Federal Register. Chapter C of 21 CFR 179 contains section 21 CFR 179,45 where all the packaging materials allowed to be used for the post-packaging irradiation treatment are listed. Many packaging materials based on paper, polymers, adjuvants, and coatings can be treated at a maximum dose of 10 kGy, except for different indications based on the product to be treated with a specific source of ionizing radiation (Komolprasert 2016).

6 Conclusion and Future Outlook

In recent years, there has been a slow increase in the sale of irradiated food products in many countries, with no harm to consumers. However, the guarantee of the safety of irradiated foods still arouses some skepticism in the food sector and especially among consumers. Given the numerous researches reported in the scientific literature and given the approval by authoritative bodies such as the WHO, the FAO, and the IAEA, there are the preconditions for the irradiation technology to be considered by the food industry a valid and efficient strategy for increasing shelf life and maintaining the sensory and nutritional quality of food. At the same time, it is necessary to sensitize consumers about the real benefits and limitations of technology, so as not to consider irradiation a dangerous method for sanitizing food.

Developing nations in Southeast Asia and European Union are increasingly expanding the idea of treating foods with γ-rays, X-rays, and electron beams; therefore, it is expected that the field of general public opinion, by means of which to dispel some clichés such as the association of irradiated foods with the onset of serious diseases. Many aspects aimed at educating consumers on issues concerning irradiation as a safe technology will have to be studied in depth. Some of these may be: radioactivity following treatment; the types of food that can be subjected to irradiating treatments; the extension of the shelf life compared to traditional heat treatment; the lack of concrete evidence that confirms the onset of cancer or degenerative diseases related to the consumption of irradiated foods; safety approved by higher bodies. The development of a broader, multidisciplinary, and careful discussion on all the aspects concerning sanitization by irradiation will certainly increase industrial interest and encourage greater consumer awareness, stimulating regulatory authorities to guarantee the safety of irradiated products and remove any unjustified obstacle to their use.

One of the most challenges of X-rays in terms of shelf life extension is their consequent economic and environmental impact. The irradiated foods with a longer shelf life show on one hand the great economic advantage to timely reach geographically distant markets and on the other allow promoting reduction of food waste because irradiated food can remain on the shelf of the supermarket for a longer time than non-irradiated food, thus increasing the probability of purchasing the product. Therefore, more conscious efforts need to be made by modern sustainable food technology. The broad diffusion of knowledge from the scientific world about the concrete advantages, the investment costs, and the impact on the future state of the environment would really promote the valorization of irradiation techniques at industrial level.

References

Achenson, D. and Steele, J.H. (2001). Food irradiation: a public health challenge for the 21st century. *Clinical Infectious Diseases* 33 (3): 376–377.

Alonso, M., Palou, L., Ángel del Río, M., and Jacas, J.-A. (2007). Effect of X-ray irradiation on fruit quality of clementine mandarin cv. 'Clemenules'. *Radiation Physics and Chemistry* 76 (10): 1631–1635.

Anon (2003). Irradiation of tropical fruits (A443). *World Food Regulation Review* 3: 3.

ANZFSC (2016). *Australia New Zealand Food Standards Code*. Standard 1.5.3. Irradiation of Food. ID F2016C00171.

Barba, F.J., Orlien, V., Mota, M.J. et al. (2016). Implementation of emerging technologies. In: *Innovation Strategies in the Food Industry: Tools for Implementation*, 1e (ed. C. Galanakis), 117–148. Academic Press https://doi.org/10.1016/B978-0-12-803751-5.00007-6.

Barba, F.J., Roohinejad, S.R., Ishikawa, K.I. et al. (2020). Electron spin resonance as a tool to monitor the influence of novel processing technologies on food properties. *Trends in Food Science and Technology* 100: 77–87.

Barbosa-Canovas, G.V. and Bermúdez-Aguirre, D. (2010). Chapter 16 - Novel food processing technologies and regulatory hurdles. In: *Ensuring Global Food Safety* (eds. C.E. Boisrobert, A. Stjepanovic, S. Oh and H.L.M. Lelieveld), 281–288. Academic Press.

Bashir, B., Swer, T.L., Prakash, K.S., and Aggarwal, M. (2017). Physico-chemical and functional properties of gamma irradiated whole wheat flour and starch. *LWT Food Science and Technology* 76 (a): 131–139.

Bhaskaram, C. and Sadasivan, G. (1975). Effects of feeding irradiated wheat to malnourished children. *American Journal of Clinical Nutrition* 23 (2): 130–135.

Boylston, T.D., Reitmeier, C.A., Moy, J.H. et al. (2002). Sensory quality and nutrient composition of three Hawaiian fruits treated by X-irradiation. *Journal of Food Quality* 25: 419–433.

CAC (2003a). Codex Alimentarius Commission. General Principles of Food Hygiene (CAC/RCP 1-1969, Rev 4-2003). Codex Alimentarius, FAO/WHO, Rome.

CAC (2003b). Codex Alimentarius Commission. General Standard for Irradiated Foods (CODEX STAN 106-1983, Rev.1-2003). Codex Alimentarius, FAO/WHO, Rome.

CAC (2010). Codex Alimentarius Commission. General Standard for the Labelling of Pre-packaged Foods (CODEX STAN 1-1985, Rev. 7-2010). Codex Alimentarius, FAO/WHO, Rome.

CFIS, 2014. Canadian Food Inspection Service. Irradiated Food. Guidance Document.

Chaine, A., Levy, C., Lacour, B. et al. (2012). Decontamination of sugar syrup by pulsed light. *Journal of Food Protection* 75 (5): 913–917.

Cho, G.-L. and Ha, J.-W. (2019). Application of X-ray for inactivation of food borne pathogens in ready-to-eat sliced ham and mechanism of the bactericidal action. *Food Control* 96: 343–350.

Diehl, J.F. (1995). Safety of irradiated food. In: (eds. O.R. Fennema, M. Karel, G.W. Sanderson, et al.), 2e 254. New York: Marcel Dekker.

Diehl, J.F. (2002). Food irradiation – past, present and future. *Radiation Physics and Chemistry* 63 (3–6): 211–215.

Elias, P.S. and Cohen, A.J. (1983). *Recent Advances in Food Irradiation*, 129–147. Amsterdam: Elsevier Biomedical.

EU (1999). European Union. Directive 1999/2/EC of the European Parliament and of the Council. Concerning food and food ingredients treated with ionizing radiation.

Franz, R. and Welle, F. (2004). Chapter 15 – Effects of ionizing radiation on the migration behavior and sensory properties of plastic packaging materials. In: *Irradiation of Food and Packaging*, ACS Symposium Series, vol. 875, 236–261.

FSANZ, Food Safety Australia New Zealand (2012). Risk and Technical Assessment Report – Application A1069. Irradiation of Tomatoes and Capsicums.

Haff, R.P. and Toyofuku, N. (2008). X-ray detection of defects and contaminants in the food industry. *Sensing and Instrumentation for Food Quality and Safety* 2: 262–273.

Hajare, V.N., Dhokane, V.S., Shashidhar, R. et al. (2006). Radiation processing of minimally processed carrot (*Daucus carota*) and cucumber (*Cucumis sativus*) to ensure safety: effect on nutritional and sensory quality. *Journal of Food Science* 71 (3): 198–203.

Hashisaka, A.E., Matches, J.R., Batters, Y. et al. (1990). Effects of gamma irradiation at -78 °C on microbial populations in dairy products. *Journal of Food Science* 55 (5): 1284–1289.

Hernàndez-Hernàndez, H.M., Moreto-Villet, L.M., and Villanueva-Rodrìguez, S.J. (2019). Current status of emerging food processing technologies in Latin America: novel non-thermal processing. *Innovative Food Science & Emerging Technologies* 58: 102233.

Hui, Y.H. (2012). Part VII: shelf-stable processed meat products. In: *Handbook of Meat and Meat Processing*, 2e, 769–790. CRC Press.

ICFMH (1982). International Committee on Food Microbiology & Hygiene of the International Union of Microbiological Societies Draft Report on Microbiological Health Hazards in Irradiated Food.

Jaiswal, A.K. (2016). *Food Processing Technologies: Impact on Product Attributes*. Boca Raton, FL: CRC Press.

Jambrak, A.R., Šimunek, M., Evačić, S. et al. (2018). Influence of high power ultrasound on selected moulds, yeasts and *Alicyclobacillus acidoterrestris* in apple, cranberry and blueberry juice and nectar. *Ultrasonics* 83: 3–17.

JECFI, Joint FAO/IAEA/WHO Expert Committee on Food Irradiation (1981). *Whole-Someness of Irradiated Food*, Technical Report Series no. 659. Geneva, Switzerland: World Health Organization.

Jeong, S., Marks, B.P., Ryser, E.T., and Moosekian, S.R. (2010). Inactivation of *Escherichia coli* O157:H7 on lettuce using low-energy X-ray irradiation. *Journal of Food Protection* 73 (3): 547–571.

Jeong, S., Marks, B.P., Ryser, E.T., and Harte, J.B. (2012). The effect of X-ray irradiation on *Salmonella* inactivation and sensory quality of almonds and walnuts as a function of water activity. *International Journal of Food Microbiology* 153 (3): 365–371.

Johnston, D.E. and Stevenson, M.H. (1990). Food irradiation and the chemistry United Kingdom. *Royal Society of Chemistry* 22 (24): 173.

Khan, M.K., Ahmad, K., Hassan, S. et al. (2018). Effect of novel technologies on polyphenols during food processing. *Innovative Food Science & Emerging Technologies* 45: 361–381.

Khouryieh, H. (2020). Novel and emerging technologies used by the U.S. food processing industry. *Innovative Food Science & Emerging Technologies* 67: 102559.

Komolprasert, V. (2013). Chapter 9 – irradiation of packaging materials in contact with food: an update. In: *Food Irradiation Research and Technology*, 2e, vol. 4, 147–168. Ames, IA: Wiley-Blackwell Publishing.

Komolprasert, V. (2014). Evaluating packaging materials for use during the irradiation of prepackaged food. Food additives and packaging. *American Chemical Society* 1162: 119–126.

Komolprasert, V. (2016). Packaging food for radiation processing. *Radiation Physics and Chemistry* 129: 35–38.

Kramer, B., Wunderlich, J., and Muranyi, P. (2015). Pulsed light induced damages in *Listeria innocua* and *Escherichia coli*. *Journal of Applied Microbiology* 119 (4): 999–1010.

Krisko, A. and Radman, M. (2013). Biology of extreme radiation resistance: the way of *Deinococcus radiodurans*. *Cold Spring Harbor Perspectives in Biology* 5 (a): 012765.

Lacivita, V., Mentana, A., Centonze, D. et al. (2019). Study of X-ray irradiation applied to fresh dairy cheese. *LWT Food Science and Technology* 103: 186–191.

Lung, H.-M., Cheng, Y.-C., Chang, Y.-H. et al. (2015). Microbial decontamination of food by electron beam irradiation. *Trends in Food Science and Technology* 44 (1): 66–78.

Mahmoud, B.S.M. (2009a). Reduction of *Vibrio vulnificus* in pure culture, half shell and whole shell oysters (*Crassostrea virginica*) by X-ray. *International Journal of Food Microbiology* 130 (2): 135–139.

Mahmoud, B.S.M. (2009b). Effect of X-ray treatments of inoculated *Escherichia coli* O157:H7, *Salmonella enterica*, *Shigella flexneri* and *Vibrio parahaemolyticus* in ready-to-eat shrimp. *Food Microbiology* 26 (8): 860–864.

Mahmoud, B.S.M. (2009c). Inactivation effect of X-ray treatments on *Cronobacter* species (*Enterobacter sakazakii*) in tryptic soy broth, skim milk, low-fat milk and whole-fat milk. *Letters in Applied Microbiology* 49 (5): 562–567.

Mahmoud, B.S.M. (2010a). Effects of X-ray radiation on *Escherichia coli* O157:H7, *Listeria monocytogenes*, *Salmonella enterica*, and *Shigella flexneri* inoculated on shredded iceberg lettuce. *Food Microbiology* 27 (8): 109–114.

Mahmoud, B.S.M. (2010b). The effects of X-ray radiation on *Escherichia coli* O157:H7, *Listeria monocytogenes*, *Salmonella enterica* and *Shigella flexneri* inoculated on whole Roma tomatoes. *Food Microbiology* 27 (8): 1057–1063.

Mahmoud, B.S.M. and Burrage, D.D. (2009). Inactivation of *Vibrio parahaemolyticus* in pure culture, whole live and half shell oysters (*Crassostrea virginica*) by X-ray. *Letters in Applied Microbiology* 48 (5): 572–578.

Mahmoud, B.S.M., Bachman, G., and Linton, R.H. (2010). Inactivation of *Escherichia coli* O157:H7, *Listeria monocytogenes*, *Salmonella enterica*, and *Shigella flexneri* on spinach leaves by X-ray. *Food Microbiology* 27 (1): 24–28.

Mahumud, B.S.M. (2012). Control of *Listeria monocytogenes* and spoilage bacteria on smoked salmon during storage at 5 °C after X-ray irradiation. *Food Microbiology* 32 (2): 317–320.

Miller, R.B. (2003). Food irradiation using bremsstrahlung X-rays. *Radiation Physics and Chemistry* 68 (6): 963–974.

Miller, R.B. (2005). *Electronic Irradiation of Foods: An Introduction to the Technology*, Food Engineering Series. New York, NY, U.S.A.: Springer Science Business Media.

Misra, N.N., Koubaa, M., Roohinejad, S. et al. (2017). Landmarks in the historical development of twenty first century food processing technologies. *Food Research International* 97: 318–339.

MOHM (2011) Ministry of Health Malaysia. Regulation 13, Food Irradiation Regulations 2011. PU(A) 143/2011.

Monk, D.H. (1994). Subject area preparation of secondary mathematics and science teachers and student achievement. *Economics of Education Review* 13 (2): 125–145.

Moosekian, S.R. and Ryser, R.T. (2011). Enhanced resistance of sanitizer-injured *E. coli* O157:H7 on baby spinach during X-ray irradiation. *Presented at Annual Meeting International Association for Food Protection*, Milwaukee, WI (31 July – 3 August) (Abstract).

Moosekian, S.R., Jeong, S., Marks, B.P., and Ryser, E.T. (2012). X-ray irradiation as a microbial intervention. *Annual Review of Food Science and Technology* 3: 493–510.

Morehouse, K.M. (2002). Food irradiation-US regulatory considerations. *Radiation Physics and Chemistry* 63 (3–6): 281–284.

Moseley, M.E., Kucharczyk, J., Mintorovitch, J. et al. (1990). Diffusion-weighted MR imaging of acute stroke: correlation with T2-weighted and magnetic susceptibility-enhanced MR imaging in cats. *American Journal of Neuroradiology* 11 (3): 423–429.

Niemira, B.A., Sommers, C.H., and Fan, X. (2002). Suspending lettuce type influences recoverability and radiation sensitivity of *Escherichia coli* O157:H7. *Journal of Food Protection* 65 (9): 1388–1393.

Olatunde, O.O. and Benjakul, S. (2018). Non-thermal processes for shelf-life extension of seafoods: a revisit. *Comprehensive Reviews in Food Science and Food Safety* 17 (4): 892–904.

Pan, Y., Sun, D.-W., and Han, Z. (2017). Applications of electromagnetic fields for non-thermal inactivation of microorganisms in foods: an overview. *Trends in Food Science and Technology* 64: 13–22.

Paquette, K.E. (2004) Chapter 12 – Irradiation of prepackaged food: evolution of the U.S. Food and Drug. Administration's regulation of the packaging materials. Irradiation of Food and Packaging. ACS Symposium Series 875, 182–202.

Ravindran, R.R. and Jaiswal, A.K. (2019). Wholesomeness and safety aspects of irradiated foods. *Food Chemistry* 285: 363–368.

Ricciardi, E.F., Lacivita, V., Conte, A. et al. (2019). X-ray irradiation as a valid technique to prolong food shelf life: the case of ricotta cheese. *International Dairy Journal* 99: 104547.

Ricciardi, E.F., Plazzotta, S., Conte, A., and Manzocco, L. (2020). Effect of pulsed light on microbial inactivation, sensory properties and protein structure of fresh ricotta cheese. *LTW Food Science and Technology* 139: 110556.

Rifna, E.J., Singh, S.K., Chakraborty, S., and Dwivedia, M. (2019). Effect of thermal and non-thermal techniques for microbial safety in food powder: recent advances. *Food Research International* 126: 108654.

Roberts, P.B. (2016). Food irradiation: standards, regulations and world-wide trade. *Radiation Physics and Chemistry* 129: 30–34.

Robertson, C.B., Andrews, L.S., Marshall, D.L. et al. (2006). Effect of X-ray irradiation on reducing the risk of Listeriosis in ready-to-eat vacuum-packaged smoked mullet. *Journal of Food Protection* 69 (7): 1561–1564.

Shahbaz, H.M., Akram, K., Ahn, J.-J., and Kwon, J.-H. (2016). Worldwide status of fresh fruits irradiation and concerns about quality, safety, and consumer acceptance. *Critical Reviews in Food Science and Nutrition* 56 (11): 1790–1807.

Song, B.-S., Choi, S.-J., Jin, Y.-B. et al. (2014). A critical review on toxicological safety of 2-alkylcyclobutanones. *Radiation Physics and Chemistry* 103: 188–193.

USFDA (2015). Food and Drug Administration of the U.S.A. Irradiation in food production, processing and handling; final rule. 21 CFR 179.26.

Vicente, A. and Castro, I. (2007). Novel thermal processing technologies. In: *Advances in Thermal and Non-Thermal Food Preservation*, 99–130. Iowa, USA: Blackwell Publishing.

WHO (1981). *Wholesomeness of Irradiated Food*. Geneva: World Health Organization. WHO technical reports series 659 (34).

Wood, O.B. and Bruhn, C.M. (2000). Position of the American dietetic association: food irradiation. *Journal of the American Dietetic Association* 100 (2): 246–253.

Yuk, H.-G., Geveke, D.J., and Zhang, H.Q. (2010). Efficacy of supercritical carbon dioxide for non-thermal inactivation of *Escherichia coli* K12 in apple cider. *International Journal of Food Microbiology* 138 (1–2): 91–99.

Zhang, W. and Jiang, W. (2019). UV treatment improved the quality of postharvest fruits and vegetables by inducing resistance. *Trends in Food Science and Technology* 92: 71–80.

6

Ultraviolet Light

Sandra N. Guerrero[1,2], *Mariana Ferrario*[1,2], *Marcela Schenk*[1,2], *Daniela Fenoglio*[1,2], *and Antonella Andreone*[1,2]

[1] *Universidad de Buenos Aires, Facultad de Ciencias Exactas y Naturales, Departamento de Industrias, Buenos Aires, Argentina*
[2] *CONICET–Universidad de Buenos Aires, Instituto de Tecnología de Alimentos y Procesos Químicos (ITAPROQ). Pabellón de Industrias, Buenos Aires, Argentina*

1 Introduction

Increasing consumers' awareness for food products and beverages processed by mild and environmentally friendly technologies continues to be on trend (Bang et al. 2017). Due to globalization, people are more connected and extremely knowledgeable about those healthy foods that have been processed using gentle technologies, with lower levels of fats, salt, sugar, and/or the partial or complete removal of synthesized additives. All-natural products that were also convenient (e.g. on-the-go, ready-to-eat, and/or ready-to-cook) and with long shelf lives are being highly demanded. More than ever before, consumers are driving their purchase decisions based on the type, effects, and extent of the food processing (Sloan 2018). These consumer-driven innovations challenge the industry and academia, requiring them to meet these expectations by developing safe products of higher quality.

The conventional pasteurization generally involves the use of high temperature, between 73 and 138 °C for a few minutes or seconds. This traditional preservation treatment is normally used for the destruction of all disease-causing microorganisms and the elimination or reduction in the number of spoilage agents. It ensures the safety and extends the shelf life of the food product that is being treated. However, it often leads to undesirable detrimental changes in the sensory and nutritional quality, thus making such alternative unviable to preserve natural fresh-like food products and beverages (Wibowo et al. 2015). A wide range of intervention technologies, which involve minimal processing, have emerged in the last two decades for the development of wholesome food products. The minimal processing concept includes a great diversity of alternatives for high-quality products, which may be: *minimally processed*, *with invisible processing*, *carefully processed*, *partially processed*, and *high moisture shelf-stable*. Its application implies a better retention of

Electromagnetic Technologies in Food Science, First Edition. Edited by Vicente M. Gómez-López and Rajeev Bhat.

product flavor, texture, color, and nutrient content, compared with conventional thermal treatments (Alzamora et al. 2016a). Only a few of these novel technologies are considered already emerged and have achieved a commercialization stage, fulfilling both technical and economic feasibility requirements. High hydrostatic pressure (HHP) is a good example, with an ever-widening list of commercial processors offered in the food and beverage market and used worldwide (Putnik et al. 2020). Others are currently considered as "emerging," for example, the light-based technologies (Guerrero et al. 2016), except for the use of UV-C for the treatment of water and clear liquid foods, which is considered as emerged (Koutchma et al. 2009). For these technologies, significant advancements have been reached, such as the system validation and the development of a variety of laboratory and pilot-scale devices. The third category includes those technologies whose concept and inactivation mode have been reported for a variety of food systems and devices, mainly on a laboratory scale. These technologies, for example, high-intensity ultrasound (US) (Chen et al. 2020) and cold plasma (Pan et al. 2019), are still considered in a less developed stage.

Among various light-based technologies, short-wave ultraviolet light (UV-C) has shown increasing potential as an alternative nonthermal technology for food processing. It has been approved by the US FDA for the sterilization of water and as alternative treatment of thermal pasteurization for the reduction of human pathogens and other microorganisms in fruit juices. The FDA requires the use of turbulent flow with a minimum Reynolds number of 2200, in a transparent tube and low-pressure mercury lamps emitting a monochromatic light signal with at least 90% of UV light at 253.7 nm (US FDA 2019). The UV dose required for pathogen reduction is not specified in the petition as it will depend on the type of juice and the microbial load. The design of the UV-C system is also crucial as it will determine the flow rate, treatment time, and the number of lamps. In 2003, for its part, Canada approved the use of a commercial UV unit (CiderSure 3500 UV, FPE Inc., NY, USA) for apple juice/cider treatment (Health Canada 2003). In 2006, the National Advisory Committee on Microbiological Criteria for Foods (NACMCF) included UV-C and its combinations with other traditional or emerging factors as alternative methods to the traditional pasteurization (NACMCF 2006). However, the UV-C technology is considered by the European Union (EU) as an irradiation process, and consequently, the legislation is not harmonized, thus being authorized by-product. For example, the European Food Safety Authority (EFSA) has recently approved the use of UV-C milk processing targeted to all the consumers with the exclusion of infants (up to one year of age). The UV-C-treated milk does not rise safety concerns, being similar in quality to the non-UV-treated milk except for an increase in the vitamin D3 content, promoted by the UV-C treatment (EFSA 2016).

When a food system is treated by UV-C, several photochemical reactions take place, and consequently, microbial inactivation proceeds due to the absorbed photons by the molecules. A photochemical reaction is promoted when those photons have enough energy. As a result, pyrimidine dimers are formed in the DNA and RNA of the microorganisms that are being irradiated, which impedes the process of cell replication (Biancaniello et al. 2018).

Many environmental and optical characteristics relative to the sample (a_w, pH, composition, viscosity, turbidity, UV absorbance, UV transmittance) and also those referred to the UV system (flow type, temperature, time of treatment, lamp power and efficiency, UV doses, design, or arrangement) affect the inactivation effectiveness and should be properly monitored (Müller et al. 2015). Compared with other light-based technologies, as pulsed light (PL) or light emitting diodes (LED), UV-C involves the use of low energy levels, in the spectra range from 200 to 280 nm with a peak at 253.7 nm, which is the primary emission wavelength of a low-pressure UV lamp (Bolton 2020). UV-C is a cost-effective alternative for small processing operations, as adjunct to other traditional or emerging processing technologies such as pasteurization (Ansari et al. 2019) or cold processing (Lemoine et al. 2010). In fact, the estimated cost of an UV-C reactor is close to $10 000–15 000, which represents half the cost of thermal pasteurizers (Shah et al. 2016), whereas HHP machinery can range in price from $300.000 to 4.0 million per machine (Yamamoto 2017). It has been used for the treatment of water in the brewery and wine industries (Mezui and Swart 2010) and/or other ingredients such as concentrated whey, sugar syrups, or liquid egg (Mahmoud and Ghaly 2004; Chaine et al. 2012; Holck et al. 2018). It is also currently considered as a good shelf-life extensor (ESL), as it well preserves the organoleptic and nutritional quality of many foods and beverages (Unluturk and Atilgan 2015). In reference to the microbial targets, it has been demonstrated that UV-C effectively destroys most of foodborne pathogens, molds and yeasts, and many viruses (Cairns 2006). Interestingly, this technology has minimal energy consumption and lower maintenance and installation costs, compared with other emerging technologies such as HHP or pulsed electric fields (PEF) (Delorme et al. 2020).

Moreover, UV-C does not leave any residue or involve the use of any chemical and is easily manageable for diverse scale-up and in-line operations (Koutchma 2009). Photoreactivation after UV-C processing may occur due to exposition of the food product to visible light, requiring storage in dark under refrigeration. However, low penetration depth is its main disadvantage, because due to material interference, it prevents UV-C light from effectively reaching microbial cells, thereby hindering the disinfection process (Fenoglio et al. 2020a). To validate the industrial rational use of the UV processing of different juices and beverages, several aspects should be studied to achieve a better design of the product/process, based on quantitative information. The adequate UV dose to be used in industrial operations should be properly determined considering the compliance of certain relevant goals, which can be summarized as follows: significant reduction of native microbiota and spoilage agents; 5-log-reductions of the most resistant pathogens; sensory, nutritional, physicochemical quality studies and enzyme inactivation (US FDA 2019).

Combining emerging technologies with conventional preserving ones or with other novel techniques to interfere with the homeostatic mechanisms of microorganisms has been successfully explored in the past years (Guerrero et al. 2017; García Carrillo et al. 2017; Ferrario et al. 2020). Notwithstanding, until now there is not a summary of the use of UV-C in a hurdle approach for food preservation. The present review analyzes the combination of UV-C with heat, novel technologies, natural antimicrobials, and other sanitizers in order to analyze and compare which conditions led to a synergistic effect on the inactivation of microorganisms while retaining nutritional quality.

2 Characterization of UV-C Dose

When it comes to the use of the UV-C light as a technology for preserving liquid foods such as juices and beverages with low to zero UV transmittance, the knowledge and characterization of the reactor design, as well as the implemented dose, are the most relevant aspects to be considered to determine process effectiveness. In this regard, as the information relative to the unit and type of dose is not harmonized, there exists today a wide range of UV-C doses and units reported in literature. For a given UV-C processing, four different types of energy should be considered. The *total applied UV dose* (E_{UVC}) is the energy that depends on the total output power of the UV source (P_{UV}, W), the average residence time of the fluid (t), and the contact area (A) and is generally expressed in Joules per unit area ($J\,cm^{-2}$) according to Eq. (1). Most of the studies in literature report this dose calculated based on the information provided by the lamp manufacturer:

$$E_{UVC} = \frac{P_{UV}}{A} \cdot t \tag{1}$$

The *incident UV fluence* (H_i, $J\,cm^{-2}$) is considered as the energy incident at the product surface and depends on the incident irradiance (I_0, $W\,cm^{-2}$), which can be measured with a radiometer at a given distance from the UV source. This value should be checked several times during the lamp lifetime. H_i is the dose generally reported for batch UV operations and does not depend on the characteristics of the food material that is being treated. It can be derived from the following:

$$H_i = I_0 \cdot t \tag{2}$$

The *absorbed and scattered UV dose* (H_r, $J\,cm^{-2}$) represents the total UV energy absorbed and scattered by the fluid constituents and is most commonly reported for batch operations but for continuous-flow devices, as well. It depends on the design of the UV system and the optical properties of the fluid. H_r can be calculated as a function of the exposure time (t) according to the following equation (Koutchma et al. 2007):

$$H_r = I_r \cdot t \tag{3}$$

where I_r represents the absorbed irradiance or fluence rate ($W\,cm^{-2}$). Light penetration decreases with the distance to the lamp and fluid absorption coefficient:

$$I_r = I \frac{r}{r_0} \exp^{a \cdot (r - r_0)} \tag{4}$$

with a, the fluid abpsortion coeffcient (cm^{-1}); r, the radial distance (cm) from a given point where the irradiance is measured to the center of the lamp; and r_0, the lamp radius (cm).

The *delivered or germicidal UV dose* (H_d) represents the remaining energy that is not absorbed or scattered by the fluid constituents, thus being available for microbial inactivation. H_d is usually estimated by actinometry through the selection of an adequate pair of chemicals (e.g. iodate/iodide), which react when exposed to UV light (Koutchma et al. 2016). To establish a specific UV process, it is necessary to clearly determine the above four doses both on a laboratory and on a pilot scale.

Although there is not any standard approach, a bio-dosimetric estimation of the inactivation rate is recommended to validate the UV processing. The bio-dosimetry approach estimates, through the use of the injection of a pulse of a microbial suspension as a tracer, the inactivation rate required for achieving 1 log-reduction, D_{10}UV, based on the knowledge of the distributions in the device of both the UV dose and the mean residence time in the UV-C unit (Koutchma et al. 2009). Considering mainly the two target microbial groups to be inactivated in several liquid foods (milk, liquid egg) and beverages (juices, soft drinks), pathogens (e.g. *Escherichia coli; Salmonella* spp, *Listeria monocytogenes*, etc.), and spoilage microorganisms (molds and yeasts, *Alicyclobacillus acidoterrestris*, *Lactobacilli*), the reported $D_{10}UV$ values are significantly higher for the spoilage agents ($D_{10}UV = 10 - 40\,mJ\,cm^{-2}$), compared with pathogenic or surrogate bacteria ($D_{10}UV = 2 - 10\,mJ\,cm^{-2}$). In particular, mold spores exhibited the highest inactivation rates (Pan et al. 2004; Gündüz and Korkmaz 2019). Then, for a given liquid food, the reduction equivalent UV dose (RED) is calculated for a pathogen or surrogate, by multiplying its respective D_{10} UV value by the specific log reduction (SLR) required by the US FDA. The product acidity and composition, apart from other environmental factors, constrain the UV resistance. Therefore, the SLR value for pathogens must be 5 (e.g. juices, milk) or 7 (e.g. liquid egg) log-reductions (Koutchma et al. 2016). In the case of considering the germicidal dose for inactivating other contaminating agents, H_d can be obtained by multiplying its $D_{10}UV$ value by the respective log-reductions achieved in each medium. There is a wide range of reported doses for single UV operations, according to whether the devices were commercial ($1.4–7\,kJ\,L^{-1}$) or laboratory-scale and pilot-scale systems (up to $187\,kJ\,L^{-1}$), probably due to their less efficiency, compared with the commercial units (Koutchma et al. 2016). For laboratory-scale continuous operations, doses ranging from 1240 to $53\,100\,mJ\,cm^{-2}$ were reported. In addition, they may differ depending on the selection of batch or continuous operations, or the type of reported dose, with the total applied UV (E_{UVC}) and the incident UV (H_i) being the most common doses informed.

Figure 1 exemplifies for an orange-tangerine juice blend (pH 3.9; 13.1 °Brix, A_{254nm} (cm^{-1}) = 0.39; UV transmittance, UVT (%) = 40.7; turbidity: 1670 NTU, particle size: 630 000 nm), and an isotonic drink (pH 3.5; 8.5 °Brix, A_{254nm} (cm^{-1}) = 0.04; UVT (%) = 91.5; 114 NTU; particle size: 706 nm), the *E. coli* inactivation dependence on the different types of UV fluence, determined in a laboratory-scale flow-through device, provided with two serially connected UV-C lamps (30 W), which operated under laminar flow conditions (Reynolds number up to 512). The delivered fluence (H_d) was calculated using the iodide/iodate actinometrical pair (up to $985\,mJ\,cm^{-2}$) according to the method proposed by (Rahn et al. 2002), which was adapted to recirculation mode (Fenoglio et al. 2020a). The decimal reduction UV dose, $D_{10}UV$, which considers both flow characteristics and UV dose distribution, was calculated according to Koutchma and Parisi (2004) by the bio-dosimetry approach, using an *E. coli* strain as a tracer ($3.1\,mJ\,cm^{-2}$). The germicidal delivered dose ($H_{d,germ}$) was estimated as the product of $D_{10}UV$ by the corresponding SLR value at each treatment time (Figure 1).

H_r and I_r were calculated according to Eqs. (3) and (4), respectively. As it can be observed in Figure 1, a typical biphasic profile of the inactivation curves, with upward concavity and tailing effect, was obtained. This profile indicates the existence of resistant members within the bacterial population to this nonthermal stressor used alone (Schenk et al. 2008). The single UV-C treatment was significantly more effective to inactivate *E. coli* in the clearer beverage (isotonic drink), compared with the juice blend, which had higher turbidity and absorbance. The UV transmittance of the matrix directly influenced the dose delivery and the inactivation effectiveness. Thus, high UV transmittance will favor photon hits against

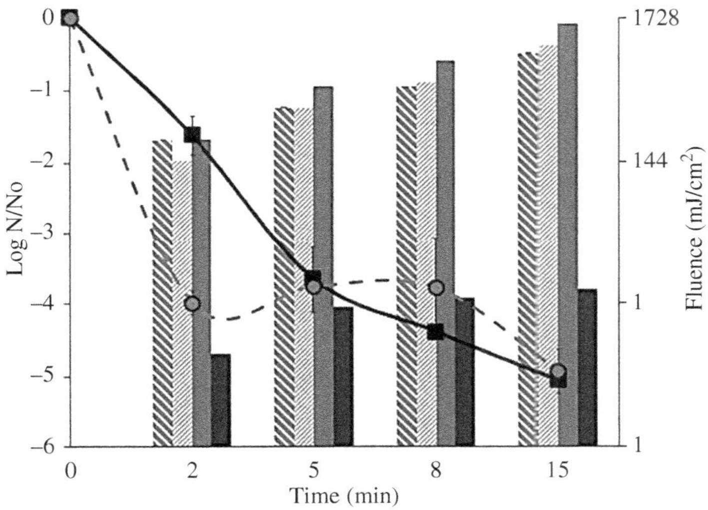

Figure 1 Inactivation of *E. coli* ATCC 25922 in an isotonic drink (–○–) and a tangerine-orange juice blend (–■–) treated by UV-C light (985 mJ cm^{-2}, 20 °C) in a laboratory-scale annular UV-C reactor according to different values of applied (■, H_r), absorbed (▨, I_r), and delivered (H_d, determined by actinometry (▧) or bio-dosimetry (■)) fluences.

the microbial cells (Koutchma 2009). Despite the H_r and I_r or H_d fluences increased with treatment time (Figure 1), the germicidal UV-C delivered dose ($H_{d,\text{germ}}$), measured using *E. coli* as a tracer in the flow, remained almost constant after five minutes of treatment, which evidenced the limited efficacy at the moment of inactivating the most resistant members within the microbial population and the scarce contribution of additional treatment time when using single UV-C. For turbid liquid foods with low or close to zero UV-C transmittance, the most relevant factors to determine the process performance will be the flow pattern, the design of the UV-C system, and its mixing efficiency. These factors significantly affect the delivered UV-C dose (Artichowicz et al. 2020). The UV-C devices are nonideal systems, characterized by volume elements represented by a nonuniform distribution of residence times (RTD). Moreover, they display a nonuniform distribution of fluence rates within the UV system, limited by the absorption of UV-C photons and by the reactor design (Koutchma and Parisi 2004; Müller et al. 2017). Consequently, the interaction between processing parameters and product characteristics will determine the effective dose distribution and the type of flow, fact that will directly affect how UV-C photons reach the microbial cells. As it can be seen in Figure 1, significantly less UV-C effectiveness on *E. coli* inactivation was determined in the juice with the most complex characteristics, thus requiring higher UV-C doses to achieve an equivalent SLR (Figure 1). Therefore, one alternative to increase UV-C effectiveness in liquid food products, counteracting or at least decreasing the effect of unfavorable characteristics, is the modification of the UV-C unit design. In this regard, it is essential to ensure that all the liquid food particles will receive the minimum UV-C treatment that guarantees a proper product pasteurization, to overcome the poor dose uniformity and low penetration depth (Koutchma et al. 2009). Continuous UV-C systems reported in literature are typically coiled tube (Koutchma et al. 2007),

Taylor–Couette, and annular flow-through reactors (Müller et al. 2017). UV-C reactors displaying thin-film flow significantly diminish UV-C path length, thus favoring photons contact with microbial cells. However, this type of reactor usually shows a widespread dose distribution (Koutchma 2009). The Taylor–Couette reactor overcomes this drawback (Gayán et al. 2014a). One enhanced commercial version of the laminar thin-film reactor is the CiderSure UV reactor (FPE; Macedon, NY, USA) commercially used for juice pasteurization, which promotes radial mixing by pumping the fluid through different serially connected chambers (Health Canada 2003). This reactor guarantees the 5 log-reductions required by US FDA for various *E. coli* O157:H7 strains, surrogates and for *Cryptosporidium parvum* in apple cider treated by UV-C light (14.3 mJ cm^{-2}) (Zhao et al. 2015). Other strategy is the promotion of turbulence within the flow, as for example, the Dean flow reactor, which presents a coiled tube design that promotes the generation of vortexes and swirls in the flow, which increases UV-C effectiveness, even under laminar flow conditions (Fenoglio et al. 2020b). The UV-C treatment efficacy can also be increased using additional factors in combination with UV light under a hurdle approach to achieve the desired product safety while maintaining or even improving its nutritional, sensory, and physicochemical quality (Guerrero et al. 2017).

The characterization of UV-C dose is highly significant to characterize a given processing, which enables a proper comparison of experimental data among different studies. Unfortunately, many researches only report the *incident dose*, thus the information derived from them is referred mostly to what happens at the surface of the matrix that is being treated. This approach does not take into consideration the phenomenon that occurs when UV-C photons penetrate the fluid and reach microbial cells. Therefore, to better characterize the UV-C processing, it is desirable to measure the absorbed dose and also perform a bio-dosimetry studies, as it was described in detail above.

3 Rational Use of the Hurdle Approach in the Design of Food Preservation Technologies

The degree of change in environmental conditions will determine whether microbial populations are injured, lose their viability, or express adaptive mechanisms that enable them to survive or even grow in unfavorable environments (Beales 2004). In foods and beverages preserved by combined methods that use sublethal levels of different stresses, the active homeostasis of vegetative cells and the passive refractory homeostasis of spores are disturbed by a combination of gentle antimicrobial factors at several sites or in a cooperative manner (Guerrero et al. 2017).

Many practical innovations in food and beverage preservation using gentle techniques use the hurdle or barrier concept as the basic principle of product design. From the practical standpoint, one of the most difficult aspects associated to the hurdle technologies design is the selection of stressors. The terms "additive," "synergistic," and "antagonistic" are commonly used to describe possible stressors interactions. Additivity refers to the sum of the effects of each hurdle acting individually. Synergism corresponds to a combined effect significantly greater than the sum of individual results, while an antagonistic effect is the opposite, i.e. the efficacy of the combined stressors is reduced compared with the sum of the effects of the individual stressors (Leistner 2010). In particular, the combination of nonthermal methods with other food intervention technologies may further increase the observed lethal effects, thus reducing the severity of nonthermal treatment required to achieve a desired level of microbial inactivation and/or prevent the proliferation of survivors.

The overall preservation effect will be as one of the following: (i) greater than individual effects but less than additive, (ii) additive, (iii) more than additive, (iv) synergistic, (v) antagonistic, or (vi) indifferent. In general, in terms of food safety and quality, mainly the effect (iv), but also the effects (iii), (ii), and (i) will be pursued (Abdullahi et al. 2016). The hurdle concept exploits the synergistic and/or additive interactions between sublethal stressors, thus contributing to reduction of the negative effects on product quality, the energy input, and the treatment intensities required to guarantee product safety.

In the growth domain, there is currently a consensus that a more effective preservation is obtained if a multi-target design is selected to inhibit microbial growth instead of choosing those stress factors with the same target. Therefore, the attack of various cellular targets would exert a synergistic effect by inducing the microbial cells to simultaneously activate every possible repair mechanism. However, there is not a final consensus for the establishment of synergic or additive effects, and in fact, the idea of multi-target preservation has been challenged considering that the combined use of stressors with the same target would improve the chances of causing irreversible damage. On the contrary, the use of stressors with different attack spots could inflict sublethal damage in several parts of the microorganism, but would fail to inactivate it (Alzamora et al. 2016b). The hurdle approach may be extended to a broader concept than just the homeostasis interference by additive or synergistic hurdles on the same microorganism, but also as the selective use of those preservation factors that may be effective against a specific organism or group of organisms, while not against others (Alzamora et al. 2016b). The number, type, and intensity of hurdles should be proposed according to the type of food product, pathogens and spoilage agents, impact of hurdles on quality, required shelf life, technical compatibility, and/or available processing/storage infrastructure (López-Gómez et al. 2009).

The multiple disruption of microbial homeostasis has been used or suggested for food preservation purposes in different arrangements

(i) using simultaneously two or more hurdles to avoid growth of spoilage and pathogenic microorganisms; and (ii) using in a simultaneous or sequential way, one or more hurdles to inactivate, injure, or physically remove some microorganisms, and then, in a sequential mode, one or more stressors to prevent survival/proliferation of the remaining refractory or sublethally damaged cells. The order in which the sequential hurdles or stressors are applied constrains the inactivation effectiveness (Alzamora et al. 2016b).

A proper design of combined intervention strategies should be thought with a multidisciplinary engineering perspective. That is, the use of certain tools applied to better understand the microbial response to the different stresses and consequently, optimize the product/process design. As a mode of example, these tools may encompass the study of the microbial inactivation behavior through predictive microbiology, growth dynamics during product storage, and an overall knowledge of microbial damage mechanisms through different tools such as flow cytometry, transmission electron microscopy (TEM), etc. (López-Gómez et al. 2009). In this regard, it is of utmost importance to predict the possibilities for the survival of a given microbial population exposed to the proposed treatment as affected by different factors. In the development of novel preservation technologies, a predictive approach is crucial to adequately select the most relevant factors and corresponding levels, from a statistical point of view (Fenoglio et al. 2020a). An accurate model prediction of survival curves would be beneficial to the food industry as it would allow to select the optimum combinations of lethal agents and environmental factors as well as the proper exposure times to obtain the desired levels of inactivation while minimizing the production costs and maintaining a maximum degree of sensory and nutrient quality. Unlike traditional preservation technologies (e.g. thermal treatments), when using emerging factors, the physiological responses of microbial cells are very complex and sometimes not fully understood. Therefore, a deep knowledge of the induced damage and the microbial response will collaborate in a better design of the proposed novel processing.

Unlike what happens with the use of single UV-C, no chapters and reviews dealing with the use of combined UV-C treatments for food preservation are available in literature. This chapter is aimed at providing a broad spectrum of some recent UV-C light–based hurdle combinations explored for the preservation of different foods and beverages. It highlights some areas of study to fully exploit the potential of the hurdle concept in the design and optimization of UV-C-based processes.

3.1 UV-C Light–based Hurdle Combinations

When employing a nonthermal stressor as UV-C, microorganisms can be sublethal or permanently damaged. During food product storage, when sublethal damage occurs, microbial cells will be capable of repairing their functional structures and growing, which potentially compromises product safety. Additionally, microorganisms may survive to the preservation treatment when the access of the killing agent to the microbial cells is impeded, as it occurs for UV-C-treated products having unfavorable optical properties (Fenoglio et al. 2020a). Some targeted UV-C-based applications including the use of other stressors such as mild heat, natural antimicrobials addition, use of sanitizers c, and/or other nonthermal technologies have been proposed for the inactivation of microorganisms and/or enzymes and/or for improving product quality with significant results and are described below.

3.1.1 Heat

The assistance of UV-C with heat (UV-C/H) is the most extensively investigated combination when considering a hurdle approach, probably because it is one of the most easily implemented at industrial level. Table 1 summarizes the experimental conditions and microbial inactivation achieved in some recent studies involving the use of UV-C combined with heat. Due to the vast information available in literature regarding this combination, the most representative examples are featured in the table. Most of the thermal treatments applied simultaneously or consecutively in the collected studies were in the range of mild heat conditions, from 40 to 65 °C. However, combinations of UV-C and heat with temperature values in the range from 75 to 121 °C (flash pasteurization) were also reported (Castro et al. 2019; Gayán et al. 2013; Kaya et al. 2015; Sommers et al. 2009; Vásquez-Mazo et al. 2019). A large variety of pathogenic and spoilage microorganisms, spores, and native microbiota were assessed in solid matrices and in clear and turbid liquid foods, and significant inactivation results were obtained (Table 1). Regarding the UV-C treatment applied, lab-scale devices were generally utilized, with the annular thin-film reactor being the most widely used design for the treatment of liquid matrices while bench-top UV-C devices were employed for the treatment of pepper powder (Cheon et al. 2015), strawberries (Pan et al. 2004) and frankfurters (Sommers et al. 2009), and some liquid foods. Fenoglio et al. (2020b) used a pilot-scale UV-C treatment device based on the Dean vortex technology, using the assistance of mild heat (50 °C), for the treatment of various clear and turbid juices under turbulent flow.

In Table 1, the reported UV-C doses were mainly informed as energy/area (mJ cm^{-2}), covering a broad range from 1 to 4080 mJ cm^{-2}, but also some authors informed the dose as energy/volume (J ml^{-1}) in a range from 0.4 to 27.1 J ml^{-1}. Most of them only informed the incident dose, and only a few measured the delivered UV-C dose (Table 1). Despite very few exceptions, those authors who have studied the inactivation dynamics during single UV-C or combined with heat treatments concur on a nonlinear profile of the inactivation curves. Discrepancies can be found regarding the inactivation curve profile, which appears to be highly dependent on the considered media, the microbial target, and/or the combined process design. The UV-C and UV-C/H inactivation curves generally displayed upward (Fenoglio et al. 2020b) or downward concavity, with shoulder and/or tail (Fenoglio et al. 2020b; García Carrillo et al. 2017; Gayán et al. 2012a, 2013, 2014b, 2016; Gouma et al. 2020). In general, the use of mild heat as an additional stressor eliminated or reduced the shoulder and/or tended to linearize the inactivation curve. The tailing effect has been attributed to multiple factors such as presence of more resistant cells within the microbial population and uneven UV-C dose distribution in the reactor (García Carrillo et al. 2017; Kaya and Ünlütürk 2019).

Shoulders were mostly observed in juices with a high absorption coefficient regardless of their turbidity (Gayán et al. 2016; García Carrillo et al. 2017). The existence of shoulder is frequently associated to an imbalance between DNA damage and repair. The multiple-hit target theory proposes that various photon events must hit a given number of microbial cells until the number of induced injuries surpasses the capacity of DNA repairing (Gouma et al. 2015b).

Overall, there was not common agreement about the best model for predicting microbial inactivation when the combined treatments involving UV-C and heat were applied. This probably occurred due to the variety of inactivation profiles obtained in the collected studies, which mainly depended on the considered targeted microorganism and the optical characteristics of the suspending media, but also on other multiple factors such as the temperature, reactor design, and process arrangement (simultaneous or consecutive, and in the last, if UV-C was applied as first or second stressor). The nonlinear inactivation curves displaying shoulder and/or tail, observed for the UV-C and UV-C/H curves, were described by a variety of models. The models used in the cited examples to describe the inactivation achieved by single and combined UV-C treatments were the Weibull, Coroller, Biphasic (Fenoglio et al. 2020b), log-linear (Gabriel et al. 2018), and log-linear plus-shoulder model proposed by Geeraerd et al. (2000) (García Carrillo et al. 2017; Gayán et al. 2012c, 2013, 2014b).

For the combined UV-C/H, fewer studies were evaluated in solid matrices compared with liquid food and beverages (Table 1). The overall inactivation effects corresponding to

Table 1 Recent experimental procedures involving the use of UV-C light combined with heat. Evaluation of microbial inactivation effectiveness in liquid food, model solutions, and culture media.

Item[a]	Experimentals/remarks			Microorganism[b]	Log (N/N0)[c]	Effect[d]	References
Simulated gastric fluid (0.1% peptone, 0.85% NaCl)	Consecutive	Benchtop UV-C (30–270 s, Incident dose: 17–500 mJ cm^{-2})	Heat (45–80 °C)	*E. coli* O26 (wild-type and ATCC BA-2196)	2.0–7.7	Additive	Castro et al. (2019)
Dried red pepper powder	Benchtop double-sided UV-C (10-cm distance, incident dose)	2040 mJ cm^{-2}, 5 min	Mild heat (35–65 °C)	Cocktail of 3-strain *E. coli* O157:H7	0.18–1.57	Additive (65 °C)	Cheon et al. (2015)
		4080 mJ cm^{-2}, 10 min		Cocktail of 2-strain *S.* Typhimurium	0.41–1.76		
				Cocktail of 3-strain *E. coli* O157:H7	0.73–2.88	More than additive (45–65 °C)	
				Cocktail of 3-strain *S.* Typhimurium	0.94–3.06		
Carrot-orange juice (pH: 4.2, 7.6 °Brix)	Thin-film lab-scale UV-C (delivered dose: 980 mJ cm^{-2}, laminar flow)		Mild heat (50 °C)	Native microbiota	TAM: 5.5	Additive	Ferrario et al. (2018)
					Y&M: 2.7		
					TC: 5.7		
Coconut liquid endosperm (pH: 5.15, 4.4 °Brix)	Benchtop UV-C (incident dose: 120 mJ cm^{-2}, 1.9 cm sample thickness)		Mild heat (55–63 °C)	Cocktail of 5-strain *E. coli* O157:H7	5.9	Additive	Gabriel et al. (2018)
				Cocktail of 7-strain *S. enterica*	5.6		
				Cocktail of 2-strain *L. monocytogenes*	6.2		
Peptone water (0.1%)	Annular thin-film lab-scale UV-C (2 lamps, delivered dose: 980 mJ cm^{-2}, laminar flow)		Mild heat (50 °C)	*S. cerevisiae* (KE 162)	4.7	Additive	García Carrillo et al. (2018)

Carrot-orange juice (pH: 3.8; 10.6 °Brix; 7667 NTU; α: 0.32 cm^{-1})	Annular thin-film lab-scale UV-C (2 lamps, delivered dose: 980 mJ cm^{-2}, laminar flow)		Mild heat (40, 45, 50 °C)	*S. cerevisiae* (KE 162)	3.4 (45 °C)	Additive	García Carrillo et al. (2017)
					3.5 (50 °C)		
				E. coli (ATCC 35218)	6.0 (50 °C)	Synergistic	
				P. fluorescens (ATCC 49838)	5.3 (45 °C)	Synergistic	
					6.0 (50 °C)		
Orange juice (pH: 3.5)	Annular thin-film lab-scale UV-C (8 lamps, incident dose: 27.1 J ml^{-1})		Mild heat (52.5 °C)	*S. aureus* (CECT 4465, CECT 976, CECT 4466, and CECT 435)	4.4	Synergistic	Gayán et al. (2014b)
Apple juice (pH: 3.3)					4.5		
Vegetable broth (pH: 5.9)	Annular thin-film lab-scale UV-C (8 lamps, incident dose: 13.6 J ml^{-1})		Mild heat (55 °C)		5.5		
Chicken broth (pH: 5.2)					5.8		
Citrate-phosphate buffer (pH: 7.0; α: 8.8–17 cm^{-1}; <5 NTU)	Simultaneous	Annular thin-film lab-scale UV-C (8 lamps, incident dose: 27.1 J ml^{-1})	Mild heat (60 °C)	Spores of *B. coagulans* (STCC 4522)	1.5 (17 cm^{-1})	Synergistic	Gayán et al. (2013)
	Consecutive	Annular thin-film lab-scale UV-C (8 lamps, incident dose: 27.1 J ml^{-1})	Heat (100–110 °C)		5.0 (8.8 cm^{-1})	Additive	
		Annular thin-film lab-scale UV-C (8 lamps, 27.1 J ml^{-1})	Mild heat (60 °C)		3.5	Indifferent	
					1.26		

(*Continued*)

Table 1 (Continued)

Item[a]	Experimentals/remarks			Microorganism[b]	Log (N/N0)[c]	Effect[d]	References
Apple juice (pH: 3.6, α: 24.9 cm^{-1})	Annular thin-film lab-scale UV-C (8 lamps, incident dose: 3.9 J ml^{-1})	Dimethyl dicarbonate (DMDC) 25, 50, 75 mg L^{-1}	Mild heat (50–55 °C)	*E. coli* (ATCC 4201)	>6.0	Synergistic	Gouma et al. (2015a)
McIlvaine citrate-phosphate buffer (pH: 3.0–7.0; α: 6.6–23.7 cm^{-1})	Simultaneous	Annular thin-film lab-scale UV-C (8 lamps, incident dose: 19.94 J ml^{-1})	Mild heat (55 °C)	*Salmonella* Typhimurium (STCC 878)	5.09	Synergistic	Gayán et al. (2012a)
	Consecutive	Annular thin-film lab-scale UV-C (8 lamps, incident dose: 19.94 J ml^{-1})	Mild heat (55 °C, 2.23 min)		2.34	Less synergistic	
		Annular thin-film lab-scale UV-C (8 lamps, incident dose: 19.94 J ml^{-1})	Mild heat (55 °C, 2.23 min)		1.69	Additive	
Apple juice (pH: 4.0, 13.8 °Brix)	Annular thin-film lab-scale UV-C (8 lamps, incident dose: 27.10 J ml^{-1})		Mild heat: 40 °C	*E. coli* (STCC 4201)	1.0	Indifferent	Gayán et al. (2012b)
			50 °C		2.0	Additive	
			52.5 °C		3.0	Additive-More than additive	
			55 °C		5.5	Synergistic	
			57.5 °C		>7.0	Synergistic	
			60 °C		>7.0	Indifferent to additive	
			55 °C	Cocktail 5-strain *E. coli*	6.0	—	

Orange juice (pH: 2.8, 10.1 °Brix; 51.52 cm^{-1}; 3075 NTU)	Annular thin-film lab-scale UV-C (8 lamps, incident dose: 27.10 $J ml^{-1}$)	Mild heat: 50 °C	*E. coli* (STCC 4201)	3.0	Synergistic	Gayán et al. (2012c)
		52.5 °C		3.0		
		55.0 °C		6.5		
		57.5 °C		7.0		
		60 °C		7.0		
Apple juice (pH: 3.3; α: 25.54 cm^{-1}; 3.34 NTU)	Annular thin-film lab-scale UV-C (8 lamps, incident dose: 27.1 $J ml^{-1}$)	40 °C	*E. coli* 0157:H7 (phage type 34)	1.0	Additive	Gayán et al. (2014b)
		50 °C		1.5		
		52.5 °C		2.0	Synergistic	
		55 °C		3.5		
		57.5 °C		5.0		
		60 °C		6.0	Additive	
		62.5 °C		6.0		
Vegetable broth (VB, pH: 5.8)	Annular thin-film lab-scale UV-C (8 lamps, incident dose: 27.1 $J ml^{-1}$)	50 °C	*E. coli* (STCC 4201)	3.5 (VB)/5.5 (CB)	10% synergism	Gayán et al. (2016)
		52.5 °C		4.0 (VB)/6.0 (CB)	10% synergism	
		55 °C		5.5 (VB)/6.0 (CB)	20% synergism	
Chicken broth (CB, pH: 5.3)		57.5 °C		7.0 (VB)/6.0 (CB)	20% synergism	
		60 °C		7.0 (VB)/6.0 (CB)	0% synergism	
		62.5 °C		7.0 (VB)/6.0 (CB)	0% synergism	

(*Continued*)

Table 1 (Continued)

Item[a]	Experimentals/remarks		Microorganism[b]	Log (N/N0)[c]	Effect[d]	References
Carrot juice (pH: 6.5)	Annular thin-film lab-scale UV-C (8 lamps, incident dose: 3.9 J ml^{-1})	50 °C	*E. coli* O157:H7 (phage type 34)	4.3	Synergistic	Gouma et al. (2020)
		60 °C	TAM	5.3		
			Y&M	4.0		
			LAB	4.5	Indifferent	
				2.8		
Apple juice (pH: 3.6, α: 24.9 cm^{-1})	Annular thin-film lab-scale UV-C (8 lamps, incident dose: 3.7 J ml^{-1})	45.0 °C	*S. cerevisiae* (STCC 1172)	0.9	Indifferent	Gouma et al. (2015b)
		50.0 °C		0.9		
		52.5 °C		1.1	Less than additive	
		55.0 °C		2	More than additive	
		57.5 °C		5	Synergistic	
		60.0 °C		6	Indifferent	
Lemon-Melon juice (pH: 4.0)	Annular thin-film lab-scale UV-C (7 lamps, 0.4–2.8 J ml^{-1})	Heat: 55 °C	*E. coli* K12 (ATCC 25253)	$D = 299$ s	—	Kaya et al. (2015)
		65 °C		$D = 128$ s		
		75 °C		$D = 5.3$ s		
		85 °C		$D = 0.99$ s		
Verjuice (pH: 2.6, 4.3 °Brix)	Annular thin-film lab-scale UV-C (7 lamps, 570 mJ cm^{-2})	Mild heat: 47.0 °C	*S. cerevisiae* (NRRL Y-139)	3.59	Synergistic	Kaya and Ünlütürk (2019)
		50.5 °C		5.16		

Stainless steel coupons	Benchtop UV-C (without specifying dose)	15 min	Superheated steam (up to 5 min)	200 °C	Cocktail of 3-strain *B. cereus* spores	2.30	Additive	Kim et al. (2020)
		30 min		250 °C		2.48		
		60 min		300 °C		2.98		
	Superheated steam (up to 5 min)	200 °C	Benchtop UV-C (without specifying dose)	15 min		2.73		
		250 °C		30 min		2.70		
		300 °C		60 min		3.06	More than additive	
							Synergistic	
Grapefruit juice (9.6 °Brix, ∝: 49.5 cm^{-1})	Teflon coiled lab-scale UV-C (delivered dose: 39.6 J/L)			Mild heat (65 °C)	TAM	4.99	Indifferent	La Cava and Sgroppo (2019)
					Y&M	5.24		
10 mM phosphate buffer (pH 7.2)	Consecutive	Benchtop UV-C (incident dose: 1.5 J cm^{-2}, 14-cm distance)		Mild heat (40, 43, 45 and 48 °C, 3, 5, 10 or 15 min)	*Botrytis cinerea* spores (MUCL 18864)	6.0	Additive. Higher inactivation with heat as a first hurdle	Marquenie et al. (2002)
		Mild heat (39, 41, 43, and 45 °C, 3, 5, 10 or 15 min)		Benchtop UV-C (incident dose: 1.5 J cm^{-2}, 100-µl inoculum 14-cm distance)	Spores of *M. fructigena* (CBS 101499)		Additive. Higher inactivation with UV-C as a first hurdle	
Orange juice (pH: 3.9, 11.3 °Brix,α 71.9 cm^{-1})	Consecutive	Benchtop UV-C (6300 mJ cm^{-2}, 1.9 cm thickness, 21-cm distance)		Mild heat (53 °C)	Cocktail of 7-strain *E. coli* O157:H7	3.98	Less than additive	Pagal and Gabriel (2020)
					TAM	1.96	—	
					Y &M	1.28		
					LAB	0.86		

(*Continued*)

Table 1 (Continued)

Item[a]	Experimentals/remarks				Microorganism[b]	Log (N/N0)[c]	Effect[d]	References
		Benchtop UV-C (Incident dose: 6300 mJ cm^{-2}, 1.9 cm thickness, 21-cm distance)	Mild heat (53 °C)		Cocktail of 7-strain *E. coli* O157:H7	4.69	Additive	
					TAM	2.99	—	
					Y&M	1.83	More than additive	
					LAB	2.56	—	
Strawberries (pH: 3.3)	Benchtop UV-C (incident dose: 4.1 kJ m^{-2} – 2.3 cm distance from lamp)		Mild heat (45 °C, 3 h in air oven)		*B.cinerea* (ICFC377/00-F18)	0% RSG	—	Pan et al. (2004)
					R. stolonifer; (ICFC364/00 -F8)	30% RSG	Additive	
Frankfurters	Benchtop UV-C (20-cm distance, incident dose)	1 mJ cm^{-2}	Flash pasteurization (121 °C steam)	0.75 s	Cocktail of 3-strain *L. innocua*	3.19	Additive	Sommers et al. (2010)
		2 mJ cm^{-2}		1.5 s		3.47		
		4 mJ cm^{-2}		3 s		3.89		
OT (pH: 3.5; 12.4 °Brix; A_{254nm}: 0.68) OBMKS (pH:3.7; 13.9 °Brix; A_{254nm}: 0.37) juices	Pilot-scale Dean-flow UV-C (delivered dose: 390 mJ cm^{-2}, turbulent flow)		Mild heat (50 °C)		*E. coli* (ATCC 25922)	OT: 6.3 OBMKS: 6.6	Additive	Fenoglio et al. (2020b)
					L. plantarum (ATCC 8014)	OT: 5.5 OBMKS: 6.7		
					S. cerevisiae (KE 162)	OT: 4.9 OBMKS: 4.6		
Raw milk (pH: 6.75)	Thin-film lab-scale UV-C (delivered dose: 980 mJ cm^{-2}, laminar flow)		Heat (85 °C, 20 min, 400-mmHg vacuum)		TAM	3.5	—	Vásquez-Mazo et al. (2019)
					TC	3.9		

[a] OBMKS, orange/banana/mango/kiwi/strawberry juice blend; OT, orange/tangerine juice blend.
[b] LAB, lactic acid bacteria; TAM, total aerobic mesophiles; TC, total coliforms; Y&M, yeasts and molds; *Escherichia coli; Salmonella* Typhimurium*; Salmonella enterica; Listeria monocytogenes; Saccharomyces cerevisiae; Pseudomonas fluorescens; Staphylococcus aureus; Bacillus coagulans; Botrytis cinerea, Monilia fructigena; Rhizopus stolonifera; Listeria innocua; Lactobacillus plantarum.*
[c] D, decimal reduction time in seconds; % RSG, percentage of reduction on spore germination.
[d] Single effects were not determined.

the combined procedures were derived from the information provided in the different studies. For instance, additive inactivation effects were determined for cocktails of three strains of *E. coli* O157:H7 and *Salmonella* Typhimurium in dried red powder pepper treated by UV-C (2040 mJ cm^{-2}) assisted by heat (65 °C). When doubled the UV-C dose, more than additive effects were determined, even at a temperature value as low as 45 °C (Cheon et al. 2015) (Table 1). Similarly, additive inactivation effects were observed for *Rhizopus stolonifer* and *Listeria innocua* in strawberries and frankfurters treated by UV-C combined with mild heat (45 °C) or flash pasteurization (121 °C, up to 3 seconds), respectively (Pan et al. 2004; Sommers et al. 2009) (Table 1). Concerning the inactivation of clear and turbid liquid food, beverages, and model media, synergistic effects were reported for several experimental conditions when applied UV-C/H treatments in a restricted temperature range from 45 to 57.5 °C (Table 1). In these studies, when high temperatures were used in combination with UV-C, the overall inactivation effect was described as inherently thermal, while the use of temperature values below 45 °C was reported as the least effective, hardly affecting the UV-C lethality, thus rendering in additive inactivation effects (García Carrillo et al. 2017; Marquenie et al. 2003; Gayán et al. 2012b).

When a range of temperature conditions were studied, the optimal values for achieving synergism were determined. For example, Gayán et al. (2012c) investigated the inactivation of *E. coli* STCC 4201 in commercial orange juice treated by UV-C/H (40–60 °C) and reported synergistic effects only when 50 °C was applied (Table 1). García Carrillo et al. (2017) observed synergism for the inactivation of *E. coli* and *Pseudomonas fluorescens* in a turbid carrot–orange juice blend subjected to UV-C/H (50 °C) (Table 1). With regard to the inactivation of *Saccharomyces cerevisiae,* a synergistic inactivation was also reported by Kaya and Ünlütürk (2019) in verjuice treated by UV-C/H (47.0 and 50.5 °C). Gayán et al. (2012b) studied the inactivation of *E. coli* STCC 4201 in commercial apple juice by UV-C/H (40–60 °C) and reported synergistic effects only when 50 °C was applied (Table 1). Furthermore, (Gouma et al. 2015b) evaluated the response of *S. cerevisiae* in apple juice treated by UV-C/H (45–60 °C). They observed synergistic effects only when a temperature of 57.5 °C was used. In a different study, Gayán et al. (2014c) reported synergistic inactivation of *E. coli* in commercial apple juice processed by UV-C/H (50, 55, and 57 °C). However, when higher temperatures were applied (60–62.5 °C), additive effects were recorded (Table 1).

Some of the UV-C/H treatments summarized in Table 1 that were performed in a range of temperature values from 45 to 60 °C reported a nonlinear relationship with downward concavity between the temperature used in the combined treatment and the parameter 4D (or 5D), with D being the decimal reduction time corresponding to the UV-C/H treatment. On the other hand, for single thermal treatments, a first-order dependence of the thermal death time (TDT) with temperature was observed (Gayán et al. 2014b, 2016; Gouma et al. 2015b). However, in general, when the temperature assisting the UV-C treatment was equal to or higher than 60 °C, the combined UV-C/H and single H inactivation curves were overlapped. This result suggested a thermal inactivation effect that strongly restricted the synergistic effect on the inactivation observed at lower temperatures.

The process arrangement or design also influenced the microbial response to the combined UV-C/H treatments. For instance, Gayán et al. (2012a) evaluated the inactivation of *S.* Typhimurium in McIlvaine citrate-phosphate buffer subjected to UV-C/H (50–60 °C) applied simultaneously or consecutively. Outstandingly, they observed synergistic effects when the hurdles were applied simultaneously in a specific temperature range from 50 to 55 °C, with negligible thermal inactivation effects. When the temperature increased up to 60 °C, synergistic effects disappeared. In addition, when the stressors were applied in the sequence UV-C followed by H, synergistic effects were observed to a lesser extent. On the same trend, Gayán et al. (2013) observed synergistic effects for the inactivation of *B. coagulans* spores in a model system when UV-C was applied simultaneously with mild heat (60 °C), in contrast to the consecutive arrangement (Table 1). The authors explained that the increased synergistic response observed when the simultaneous arrangement was used may be related to an additional damage that arises from the interaction of sublesions induced by both stressors, which was more effective than when they were applied sequentially.

With the idea of finding a pattern in the inactivation response, the *log synergism* was calculated, when possible, from data collected from Table 1 and is displayed in Figure 2. This parameter was estimated according to Equation 5 as the difference between the log reductions achieved after combined treatment (log UVC/*H*) and the sum of the log reductions yielded after exposure to the single treatments (log UVC and log *H*) (Bang et al. 2017):

$$\log \text{synergism} = \left(\log \text{UVC}/H - \left(\log \text{UVC} + \log H\right)\right) \tag{5}$$

Theoretically, a synergistic effect will be assumed by any positive value as derived from Eq. (5). Nevertheless, a strong synergistic effect may be confirmed when *log synergism* values are higher than or equal to 1. While antagonistic to indifferent effects were considered when negative to zero values were obtained, respectively. As noted from the analysis of Figure 2, a common synergistic profile was observed in a wide variety of microorganisms and liquid treatment media, when the temperature simultaneously assisting the UV-C processing was in the range from 50 to 57.5 °C. Conversely, only additive effects were

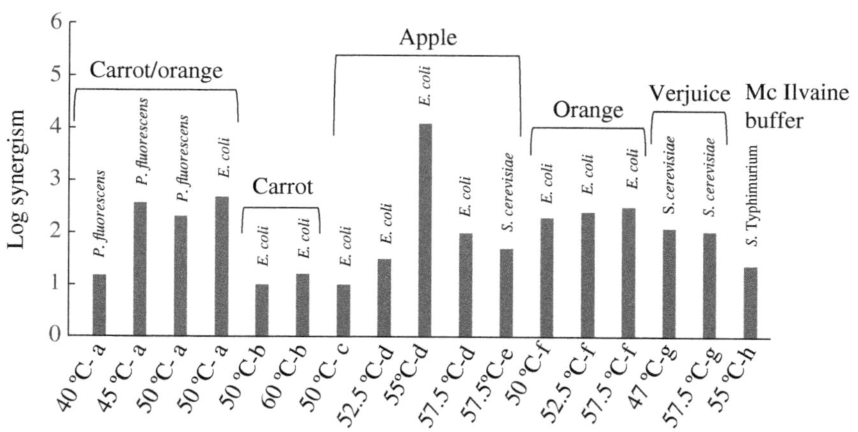

Figure 2 *Log synergism* estimated according to Eq. (5) from data reported in literature for *Pseudomonas fluorescens*, *Escherichia coli*, *Saccharomyces cerevisiae*, and *Salmonella* Typhimurium in a variety of media submitted to UV-C assisted by mild heat treatments. (a) García Carrillo et al. (2017); (b) Gouma et al. (2020); (c) Gouma et al. (2015a); (d) Gayán et al. (2012b); (e) Gouma et al. (2015b); (f) Gayán et al. (2012c); (g) Kaya and Ünlütürk (2019); (h) Gayán et al. (2012a).

obtained in solid media. In general, *P. fluorescens* seems to be more sensitive than *E. coli*, as it showed synergistic inactivation effects, even at a very low temperature (40 °C) assisting the UV-C processing, whereas *E. coli* displayed synergism when the UV-C treatment was performed at 50 °C.

The *log synergism* for different strains of *E. coli* raised either by increasing the temperature value from 50 up to 57.5 °C or by decreasing the turbidity of the juice. For instance, Gouma et al. (2020) obtained a *log synergism* value of 1.1 when processing *E. coli* by UV-C assisted by mild heat (60 °C) in a turbid (9148 NTU) carrot juice, whereas *log synergism* values in the range from 2.3 to 2.5 were estimated for *E. coli* in an orange juice with lower turbidity (4460 NTU) treated by UV-C assisted by mild heat at lower temperature (50.0–57.5 °C) (Gayán et al. 2012c). Outstandingly, a *log synergism* value of 4.1 was recorded for *E. coli* treated by UV-C/H at 55 °C in an apple juice (Gayán et al. 2012b). With regard to *S. cerevisiae*, *log synergism* values of 2.0 and 2.1 were estimated in a clear verjuice (38 NTU) processed by UV-C/H at 47.0 and 57.5 °C, respectively (Kaya and Ünlütürk 2019).

With regard to the quality parameters that were affected by the preservation treatments involving UV-C and heat (UV-C/H), as a mode of example, Figure 3 shows the percentage of retention of the total polyphenol content (TPC), total antioxidant activity (TAA), and ascorbic acid content for different fruits, vegetables, and juices treated by UV-C/H at different temperature values. Even though synergistic microbial inactivation effects were reported when UV-C/H was applied in the range from 50 to 60 °C, under different arrangements, a decrease in the retention of bioactive compounds was observed when temperature values higher than 55 °C were used, regardless of the considered UV-C dose (Figure 3). In particular, when UV-C (39 600 mJ L^{-1}) was performed at 65 °C in a coiled-tube reactor for the treatment of grapefruit juice (9.6 °Brix, absorption coefficient (α): 49.5 cm^{-1}), retention values of 86%, 84%, and 37% were determined for TPC, TAA, and ascorbic acid, respectively (La Cava and Sgroppo 2019). Similarly, a TPC retention of 82% was observed in fresh-cut pomegranate arils treated by benchtop UV-C (4540 mJ cm^{-2}; 15-cm distance) combined with mild heat (55 °C, 30 seconds in hot water) and subsequently packaged in high-oxygen (90 kPa O_2) modified atmosphere (Maghoumi et al. 2013). Even though indifferent to synergistic enzyme inactivation effects were determined, the overall reduction percentages for catalase (10%) and polyphenol oxidase, PPO (14.6%) activities were low. When the temperature that assisted UV-C was 55 °C, a retention of ascorbic acid higher than 80% and higher enzyme inactivation were generally observed. For example, 83% of ascorbic acid was retained while 39% of PPO was inactivated in an apple juice (pH: 4.0; 13.8 °Brix) (Gayán et al. 2012b), whereas 83% of ascorbic acid retention and 64% of pectin methylesterase (PME) inactivation were achieved in an orange juice (pH: 2.8; 10.1 °Brix) (Gayán et al. 2012c) subjected to UV-C (27.1 mJ L^{-1}) assisted by mild heat (55 °C). The order in which the hurdles were applied strongly influenced the retention of the quality parameters. For example, ascorbic acid retention values of 87% (UV-C as first hurdle) and 96% (mild heat as first hurdle) were observed in an orange juice when benchtop UV-C (8066 mJ cm^{-2}; 21-cm distance from the lamp) was combined in sequence with mild heat (53 °C, 400 seconds) (Pagal and Gabriel 2020).

When the hurdles were simultaneously applied using a lower UV-C dose (807 mJ cm^{-2}; 53 °C), 93% of ascorbic acid retention was observed. On the contrary, when UV-C was combined with mild heat under different arrangements, at lower temperature values, the quality parameters TPC and TAA were almost totally retained (99–100%) in a carrot–orange

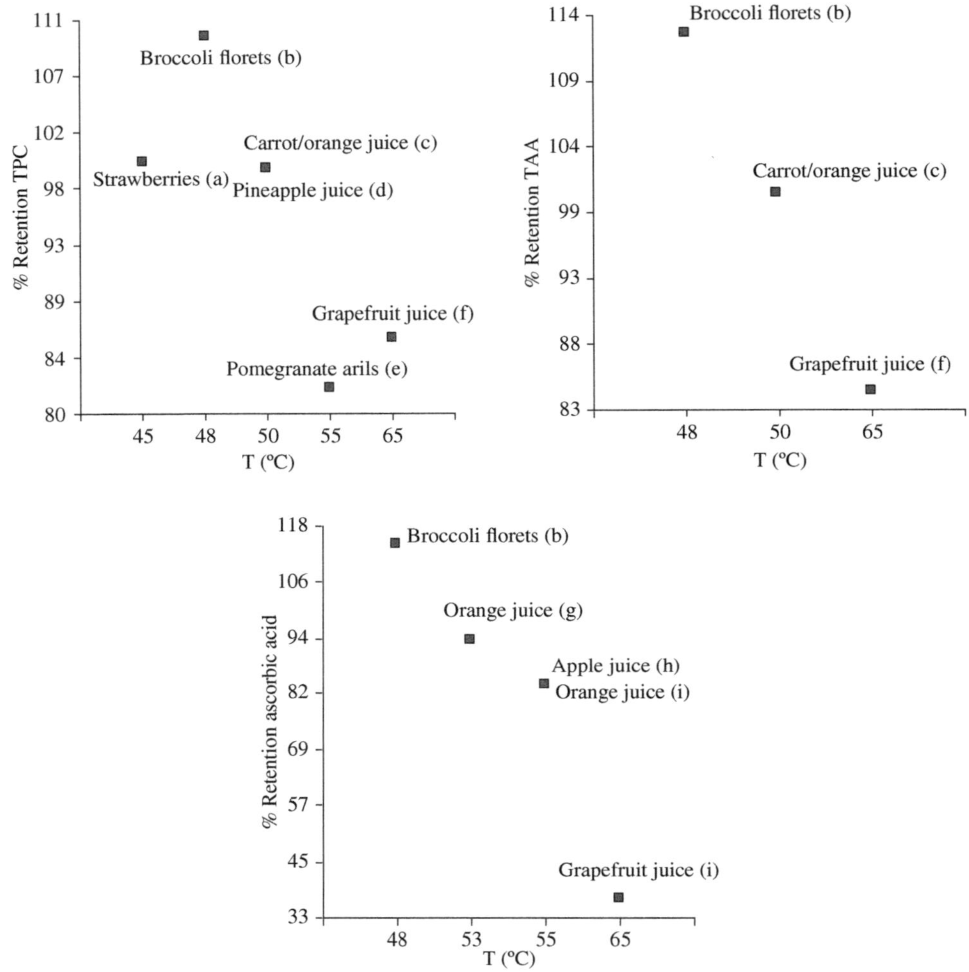

Figure 3 Percentage retention (%) of total polyphenol content (TPC), total antioxidant activity (TAA), and ascorbic acid of different vegetables, fruits, and juices subjected to UV-C/H treatments. (a) 690 mJ cm^{-2} (Pan et al. 2004); (b) Lemoine et al. (2010); (c) 980 mJ cm^{-2} (Ferrario et al. 2018); (d) 7.55 mJ cm^{-2} (Sew et al. 2014); (e) 454 mJ cm^{-2} (Ferrario et al. 2011); (f) 39 600 mJ L^{-1} (La Cava and Sgroppo 2019); (g) 706 mJ cm^{-2} (Pagal and Gabriel 2020); (h) 27.1 J ml^{-1} (Gayán et al. 2012b); (i) 27.1 J ml^{-1} (Gayán et al. 2012c).

(annular thin-flow reactor, 985 mJ cm^{-2}, 50 °C) (Ferrario et al. 2018) and pineapple (7.55 mJ cm^{-2}; 50 °C) (Sew et al. 2014) juices and in strawberries (benchtop UV-C, 690 mJ cm^{-2}; 30-cm distance; 45 °C/3 hours; consecutive treatment) (Pan et al. 2004). Outstandingly, Lemoine et al. (2010) applied a benchtop UV-C treatment (800 mJ cm^{-2}, 30-cm distance from lamp) to broccoli florets and observed an increase of 10, 13, and 14% in TPC, TAA, and ascorbic acid, respectively after treatment without observing changes in catalase and peroxidase activities (Figure 3). In general, sensory panel evaluations determined no differences between the UV-C/H combined treated and untreated samples, regarding the overall appearance, acceptability, and color (Pagal and Gabriel 2020). In the same trend, no texture modifications or even positive texture changes were described, compared with the single treatments. For instance, Vásquez-Mazo et al. (2019) observed a syneresis reduction of 60% and higher cohesiveness and elasticity values in yogurt elaborated from milk simultaneously treated by UV-C (1060 mJ cm^{-2}, thin-flow device) assisted by vacuum (400 mmHg) and heat (H, 85 °C) and stored under refrigeration during 20 days, compared with the single UV-C and H treatments. Similarly, Cruz et al. (2016) observed an increase in firmness (125%) when Portuguese cabbage was treated by benchtop UV-C (198 mJ cm^{-2}; 21.5-cm distance) followed by water blanching (80 °C; 90 seconds), compared with a single heat treatment. A synergistic fourfold reduction of peroxidase activity (90% reduction) was also observed for the combined treatment.

A proper design for combined intervention strategies uses a multidisciplinary engineering perspective, as explained above (López-Gómez et al. 2009). Certain tools such as predictive microbiology, growth dynamics study during product storage, and induced damage studies by flow cytometry and/or transmission electron microscopy (TEM), etc. should be explored. Figure 4a shows, as a mode of example, the inactivation profile of *Candida parapsilosis* ATCC 22019 and *Saccharomyces cerevisiae* KE 162 in a turbid carrot–orange (pH: 4.2; 7.8 °Brix) juice blend treated by single UV-C (985 mJ cm^{-2}, 25 °C, laboratory-scale thin-film annular reactor, laminar flow), single mild heat (H, 50 °C), and the corresponding combined treatment (UV-C/H).

An important reduction of up to 4.7 log cycles and complete inactivation of *S. cerevisiae* and *C. parapsilosis*, respectively were achieved after applying the combined UV-C/H treatment. The combinations of these single treatments showed additive effects for both yeast strains.

With regard to the microbial inactivation mechanism of the UV-C assisted by mild heat treatments, there is some evidence derived from flow cytometry studies that the exposure to the combined UV-C/H treatment compromises membrane permeability and reduces enzyme activity. The use of flow cytometry has proved to be useful for the assessment of microbial physiological status on a single cell level analyzing large populations of cells in a very short time. The physiological state of treated cells can be monitored by using fluorescent dyes that are targeted on specific cellular structures (García Carrillo et al. 2020). In reference to the physiological state of the cells, there is evidence that the assistance of heat to the UV-C treatment avoids the existence of double-stained cells. To exemplify this, the induced damage of *C.parapsilosis* ATCC 22019 and *S. cerevisiae* KE162 in a carrot–orange (pH: 4.2; 7.8 °Brix; 7667 NTU) juice blend as affected by single UV-C (985 mJ cm^{-2}, <25 °C), single H (50 °C), and the combined treatment UV-C/H was analyzed in a recent study. A double-staining strategy using fluorescein diacetate and propidium iodide was applied to study changes in esterase activity and membrane permeability after exposure to single or combined treatments (Figure 4b,c). Additionally, Figure 5 shows the corresponding density plots depicting the enzyme activity and membrane damage before and after exposure to the single and combined treatments. The percentage of cells located in each gate can be visualized in the four edges of each plot. All treatments induced a shift of cells from gate #1 (cells with esterase activity and intact membrane) to gate #4 (cells with compromised membrane) (Figure 5). Therefore, all the treatments disrupted the integrity of the cell membrane, allowing PI to penetrate the cell. The fastest shift to the death gate (#4) was more pronounced in the case of the combined UV-C/H treatment, reaching up to 97–98% of PI-stained cells. In contrast, only 0.0–17.5% and 47.6–67.8% were achieved after single H and

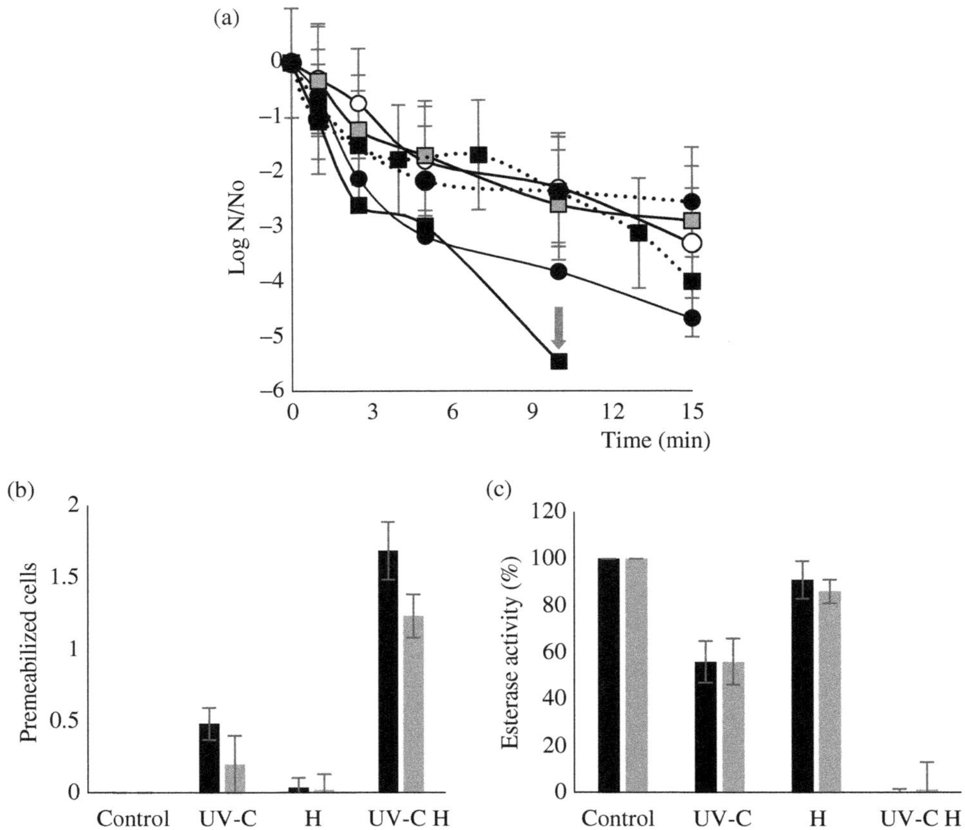

Figure 4 (a) Survival curves of *Candida parapsilosis* ATCC 22019 (■) and *Saccharomyces cerevisiae* KE 162 (●) in a carrot–orange juice blend treated by single UV-C (■, ○), single mild heat at 50 °C (…••■•••…,…••●•••…), and UV-C assisted by mild heat at 50 °C (–■–, –●–) (delivered maximum dose: 980 mJ cm^{-2}), determined by the plate count method. Changes in membrane permeability (b) and esterase activity (c) of *C. parapsilosis* (■) and *S. cerevisiae* (■) cells. Permeabilized cells determined by PI uptake (Log No_t − Log N_t) and relative changes of esterase activity as a function of treatment: untreated cells (Control), cells treated with UV-C (985 mJ cm^{-2}), cells treated with single mild heat (H) and UV-C assisted by mild heat (UV-C/H). The results are means on data from three independent experiments with error bars indicating standard deviation (I).

UV-C treatment, respectively (Figure 5). Therefore, single UV-C and UV-C/H induced, for both strains, significant esterase activity loss and membrane permeabilization, which exposed a synergistic inactivation behavior when the combined treatment was applied (Figure 5a,b). Exceptionally, in the case of *S. cerevisiae* cells exposed to single treatments, an important fraction (7.8–8.5%) of double-stained cells was recorded, which suggested the presence of sublethally damaged cells. The higher fraction of permeabilized cells observed for the combined treatment, compared with single UV-C and H, is in accordance with the higher inactivation observed by the plate count method illustrated in Figure 4a. In the same study, TEM images corresponding to *S. cerevisiae* KE 162 cells in the carrot–orange juice blend were obtained to further analyze the induced damage. As shown in Figure 6a–b,

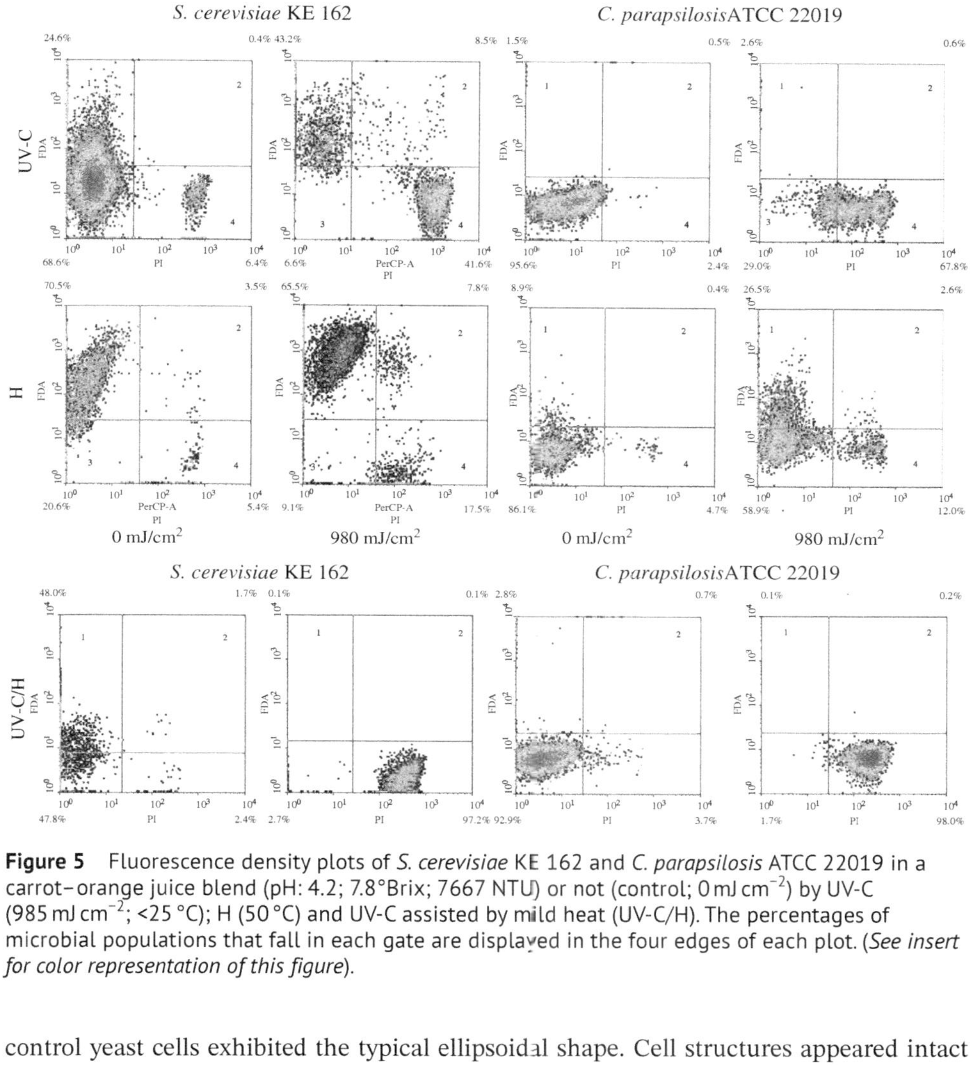

Figure 5 Fluorescence density plots of *S. cerevisiae* KE 162 and *C. parapsilosis* ATCC 22019 in a carrot–orange juice blend (pH: 4.2; 7.8°Brix; 7667 NTU) or not (control; 0 mJ cm^{-2}) by UV-C (985 mJ cm^{-2}; <25 °C); H (50 °C) and UV-C assisted by mild heat (UV-C/H). The percentages of microbial populations that fall in each gate are displayed in the four edges of each plot. (*See insert for color representation of this figure*).

control yeast cells exhibited the typical ellipsoidal shape. Cell structures appeared intact displaying high electronic density, a smooth cell wall without discontinuities, and undamaged organelles such as mitochondria (M), vacuoles (V), bud vacuoles (BV), cell nucleus (N), lysosomes (L), endoplasmic reticulum (ER), and storage granules (SG). The thermal treatment (H) induced plasmatic membrane alterations. However, the cell wall looked intact (Figure 6c). Moreover, an unnatural more rounded shape was recorded and the cytoplasm was less differentiated, when compared with control cells (Figure 6c). Similarly, the UV-C-treated cells showed less differentiated cytoplasm with preserved cell wall but certain membrane deformation with discontinuities (Figure 6d). Moreover, organelles such as lysosome and mitochondria appeared deformed (Figure 6d), whereas, the UV-C/H treatment induced several changes at different levels, reflecting the higher severity of the combined treatment, compared with the single procedures. For instance, a discontinuity of plasma membrane with leakage of content (Figure 6e,f) and separation of the plasmatic

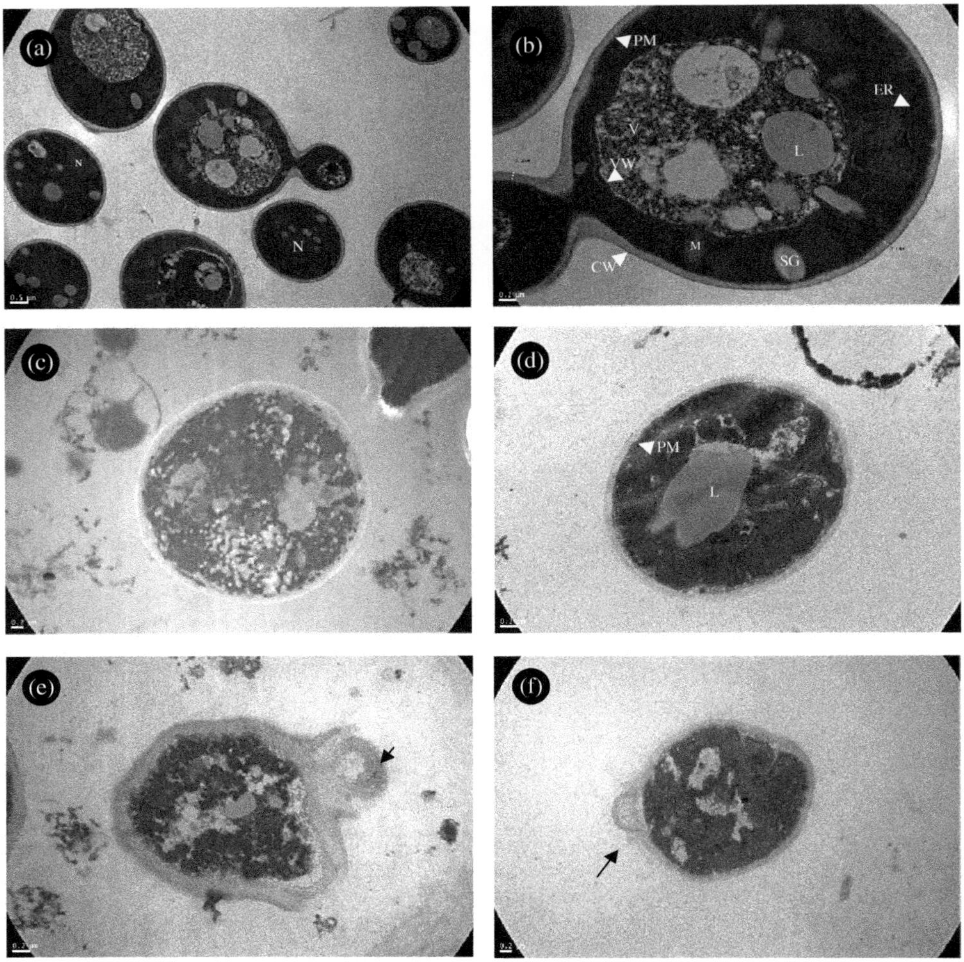

Figure 6 TEM images of *S. cerevisiae* KE 162. (a, b) Untreated cells and cells treated by: (c) H (50 °C, 15 min); (d) UV-C (985 mJ cm^{-2}); (e, f) UV-C/H (985 mJ cm^{-2} 50 °C). (a) Group of intact cells having round shape, round nucleus, and smooth membrane; (b) Intact yeast cell showing budding ascus with well-defined organelles, cell wall, and plasmatic membrane; (c) Cell with undistinguishable intracellular content; (d) Cell with disrupted inner structure; (e) Deformed cell with disrupted inner structure; broken cell wall and plasmalemma; inner material radiating from the cell (arrow). Scale: (b–f) 0.2 μm; (a) 0.5 μm. Mitochondria (M), nucleus (N), cell wall (CW), plasma membrane (PM), vacuole (V), vacuole wall (VW), lysosome (L), storage granule (SG), endoplasmic reticulum (ER).

membrane from the cell wall were observed (Figure 6e,f; black arrow). Moreover, disruption of organelles and generalized rupture of membranes were also detected, without any possibility of distinguishing cell structures (Figure 6e,f).

Overall, there is a vast number of studies assessing the impact of different treatments involving UV-C assisted by heat on microbial inactivation and physicochemical properties of the treated media. The combinations of UV-C and mild heat with temperature values ranging from 50 to 55 °C were the most successful in achieving synergistic inactivation effects while preserving food quality. Moreover, these combinations induced more severe damage on treated yeasts cells compared with single treatments, as well as it ensured the absence of viable non-culturable cells in a turbid juice.

3.1.2 UV-C Combined with Other Novel Technologies

There is a wide range of novel technologies that have been increasingly used under hurdle approach in the last two decades, causing microbial inactivation at sublethal temperatures and additionally extending the product shelf life. The use of these technologies is proposed for food and beverage preservation, without the need for severe heating, expecting that they cause minimal damage to texture, flavor, and nutritional quality of a great variety of products (Zhang et al. 2018). Overall, they are effective at inactivating mostly vegetative cells and spores to a lesser extent as they are far more resistant. Many of these traditional or emerging technologies affect, through multiple mechanisms that do not have one specific cell target, the individual constituents, structures, molecules, and reactions. In general, the primary cause of the cell death provoked by many of these nonthermal agents is still not clearly defined, and this lack of accuracy attempts against the intelligent selection of stressors that should be promoted to surpass microbial resistance. Some of these agents that have been more recently studied to be used in combination with UV-C are as follows: PEF (Walkling-Ribeiro et al. 2008; Palgan et al. 2011); alkaline (AEW) and neutral (NEW) electrolyzed water (Jemni et al. 2014; Mikš-Krajnik et al. 2017; Martínez-Hernández et al. 2013); US (Gabriel 2012, 2015; Char et al. 2010; Tremarin et al. 2017; Mikš-Krajnik et al. 2017; Esua et al. 2018; Şengül et al. 2011); nonthermal atmospheric plasma (NTAP) (Colejo et al. 2018) and ohmic heating (Kim et al. 2019), among others.

In reference to high-intensity US, injury or disruption of microbial cells is widely attributed to the *cavitation* phenomenon, that is, the rupture of liquids when high-intensity US is applied (e.g.: intensities higher than $1\,\mathrm{W\,cm^{-2}}$ and frequencies between 18 and 100 kHz), and the effects produced by the motion of the cavities or bubbles, which is denominated "*stable*" cavitation. The bubbles can undergo relatively stable, low-energy oscillations, provoking a flux or stream in their immediate surroundings (microstreaming effect) that could shear and disrupt cellular membranes or break cells (Chemat et al. 2011). When US is applied assisted by heat and pressure, it is called manothermosonication (MTS) (Raso and Barbosa-Cánovas 2003). Recent studies have demonstrated that the inactivation effectiveness achieved by US is largely dependent on the specific microorganism. It displays a disruptive action over several targets within the microbial cell such as the cell wall, cytoplasmic membrane, DNA, other internal cell structures, and the outer membrane, with the cytoplasmic membrane not being the primary target (Ananta et al. 2005; Ferrario and Guerrero 2017). Regarding the use of PEFs, the PEF treatment involves the application of short pulses (1–10 μs) with high-intensity electric field (typically $20\text{–}80\,\mathrm{kv\,cm^{-1}}$) to fluid foods placed between two electrodes. The mechanism of microbial inactivation is well-known and is based on the formation of pores in the cellular membrane, which can generate cellular lysis with leakage of extracellular components due to the induced electric field. For high levels of electric field intensity, this phenomenon is irreversible, which renders into cellular death. However, this innovative technology demonstrated to be more effective for food preservation when it is employed under hurdle approach (Mosqueda-Melgar et al. 2008). Ohmic heating involves the use of electric current, which is directly dissipated into foods. Food components become the elements of an electric circuit, through which alternating current flow generates heat in a short period of time, depending on their intrinsic electrical resistance. This technology has the potential for achieving rapid and uniform heating in foods, providing microbiologically safe and high-quality products (Kumar 2018). Microfiltration (MF) is a membrane separation process that has been recently used to physically remove microbial contaminants and suspended solids from diverse beverages such as beer, juices, and wine (Zhao et al. 2015). With regard to the use of ozone and diverse types of electrolyzed water, these emerging technologies were described as suitable, due to its bactericidal activity, for the control and/or prevention of microorganisms in a variety of foods and beverages. The oxidative stress provoked by reactive oxygen species causes an imbalance between intracellular oxidant concentration, cellular antioxidant protection, and oxidative change of lipids of membrane, proteins, and DNA repair enzymes (Zheng et al. 2014; García Loredo et al. 2015). On the other hand, infrared (IR) radiant heating offers significant advantages over convection and conduction heating. It displays higher heat transfer capacity and high energy efficiency, because it heats directly the product, without the influence of the surrounding air. IR radiation transfers thermal energy in the form of electromagnetic waves and is classified into three regions, near IR (NIR; 0.76–2 mm), medium IR (MIR; 2–4 mm), and far IR (FIR; 4–1000 mm) (Ha and Kang 2015). More recently, the use of NTAP has been investigated for preservation purposes. This technology induces gas ionization by the application of enough energy, generally involving electric discharges at atmospheric pressure and room temperature. Microbial inactivation is achieved due to the interaction between excited atoms and molecules, cations and anions free radicals, electrons and ultraviolet radiation with the microorganisms (Colejo et al. 2018).

Table 2 shows, by way of example, some binary and ternary UV-C-based combinations based on the use of novel nonthermal agents, which have been more recently suggested for the microbial inactivation step, as part of an overall preservation strategy that was usually complemented with posterior chilling or modified atmosphere packaging (MAP) during storage. The factors involved in the preservation processing technologies proposed exerted diverse modes of action toward microbial inactivation, different from those described for UV-C.

Combined preservation processes including emerging nonthermal agents are gaining popularity most quickly in fruit derivatives, probably due the inherent low pH of this type of food material, hurdle that contributes to an overall preservation strategy (Alzamora et al. 2016b). As displayed in Table 2, fruit and vegetable derivatives are the preferred matrices, although salmon (Mikš-Krajnik et al. 2017; Colejo et al. 2018); ham (Ha and Kang 2015); powered infant formula (Ha and Kang 2014) and milk (Şengül et al. 2011) were other food matrices used. According to Table 2, a wide range of UV-C doses was used, mainly measured as incident doses covering a wide range from 0.45 to 3100 mJ cm^{-2}. Solid food matrices were treated by UV-C light using benchtop reactors (Colejo et al. 2018; Choi et al. 2015; Jemni et al. 2014; Kim and Hung 2012; Mikš-Krajnik et al. 2017; Maghoumi et al. 2012; Martínez-Hernández et al. 2013). Similarly, the studied liquid food products were mainly treated in benchtop units (Tremarin et al. 2017; Kim et al. 2019; Gabriel 2012, 2015; Walkling-Ribeiro et al. 2008), but also in annular flow devices (Palgan et al. 2011; Char et al. 2010; Zhao et al. 2015), and in one exceptional case, cow milk was treated using an

Table 2 Recent experimental procedures involving the use of UV-C light combined with other novel technologies. Evaluation of microbial inactivation effectiveness in different media.

Item	Experimentals/remarks[a]			Microorganism[b]	Log (N/N0)[c]	Effect[d]	References
Apple cider (pH:3.7–3.8; 12.8–14.4 °Brix)	Consecutive	MF (1.4 μm; 10°C; 155 kPa)	Annular flow UV-C (incident dose: 1.75 mJ cm^{-2})	*E. coli* (ATCC 25922)	8.0	HI/single MF	Zhao et al. (2015)
				Cocktail 5-strain *E. coli* O157:H7	7.7	HI/single MF	
				A. acidoterrestris VF (cells)	5.7	HI/single MF	
				A. acidoterrestris VF (spores)	5.7	HI/single MF	
				C. parvum (oocysts)	ND[2] (5.3)	Indifferent	
Apple juice (pH: 3.5; 11.8 °Brix)	Consecutive	Benchtop UV-C (30 W; 30 min; 26-cm distance)	PEF (46/58 °C 40 kV cm^{-1}; 100 μs)	*S. aureus* (SST 2.4)	9.5	Additive	Walkling-Ribeiro et al. (2008)
Smoked salmon		Benchtop UV-C (incident dose: 500 mJ cm^{-2})	NTAP (8 min; 1 atm; <40°C)	*S.* Typhimurium (CECT 443)	1.3	Additive	Colejo et al. (2018)
				S. Enteritidis (CECT 4300)		More than additive	
				P. shigelloides (ATCC 14029)	1.0	Less than additive	
				A hydrophila (CECT 5174)	1.1	Less than additive	
				E. coli O157:H7 (ATCC 43895)	1.0	Less than additive	
				L. monocytogenes (ATCC 15313)	0.7	Less than additive	
				S. aureus (CECT 4459)	0.7	Indifferent	

(*Continued*)

Table 2 (Continued)

Item	Experimentals/remarks[a]			Microorganism[b]	Log (N/N0)[c]	Effect[d]	References
Orange juice (pH:3.4; 10°Brix)	Consecutive	US (20 kHz; 95 μs; 10 min)	Annular flow UV-C (delivered dose: 1870 mJ cm^{-2})	*E. coli* (ATCC 35218)	1.9	Indifferent	Char et al. (2010)
	Simultaneous (20 min)				3.3	Additive	
Cherry tomatoes	Consecutive	Benchtop UV-C (Incident dose: 2 kJ m^{-2})	Active MAP (5.3% CO_2;5.5% O_2/9-day storage; 4 °C)	Cocktail of 3 strain *S.* Typhimurium	2.2	Additive	Choi et al. (2015)
			Passive MAP (9-day storage; 4 °C)		1.6	Less than additive	
Tomatoes	Simultaneous	Benchtop UV-C (Incident dose: 6.46 kJ m^{-2})	US (40 kHz; 13.78 W/L)	TAM	2.3	—	Esua et al. (2018)
				Y&M	3.4		
Cloudy apple juice (pH:3.8; 11.1 °Brix)	Simultaneous	Multifrequency *Dynashock* US (28/4/100 kHz; 1-ms time interval; up to 30 min)	Benchtop UV-C (15-W lamp; 36-cm distance)	Cocktail of 5-strain *E. coli* (O157:H7)	3.2	Synergistic	Gabriel (2012)
Orange juice (pH:3.7; 11.1 °Brix)	Simultaneous	Multifrequency *Dynashock* US (28/4/100 kHz; 1-ms time interval; up to 30 min)	Benchtop UV-C (15-W lamp; 36-cm distance)	Cocktail of 5-strain *E. coli* (O157:H7)	3.7	HI/single US	Gabriel (2015)
Cloudy apple juice (pH:3.8; 11.1 °Brix)					4.5	HI/ single US	
Ham slices (90 × 90 × 2 mm)	Simultaneous	NIR (200.4 μW cm^{-2} nm; 13.5-cm distance; 70 s)	Benchtop UV-C (incident dose: 130 mJ cm^{-2}; 13.5-cm distance)	Cocktail of 3-strain *E. coli* (O157:H7)	3.6	More than additive	Ha and Kang (2015)
				Cocktail of 3-strain *S.* Typhimurium	4.2	More than additive	
				Cocktail of 3-strain *L.monocytogenes*	3.4	More than additive	

Powdered infant formula (a_w: 0.42)	Simultaneous 23-rpm mixing	NIR (141.8 μW cm^{-2} nm; 13.5-cm distance; 420 s)	Benchtop UV-C (incident dose: 1100 mJ cm^{-2}; 13.5-cm distance)	Cocktail of 3-strain *C.sakazakii*	2.8	Synergistic	Ha and Kang (2014)
Date palm	Consecutive	OW (ORP: 390 mv; O_3: 0.55 mg l^{-1}; 2 min; 15 °C)	Benchtop UV-C (incident dose: 622 mJ cm^{-2})	TAM	0.9	HI/single UV-C	Jemni et al. (2014)
				Y&M	1.6		
				TC	0.8		
		NEW (pH: 7.0; ORP: 870 mv; free Cl: 100 mg l^{-1}; 2 min; 15 °C)		TAM	1.1		
				Y&M	1.3		
				TC	0.5		
		AEW (pH:11.3; ORP: -880 mv; free Cl: 1.83 mg l; 2 min; 15 °C)		TAM	0.9		
				Y&M	1.2		
				TC	0.8		
Blueberries calyx	Consecutive	Ozone (4000 mg l^{-1}; 1 min)	Benchtop UV-C (Incident dose: 954 mJ cm^{-2}; 0.9-cm distance)	Cocktail of 5-strain *E. coli* 0157:H7	3.1	Additive	Kim and Hung (2012)
		Benchtop UV-C (954 mJ cm^{-2}; 0.9-cm distance)	Ozone (4000 mg l^{-1}; 1 min)		3.5	Antagonist	
Buffered peptone water (pH:7.2)	Simultaneous	Ohmic heating (13.4 Vrms cm^{-1}; 0.05 duty ratio; 500 Hz; 50s; 60 °C)	Benchtop UV-C (Incident dose: 45.6 mJ cm^{-2})	Cocktail of 3-strain *E. coli* 0157:H7	2.5	Synergistic	Kim et al. (2019)
				Cocktail of 3-strain *S.* Typhimurium	2.0		
				Cocktail of 3-strain *L. monocytogenes*	1.8		

(*Continued*)

Table 2 (Continued)

Item	Experimentals/remarks[a]				Microorganism[b]	Log (N/N0)[c]	Effect[d]	References
Tomato juice (pH:3.6; 11.8°Brix)	Simultaneous	Ohmic heating (13.4 Vrms cm^{-1}; 0.05 duty ratio; 500 Hz; 210 s; 63 °C)		Benchtop UV-C (Incident dose: 191.5 mJ cm^{-2})	Cocktail of 3-strain *E. coli* 0157:H7	3.8		
					Cocktail of 3-strain *S.* Typhimurium	2.2		
					Cocktail of 3-strain *L. monocytogenes*	2.7		
Raw salmon fillets (10 g; 4×2×1 cm)	Consecutive	Benchtop UV-C (Incident dose: 3100 mJ cm^{-2})		US (45Hz; 1 min)	Cocktail of 3-strain *L. monocytogenes*	0.8		
					TAM/TC	0.6/0.6	Synergistic/ additive	Mikš-Krajnik et al. (2017)
					Y&M	0.4	Synergistic	
				AEW (pH:2.6; ORP: 11 490 mV; free chlorine: 65 ppm; 1 min)	Cocktail of 3-strain *L. monocytogenes*	0.6	Less than additive	
					TAM/TC	0.4/0.1	Synergistic/ antagonist	
					Y&M	—	Synergistic	
				US (45 Hz; 1 min) + AEW (pH:2.6; ORP: 11490 mV; free chlorine: 65 ppm; 1 min)	Cocktail of 3-strain *L. monocytogenes*	0.8	Less than additive	
					TAM/TC	0.6/0.5	Synergistic/ additive	
					Y&M	—	Indifferent	
Fresh-cut pomegranate arils	Consecutive	Hot water immersion (55 °C; 30 s)	Benchtop UV-C (incident dose: 0.45 mJ cm^{-2})	Passive MAP (14-day storage; 5 °C)	TAM	0.1	Indifferent	Maghoumi et al. (2012)
					Yeasts	1.0	Less than additive	
					Molds	0.9	Less than additive	

		Benchtop UV-C (incident dose: 0.45 $mJ\,cm^{-2}$		High oxygen MAP (100 $kPaO_2$; 14-day storage; 5 °C)	TAM	0.7	Less than additive	
					Yeasts	1.3	Less than additive	
					Molds	0.5	Indifferent	
		Hot water immersion (55 °C; 30 s)	Benchtop UV-C (incident dose: 0.45 $mJ\,cm^{-2}$)	High oxygen MAP (100 $kPaO_2$;14-day storage; 5 °C)	TAM	2.2	Additive	
					Yeasts	1.0	Less than additive	
					Molds	2.2	Additive	
Kailan-hybrid broccoli	Consecutive	NEW (100 $mg\,l^{-1}$ Cl_2; pH:7.0; ORP: 900 mV)		Benchtop UV-C (incident dose: 600 $mJ\,cm^{-2}$; 19-day storage; 5 °C)	TAM	1.8	Indifferent	Martínez-Hernández et al. (2013)
					Psychrophilic	2.2	Antagonist	
					Y&M	3.2	Synergistic	
		UV-C (incident dose: 600 $mJ\,cm^{-2}$)		High oxygen MAP (90 $kPaO_2$;19-day storage; 5 °C)	TAM	1.6	Indifferent	
					Psychrophilic	3.0	Indifferent	
					Y&M	0.4	Indifferent	
	Consecutive	NEW (100 $mg\,l^{-1}$ free Cl_2; pH:7.0; ORP: 900 mV)	UV-C (incident dose: 600 $mJ\,cm^{-2}$)	High oxygen MAP (90 $kPaO_2$;19-day storage; 5 °C)	TAM	1.7	Indifferent	
					Psychrophilic	2.9	Indifferent	
					Y&M	3.2	Synergistic	
Milk (pH: 6.4; 3% fat)	Simultaneous	US (24 kHz; < 30 °C; 120 μm)		Immersed UV-C system (incident dose: 11.8 $mJ\,cm^{-2}$)	TAM/TC	4.8/5.3		Şengül et al. (2011)
Apple juice (pH: 3.3; 11°Brix)	Consecutive	US (35 kHz; 120–480 W; 5 min)		Benchtop UV-C (Incident dose: 2010 $mJ\,cm^{-2}$)	*A.acidoterrestris* spores (CCT 4384)	5.0	Additive	Tremarin, et al. (2017)

(*Continued*)

Table 2 (Continued)

Item	Experimentals/remarks[a]			Microorganism[b]	Log (N/N0)[c]	Effect[d]	References
Apple/cranberry juice blend (90 : 10)	Consecutive	Annular flow UV-C (incident dose: 5.3 mJ cm^{-2})	PEF (18 Hz; 34 kV cm^{-1}; 93 μs)	*E. coli* K 12 (DSM 1607)	6.0	Additive	Palgan et al. (2011)
				E. coli K 12 (DSM 1607)	5.9		
			MTS (20 kHz; 23 μm; 400 kPa; −15 °C)	*P. fermentans* (DSM 70090)	6.0		
				P. fermentans (DSM 70090)	5.9		

[a] US (Ultrasound); NIR Near-Infrared Heating; MF (Microfiltration); PEF (Pulsed Electric Fields); NTAP (Nonthermal Atmospheric Plasma); HPEF (High intensity Pulsed Electric Fields); MAP (Modified Atmosphere Packaging); NEW (neutral electrolyzed water; OW: ozonized water; AEW: alkaline electrolyzed water; ORP (Oxidation/reduction potential); MTS (Manothermosonication)

[b] *Escherichia coli*; *Alicyclobacillus acidoterrestris*; *Cryptosporidium parvum*; *Staphylococcus aureus*; *Salmonella* Typhimurium ; *Salmonella* Enteritidis; *Listeria monocytogenes; Listeria innocua.*

[c] ND: the combined effect could not be determined due to the achievement of complete inactivation; HI/: Higher inactivation than, Increased/MF: Increased inactivation compared with single MF, Increased/UV-C: Increased inactivation compared with the single UV-C treatment.

[d] HI/: Higher inactivation than; Increased/MF: Increased inactivation compared to single MF; Increased/UV-C: Increased inactivation compared to the single UV-C treatment.

immersed UV-C system, similar to those commonly used for water treatment (Şengül et al. 2011). A broad spectrum of combined effects from indifferent to synergistic inactivation, was found, which mainly depended on the considered food product. For example, Gabriel (2012 and 2015) evaluated the inactivation of a cocktail of different strains of *E. coli* O157:H7 in apple and orange juices treated by multifrequency *Dynashock* US (28/45/100 kHz, 1-ms time interval) combined with UV-C using the same experimental conditions. They found that the inactivation response to the combined treatment varied from higher than single US to synergistic and mainly depended on the evaluated fruit juice (Table 2). In other cases, the overall effect of the proposed combined processing was strongly dependent on the target microorganism. For instance, Mikš-Krajnik et al. (2017) found a less than additive effect when combining acidic electrolyzed water (AEW) for the treatment of raw salmon fillets contaminated with a three-strain cocktail of *L. monocytogenes*. In contrast, a synergistic effect in the case of total aerobic mesophiles and yeasts and molds while an antagonistic effect in the case of total coliforms were determined (Table 2). The shape of the inactivation curve, when studied, showed dependence on the considered treatment. For instance, in the studies reported by Gabriel (2012, 2015) previously mentioned, the authors reported single inactivation curves that displayed a biphasic behavior with a significant shoulder at the beginning of the treatment (Table 2). However, when the US treatment was assisted by UV-C (1 lamp, 15 W), the mentioned shoulder completely disappeared in the case of the apple juice but remained in the case of the orange juice. This response was attributed to the lower UV transmittance of the orange juice compared with the apple juice that impeded the light from reaching the target microorganisms, thus decreasing the efficacy of the UV-C treatment. In the same fashion, (Char et al. 2010) evaluated the inactivation of *E. coli* in an orange juice treated by US (20 kHz) and UV-C (1870 $mJ\,cm^{-2}$) under both consecutive and simultaneous arrangements. They reported that the inactivation curves obtained by single US exhibited a first-order behavior while those that corresponded to single UV-C had a clear upward concavity. This inactivation profile suggested that microbial inactivation was higher during the first minutes of the annular flow UV-C treatment while some more resistant members within the microbial population were harder to inactivate (Table 2).

Some authors incorporated the use of MAP to contribute to extend product shelf life during cold storage. For example, (Martínez-Hernández et al. 2013) treated Kailan-hybrid broccoli with NEW (100 $mg\,l^{-1}$ free Cl_2) followed by UV-C (600 $mJ\,cm^{-2}$) and stored the product in a high-oxygen MAP (100 kPa O_2; 5 °C). They observed, after 19-day storage, that those systems including high-oxygen MAP packaging succeeded in preventing yeast and mold growth and the recovery of psychrophilic bacteria (Table 2). The same results were obtained by Maghoumi et al. (2012), who combined UV-C (0.45 $mJ\,cm^{-2}$) with hot water (55 °C) or with high-oxygen MAP (100 kPa O_2; 14 days; 5 °C) or a ternary approach combined with UV-C involving both factors, hot water and high-oxygen MAP, consecutively. They concluded that only when combining the three stressors, total mesophilic aerobes and molds were significantly prevented from growth. On the contrary, when a passive MAP packaging was used, total aerobic aerobes were completely recovered (Table 2).

Hurdle strategies involving the use of UV-C and physical methods such as US, PEF, MF, and MTS did not significantly modify the pH, titratable acidity, soluble solids, TPC TAA values, color, or texture of the treated samples. For instance, Walkling-Ribeiro et al. (2008)

assessed the effect of UV-C (30-W lamp, 26-cm distance; 30 minutes) combined with PEF (36–40 kV cm^{-1}, 75–100 μs, 43–46 °C) for the treatment of apple juice (pH: 3.5). They did not observe modifications on the soluble solids, conductivity, pH, nor color of the juice. On the contrary, they reported a decrease of 44% in the content of ascorbic acid and TPC values. Similarly, Caminiti et al. (2011) reported no changes in pH, °Brix, TPC, TAA, color, and nonenzymatic browning of an apple–cranberry juice blend (90 : 10 v/v, pH: 3.2, 7 °Brix) subjected to UV-C (annular thin-film reactor, 5300 mJ cm^{-2}) combined with lab-scale PEF (34 kV cm^{-1}, 18 Hz, 93 μs) or MTS (20 kHz; 23 μm, 4 bar, 58 °C). However, they observed a 19–24% of loss in the anthocyanin cyanidin-3-glucoside content after both treatments. Similarly, Choi et al.(2015) evaluated the effect of UV-C (1000 mJ cm^{-2}) and passive or active modified atmosphere (5.3% CO_2 + 5.5% O_2) stored at 4 and 20 °C of cherry tomatoes (*Lycopersicon esculentum* L. species; Beta Tiny), without observing changes in color, firmness, nor lycopene content, immediately after treatment or during storage at both studied temperatures. On the same trend, Kim et al. (2019) reported no significant changes in color nor lycopene content of a tomato juice exposed to a combined treatment involving UV-C (191.5 mJ cm^{-2}) and ohmic heating (13.4 V_{rms} cm, 63 °C). On the contrary, Caminiti et al. (2012) observed a slight decrease in the TPC (7%) and TAA (8%) values of a carrot–orange juice blend (pH: 3.8) treated by UV-C (10.6 J cm^{-2}) and MTS (20 kHz, 63 °C, 400 kPa). Outstandingly, an increase in the bioactive compounds' extractability was observed in some cases after combinations of UV-C and US. For instance, Santhirasegaram et al. (2015) obtained an increase in the carotenoid (15%), flavonoid (35%), TPC (37%), and TAA values of Chokanan mango juice (pH: 4.6) treated by UV-C (353 mJ cm^{-2}) and US (40 kHz). Additionally, in agreement with the previous studies, they did not observe changes in the soluble solids, pH, and titratable acidity. Similarly, Esua et al. (2018) determined an increase in TPC (55%) and TAA (36%) after subjecting tomatoes to UV-C (72–108 mJ cm^{-2}) and US (40 kHz; 80–1200 μs). Using a different strategy, Jemni et al. (2014) studied the influence of a combined treatment involving UV-C (622 mJ cm^{-2}) and ozonated water (0.5 mg L^{-1}, 2 minutes, 15 °C), AEW (pH 11.3, 2 minutes) and NEW (pH 11.3, 2 minutes) on certain quality parameters of Date fruits (*Deglet Nour cv.*). Even though they did not observe changes in pH, titratable acidity, and TPC for all the combinations, all treatments prevented from weight loss (67–68%) and increased firmness (48–69%) compared with untreated samples.

In general terms, the effect of UV-C combined with other novel intervention technologies is highly dependent on the target microorganism and the food matrix, thus it seems unlikely to generalize about the best conditions and/or combinations to be used. However, it can be inferred from the examples commented above that some smart combinations can be proposed to obtain products preserved by more sustainable technologies that can guarantee similar or even higher nutritional quality than the thermally treated ones.

3.1.3 UV-C Combined with the Addition of Natural Antimicrobials

The use of natural antimicrobials is gaining popularity as a replacement of synthetic preservatives, which are highly associated to certain health issues (e.g. allergies). Additionally, its use may provide a great diversity of flavors to better attract consumer's attention. However, the evaluation of the antimicrobial capacity of some additives, used in combination or not, in buffers, culture media, or water is, generally, a poor measure of its performance in more sophisticated food systems. This concept acquires a higher relevance in the case of natural antimicrobials, as the extrapolation is more troubled. In many cases, the interactions with food components such as lipids, aldehydes, proteins, and many macromolecules, and with other stressors, are the responsible for reducing the observed antimicrobial activity (Gutierrez et al. 2009). Therefore, microbial inhibition or inactivation in real food products demands significantly higher antimicrobial concentrations than in laboratory media and frequently, it will be above the acceptable sensory thresholds. This issue may be overcome using a hurdle approach that includes the use of one or more natural antimicrobials having different targets. When more than one antimicrobial is chosen to be combined with a given traditional or emerging technology, the antimicrobial combinations can be selected based on their effectiveness against those target microorganisms that are resistant to the antimicrobials used on a single base (Ferrario et al. 2020). Four mechanisms have been proposed for describing those antimicrobial interactions that produce synergism: (i) inhibition of protective enzymes, (ii) sequential inhibition of a given biochemical pathway, (iii) use of cell-wall active agents to enhance the uptake of other antimicrobials, and (iv) combinations of cell-wall active agents (Schenk et al. 2018). Additionally, antimicrobial combinations can be designed to provide a broad spectrum of possibilities in the preservation of foods and sensory acceptance, as well as to find synergistic combinations. Nonetheless, the knowledge of the specific mode of action of naturally occurring agents on microbial metabolic activities still needs more research, even when they were the only stressors involved in a food processing design.

Relatively less work has been published to date on the combined action of mixtures of natural occurring antimicrobials and UV-C light for the treatment of solid and liquid food products, in comparison to literature referred to the single inactivation effects derived from the use of these antimicrobials. For instance, Sommers et al. (2010) studied a binary consecutive treatment, which involved UV-C surface decontamination in a rotary food conveyor hood (500 mJ cm^{-2}) and the subsequent immersion in a solution (5%) of the GRAS (generally recognized as safe) antimicrobial Lauric–Arginate–Ester (LAE) for a few seconds. They informed that decombined treatment inactivated between 2.3 and 2.8 log cycles immediately after treatment and between 3.6 and 4.1 log cycles after 12 weeks of refrigerated storage (10 °C) of freshly manufactured frankfurters inoculated with three-strain cocktails of *Listeria monocytogenes*, *Salmonella enterica*, and *Staphylococcus aureus*. These results demonstrated a higher effectiveness of the combined treatment, compared with the single ones. Furthermore, no changes in some color and texture parameters of the frankfurters were determined, compared with the untreated system. Similarly, Ferrario et al. (2011) reported that the incorporation of 1500-ppm vanillin plus 50–100 ppm of citral to an orange juice treated by UV-C light (2500 mJ cm^{-2}) in a laboratory-scale annular thin-flow reactor delayed during 13 days of storage (5 °C) the recovery of *E. coli* ATCC 35218, *Listeria innocua* ATCC 33090, and *Zygosaccharomyces bailii* NRRL 7256, which were refractory to the single UV-C treatment. Similarly, Tawema et al. (2016) studied *Listeria monocytogenes* (HPB 2812 serovar 1/2a), *E. coli* O157: H7 (EDL933), and total yeast and mold inactivation achieved in fresh-cut cauliflower treated by UV-C (500–1000 mJ cm^{-2}) with the addition of two formulations of some natural antimicrobials (lemongrass, citral, lactic acid, and oregano) in different ratios and stored during 14 days (5 °C). These authors reported an increase of up to 0.66–3.27 log reductions for the combined treatment compared with the UV-C-treated samples. The evaluation of other antimicrobials in combination with UV-C was also reported by Choi et al. (2017), who found additive inactivation of natural microbiota in dongchimi juice added with 0.1% grape seeds extract and treated by UV-C (600 mJ cm^{-2}). The combined treatment led up to 2.2 and 3.0 log cycles of reduction, without altering the titratable acidity and pH of the juice.

Less attention has been given to the study of the inactivation mechanisms of UV-C combined with the addition of natural antimicrobials. Park et al. (2018b) used a combination of UV-C (120 mJ cm^{-2}) and an emulsion composed of 0.02% cinnamon bark oil plus a cationic surfactant (0.002% cetylpyridinium chloride) for washing fresh-cut red chard. They observed additive effects for the inactivation of two-strain cocktails of *L. monocytogenes* or *S.* Typhimurium, achieving up to 1.3–1.4 additional log reductions, respectively. Furthermore, they reported the existence of membrane damage evaluated by scanning electron microscopy. Furthermore, they did not observe changes compared with control samples in color, appearance, odor, texture, TPC, nor overall acceptability. Recently, Ferrario et al. (2020) observed a synergistic response in the inactivation of *E. coli* ATCC 25922, *L. plantarum* ATCC 8014, and *S. cerevisiae* KE 162 in a turbid orange–banana–mango–kiwi–strawberry juice blend (pH: 3.7; 13.9 °Brix) processed in a pilot-scale coiled tube UV-C reactor (390 mJ cm^{-2}) loaded with encapsulated vanillin (1000 ppm) and citral (100 ppm), after which significant inactivation was achieved (7.6; 5.6 and 6.4 log reductions, respectively). When an orange–tangerine juice with worst optical characteristics was tested (pH: 3.5; 12.4 °Brix), a synergistic response was observed only for *E. coli* ATCC 25922, which was the most sensitive assayed strain.

Although the use of naturally occurring antimicrobials combined with UV-C has been less explored, the results reported so far show the potential of this hurdle approach. On the one hand, the investment is much lower than that required for the use of other technologies mentioned in this chapter. On the other hand, it would lead to products preserved by more natural means compared with the use of chemical additives, opening a diversity of products, specially designed for those adventurous consumers interested in food and beverages with distinctive flavor notes.

3.1.4 UV-C Combined with Sanitizers

UV-C irradiation has been applied in combination with a great variety of chemical agents, mainly in consecutive arrangement, but also simultaneously to inactivate pathogenic or spoilage microorganisms in different model systems, foods, and beverages. Table 3 illustrates the inactivation achieved immediately after UV-C exposure in a hurdle strategy combined with diverse sanitizers, mostly applied to fruit and vegetables such as strawberries (Kim et al. 2010), pear (Schenk et al. 2012), lettuce (Lee et al. 2012), among others, but also apple juice (Gouma et al. 2015b), fish (Lee et al. 2019), and stainless steel chips (Ha and Ha 2010, 2011). Mainly benchtop UV-C devices were used, almost always reporting the incident UV-C dose measured with a radiometer. A broad spectrum of UV-C doses was reported, in a range from 30 to 4800 mJ cm^{-2}, with most of the research studies using doses around 500 mJ cm^{-2} (Table 3). The chemical solutions used for decontamination were generally applied by immersion of the fruit or vegetable produce before or after the UV-C treatment. However, the modalities of spray (Lee et al. 2019) and fine mist (Hadjok et al. 2008) were also employed. The selected sanitizers were those traditionally used in the industry for the microbial decontamination of fruits and vegetables during the washing step

Table 3 Recent experimental procedures involving the use of UV-C light combined with sanitizers.

Item	Experimentals/remarks			Microorganism[a]	Log (N/N0)[b]	Effect[c]	References
Buckwheat sprouts (13 cm)	Consecutive	Blend 1 : 10 (w/v): ClO_2 ($100\,mg\,l^{-1}$) + Fumaric acid (0.3%) (pH: 2.5; 5 min)	Benchtop UV-C (incident dose: $200\,mJ\,cm^{-2}$; 18-cm distance)	*S.* Typhimurium (ATCC 14028 and KCTC 2514)	2.3	—	Chun and Song (2013)
				E. coli O157:H7 (NCTC 12079)	3.0		
				Native microbiota (TAM)	1.8		
Yam (20/25 cm × 5 cm)	Consecutive	ClO_2 (50 ppm; 10 min)	Benchtop UV-C (incident dose: $500\,mJ\,cm^{-2}$; 18-cm distance)	Native microbiota	2.3 (Y&M)	Less than additive	Chun et al. (2013)
					3.2 (TAM)	Less than additive	
					3.0 (TC)	Synergistic	
Apple juice (pH: 3.6; α: $24.9\,cm^{-1}$)	Simultaneous	Dimethyl dicarbonate ($75\,mg\,l^{-1}$; 3.6 min)	Annular thin-film UV-C (incident dose: $3.9\,J\,ml^{-1}$)	*E. coli* (STCC 4201)	2.9	Synergistic	Gouma et al. (2015a)
Stainless steel chips (0.2 cm × 0.2 cm)	Consecutive	Ethanol (50% v/v; 5 min) (E) NaClO (200 ppm; 2 min) (Cl)	Benchtop collimated-beam UV-C (incident dose: $504\,mJ\,cm^{-2}$)	*B. cereus* (F4810/72)	5.5 (E)/ 4.6 (Cl)	Synergistic	Ha and Ha, (2010, 2011)
				C. sakazakii (KCTC 2949)	6.4 (E)/ 4.9 (Cl)		
				S. aureus (ATCC 35556)	3.7 (E)/ 3.8 (Cl)		
				S. Typhimurium (NO/NA)	4.9 (E)/ 3.9 (Cl)		
				E. coli (ATCC 10536)	5.2(E)/ 4.1 (Cl)		

(*Continued*)

Table 3 (Continued)

Item	Experimentals/remarks			Microorganism[a]	Log (N/N0)[b]	Effect[c]	References
Oyster mushroom (2×2 cm)	Consecutive (20 °C, 5-min chemical contact)	Ethanol (70% v/v)	Benchtop collimated-beam UV-C (incident dose: 504 mJ cm^{-2})	*B. cereus* (F4810/72)	5.0	Synergistic	Ha et al. (2011)
		H_2O_2 (2000 ppm)			3.7		
		NaClO (200 ppm)			3.6		
Oyster mushroom (2 cm × 2 cm)	Consecutive (20 °C, 5-min chemical contact)	Ethanol (70% v/v)	Benchtop collimated-beam UV-C (incident dose: 504 mJ cm^{-2})	*S. aureus* (ATCC 35556)	4.1	Synergistic	Ha et al. (2011)
		H_2O_2 (2000 ppm)			2.8		
		NaClO (200 ppm)			3.3		
Strawberries (*Fragaria* × *ananassa* Duch.) (18 g)	Consecutive (20 °C, 5-min chemical contact)	ClO_2 (50 mg l^{-1}; pH: 3.0)	Benchtop UV-C (incident dose: 500 mJ cm^{-2}; 17.5-cm distance)	TAM	2.1	—	Kim et al. (2010)
				Y&M	1.9		
		Fumaric acid (0.5%; pH 2.3)		TAM	2.3		
				Y&M	2.0		
Plums (20 g)	Consecutive	ClO_2 (15 ppm; 20 min)	Benchtop UV-C (incident dose: 1000 mJ cm^{-2}; 17.5-cm distance)	Cocktails of 3-strain *L. monocytogenes* or 2-strain *E. coli* O157:H7	3.3–4.0	More than additive	Kim and Song (2017)
Romaine lettuce (L) (2 cm × 2 cm) or Kale (K) (2 cm × 2 cm)	Consecutive	ClO_2 (50 µl l^{-1}; 5 min)	UV-C (incident dose: 1000 mJ cm^{-2}; 18-cm distance)	Cocktail of 3-strain *S.* Typhimurium	>6.1 (K)–6.3 (l)	More than additive (K)/ synergistic (L)	Lee et al. (2012)
				E. coli O157:H7 (NCTC 12079)	>6.2(K)–6.3(l)		
Gwamegi (3.0 cm × 4.0 cm × 0.5 cm)	Consecutive	Benchtop collimated-beam (incident dose: 4800 mJ cm^{-2})	Ethanol (70%; 2 min)	*B. cereus* (F4810/72)	2.7	—	Lee et al. (2019)
				Cocktail of 4-strain *E. coli*	3.0		

Kailan-hybrid broccoli florets (15–18 cm)	Consecutive	Peroxyacetic acid (100 mg l^{-1}; pH: 5.3; 5 °C; 2 min)	Benchtop UV-C (incident dose:50 mJ cm^{-2}; 15-cm distance)	*E. coli* (CETC 7619)	3.0	Additive	Martínez-Hernández et al. (2013)
				S. enteriditis (CETC 4300)	3.2		
Strawberries (*Seolhyang* 15 g)	Consecutive	Fumaric acid (0.5% w/v; 5 min)	Benchtop UV-C (incident dose:500 mJ m^{-2})	Native microbiota	1.6 (TAM)	—	Shin et al. (2011)
					1.6 (Y&M)		
Strawberry (*Goha;* 10 g)	Consecutive	ClO_2 (50 ppm; 5 min)	Benchtop UV-C (incident dose:500 mJ m^{-2})	Native microbiota	1.5 (TAM)	—	Shin et al. (2012)
					1.9 (Y&M)		
Spinach leaves (S) (5 cm × 2 cm) Tomato (T) (5 cm × 2 cm)	Simultaneous (20 min; 22 °C)	Benchtop UV-C (incident dose: 85 mJ cm^{-2})	ClO_2 gas (5 ppmV)	Cocktail of 3-strain *L. monocytogenes*	3.1 (S)/2.7 (T)	Additive	Park et al. (2018a)
				Cocktail of 3-strain *S.* Typhimurium	3.7 (S)/4.3 (T)		
				Cocktail of 3-strain *E. coli* O157:H7	4.4 (S)/4.8 (T)		
Papaya (0.7 cm × 3.5 cm × 2.5 cm; pH:5.5)	Consecutive (25 °C; 1-min contact)	Calcium lactate (1% w/v) + ascorbic acid (0.5% w/v) + malic acid (1.5% w/v)	Benchtop UV_C (incident dose: 576 mJ cm^{-2}; 15-cm distance)	*L. monocytogenes* (CVCM 449)	2.5	Synergistic	Raybaudi-Massilia et al. (2013)
				S. Poona (CVCM 1921)	3.8		
Cherry tomatoes	Consecutive	ClO_2 (10 ppm; 5 min)	Benchtop UV (incident dose: 500 mJ cm^{-2}; 18-cm distance)	*E. coli* O157:H7 (NCTC 12079)	$^{>}$5.9–6.2	—	Song et al. (2011)
				Cocktail of 3-strain *S.* Typhimurium	$^{>}$6.2		

(*Continued*)

Table 3 (Continued)

Item	Experimentals/remarks			Microorganism[a]	Log (N/N0)[b]	Effect[c]	References
Whole plum tomatoes	Consecutive (20 °C, 2-min chemical contact)	Benchtop UV-C (incident dose: 60 mJ cm^{-2}; 10-cm distance)	Chlorinated water (200 ppm)	Y&M	1.4	Additive	Mukhopadhyay et al. (2015)
			Aqueous ozone (1.5 ppm)		2.5		
			H_2O_2 (3% v/v) + 0.02 mM EDTA + nisin (20 mg ml^{-1})		3.1		
			Lactic acid (1% w/v)		2.6		
			Citric acid (1% w/v)		3.2		
			Lactic acid + citric acid (1% w/v)		˃4.6		
Pear discs (3-cm diameter; pH: 4.2)	Consecutive	H_2O_2 (3% v/v; pH: 3.0; 5 min)	Benchtop UV-C (delivered dose:370 mJ cm^{-2}; 10-cm distance)	*E. coli* (ATCC 11229)	3.6	Higher than single UV-C	Schenk et al. (2012)
				L. innocua (ATCC 33090)	2.4		
				Z. bailii (NRRL 7256)	2.7		
Iceberg lettuce leaves	Simultaneous (50 °C, 30-s chemical contact)	H_2O_2 (1.5% v/v; fine mist)	Benchtop UV_C (incident dose: 38 mJ cm$^{-2;}$ 35-cm distance)	*S.* Montevideo (P2)	4.1	—	Hadjok et al. (2008)
				P. fluorescens (P30)	4.2		
				E. coli O157:H7	3.9		

Evaluation of microbial inactivation effectiveness in different media.
[a] *Salmonella* Typhimurium; *Escherichia coli; Bacillus cereus; Cronobacter sakazakii; Staphylococcus aureus ; Listeria monocytogenes; Salmonella enterica; Listeria innocua; Zygosaccharomyces bailii; Salmonella* Montevideo; *Pseudomonas fluorescens.*
[b] TAM: total aerobic mesophiles; TC: total coliforms; Y&M: yeasts and molds.
[c] Type of combined effect determined according Eq. (5) when possible; –: single effects were not determined.

(e.g. chlorine derivatives; acids; ozone; etc.) applied prior to other industrial unit operations or commercialization. Other emerging sanitizers, such as mixtures of chemicals involving the additional use of calcium lactate, malic acid (Raybaudi-Massilia et al. 2013), or nisin, a natural antimicrobial, (Mukhopadhyay et al. 2015) were also employed.

With regard to the assessment of nutritional and sensory quality of food products subjected to UV-C combined with sanitizers, in general, no changes in color, pH, titratable acidity, texture, TAA, TPC, overall appearance, and odor were observed in treated compared with control samples. For instance, the combination of UV-C (4800 mJ cm^{-2}) and ethanol (70%, 2 minutes) did not alter the color and pH of fresh-cut gwamegi (Lee et al. 2019). However, Schenk et al. (2012) reported a decrease in hardness of pear discs treated by benchtop UV-C (370 mJ cm^{-2}) and H_2O_2 (3%, 5 minutes).

In reference to the microbial challenge studies summarized in Table 3, the inactivation of different pathogens, surrogates, and native microbiota was analyzed immediately after treatment (Table 3). In some studies, growth dynamics along product refrigerated storage was evaluated, generally observing between 1 and 2 log cycles of recovery (data not shown). Only two of the collected studies evaluated the inactivation kinetics during the application of the combined treatment without mathematically characterizing the inactivation curves. Gouma et al. (2015a) and Raybaudi-Massilia et al. (2013) evaluated the inactivation of *E. coli* or *L. monocytogenes* and *S. enterica* in apple juice and papaya, respectively, subjected to UV-C combined with sanitizers in different arrangements. Both researches reported inactivation curves with downward concavity and presence of a significant shoulder after single UV-C exposure, whereas the shoulder length was shortened when the combined treatment was applied (data not shown). This type of response in the case of single UV-C treatments and the decrease of the shoulder length when an additional stressor is introduced has been previously commented in this chapter. The hurdle strategies summarized in Table 3 produced additive to synergistic inactivation effects when applied ClO_2, NaClO, ethanol (50% or 70%), H_2O_2, or calcium lactate combined with ascorbic and malic acid or dimethyl decarbonate. Outstandingly, between 2.8 and 6.4 log reductions of single cultures or representative cocktails of target inoculated microorganisms were achieved. For instance, Lee et al. (2012) examined the impact of a combined treatment, which involved the use of ClO_2 (50 μl L^{-1}; 5 minutes) followed by UV-C (1000 mJ cm^{-2}) on the inactivation of cocktails of *S.* Typhimurium and *E. coli* O157:H7 in lettuce and kale leaves. They reported synergistic and more than additive inactivation effects for lettuce and kale respectively, achieving more than 6.0 log reductions (Table 3) while no changes in color, odor, nor overall appearance were detected, compared with untreated samples. In the same trend, Ha et al. (2011) obtained a synergistic inactivation of *B. cereus* spores (4–5 log reductions) on fresh-cut oyster mushrooms after the exposure to a consecutive treatment, which combined ethanol (70% v/v) or H_2O_2 (2000 ppm) or NaClO (200 ppm) followed by UV-C (504 mJ cm^{-2}) (Table 3). On the other hand, Park et al. (2018a) studied the simultaneous combination of UV-C (85 mJ cm^{-2}) with ClO_2 on different strains of *S.* Typhimurium, *L. monocytogenes*, and *E. coli* O157:H7 on spinach and tomatoes and reported an additive inactivation effect when the combined treatment was applied (Table 3), while no changes in color and texture were determined. The effect of UV-C treatments combined with sanitizers on native microbiota inactivation was less studied achieving in general, between 1.5 and 3.0 log reductions. For instance, Chun et al. (2013) obtained less-than-additive inactivation effects in yam treated by a consecutive treatment involving ClO_2 and UV-C. Exceptionally, they observed a synergistic response for the reduction of total coliforms (Table 3). In addition, no significant changes in weight loss and lightness were found. Shin et al. (2011) and Shin et al. (2012) examined the decontamination effects on the native microbiota of strawberries consecutively treated by fumaric acid (0.5%, 5 minutes) or ClO_2 (50 ppm, 5 minutes) followed by UV-C (500 mJ cm^{-2}). Both consecutive treatments induced 1.5–1.6 log reductions of total aerobic mesophiles while 1.6–1.9 log reductions of yeast and mold populations were achieved (Table 3). They reported no changes in color, appearance, firmness, odor, nor overall acceptability, compared with control samples. However, as the single treatments were not assayed, it was not possible to determine the type of the combined inactivation effect.

Some successful combined technologies involving the use of sanitizers and UV-C are mentioned in this section. These approaches greatly reduced the concentration of chemicals traditionally employed by industrial processors, which represents a significant advantage for consumer health. Moreover, they achieved moderate-to-high microbial inactivation while demonstrated not to significantly alter food quality.

4 Conclusions and Future Perspectives

There is no doubt that the implementation of UV-C-based technologies may play a significant role in food preservation. Most of the combined technologies mentioned in this chapter were able to achieve the microbial inactivation required by the FDA and UE regulations, while well preserved the food quality of different foods and beverages submitted to those treatments. All these UV-C-based combinations intended to produce better-quality products, in a more sustainable way, by reducing or eliminating the use of chemicals, or using milder and/or green technologies and in many cases, with low investment. As stated in this chapter, despite few exceptions, UV-C light and its derived combined intervention procedures, have not yet found their way into widespread commercial use for preservation purposes. Therefore, the future challenges associated with their research, development, implementation, and commercialization are still numerous.

Several key factors that influence UV-C processing effectiveness remain unsolved. To start with, there is a lack of systematic studies about safety and quality of the food products processed by UV-C, used alone or combined with other factors. The need for a better knowledge and unified criteria when studying novel technologies has been highlighted in this chapter, as it has been discussed that the success of each combination is highly dependent on the type of product that is being treated and of the treatment conditions. More information about the inactivation effect on microbial spores, other pathogens, and spoilage agents, immediately after treatment and during storage is demanded. Many tools, such as predictive microbiology, could cooperate in achieving this goal. In addition, further research with regard to the impact of these hurdle technologies upon nutritional, sensory, and functional food stability, immediately after treatment and along product shelf life, is required. For example, studies from literature regarding the effect of UV-C on toxins produced by microorganisms (e.g. micotoxins) or certain food bioactive compounds such as enzymes or vitamins are still scarce.

Other stumbling block is the need for an establishment of equivalent processing conditions and/or comparisons among the different UV-C systems. This drawback arises from the lack of standardization in the evaluation of the UV-C dose and from the great diversity of the proposed UV-C devices and processes. A proper characterization of the UV-C dose in the reactor will allow to determine the best one for each combined preservation process, to validate it and to develop proper devices for an effective treatment. Furthermore, this will avoid overprocessing due to nonuniformity issues Overall, the proper knowledge of UV-C dose and dose distribution in the reactor enables to estimate the pros and cons for UV-C industrial adoption, based on the specific product requirements.

All the future goals could be better fulfilled by a more extensive look of the hurdle concept, which encompasses multiple disciplines (microbial ecology, chemistry, sensory, etc.) with a food safety engineering perspective, to address microbial and chemical safety challenges. As shown in this chapter, most of the binary and ternary nonthermal combinations reported in literature have shown so far additive inactivation effects more than synergistic ones. Even when synergistic effects on microbial inactivation were observed, not always the proposed technology was convenient. For instance, despite the combination of UV-C and mild heat in the range from 55 to 60 °C yielded synergistic inactivation effects, a negative impact on the nutritional quality of the treated food products was determined after the exposure to these combinations of hurdles. On the contrary, when the temperature assisting the UV-C treatment was in the range from 50 to 55 °C, more than additive to synergistic inactivation effects were observed for some relevant microorganisms, without altering the physicochemical parameters of the food products, and consequently, there is a niche of opportunities in this field to further explore.

Ultimately, the intelligent finding of other useful UV-C-based combinations that may guarantee product safety and achieve reasonable product shelf life and process efficiency is suggested. Additional factors should be evaluated as, for example, the proper packaging that should maintain or even reinforce the overall quality of the food product and in the case of beverages, the possible incorporation of oxygen during the filling process.

On the positive side, the information provided in this chapter evidences that there is a good opportunity for a variety of combined preservation UV-C-based approaches, which could be capitalized on the biochemical and cellular damage of the microbial populations caused by the involved stressors. Therefore, the implementation of UV-C under a hurdle approach gives the opportunity to produce value-added products implementing a low-cost and low-investment technology, even for those products with unfavorable optical characteristics, where the use of single UV-C fails to provide an effective microbial inactivation.

Acknowledgments

The authors want to thank the Universidad de Buenos Aires, CONICET, ANPCyT of Argentina, and BID for financial support.

References

Abdullahi, N., Ariahu, C., and Abu, J. (2016). Critical review on principles and applications of hurdle technology in food preservation. *Annals. Food Science and Technology* 17 (2): 485–491.

Alzamora, S.M., López-Malo, A., Tapia, M.S., and Welti-Chanes, J. (2016a). Minimally processed foods. In: *Encyclopedia of Food and Health* (eds. P. Finglas and F. Toldrá), 767–771. New York, USA: Elsevier.

Alzamora, S., Guerrero, S., Raffellini, S. et al. (2016b). Hurdle technology in fruit processing. In: *Fresh-Cut Fruits and Vegetables Technology, Physiology, and Safety* (ed. S. Pareek), 101–138. Boca Raton, USA: CRC.

Ananta, E., Voigt, D., Zenker, M. et al. (2005). Cellular injuries upon exposure of *Escherichia coli* and *Lactobacillus rhamnosus* to high-intensity ultrasound. *Journal of Applied Microbiology 99* (2): 271–278.

Ansari, J.A., Ismail, M., and Farid, M. (2019). Investigate the efficacy of UV pre-treatment on thermal inactivation of *Bacillus subtilis* spores in different types of milk. *Innovative Food Science and Emerging Technologies 52*: 387–393.

Artichowicz, W., Luczkiewicz, A., and Sawicki, J.M. (2020). Analysis of the radiation dose in UV-disinfection flow reactors. *Water (Switzerland) 12* (1): 231. https://doi.org/10.3390/w12010231.

Bang, H.J., Park, S.Y., Kim, S.E. et al. (2017). Synergistic effects of combined ultrasound and peroxyacetic acid treatments against *Cronobacter sakazakii* biofilms on fresh cucumber. *LWT – Food Science and Technology* 84: 91–98.

Beales, N. (2004). Adaptation of microorganisms to cold temperatures, weak acid preservatives, low pH, and osmotic stress: a review. *Comprehensive Reviews in Food Science and Food Safety 3* (1): 1–20.

Biancaniello, M., Popović, V., Fernandez-Avila, C. et al. (2018). Feasibility of a novel industrial-scale treatment of green cold-pressed juices by UV-C light exposure. *Beverages 4* (2): 29.

Bolton, J. (2020). Action Spectra, A Review. IUVA News. https://iuvanews.com/stories/070317/action-spectra.shtml. Last accessed, March 2021.

Cairns, B. (2006). UV dose required to achieve incremental log inactivation of bacteria, protozoa and viruses. *IUVA News 8* (1): 38–45.

Caminiti, I.M., Noci, F., Muñoz, A. et al. (2011). Impact of selected combinations of non-thermal processing technologies on the quality of an apple and cranberry juice blend. *Food Chemistry 124* (4): 1387–1392.

Caminiti, I.M., Noci, F., Morgan, D.J. et al. (2012). The effect of pulsed electric fields, ultraviolet light or high intensity light pulses in combination with manothermosonication on selected physico-chemical and sensory attributes of an orange and carrot juice blend. *Food and Bioproducts Processing 90* (3): 442–448.

Castro, V.S., Rosario, D.K.A., Mutz, Y.S. et al. (2019). Modelling inactivation of wild-type and clinical *Escherichia coli* O26 strains using UV-C and thermal treatment and subsequent persistence in simulated gastric fluid. *Journal of Applied Microbiology 127* (5): 1564–1575.

Chaine, A., Levy, C., Lacour, B. et al. (2012). Decontamination of sugar syrup by pulsed light. *Journal of Food Protection 75* (5): 913–917.

Char, C.D., Mitilinaki, E., Guerrero, S.N., and Alzamora, S.M. (2010). Use of high-intensity ultrasound and UV-C light to inactivate some microorganisms in fruit juices. *Food and Bioprocess Technology 3* (6): 797–803.

Chemat, F., Zill-E-Huma, and Khan, M.K. (2011). Applications of ultrasound in food technology: processing, preservation and extraction. *Ultrasonics Sonochemistry 18* (4): 813–835.

Chen, F., Zhang, M., and Yang, C.H. (2020). Application of ultrasound technology in processing of ready-to-eat fresh food: a review. *Ultrasonics Sonochemistry 63*: 104953.

Cheon, H., Shin, J., Park, K. et al. (2015). Inactivation of foodborne pathogens in powdered red pepper (*Capsicum annuum L.*) using combined UV-C irradiation and mild heat treatment. *Food Control 50*: 441–445.

Choi, D.S., Park, S.H., Choi, S.R. et al. (2015). The combined effects of ultraviolet-C irradiation and modified atmosphere packaging for inactivating *Salmonella enterica* serovar Typhimurium and extending the shelf life of cherry tomatoes during cold storage. *Food Packaging and Shelf Life 3*: 19–30.

Choi, E., Park, H., Yang, H., and Chun, H. (2017). Effects of combined treatment with ultraviolet-C irradiation and grape seed extract followed by supercooled storage on microbial inactivation and quality of dongchimi. *Food Science and Technology 85*: 110–120.

Chun, H. and Song, K. (2013). The combined effects of aqueous chlorine dioxide, fumaric acid, and ultraviolet-C with modified atmosphere packaging enriched in CO_2 for inactivating existing microorganisms, *Escherichia coli* O157: H7 and *Salmonella* Typhimurium inoculated on buckwheat sprouts. *Postharvest Biology and Technology 86*: 118–124.

Chun, H., Yu, D., and Song, K. (2013). Effects of combined nonthermal treatment on microbial growth and the quality of minimally processed yam (*Dioscorea japonica* Thumb) during storage. *International Journal of Food Science and Technology 48* (2): 334–340.

Colejo, S., Alvarez-Ordóñez, A., Prieto, M. et al. (2018). Evaluation of ultraviolet light (UV), non-thermal atmospheric plasma (NTAP) and their combination for the control of foodborne pathogens in smoked salmon and their effect on quality attributes. *Innovative Food Science and Emerging Technologies 50*: 84–93.

Cruz, R.M.S., Godinho, A.I.A., Aslan, D. et al. (2016). Modelling the kinetics of peroxidase inactivation, colour and texture changes of Portuguese cabbage (*Brassica oleracea L.* var. *costata DC*) during UV-C light and heat blanching. *International Journal of Food Studies 5* (2): 180–192.

Delorme, M., Guimarães, J.T., Coutinho, N.M. et al. (2020). Ultraviolet radiation: an interesting technology to preserve quality and safety of milk and dairy foods. *Trends in Food Science and Technology 102*: 146–154.

EFSA (2016). Safety of UV-treated milk as a novel food pursuant to Regulation (EC) No 258/97. *EFSA Journal 14* (1) https://doi.org/10.2903/j.efsa 2016.4370.

Esua, O.J., Chin, N.L., and Yusof, Y.A. (2018). Simultaneous UV-C and ultrasonic energy treatment for disinfection of tomatoes and its antioxidant properties. *Journal of Advanced Agricultural Technologies 5* (3): 209–214.

Fenoglio, D., Ferrario, M., García Carrillo, M. et al. (2020a). Characterization of microbial inactivation in clear and turbid juices processed by short-wave ultraviolet light. *Journal of Food Processing and Preservation 44* (6) https://doi.org/10.1111/jfpp.14452.

Fenoglio, D., Ferrario, M., Schenk, M., and Guerrero, S. (2020b). Effect of pilot-scale UV-C light treatment assisted by mild heat on *E. coli, L. plantarum and S. cerevisiae* inactivation in clear and turbid fruit juices. Storage study of surviving populations. *International Journal of Food Microbiology* 332: 108767. https://doi.org/10.1016/j.ijfoodmicro.2020.108767.

Ferrario, M. and Guerrero, S. (2017). Impact of a combined processing technology involving ultrasound and PL on structural and physiological changes of *Saccharomyces cerevisiae* KE 162 in apple juice. *Food Microbiology 65*: 83–94.

Ferrario, M, Alzamora, S., and Guerrero, S. (2011). Use of modified Gompertz and Weibullian model to characterize microbial inactivation in orange juice processed with ultraviolet light and natural. *Proceedings of the 8th International Congress on Engineering and Food - CIBIA 8 (CD version), Lima, Perú.*

Ferrario, M., Schenk, M., García Carrillo, M., and Guerrero, S. (2018). Development and quality assessment of a turbid carrot-orange juice blend processed by UV-C light assisted by mild heat and addition of Yerba Mate (*Ilex paraguariensis*) extract. *Food Chemistry 269*: 567–576.

Ferrario, M., Fenoglio, D., Chantada, A., and Guerrero, S. (2020). Hurdle processing of turbid fruit juices involving encapsulated citral and vanillin addition and UV-C treatment. *International Journal of Food Microbiology 332*: 108811. https://doi.org/10.1016/j.ijfoodmicro.2020.108811.

Gabriel, A.A. (2012). Microbial inactivation in cloudy apple juice by multi-frequency *Dynashock* power ultrasound. *Ultrasonics Sonochemistry 19* (2): 346–351.

Gabriel, A.A. (2015). Combinations of selected physical and chemical hurdles to inactivate *Escherichia coli* O157: H7 in apple and orange juices. *Food Control 50*: 722–728.

Gabriel, A.A., Ostonal, J.M., Cristobal, J.O. et al. (2018). Individual and combined efficacies of mild heat and ultraviolet-c radiation against *Escherichia coli* O157:H7, *Salmonella enterica*, and *Listeria monocytogenes* in coconut liquid endosperm. *International Journal of Food Microbiology 277*: 64–73.

García Carrillo, M., Ferrario, M., and Guerrero, S. (2017). Study of the inactivation of some microorganisms in turbid carrot-orange juice blend processed by UV-C assisted by mild heat treatment. *Journal of Food Engineering 212*: 213–225.

García Carrillo, M., Ferrario, M., and Guerrero, S. (2018). Effectiveness of UV-C light assisted by mild heat on *Saccharomyces cerevisiae* KE 162 inactivation in carrot-orange juice blend studied by flow cytometry and transmission electron microscopy. *Food Microbiology 73*: 1–10.

García Carrillo, M., Ferrario, M., Schenk, M., and Guerrero, S. (2020). Effect of an UV-C light-based hurdle strategy for carrot-orange juice processing on *Candida parapsilosis* inactivation and physiological state: impact on juice sensory and physicochemical quality parameters. *Food and Bioprocess Technology 13* (11): 1954–1967.

García Loredo, A.B., Guerrero, S.N., and Alzamora, S.M. (2015). Inactivation kinetics and growth dynamics during cold storage of Escherichia coli ATCC 11229, *Listeria innocua* ATCC 33090 and *Saccharomyces cerevisiae* KE162 in peach juice using aqueous ozone. *Innovative Food Science and Emerging Technologies 29*: 271–279.

Gayán, E., Serrano, M.J., Raso, J. et al. (2012a). Inactivation of *Salmonella enterica* by UV-C light alone and in combination with mild temperatures. *Applied and Environmental Microbiology 78* (23): 8353–8361.

Gayán, E., Serrano, M.J., Monfort, S. et al. (2012b). Pasteurization of apple juice contaminated with *Escherichia coli* by a combined UV-Mild temperature treatment. *Food and Bioprocess Technology 6* (11): 3006–3016.

Gayán, E., Serrano, M.J., Monfort, S. et al. (2012c). Combining ultraviolet light and mild temperatures for the inactivation of *Escherichia coli* in orange juice. *Journal of Food Engineering 113* (4): 598–605.

Gayán, E., Álvarez, I., and Condón, S. (2013). Inactivation of bacterial spores by UV-C light. *Innovative Food Science and Emerging Technologies 19*: 140–145.

Gayán, E., Condón, S., and Álvarez, I. (2014a). Continuous-flow UV liquid food pasteurization: engineering aspects. *Food and Bioprocess Technology 7* (10): 2813–2827.

Gayán, E., García-Gonzalo, D., Álvarez, I., and Condón, S. (2014b). Resistance of *Staphylococcus aureus* to UV-C light and combined UV-heat treatments at mild temperatures. *International Journal of Food Microbiology 17* (172): 30–39.

Gayán, E., Torres, J.A., Ávarez, I., and Condón, S. (2014c). Selection of process conditions by risk assessment for apple juice pasteurization by UV-heat treatments at moderate temperatures. *Journal of Food Protection* 77 (2): 207–215.

Gayán, E., Serrano, M.J., Álvarez, I., and Condón, S. (2016). Modeling optimal process conditions for UV-heat inactivation of foodborne pathogens in liquid foods. *Food Microbiology 60*: 13–20.

Geeraerd, A.H., Herremans, C.H., and Van Impe, J.F. (2000). Structural model requirements to describe microbial inactivation during a mild heat treatment. *International Journal of Food Microbiology 59* (3): 185–209.

Gouma, M., Gayán, E., Raso, J. et al. (2015a). Influence of dimethyl dicarbonate on the resistance of *Escherichia coli* to a combined UV-Heat treatment in apple juice. *Frontiers in Microbiology 6*: 501.

Gouma, M., Gayán, E., Raso, J. et al. (2015b). Inactivation of spoilage yeasts in apple juice by UV-C light and in combination with mild heat. *Innovative Food Science and Emerging Technologies 32*: 146–155.

Gouma, M., Álvarez, I., Condón, S., and Gayán, E. (2020). Pasteurization of carrot juice by combining UV-C and mild heat: impact on shelf-life and quality compared to conventional thermal treatment. *Innovative Food Science and Emerging Technologies 64*: 102362.

Guerrero, S., Alzamora, S.M., and Ferrario, M. (2016). The use of pulsed light technology in a hurdle preservation strategy. In: *High Intensity Pulsed light in Processing and Preservation of Foods* (eds. G. Pataro and J. Ling), 205–228. New York, USA: Nova Publishers.

Guerrero, S., Ferrario, M., Schenk, M., and García Carrillo, M. (2017). Hurdle technology using ultrasound for food preservation. In: *Ultrasound: Advances in Food Processing and Preservation* (ed. D.B. Aguirre), 39–99. New York, USA: Elsevier Inc.: Academic Press.

Gündüz, G.T. and Korkmaz, A. (2019). UV-C treatment for the inhibition of molds isolated from dried persimmons (*Diospyros kaki L.*) and modelling of UV-C inactivation kinetics. *LWT* 115: 108451.

Gutierrez, J., Barry-Ryan, C., and Bourke, P. (2009). Antimicrobial activity of plant essential oils using food model media: efficacy, synergistic potential and interactions with food components. *Food Microbiology 26* (2): 142–150.

Ha, J.H. and Ha, S.D. (2010). Synergistic effects of ethanol and UV radiation to reduce levels of selected foodborne pathogenic bacteria. *Journal of Food Protection 73* (3): 556–561.

Ha, J.H. and Ha, S.D. (2011). Synergistic effects of sodium hypochlorite and ultraviolet radiation in reducing the levels of selected foodborne pathogenic bacteria. *Foodborne Pathogens and Disease 8* (5): 587.

Ha, J. and Kang, D. (2014). Synergistic bactericidal effect of simultaneous near-infrared radiant heating and UV radiation against *Cronobacter sakazakii* in powdered infant formula. *Applied and Environmental Microbiology 80* (6): 1858–1863.

Ha, J.W. and Kang, D.H. (2015). Enhanced inactivation of food-borne pathogens in ready-to-eat sliced ham by near-infrared heating combined with UV-C irradiation and mechanism of the synergistic bactericidal action. *Applied and Environmental Microbiology 81* (1): 2–8.

Ha, J.H., Lee, D.U., Auh, J.H., and Ha, S.D. (2011). Synergistic effects of combined disinfecting treatments using sanitizers and UV to reduce levels of *Bacillus cereus* in oyster mushroom. *Journal of Applied Biological Chemistry 54* (2): 269–274. https://doi.org/10.3839/jksabc.2011.042.

Hadjok, C., Mittal, G.S., and Warriner, K. (2008). Inactivation of human pathogens and spoilage bacteria on the surface and internalized within fresh produce by using a combination of ultraviolet light and hydrogen peroxide. *Journal of Applied Microbiology 104* (4): 1014–1024.

Health Canada. (2003). *Ultraviolet light treatment of apple juice/cider using the CiderSure 3500 - Canada.ca*. https://www.canada.ca/en/health-canada/services/food-nutrition/genetically-modified-foods-other-novel-foods/approved-products/novel-food-information-ultraviolet-light-treatment-apple-juice-cider-using-cidersure-3500.html, Last accessed, June 2021.

Holck, A.L., Liland, K.H., Drømtorp, S.M. et al. (2018). Comparison of UV-C and pulsed UV light treatments for reduction of *Salmonella, Listeria monocytogenes*, and enterohemorrhagic *Escherichia coli* on eggs. *Journal of Food Protection 81* (1): 6–16.

Jemni, M., Gómez, P., Souza, M. et al. (2014). Combined effect of UV-C, ozone and electrolyzed water for keeping overall quality of date palm. *LWT 59*: 649–655.

Kaya, Z. and Ünlütürk, S. (2019). Pasteurization of verjuice by UV-C irradiation and mild heat treatment. *Journal of Food Process Engineering 42* (5): 1–11.

Kaya, Z., Yildiz, S., and Ünlütürk, S. (2015). Effect of UV-C irradiation and heat treatment on the shelf life of a lemon-melon juice blend. Multivariate statistical approach. *Innovative Food Science and Emerging Technologies 29*: 230–239.

Kim, C. and Hung, Y.C. (2012). Inactivation of *E. coli* O157:H7 on blueberries by electrolyzed water, ultraviolet light, and ozone. *Journal of Food Science* 77 (4): 206–211.

Kim, J., Kim, H., Lim, G. et al. (2010). The effects of aqueous chlorine dioxide or fumaric acid treatment combined with UV-C on postharvest quality of *Maehyang* strawberries. *Postharvest Biology and Technology 56*: 254–256.

Kim, S.S., Park, S.H., Kim, S.H., and Kang, D.H. (2019). Synergistic effect of ohmic heating and UV-C irradiation for inactivation of Escherichia coli O157:H7, *Salmonella* Typhimurium and *Listeria monocytogenes* in buffered peptone water and tomato juice. *Food Control 102*: 69–75.

Kim, S.S., Kim, S.H., Park, S.H., and Kang, D.H. (2020). Inactivation of *Bacillus cereus* spores on stainless steel by combined superheated steam and UV-C irradiation treatment. *Journal of Food Protection 83* (1): 13–16.

Kim H.G. and Song K. (2017). Combined treatment with chlorine dioxidegas, fumaric acid, and ultraviolet-C light for inactivating Escherichia coli O157:H7 and Listeria monocytogenes inoculated on plums. Food Control 71: 371–375.

Koutchma, T. (2009). Advances in ultraviolet light technology for non-thermal processing of liquid foods. *Food and Bioprocess Technology 2* (2): 138–155.

Koutchma, T. and Parisi, B. (2004). Biodosimetry of *Escherichia coli* UV inactivation in model juices with regard to dose distribution in annular UV reactors. *Journal of Food Science 69* (1): FEP14–FEP22.

Koutchma, T., Parisi, B., and Patazca, E. (2007). Validation of UV coiled tube reactor for fresh juices. *Journal of Environmental Engineering and Science 6* (3): 319–328.

Koutchma, T., Forney, L.J., and Moraru, C.I. (2009). *Ultraviolet Light in Food Technology: Principles and Applications*. Taylor and Francis Group, Boca Raton, USA: *CRS Press*.

Koutchma, T., Popović, V., Ros-Polski, V., and Popielarz, A. (2016). Effects of ultraviolet light and high-pressure processing on quality and health-related constituents of fresh juice products. *Comprehensive Reviews in Food Science and Food Safety* 15 (5): 844–867.

Kumar, T. (2018). A review on ohmic heating technology: principle, applications and scope. *International Journal of Agriculture, Environment and Biotechnology 11* (4): 679–687.

La Cava, E. and Sgroppo, S. (2019). Combined effect of UV-C light and mild heat on microbial quality and antioxidant capacity of grapefruit juice by flow continuous reactor. *Food and Bioprocess Technology 12* (4): 645–653.

Lee, J.H., Chun, H.H., Oh, D.H. et al. (2012). Sensory and microbiological qualities of romaine lettuce and kale affected by a combined treatment of aqueous chlorine dioxide and ultraviolet-C. *Horticulture, Environment, and Biotechnology 53* (5): 387–396.

Lee, E.S., Park, S.Y., and Ha, S. (2019). Application of combined UV-C light and ethanol treatment for the reduction of pathogenic *Escherichia coli* and *Bacillus cereus* on Gwamegi (semidried Pacific saury). *Journal of Food Safety 39* (6): 1–9.

Leistner, L. (2010). Basic aspects of food preservation by hurdle technology. *International Journal of Food Microbiology 55*: 181–186.

Lemoine, M., Chaves, A., and Martínez, G. (2010). Influence of combined hot air and UV-C treatment on the antioxidant system of minimally processed broccoli (*Brassica oleracea* L. var. Itálica). *LWT 43* (9): 1313–1319.

López-Gómez, A., Fernández, P.S., Palop, A. et al. (2009). Food safety engineering: an emergent perspective. *Food Engineering Reviews 1* (1): 84–104.

Maghoumi, M., Gómez, P.A., Artés-Hernández, F. et al. (2012). Hot water, UV-C and super atmospheric oxygen packaging as hurdle techniques for maintaining overall quality of fresh-cut pomegranate arils. *Journal of the Science of Food and Agriculture 93* (5): 1162–1168.

Maghoumi, M., Gómez, P.A., Mostofi, Y. et al. (2013). Combined effect of heat treatment, UV-C and super atmospheric oxygen packing on phenolics and browning related enzymes of fresh-cut pomegranate arils. *LWT - Food Science and Technology 54* (2): 389–396.

Mahmoud, N.S. and Ghaly, A.E. (2004). On-line sterilization of cheese whey using ultraviolet radiation. *Biotechnology Progress 20* (2): 550–560.

Marquenie, D., Lammertyn, J., Geeraerd, A.H. et al. (2002). Inactivation of conidia of *Botrytis cinerea* and *Monilinia fructigena* using UV-C and heat treatment. *International Journal of Food Microbiology 74* (1–2): 27–35.

Marquenie, D., Geeraerd, A.H., Lammertyn, J. et al. (2003). Combinations of pulsed white light and UV-C or mild heat treatment to inactivate conidia of *Botrytis cinerea* and *Monilia fructigena*. *International Journal of Food Microbiology 85* (1–2): 185–196.

Martínez-Hernández, G.B., Artés-Hernández, F., Gómez, P.A. et al. (2013). Combination of electrolysed water, UV-C and super atmospheric O_2 packaging for improving fresh-cut broccoli quality. *Postharvest Biology and Technology 76*: 125–134.

Martínez-Hernández G. B., Navarro-Rico J., Gómez P.A. et al.(2015). Combined sustainable sanitising treatments to reduce Escherichia coli and *Salmonella* Enteritidis growth on fresh-cut kailan-hybrid broccoli. *Food Control* 47: 312–317.

Mezui, A.M. and Swart, P. (2010). Effect of UV-C disinfection of beer - sensory analyses and consumer ranking. *Journal of the Institute of Brewing 116* (4): 348–353.

Mikš-Krajnik, M., Feng, L., Bang, W., and Yuk, H. (2017). Inactivation of *Listeria monocytogenes* and natural microbiota on raw salmon fillets using acidic electrolyzed water, ultraviolet light or/and ultrasounds. *Food Control 74*: 54–60.

Mosqueda-Melgar, J., Raybaudi-Massilia, R.M., and Martín-Belloso, O. (2008). Non-thermal pasteurization of fruit juices by combining high-intensity pulsed electric fields with natural antimicrobials. *Innovative Food Science and Emerging Technologies 9*: 328–340.

Mukhopadhyay, S., Ukuku, D.O., and Juneja, V.K. (2015). Effects of integrated treatment of nonthermal UV-C light and different antimicrobial wash on *Salmonella enterica* on plum tomatoes. *Food Control 56*: 147–154.

Müller, A., Günthner, K., Stahl, M. et al. (2015). Effect of physical properties of the liquid on the efficiency of a UV-C treatment in a coiled tube reactor. *Innovative Food Science and Emerging Technologies 29*: 240–246.

Müller, A., Orlowska, M., Knörr, M. et al. (2017). Actinometric and biodosimetric evaluation of UV-C dose delivery in annular, Taylor-Coutte and coiled tube continuous systems. *Food Science and Technology International 23* (3): 222–234.

NACMCF (2006). Requisite scientific parameters for establishing the equivalence of alternative methods of pasteurization. *Journal of Food Protection 69* (5): 1190–1216.

Pagal, G.A. and Gabriel, A.A. (2020). Individual and combined mild heat and UV-C processes for orange juice against *Escherichia coli* O157:H7. *LWT 126*: 109295.

Palgan, I., Caminiti, I.M., Muñoz, A. et al. (2011). Combined effect of selected non-thermal technologies on *Escherichia coli* and *Pichia fermentans* inactivation in apple and cranberry juice blend and on product shelf life. *International Journal of Food Microbiology 151* (1): 1–6.

Pan, J., Vicente, A., Martínez, G. et al. (2004). Combined use of UV-C irradiation and heat treatment to improve postharvest life of strawberry fruit. *Journal of the Science of Food and Agriculture 84* (14): 1831–1838.

Pan, Y., Cheng, J.H., and Sun, D.W. (2019). Cold plasma-mediated treatments for shelf life extension of fresh produce: a review of recent research developments. *Comprehensive Reviews in Food Science and Food Safety 18* (5): 1312–1326.

Park, S.H., Kang, J.W., and Kang, D. (2018a). Inactivation of foodborne pathogens on fresh produce by combined treatment with UV-C radiation and chlorine dioxide gas, and mechanisms of synergistic inactivation. *Food Control 92*: 331–340.

Park, J.B., Kang, J.H., and Song, K.B. (2018b). Combined treatment of cinnamon bark oil emulsion washing and ultraviolet-C irradiation improves microbial safety of fresh-cut red chard. *LWT. Food Science and Technology* 93: 109–115.

Putnik, P., Pavlić, B., Šojić, B. et al. (2020). Innovative hurdle technologies for the preservation of functional fruit juices. *Foods* 9: 699.

Rahn, R.O., Gerstenberg, H.M., and Vavrina, G.A. (2002). Dosimetry of ionizing radiation using an iodide/iodate aqueous solution. *Applied Radiation and Isotopes* 56: 525–534.

Raso, J. and Barbosa-Cánovas, G. (2003). Nonthermal preservation of foods using combined processing techniques. *Critical Reviews in Food Science and Nutrition 43* (3): 265–285.

Raybaudi-Massilia, R., Calderón-Gabaldón, M.I., Mosqueda-Melgar, J., and Tapia, M.S. (2013). Inactivation of *Salmonella enterica* ser. *Poona* and *Listeria monocytogenes* on fresh-cut

"Maradol" red papaya (*Carica papaya L*) treated with UV-C light and malic acid. *Journal of Consumer Protection and Food Safety 8* (1–2): 37–44.

Santhirasegaram, V., Razali, Z., and Somasundram, C. (2015). Effects of sonication and ultraviolet-C treatment as a hurdle concept on quality attributes of Chokanan mango (*Mangifera indica L.)* juice. *Food Science and Technology International 21* (3): 232–241.

Schenk, M., Guerrero, S., and Alzamora, S.M. (2008). Response of some microorganisms to ultraviolet treatment on fresh-cut pear. *Food and Bioprocess Technology 1* (4): 384–392.

Schenk, M., García Loredo, A., Raffellini, S. et al. (2012). The effect of UV-C in combination with H_2O_2 treatments on microbial response and quality parameters of fresh cut pear discs. *International Journal of Food Science and Technology 47* (9): 1842–1851.

Schenk, M., Ferrario, M., and Guerrero, S. (2018). Antimicrobial activity of binary and ternary mixtures of vanillin, citral, and potassium sorbate in laboratory media and fruit purées. *Food and Bioprocess Technology 11* (2): 324–333.

Şengül, M., Erkaya, T., Başlar, M., and Ertugay, M.F. (2011). Effect of photosonication treatment on inactivation of total and coliform bacteria in milk. *Food Control 22* (11): 1803–1806.

Sew, C.C., Mohd Ghazali, H., Martín-Belloso, O., and Noranizan, M.A. (2014). Effects of combining ultraviolet and mild heat treatments on enzymatic activities and total phenolic contents in pineapple juice. *Innovative Food Science and Emerging Technologies 26*: 511–516.

Shah, N.N.A.K., Shamsudin, R., Rahman, R.A., and Adzahan, N.M. (2016). Fruit juice production using ultraviolet pasteurization: a review. *Beverages 2* (3).

Shin, Y., Jang, S., Song, H. et al. (2011). Effects of combined fumaric acid-UV-C treatment and rapeseed protein-gelatin film packaging on the postharvest quality of *Seolhyang* strawberries. *Food Science and Biotechnology 20* (4): 1161–1165.

Shin, Y., Song, H., and Song, K. (2012). Effect of a combined treatment of rice bran protein film packaging with aqueous chlorine dioxide washing and ultraviolet-C irradiation on the postharvest quality of *Goha* strawberries. *Journal of Food Engineering 113*: 374–379.

Sloan, E. (2018). Top 10 functional food trends. *Food Technology 72* (4): 27–43.

Sommers, C.H., Geveke, D.J., Pulsfus, S., and Lemmenes, B. (2009). Inactivation of *Listeria innocua* on frankfurters by ultraviolet light and flash pasteurization. *Journal of Food Science 74* (3): 3–6.

Sommers, C.H., Scullen, J.J., and Sites, J.E. (2010). Inactivation of foodborne pathogens on frankfurters using ultraviolet light and gras antimicrobials. *Journal of Food Safety 30* (3): 666–678.

Song, H.J., Choi, D.W., and Song, K.B. (2011). Effect of aqueous chlorine dioxide and UV-C treatment on the microbial reduction and color of cherry tomatoes. *Horticulture Environment and Biotechnology 52* (5): 488–493.

Tawema, P., Han, J., Vu, K.D. et al. (2016). Antimicrobial effects of combined UV-C or gamma radiation with natural antimicrobial formulations against *Listeria monocytogenes, Escherichia coli* O157: H7, and total yeasts/molds in fresh cut cauliflower. *LWT - Food Science and Technology 65*: 451–456.

Tremarin, A., Brandão, T.R.S., and Silva, C.L.M. (2017). Application of ultraviolet radiation and ultrasound treatments for *Alicyclobacillus acidoterrestris* spores inactivation in apple juice. *LWT. Food Science and Technology 78*: 138–142.

Unluturk, S. and Atilgan, M.R. (2015). Microbial safety and shelf life of UV-C treated freshly squeezed white grape juice. *Journal of Food Science 80* (8): M1831–M1841.

US FDA (2019). Ultraviolet radiation for the processing and treatment of food. https://www.accessdata.fda.gov/scripts/cdrh/cfdocs/cfcfr/cfrsearch.cfm?fr=179.39, Last accessed, March 2021

Vásquez-Mazo, P., García Loredo, A., Ferrario, M., and Guerrero, S. (2019). Development of a novel milk processing to produce yogurt with improved quality. *Food and Bioprocess Technology 12* (6): 964–975.

Walkling-Ribeiro, M., Noci, F., Cronin, D.A. et al. (2008). Reduction of *Staphylococcus aureus* and quality changes in apple juice processed by ultraviolet irradiation, pre-heating and pulsed electric fields. *Journal of Food Engineering 89* (3): 267–273.

Wibowo, S., Grauwet, T., Santiago, J.S. et al. (2015). Quality changes of pasteurised orange juice during storage: a kinetic study of specific parameters and their relation to colour instability. *Food Chemistry 187*: 140–151.

Yamamoto, K. (2017). Food processing by high hydrostatic pressure. *Bioscience, Biotechnology, and Biochemistry 81* (4): 672–679.

Zhang, Z.-H., Wang, L.-H., Zeng, X.A. et al. (2018). Non-thermal technologies and its current and future application in the food industry: a review. *International Journal of Food Science and Technology 54* (1): 1–13.

Zhao, D., Barrientos, J.U., Wang, Q. et al. (2015). Efficient reduction of pathogenic and spoilage microorganisms from apple cider by combining microfiltration with UV treatment. *Journal of Food Protection 78* (4): 716–722.

Zheng, W., Zhao, Y., Xin, H. et al. (2014). Airborne particulate matter and culturable bacteria reduction from spraying slightly acidic electrolyzed water in an experimental aviary laying-hen housing chamber. *Transactions of the ASABE 57* (1): 229–236.

7

Visible Light

Laura M. Hinds[1,2], *Mysore L. Bhavya*[3], *Colm P. O'Donnell*[2], *and Brijesh K. Tiwari*[1,2]

[1] *Food Chemistry and Technology Department, Teagasc Food Research Centre, Dublin, Ireland*
[2] *School of Biosystems and Food Engineering, University College Dublin, Dublin, Ireland*
[3] *Food Engineering Department, CSIR-Central Food Technological Research Institute, Mysuru, India*

1 Introduction

With the global population is expected to rise from 7.7 billion to 8.5 billion in the next decade, with a further increase to 10 billion by 2050, the food industry will be required to adapt to facilitate an increasing demand for food (United Nations, Department of Economic and Social Affairs, Population Division 2019). Further to this, the processing technologies that will be utilized must be sustainable and preferably would not involve the use of dangerous chemicals in the food processing environment. Light-based technologies have been gaining more and more attention in recent years due to their wide applications in the field of food science while also being a potential sustainable non-thermal technology (D'Souza et al. 2015; Bantis et al. 2018; Prasad et al. 2020).

Light is one of the most crucial environmental factors for plants; this is because it provides the source of energy for plant life. Considering this, it is unsurprising that plants have adopted the ability to sense multiple parameters of ambient light signals, including light quantity (fluence), quality (wavelength), direction, and duration. Light signals are perceived through at least four distinct families of photoreceptors, such as phytochromes, cryptochromes, phototropins, and unidentified ultraviolet B photoreceptors (Jiao et al. 2007). Since plants respond to various types of light in various ways, this phenomenon can be used to facilitate optimal growth conditions and improved postharvest processing which subsequently could also reduce food waste. While this technology is still in preliminary stages, there are significant opportunities for visible light-based technologies in the food industry including horticulture, postharvest treatments, aquaculture, and decontamination.

Visible light is electromagnetic radiation which is detectable through our sense of sight. The wavelength range of visible light is about 400–700 nm, with some describing ranges extending these values as there are no agreed limits to the visible spectrum (Sliney 2016). Visible, or

Electromagnetic Technologies in Food Science, First Edition. Edited by Vicente M. Gómez-López and Rajeev Bhat.

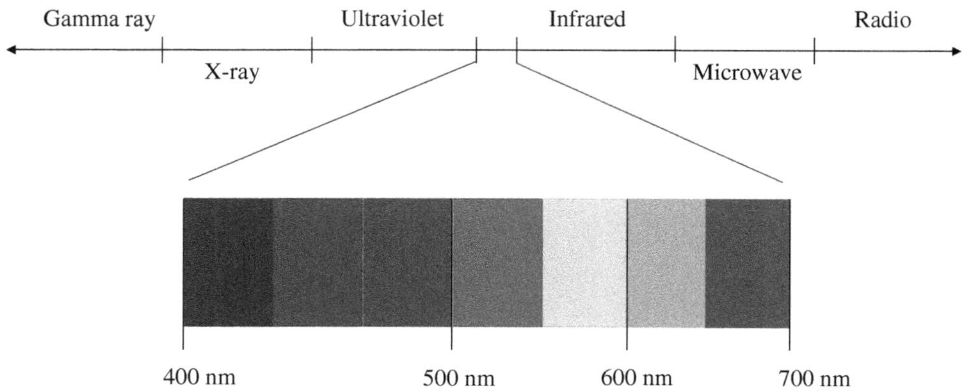

Figure 1 Visible light colors/wavelength within the electromagnetic spectrum. (*See insert for color representation of this figure*).

Table 1 Semiconductor materials commonly used in light emitting diodes across the visible spectral ranges.

Color/wavelength (nm)	Semiconductor (p, n)	QW material (active layer)
Blue/green (400–570)	GaN	InGaN
Yellow (570–610)	GaAsP	InGaAlP
Red (610–760)	GaInP	InGaAlP
	AlGaAs	

white, light but comprises various "colors" of light (Figure 1). Light interacts with different biological materials differently, giving rise to photobiological research, which has highlighted the various applications of each color in the food science sector. There are numerous sources that can generate visible light either as white light or as individual colors. Most studies in this field are looking to light-emitting diodes (LEDs) as they emit a narrow spectrum of light allowing illumination of food with different colors both individually and in various combinations. LEDs emit light through a process called electroluminescence, in which the type (color) of light is dependent on the material used in the production of the LED (Table 1).

The objective of this chapter is to provide a framework for visible light technologies and their applications in the food industry. The applications of LED technology will be discussed; however, it should be noted that this field of research is still in its infancy and there is no consistent methodology and therefore makes directly comparing various studies to make conclusions quite difficult.

2 Sources

There are various sources that can artificially produce light in the visible region, either as white light or as individual colors with near monochromatic output. Among these sources are fluorescent, high-pressure sodium, metal halide, incandescent lights. However, these traditional sources are characterized by broad spectral power distribution and have limited control over the emissions of ultraviolet or infrared radiation. From a sustainability and versatility perspective, LEDs are significantly superior to other sources (D'Souza et al. 2015). There are various advantages of LEDs over the conventional sources such as being environmentally friendly and sustainable, cost-effective/low investment, low energy consumption, long lifespan, low maintenance, highly durable, and safe operation (D'Souza et al. 2015). Further, due to low emission of radiant heat (infrared light), LEDs are known for reducing the undesirable deterioration in food (Mitchell et al. 2012).

LEDs are two-terminal semiconductor devices, which emit light under forward bias conditions. Different semiconductor materials emit different colors (or "wavelengths") of light; the emission wavelengths are dependent on the semiconductor band gap of the material (Table 1). Compound semiconductor materials, also known as III–V materials (because they are made up of group III and group V elements from the periodic table), are generally direct band gap materials and therefore efficient in the conversion of electrical current into light. These III–V materials form the backbone of modern LED technology. Positively charged carriers (holes) in the p-type layer are driven toward the active (junction) layer where they recombine with electrons which are driven by the same bias from the n-type layer in the opposite direction toward the junction. When the electron and hole meet, they recombine under emission of a photon which carries away the energy that is released upon recombination. The higher the band gap of the material, the higher the energy of the emitted photon, and the shorter the wavelength of the emitted light. In practical devices, there is usually a very thin layer of a material with a lower band gap at the semiconductor junction. This thin quantum well speeds up the recombination process and makes it more efficient. The band gap of the quantum well then sets the emission wavelength. By varying the III–V composition of the quantum well, the emission wavelength can be tuned (Hinds et al. 2019). As LEDs can produce light with a narrow spectral output, this can facilitate the study of various colors of the visible light spectrum for applications in the food industry further and provide insight which can then optimize this technology.

3 Quantifying Light Treatment

The basic unit of light is candela which is described as the luminous power per unit solid angle emitted by a point light source in a particular direction. However, this unit is rarely used in the field of food science. Within this field however, there are numerous terms and metrics used to describe how visible light treatments are reported with units such as Einsteins, photon flux densities, and irradiance; studies focusing on ultraviolet light have predominantly used the metric fluence, fluence rate, and irradiance, with some studies reporting "dose" – which should be avoided.

While in some cases these two terms "fluence rate" and "irradiance" are synonymous, such as a one directional light source illuminating a flat surface, there are other circumstances, i.e. an annular system with multidirectional light, where a distinction must be made and in such a case fluence rate would apply. Fluence rate describes the total radiant power from all directions, and irradiance describes the radiant power from all upward/downward directions. Fluence rate describes a three-dimensional scenario, while irradiance describes a two-dimensional surface. Therefore, it will depend on the angle at which the light is applied to the sample/surface and which directions the light is coming from (Bolton et al. 2015).

While it may not be feasible to determine standard units for photobiological research in the food sector in the near future, there are key parameters that should be reported in order to gain insight into the experimental design and treatment. These parameters are wavelength of light, distance from the source, and duration of the treatment.

4 Applications of Visible Light in the Food Industry

There are countless applications of visible light across many areas within the food industry, spanning from horticulture (cultivation of crops), postharvest technology (delaying of fruit ripening and postharvest storage), and food safety (D'Souza et al. 2015; Davis and Burns 2016). The focus of this chapter will emphasize on the postharvest and food safety applications of various visible light colors on food products.

4.1 Postharvest Handling

The exposure of visible light to food after harvest induces processes which can prolong shelf life and reduces postharvest losses and therefore reduces the amount of food waste and also introduces a more cost-effective process (D'Souza et al. 2015). While visible light technologies are still quite novel, there have been many studies that have highlighted their potential. Various studies and their applications are highlighted in Table 2. Bárcena et al. (2019) investigated the potential of low-intensity light pulses (white and red light) in delaying postharvest senescence of purple kale stored at room temperature using various metrics such as chlorophyll and protein breakdown – which are indicators of senescence progression in green tissues. It was reported that approximately 62% of the initial chlorophyll content was lost in control kale leaves (stored in darkness), while only 20% of initial chlorophyll was lost after white or red light-treated (daily pulses over three days) leaves. Soluble protein content of control kale leaves diminished during storage, showing a decrease of 75% (three days at room temperature). In contrast, irradiated leaves with red light retained their protein content with respect to initial value, while white light treatment proteins decreased slightly after three days (36%) but maintained higher levels than that of the corresponding controls. These results suggest that red light pulses were more effective than white light to delay postharvest senescence of kale leaves, in terms of protein retention. However, white light can also be fine-tuned to have different properties such as cold white light or warm white light, depending on the amount of blue/white light emitted from the light source. Hasperué et al. (2016a) also observed delayed senescence and increase in sugars (glucose, fructose, and sucrose) throughout the storage (5 or 22 °C) in broccoli heads when treated with white and blue light continuously. Another study by Pola et al. (2020) reported that red LED illumination accelerated the red color development and total chlorophyll content of chili compared to blue LED and control treatment, while blue LED treatment increased total phenolic, ascorbic acid content, and antioxidant capacity in chili. Burattini et al. (2017) investigated the growth of spinach plants in experimental boxes under two white treatments each having different distributions of color. Using two metrics leaf length and leaf width, it was shown that cold white light stimulated leaf growth far more than warm white light; more specifically, plants grown under cold white light treatments showed longer and larger leaves, compared to the warm white light treatment. It was also noted that warm white light produced a better development of the photosynthetic photoreceptors; it was stated that this could be relevant for the production quality. Therefore, further studies are required to fully optimize this technology, while this study focused on warm white light which contains more yellow light. There are limited studies investigating the yellow/orange colors individually for postharvest handling applications.

Loi et al. (2019) studied the overall effect of blue, green, yellow, red, and white on the quality parameters of broccoli florets over 20 days of cold storage. An increase in chlorophyll a and b and total chlorophyll content was observed for all samples, including the control samples which were not exposed to light but a significantly higher difference in those exposed to green light at almost all time points. A higher chlorophyll content results in increased photosynthetic process which produces bioactives called photosynthates, which is desirable for consumers. Jin et al. (2015) also had positive effects of chlorophyll content in broccoli florets when exposed to green light. Green light was shown to be the most effective treatment in improving shelf life and product quality. Another study by Kim et al. (2011) investigated the effect of multiple LEDs (white, blue, green, and red) on the nutrient content of cabbages over a period of 18 days, and results showed that green and white LEDs were most effective at stimulating chlorophyll production, followed by red and blue LEDs, while blue and white LEDs were generally better at increasing vitamin C and total phenolics. Kasim and Kasim (2017) reported that blue LED illumination on lettuce leaves showed negative effects when stored at 5 °C for 21 days, whereas white LED illumination showed the best results by reducing the chlorophyll degradation, maintaining green color and also increased the total soluble content of lettuce. Even, they observed reduction in weight loss and decay rate of lettuce by red and green light treatment, respectively. An investigation by Liu et al. (2019) mentioned that blue and red LED light irradiation increased the levels of polymethoxylated flavonoids in navel oranges during six days of exposure. Further, red light was found to induce the accumulation of flavone and flavanone glycosides throughout the exposure period, thus suggesting the application of white and red light for improving the nutrition and flavor profile of the citrus fruits.

Nájera et al. (2018) studied the effect of LED light exposure on the postharvest and lycopene quality attributes of tomatoes. They observed that tomatoes exposed to LED light with high red:far red ratio had higher lycopene content by 30–60% compared to that of dark-exposed ones. The firmness and titratable acidity of LED-treated tomatoes increased even with one hour of exposure per day during the storage. A study by Dhakal and Baek (2014) investigated red and blue light as a pre-treatment, prior to storage in dark, as a potential technology to preserve mature green tomatoes postharvest to improve long-term storage. Mature green tomatoes irradiated with blue light for seven days had slower rate of color change from green to red compared with mature green tomatoes stored in darkness or treated with red light for the same duration. This highlights that blue light is also a potential tool for this technology for a different approach. Hasperué et al. (2016b) found that Brussels sprouts treated with white–blue LEDs showed delayed senescence and lower respiration rate on the outer and inner leaves. By the end of the storage (10 days at 22 °C), the sprouts also retained greenness with >10 and 1.6 times of chlorophyll in outer and inner leaves, respectively, compared to that of control samples. It was also observed that outer leaves contained higher levels of flavonoids, antioxidants compared to the inner ones. The authors indicated that the effect of visible light treatment was not only confined to the outer leaves but they also improved the quality of the inner leaves. A report by Huang et al. (2018) mentioned that LED treatment especially blue light accelerated the ripening of mature green banana compared to that of storage in dark. The blue light treatment showed faster yellowing/de-greening, increased flesh softening, and higher respiration rate and ethylene production than the red and green light. Also, they observed that there was an increase in the ascorbic acid, phenolics, and sugar content in the LED-treated bananas, which would be a potential chemical-free strategy for ripening banana in a shorter time without affecting its nutritional quality. Another study by Maroga et al. (2019) studied the influence of LEDs (eight hours illumination/day) on the functional quality of three different fresh cut sweet peppers during the storage of 14 days at 7 °C. Red (660 nm) and blue (450 nm) LED reduced the weight loss to an acceptable level of <2% in fresh-cut sweet peppers (yellow, green, and red) until 11 days of storage. It was also noticed that yellow and green sweet peppers had least change in color parameters (total color change) and quality attributes (total phenolics, ascorbic acid content, and antioxidant activity) until 7 and 11 days, respectively, when treated with red LEDs. While, it was maintained until 11 days in case of red peppers treated with blue LEDs. Even, red LEDs maintained the highest concentrations of the functional pigments and phenolics (chlorophyll, β-carotene, and lycopene) in green and yellow peppers, but blue LEDs increased it in red sweet peppers until seven days of storage. A report by Ngcobo et al. (2020) mentioned that red LED (118 $\mu mol\,m^{-2}\,s^{-1}$ at 638 nm) illumination for 48 hours enhanced color parameters in red and yellow cultivars of green mature cherry tomatoes resulting in visually redder and yellower tomatoes, respectively. In addition, the LED treatment increased the concentration of sugars, lycopene, and β-carotene in the tomato fruit without significantly affecting fruit mass and firmness at the green-mature stage.

4.2 Food Safety

The food safety applications of visible light are being investigated in different food matrices using different types of light (Table 2). LED technology has the ability to inactivate bacteria, yeast, fungi as well as bacterial spores and thus can be explored for developing an effective and safe technology to deliver hygienic food that improve the public health (Maclean et al. 2009; Kumar et al. 2017; Murdoch et al. 2013; Moorhead et al. 2016; Trzaska et al. 2017). Blue light is the most widely studied color range of the visible light spectrum. Blue region (400–500 nm) of visible light has higher inactivation effect on foodborne pathogens (Gram negative and Gram positive bacteria) than other visible range (green and red) (Ghate et al. 2013; Kumar et al. 2016). Various studies have shown that visible light can inactivate various foodborne pathogens in laboratory media as shown in Table 3. It is reported that Gram positive bacteria are more susceptible to 405 nm illumination than Gram negative bacteria (Maclean et al. 2009) Kim et al. (2016) found that *Salmonella* was more susceptible to LED illumination than *Escherichia coli* O157:H7 in phosphate-buffered saline at 4 °C. Kim et al. (2015) found 1.9, 2.1, and 0.9 $\log_{10}$ reduction of *Bacillus cereus*,

Table 2 Studies highlighting the applications of visible light technology in the food industry.

Food matrix	Color of light/sources	Treatment conditions	Findings	References
Postharvest handling				
Kale leaves	Fluorescent lamp (white with filters)	White light (1 h of daily irradiation at 20–25 μmol m^{-2} s^{-1}) Red light (1 h of daily irradiation at 20–25 μmol m^{-2} s^{-1})	Irradiated samples showed higher chlorophyll, protein, antioxidant capacity, and soluble sugar content than non-irradiated ones. However, red light pulses were more effective than white light to delay postharvest senescence of kale leaves, in terms of protein retention	Bárcena et al. (2019)
Green tomatoes	Blue and red (LEDs)	Blue or red light at 85.72 and 102.70 μeinstein $m^{-2}s^{-1}$ for seven days, respectively	Mature green tomatoes pre-treated with blue light emitted from blue light-emitting diodes for seven days developed a yellowish color and high levels of firmness, while those pre-treated with darkness or red light from red LEDs for the same period ripened and developed red color	Dhakal and Baek (2014)
Strawberry	White, blue, green, and red (LEDs)	Treatments were carried out over 18 days	Green and white LEDs were most effective at stimulating chlorophyll production, followed by red and blue	Kim et al. (2011)
Broccoli florets	Blue, green, yellow, red, and white (LEDs)	Experiments extended 20 days. The photosynthetic photon flux (PPF) level at the top of the broccoli surface was 21, 24, 27, 66, and 40 μmol m^{-2} s^{-1} for blue, green, yellow, red light, and white light, respectively	Compared to darkness green light increased the chlorophyll content; red and yellow light increased the phenolic compound content and the appearance of broccoli florets; green, red, and yellow light caused an increase in the ascorbic acid content, and all lights increased the soluble protein content	Loi et al. (2019)
Broccoli heads	White and blue (LEDs)	Continuous white and blue LED light at the distance of 9 cm with light intensity of 20 μmol m^{-2} s^{-1} and stored at 5 or at 22 °C	The white–blue LED treatment extended the shelf life and delayed senescence. Chlorophyll content doubled and even carotenoid content also increased by the end of the storage in treated samples than in the dark stored controls	Hasperué et al. (2016a)

(*Continued*)

Table 2 (Continued)

Food matrix	Color of light/sources	Treatment conditions	Findings	References
Grapes	Blue (LED)	Stored for seven to nine days at 10, 15, 20, or 25 °C with continuous irradiation with blue LED light of 445 nm and 100 μmol m^{-2} s^{-1}	Anthocyanin accumulation of berries was observed when white light + blue LED light was used with 15–25 °C treatments without any reduction in titratable acidity and berry weight	Azuma et al. (2019)
Strawberry	Blue (LED)	Illuminated with blue LED at 40 μmol m^{-2} s for 12 days at 5 °C	Total anthocyanin content in strawberry was found to be enhanced during storage when treated with blue light	Xu et al. (2014)
Broccoli	Red (LED)	Red light illumination at 50 μmol m^2 s^{-1} for five days at 20 °C	Red light inhibited yellowing and the degradation of chlorophyll, maintained the sensory appearance of broccoli, and decreased weight loss and malondialdehyde content	Jiang et al. (2019)
Food safety				
Milk	Blue LED (413 nm)	Irradiance level of 100 mW cm^{-2}. The LED probe was placed on a holder 1 mm above the surface of a 24-well plate	All bacterial species tested presented more than 5 log_{10} of inactivation within 2 h of irradiation (720 J cm^{-2})	dos Anjos et al. (2020)
Cucumbers	Light pads (464 nm)	464 nm at 6, 12, and 18 J cm^2	*Salmonella* – 464 nm light at doses of 6, 12, and 18 J cm^{-2} produced significant inactivation of the organism. Kill rates of 80.23–100% were obtained. Reductions of 75.61–96.34% were achieved for *E. coli*	Guffey et al. (2016)
Packaged hot dogs	Supraluminous diode (405 nm)	405 nm at 30, 60, and 100 J cm^2		
Broccoli sprouts	Blue, red, and far-red (LEDs) White (fluorescent lamps)	30 cm from the light source and were continuously illuminated using a photosynthetic photon flux density (PPFD) of 35 ± 2.5 μmol m^{-2} s^{-1}	After four days, far-red LED-treated sprouts reported the lowest mesophilic bacterial growth as compared to the other light regimes	Castillejo et al. (2021)
Packaged sliced cheese	Blue LEDs (405 nm)	1.00 mW cm^{-2} on the surface of packaged sliced cheese	*L. monocytogenes* and *P. fluorescens* were reduced by from 6.53 and 4.98 to 5.14 and 3.60 log_{10}, respectively	Hyun and Lee (2019)

Fresh-cut papaya	Blue LEDs (405 nm)	Illuminated by blue LED for 24–48 h (0.9–1.7 kJ cm^{-2}) at set temperatures of 4, 10, or 20 °C	*Salmonella* reduced by 1.0–1.2 log_{10} cfu cm^{-2}	Kim et al. (2017a)
Fresh-cut mango	Blue LEDs (405 nm)	Illuminated with 2.6–3.5 kJ cm^{-2} for 36–48 h at 4 and 10 °C	Inactivated *E. coli, L. monocytogenes* and *Salmonella* by 1.0–1.6 log_{10} cfu cm^{-2}	Kim et al. (2017b)
Fresh-cut pineapple	Blue LEDs (460 nm)	92.0 mW cm^{-2} at 16 °C maintained at a distance of 4.5 cm for 24 h to achieve 7950 J cm^{-2}	*Salmonella* spp. reduced by 1.7 log_{10} cfu g^{-1}	Ghate et al. (2017)
Orange juice	Blue LEDs (460 nm)	460 nm for 4.9–13.6 h at 4–20 °C	*Salmonella* reduced by 3.3–4.8 log_{10} cfu ml^{-1}	Ghate et al. (2016)
Orange juice	Blue LEDs (462 nm)	The irradiance and fluence on the surface of the orange juice were 6.34 ± 0.05 mW cm^{-2} and 70 J cm^{-2}, respectively	*E. coli* and *S. aureus* reduced by <0.5 log_{10} cfu ml^{-1}	Bhavya and Hebbar (2019)
Milk	Blue LEDs (405, 430, and 460 nm)	Irradiance at three different temperatures of 5, 10, and 15 °C	At 5 °C, 4.69 log_{10} reduction at 405 nm after 60 min, 4.11 log_{10} at 430 nm after 75 min, 3.41 log_{10} at 460 nm after 90 min	Srimagal et al. (2016)
			At 10 °C, 405, 430, and 460 nm lights showed 4.82 log_{10} reduction after 75 min, 4.24 after 75 min, and 3.64 after 90 min, respectively	
			At 15 °C showed reduction of 5.27 log_{10} after 60 min, 5.04 log_{10} after 75 min, and 4.64 after 90 min, respectively, for 405, 433, and 460 nm lights	
Almonds	Blue LEDs (405 nm)	3.4 W of collimated light, for 0–10 min at a working distance of 7 cm	Reductions up to 1.85, 2.44, 0.54, and 0.70 log_{10} cfu g^{-1} of *E. coli* K-12, *E. coli* O157:H7, attenuated *Salmonella* and pathogenic *Salmonella*, respectively	Lacombe et al. (2016)

Table 3 in vitro microbial reduction by the application of visible light.

Micro-organism	Matrix/medium	Treatment	Log_{10} reduction	References
Blue light				
B. cereus	Phosphate-buffered saline	405 ± 5 nm for 7.5 h at 4 °C	$1.9\ log_{10}\ cfu\ ml^{-1}$	Kim et al. (2015)
S. aureus			$0.9\ log_{10}\ cfu\ ml^{-1}$	
L. monocytogenes			$2.1\ log_{10}\ cfu\ ml^{-1}$	
E. coli	Maximum recovery diluent	462 ± 3 nm for 1 h at 9 and 27 °C	$<0.5\ log_{10}\ cfu\ ml^{-1}$	Bhavya and Umesh Hebbar (2019)
S. aureus				
S. aureus	Phosphate-buffered saline	405 nm for 60–90 min at 30 °C	$>5.0\ log_{10}\ cfu\ ml^{-1}$	Maclean et al. (2009)
L. monocytogenes	Phosphate-buffered saline	405 nm for 36 min at 32 °C	$3.3–3.7\ log_{10}\ cfu\ ml^{-1}$	Endarko et al. (2012)
B. cereus	Phosphate-buffered saline	405 nm for 40 min at 30 °C	$4.0\ log_{10}\ cfu\ ml^{-1}$	Maclean et al. (2013)
E. coli	Trypticase soya broth	461 nm for 10 h at 15 °C	$4.0–6.0\ log_{10}\ cfu\ ml^{-1}$	Kim and Bang (2013)
S. aureus				
P. aeruginosa	Phosphate-buffered saline	415 nm for 40 min at 25 °C	$3.5\ log_{10}\ cfu\ ml^{-1}$	Amin et al. (2016)
E. coli	Phosphate-buffered saline	405 nm for 8 h at 26 °C	$3.5–5.0\ log_{10}\ cfu\ ml^{-1}$	Murdoch et al. (2012)
L. monocytogenes				
S. Enteritidis				
E. coli	Trypticase soya broth	461 nm for 7.5 h at 15 °C	$4.9\ log_{10}\ cfu\ ml^{-1}$	Ghate et al. (2013)
		461 nm for 7.5 h at 10 °C	$5.1\ log_{10}\ cfu\ ml^{-1}$	
L. monocytogenes		461 nm for 7.5 h at 15 °C	$4.3\ log_{10}\ cfu\ ml^{-1}$	
		461 nm for 7.5 h at 10 °C	$5.2\ log_{10}\ cfu\ ml^{-1}$	
E. coli	Trypticase soya broth	461 nm for 7.5 h at 15 °C – 4.5 pH	$2.1\ log_{10}\ cfu\ ml^{-1}$	Ghate et al. (2015)
		461 nm for 7.5 h at 15 °C – 7.3 pH	$1.2\ log_{10}\ cfu\ ml^{-1}$	
		461 nm for 7.5 h at 15 °C – 9.5 pH	$4.1\ log_{10}\ cfu\ ml^{-1}$	
S. aureus methicillin-resistant (MRSA)	Tryptic soya agar	470 nm at 55, 110, 165, and 220 J cm^{-2}	92% reduction at 55 J cm^{-2} and 100% reduction at 110, 165, or 220 J cm^{-2}	Bumah et al. (2015)
S. Typhimurium	*Salmonella Shigella* agar		31% reduction at 55 or 110 J cm^{-2} and 93% reduction at 165 or 220 J cm^{-2}	

Listeria monocytogenes, and *Staphylococcus aureus*, respectively, with 405 ± 5 nm illumination for 7.5 hours until 486 J cm^{-2} at 4 °C. Antibacterial effect of 405 ± 5 nm LED on these Gram-positive bacteria was attributed to physical damage to cellular membrane rather than DNA, as the cells were more sensitive to NaCl than control cells (non-illuminated). Srimagal et al. (2016) and Kumar et al. (2016) observed that 405 nm LED illumination had more antimicrobial effect than 460 nm, since 405 nm spectrum fell in UV range which may directly cause the DNA damage. LED illumination excites the intracellular photosensitizers which absorb photon of visible light. While, returning to ground state, these molecules transfer energy to an oxygen molecule, resulting in the production of reactive oxygen species (ROS). ROS such as superoxide anion, hydrogen peroxide, hydroxyl radicals, and singlet oxygen may attack cellular components like DNA, lipids, and proteins ultimately causing bacterial death (Luksiene and Brovko 2013). ROS are known to cause damage to DNA by targeting guanine bases and forming 8-hydroxy-deoxyguanosine (8-OHdG), an oxidized derivative (Kim et al. 2015). Kim et al. (2017a) did not observe any significant difference in lipid peroxidation; however, DNA oxidation was increased by 1.7–1.8 times in *Salmonella* spp. treated with 405 nm light, suggesting that ROS might preferentially oxidize DNA rather than lipids in cell membrane. Kim et al. (2016) observed that 405 ± 5 nm LED illumination resulted in loss of bacterial membrane permeability and no DNA damage was observed; therefore, physical damage might partly be the cause for antibacterial effect. However, while there are various theories as to how blue light induces damage to microbial cells, such as the generation of ROS within the cell, this exact mechanism remains unknown (Hyun and Lee 2020b).

A study by dos Anjos et al. (2020) showed that blue light has significant effect on *S. aureus*, *E. coli*, *Pseudomonas aeruginosa*, *S.* Typhimurium, and *Mycobacterium fortuitum* cells suspended in whole milk or saline solution, achieving 5 log_{10} inactivations after 2 hours treatment. While some studies using light in the visible range to inactivate bacteria employ an exogenous photosensitizers to help facilitate the photoreaction (called photodynamic inactivation), this study reported that it relied only on endogenous compounds broadly found in pathogens, such as porphyrins and flavins, which can absorb light at specific wavelengths and, hence, are promoted to excited energy states and result in the generation of ROS which could have detrimental effects on biological materials. Another study by Guffey et al. (2016) also demonstrated that using blue light (464 nm), with a fluence of 18 J cm^{-2}, can completely eliminate *S.* Typhimurium on cucumbers, showing that there is potential for this technology in the fresh food sector.

Some studies have focused on comparing the decontamination capabilities of different wavelengths. In addition to this, the effect of using different treatment temperatures was also investigated. Ghate et al. (2017) observed that both irradiance and temperature influence the inactivation of *Salmonella* (2–5 log_{10} cfu ml^{-1}) at 460 nm. Kim et al. (2017b) showed that with 405 ± 5 nm treatment, *E. coli*, *L. monocytogenes*, and *Salmonella* spp. on mango surfaces were effectively inactivated by 97–99% at chilling temperatures, but less effective to *E. coli* and *L. monocytogenes* at room temperature. In another study by Ghate et al. (2013), the population of *E. coli*, *S.* Typhimurium, *L. monocytogenes*, and *S. aureus* decreased by 4.9, 5.0, 4.3, and 5.2 log_{10} cfu ml^{-1} after 7.5 hours under 460 nm LED at 15 °C, respectively. Ghate et al. (2015) mentioned that 461 nm illumination was more detrimental in acidic conditions than alkaline conditions for *L. monocytogenes*, while inactivation was more effective for alkaline conditions for *E. coli* and *S.* Typhimurium, suggesting the potential of LEDs in preserving acidic and alkaline foods. In a study by Castillejo et al. (2021), far-red LEDs were successful in lowering the growth of mesophilic bacteria in broccoli sprouts over four days, with blue LEDs having no effect. Conversely, in a study by Ghate et al. (2013), red LEDs (642 nm) did not show bactericidal or bacteriostatic effect at any temperature, while blue (461 nm) and green (521 nm) LEDs produced a bactericidal effect at 10 and 15 °C but not 20 °C, with the blue having a more significant effect. It should be noted that in the previously mentioned study (Castillejo et al. 2021), the treatment suspension medium was tryptic soya broth, which is nutrient-rich media for many foodborne pathogens and an increase in temperature could increase the growth rate of bacteria suspended in this media and therefore limited the effect of the light. In another study investigating similar conditions, a different medium that does not facilitate growth showed that treatments using blue LEDs were effective at both room temperature and chilling temperature (Kumar et al. 2017). Hyun and Lee (2020a) studied the effect of blue LEDs on populations of *L. monocytogenes* and *P. fluorescens* on packaged sliced cheese, and reductions of 5.14 and 3.60 $\log_{10}$, respectively, were achieved, after seven days of treatment at 4 °C.

Few reports on the effect of blue light on the physicochemical attributes of the food are available. A study by Ghate et al. (2017) reported that color of pineapple slices was reduced when it was treated with 460 nm LED which could be attributed to the absorption of light by β-carotene in pineapple. Kim et al. (2017a) noticed that there was no significant effect on color and antioxidant capacity of fresh-cut papaya illuminated at 405 ± 5 nm, whereas flavonoid content increased by 1.5–1.9 times higher than that of untreated fruit. No significant changes in the levels of lycopene, ascorbic acid, and β-carotene were observed between LED illuminated fruits, as compared to control. Kim et al. (2017b) observed that the physicochemical and nutritive qualities of fresh-cut mangoes such as ascorbic acid, β-carotene, antioxidant capacity, and flavonoids were preserved by 405 ± 5 nm LED illumination compared to non-illuminated fruits. Another study by Srimagal et al. (2016) evaluated the shelf life of 405 nm LED-treated milk. They found a shelf life of 19 hours and 9 days at room temperature and refrigerated conditions, respectively, with no significant changes in composition and physicochemical properties (content of carbohydrates, proteins, fats, moisture, total nitrogen, color, acidity, pH, and viscosity) of treated milk. A study by Bhavya and Hebbar (2019) also mentioned that there was no significant alteration in color parameters, phenolics, flavonoids, ascorbic acid, and antioxidant activity of the orange treated with blue light. The above reports indicate that further studies on optimization of LED-based processing conditions such as treatment time, light intensity, and processing temperature to minimize quality changes are required to further develop the knowledge base and upscale the technology.

Early studies with blue LED have shown potential application in food safety, especially for surface decontamination. However, prolonged exposure period was required for inactivation of foodborne pathogens. To reduce the processing time, many exogenous photosensitizers such as curcumin, chlorophyllin, hypericin, α-terthienyl, and 5-aminolevulinic acid can aid the photodynamic inactivation by blue light (D'Souza et al. 2015). Photodynamic inactivation is a novel technique used to inactivate micro-organisms, which is based on the use of photosensitizers (photoactive molecule) activated by specific wavelengths of light, including LEDs (Alves et al. 2015; Jiang et al. 2014). A few studies have been reported in the field of clinical research on the application of curcumin and blue on different pathogenic bacteria that cause dental complications (Araújo et al. 2012; Bulit et al. 2014; Leite et al. 2014). Curcumin-mediated photosensitization has significantly reduced the bacterial cells and fungal spores in vitro (Temba et al. 2016; Wu et al. 2016; Al-Asmari et al. 2017). Josewin et al. (2018) observed 2.3 and 2.2 $\log_{10}$ cfu cm^{-2} reduction of *L. monocytogenes* at 4 and 20 °C, respectively, with 100 μM sodium copper chlorophyllin-mediated blue LED illumination. In a similar manner, photosensitization-based treatment in the food industry is reported to be one of the potential hurdle technologies for decontamination of food packaging material, vegetables, and fruits (Luksiene and Brovko 2013).

A study by de Oliveira et al. (2018) investigated that combination of curcumin at 5 mg l^{-1} (diluted in ethanol) and 6 J cm^{-2} of visible light (400–800 nm) inactivated *E. coli* O157:H7 by ~4–5 $\log_{10}$ cfu ml^{-1} in fresh produce wash water. Gao and Matthews (2020) also mentioned that incubation time did not have any significant effect on the inactivation of *L. monocytogenes* in both broth and chicken skin. Another study by Aurum and Nguyen (2019) reported that when grapes were treated with 465 nm (36.3 J cm^{-2}) and curcumin at a concentration of 1.6 mM, *E. coli* reduced by 2.4 $\log_{10}$ cfu g^{-1}. Bhavya and Umesh Hebbar (2019) reported an inactivation of *E. coli* and *S. aureus* by 6 $\log_{10}$ cfu ml^{-1} when a combination of curcumin (20 μM) and blue light (13 J cm^{-2}). Luksiene and Paskeviciute (2011) noticed that a combination of sodium chlorophyllin and 400 nm LEDs reduced the *L. monocytogenes* counts by ~1.8 $\log_{10}$ cfu g^{-1} in strawberries.

Several studies have demonstrated the use of curcumin-mediated photosensitizers against a range of fungi and bacteria such as *S. aureus* (Jiang et al. 2014; Penha et al. 2017), *S.* Typhimurium (Penha et al. 2017), and *E. coli* (Haukvik et al. 2009). Haukvik et al. (2009) mentioned 3.0 $\log_{10}$ reductions in *E. coli* count when treated with 25 μM curcumin and 430 nm LED in combination. Similarly, Jiang et al. (2014) also reported 2.0 $\log_{10}$ reduction in *S. aureus* population with a treatment of curcumin (2.5 μM) and blue LED treatment (470 nm) in vitro. Penha et al. (2017) reported that a combination of curcumin (75 μM) and blue LED illumination (470 nm) reduced the pathogenic bacteria (*E. coli*, *S.* Typhimurium, *Aeromonas hydrophila*, *S. aureus*, *P. aeruginosa*) by 3.5–6.0 $\log_{10}$ cfu ml^{-1}. Aponiene et al. (2015) observed the reduction of mesophilic bacteria on fresh produce (apricots, plums, and cauliflowers), after a combination treatment with 15 μM hypericin and 585 nm LEDs for 30 minutes. Blue light-activated curcumin was shown to markedly damage the membrane permeability, resulting in cell death caused by intracellular ROS increase that could cause photodynamic killing of *S. aureus* in the presence of curcumin (Jiang et al. 2014).

ROS are generated when the photosensitizer gets excited by the light energy and returns to ground state through two pathways. In Type I mechanism, the transfer of an electron happens from the excited photosensitizer to molecular oxygen that is consequently reduced to ROS. While, in Type II mechanism, triplet oxygen gets excitation to the reactive singlet oxygen when energy transfers occur from the photosensitizer to triplet oxygen. Further, these ROS reacts with various biochemical components to form cytotoxic compounds that target cell membrane, DNA, and various enzymes leading to cell death (Luksiene and Brovko 2013).

Although there have been recently a few studies investigating the effects of combination treatment on food systems to improve antibacterial effects, the interactions of combination treatment are still not clear. Therefore, the combined effects and underlying antibacterial mechanisms need to be further studied for more efficient application of LEDs in the food industry.

5 Challenges and Limitations

The primary challenge with this technology is the penetration depth of visible light. Light interacts with different materials and tissues differently depending on the optical properties of the recipient (Thomas C. Vogelmann 1989). In some cases, only surface application is required. However, for food safety treatments, surface treatments are a limitation and it is possible hurdle technology approaches would be required to achieve desired treatments. Second, since the primary technology of interest – LEDs – is quite novel, there is still improvements to be made to electrical efficiency. Furthermore, as they are very small, large treatment systems will require a significant number of LEDs. While the cost of LEDs has improved, the cost remains quite high for those LEDs in the lower end (higher energy) of the spectrum. Finally, the advancement of this technology relies on united effective standard reporting of light treatment parameters. At present, there is no such practice and the most effective and methodology is still to be determined. However, reporting of parameters such as distance, wavelength, treatment duration, and temperature could facilitate thorough assessment of each study.

6 Conclusion

Light-based technologies are showing great promise in the food industry due to the many applications available and its sustainable chemical-free approach. Visible light is composed of numerous colors or wavelengths which interact with plants and microorganisms differently and thus can elicit responses that could increase shelf life or reduce microbial load. There are various sources that can produce visible light artificially; however, LEDs are quickly becoming a forerunner for the treatment using visible light because this technology can produce near monochromatic light and therefore allows more specific treatments. Moreover, this technology is sustainable and energy efficient. However, there is still further research required to optimize this technology.

References

Al-Asmari, F., Mereddy, R., and Sultanbawa, Y. (2017). A novel photosensitization treatment for the inactivation of fungal spores and cells mediated by curcumin. *Journal of Photochemistry and Photobiology B* 173: 301–306.

Alves, E., Faustino, M.A.F., Neves, M.G.P.M.S. et al. (2015). Potential applications of porphyrins in photodynamic inactivation beyond the medical scope. *Journal of Photochemistry and Photobiology C: Photochemistry Reviews* 22: 34–57.

Amin, R.M., Bhayana, B., Hamblin, M.R., and Dai, T. (2016). Antimicrobial blue light inactivation of *Pseudomonas aeruginosa* by photo-excitation of endogenous porphyrins: in vitro and in vivo studies. *Lasers in Surgery and Medicine* 48: 562–568.

Aponiene, K., Paskeviciute, E., Reklaitis, I., and Luksiene, Z. (2015). Reduction of microbial contamination of fruits and vegetables by hypericin-based photosensitization: comparison with other emerging antimicrobial treatments. *Journal of Food Engineering* 144: 29–35.

Araújo, N.C., Fontana, C.R., Bagnato, V.S., and Gerbi, M.E. (2012). Photodynamic effects of curcumin against cariogenic pathogens. *Photomedicine and Laser Surgery* 30: 393–399.

Aurum, F.S. and Nguyen, L.T. (2019). Efficacy of photoactivated curcumin to decontaminate food surfaces under blue light emitting diode. *Journal of Food Process Engineering* 42: e12988.

Azuma, A., Yakushiji, H., and Sato, A. (2019). Postharvest light irradiation and appropriate temperature treatment increase anthocyanin accumulation in grape berry skin. *Postharvest Biology and Technology* 147: 89–99.

Bantis, F., Smirnakou, S., Ouzounis, T. et al. (2018). Current status and recent achievements in the field of horticulture with the use of light-emitting diodes (LEDs). *Scientia Horticulturae* 235: 437–451.

Bárcena, A., Martínez, G., and Costa, L. (2019). Low intensity light treatment improves purple kale (*Brassica oleracea* var. *sabellica*) postharvest preservation at room temperature. *Heliyon* 5: e02467.

Bhavya, M.L. and Hebbar, H.U. (2019). Sono-photodynamic inactivation of *Escherichia coli* and *Staphylococcus aureus* in orange juice. *Ultrasonics Sonochemistry* 57: 108–115.

Bhavya, M.L. and Umesh Hebbar, H. (2019). Efficacy of blue LED in microbial inactivation: effect of photosensitization and process parameters. *International Journal of Food Microbiology* 290: 296–304.

Bolton, J.R., Mayor-Smith, I., and Linden, K.G. (2015). Rethinking the concepts of fluence (UV dose) and fluence rate: the importance of photon-based units – a systemic review. *Photochemistry and Photobiology* 91: 1252–1262.

Bulit, F., Grad, I., Manoil, D. et al. (2014). Antimicrobial activity and cytotoxicity of 3 photosensitizers activated with blue light. *Journal of Endodontics* 40: 427–431.

Bumah, V.V., Masson-Meyers, D.S., and Enwemeka, C.S. (2015). Blue 470 nm light suppresses the growth of *Salmonella enterica* and methicillin-resistant *Staphylococcus aureus* (MRSA) in vitro. *Lasers in Surgery and Medicine* 47: 595–601.

Burattini, C., Mattoni, B., and Bisegna, F. (2017). The impact of spectral composition of white LEDs on spinach (*Spinacia oleracea*) growth and development. *Energies* 10 (9): 1383.

Castillejo, N., Martínez-Zamora, L., Gómez, P.A. et al. (2021). Postharvest LED lighting: effect of red, blue and far red on quality of minimally processed broccoli sprouts. *Journal of the Science of Food and Agriculture* 101: 44–53.

Davis, P.A. and Burns, C. (2016). Photobiology in protected horticulture. *Food and Energy Security* 5: 223–238.

Dhakal, R. and Baek, K.-H. (2014). Short period irradiation of single blue wavelength light extends the storage period of mature green tomatoes. *Postharvest Biology and Technology* 90: 73–77.

dos Anjos, C., Sellera, F.P., de Freitas, L.M. et al. (2020). Inactivation of milk-borne pathogens by blue light exposure. *Journal of Dairy Science* 103: 1261–1268.

D'Souza, C., Yuk, H.-G., Khoo, G.H., and Zhou, W. (2015). Application of light-emitting diodes in food production, postharvest preservation, and microbiological food safety. *Comprehensive Reviews in Food Science and Food Safety* 14: 719–740.

Endarko, E., Maclean, M., Timoshkin, I.V. et al. (2012). High-intensity 405 nm light inactivation of *Listeria monocytogenes*. *Photochemistry and Photobiology* 88: 1280–1286.

Gao, J. and Matthews, K.R. (2020). Effects of the photosensitizer curcumin in inactivating foodborne pathogens on chicken skin. *Food Control* 109: 106959.

Ghate, V.S., Ng, K.S., Zhou, W. et al. (2013). Antibacterial effect of light emitting diodes of visible wavelengths on selected foodborne pathogens at different illumination temperatures. *International Journal of Food Microbiology* 166: 399–406.

Ghate, V., Leong, A.L., Kumar, A. et al. (2015). Enhancing the antibacterial effect of 461 and 521 nm light emitting diodes on selected foodborne pathogens in trypticase soy broth by acidic and alkaline pH conditions. *Food Microbiology* 48: 49–57.

Ghate, V., Kumar, A., Zhou, W., and Yuk, H.G. (2016). Irradiance and temperature influence the bactericidal effect of 460-nanometer light-emitting diodes on Salmonella in orange juice. *Journal of Food Protection* 79: 553–560.

Ghate, V., Kumar, A., Kim, M.-J. et al. (2017). Effect of 460 nm light emitting diode illumination on survival of *Salmonella* spp. on fresh-cut pineapples at different irradiances and temperatures. *Journal of Food Engineering* 196: 130–138.

Guffey, J.S., Payne, W.C., Motts, S.D. et al. (2016). Inactivation of Salmonella on tainted foods: using blue light to disinfect cucumbers and processed meat products. *Food Science & Nutrition* 4: 878–887.

Hasperué, J.H., Guardianelli, L., Rodoni, L.M. et al. (2016a). Continuous white–blue LED light exposition delays postharvest senescence of broccoli. *LWT - Food Science and Technology* 65: 495–502.

Hasperué, J.H., Rodoni, L.M., Guardianelli, L.M. et al. (2016b). Use of LED light for Brussels sprouts postharvest conservation. *Scientia Horticulturae* 213: 281–286.

Haukvik, T., Bruzell, E., Kristensen, S., and Tønnesen, H.H. (2009). Photokilling of bacteria by curcumin in different aqueous preparations. Studies on curcumin and curcuminoids XXXVII. *Pharmazie* 64: 666–673.

Hinds, L., O'Donnell, C., Akhter, M., and Tiwari, B. (2019). Principles and mechanisms of ultraviolet light emitting diode technology for food industry applications. *Innovative Food Science & Emerging Technologies* 56: 102153.

Huang, J.Y., Xu, F., and Zhou, W. (2018). Effect of LED irradiation on the ripening and nutritional quality of postharvest banana fruit. *Journal of the Science of Food and Agriculture* 98: 5486–5493.

Hyun, J.-E. and Lee, S.-Y. (2020a). Antibacterial effect and mechanisms of action of 460–470 nm light-emitting diode against *Listeria monocytogenes* and *Pseudomonas fluorescens* on the surface of packaged sliced cheese. *Food Microbiology* 86: 103314.

Hyun, J.-E. and Lee, S.-Y. (2020b). Blue light-emitting diodes as eco-friendly non-thermal technology in food preservation. *Trends in Food Science & Technology* 105: 284–295.

Jiang, Y., Leung, A., Hua, H. et al. (2014). Photodynamic action of LED-activated curcumin against *Staphylococcus aureus* involving intracellular ROS increase and membrane damage. *International Journal of Photoenergy* 2014: 637601.

Jiang, A., Zuo, J., Zheng, Q. et al. (2019). Red LED irradiation maintains the postharvest quality of broccoli by elevating antioxidant enzyme activity and reducing the expression of senescence-related genes. *Scientia Horticulturae* 251: 73–79.

Jiao, Y., Lau, O.S., and Deng, X.W. (2007). Light-regulated transcriptional networks in higher plants. *Nature Reviews Genetics* 8: 217–230.

Jin, P., Yao, D., Xu, F. et al. (2015). Effect of light on quality and bioactive compounds in postharvest broccoli florets. *Food Chemistry* 172: 705–709.

Josewin, S.W., Kim, M.-J., and Yuk, H.-G. (2018). Inactivation of *Listeria monocytogenes* and *Salmonella* spp. on cantaloupe rinds by blue light emitting diodes (LEDs). *Food Microbiology* 76: 219–225.

Kasim, M.U. and Kasim, R. (2017). While continuous white LED lighting increases chlorophyll content (SPAD), green LED light reduces the infection rate of lettuce during storage and shelf-life conditions. *Journal of Food Processing and Preservation* 41: e13266.

Kim, S.-H. and Bang, W.-S. (2013). Effects of the 461-nm LED light and combination with acid stress treatment on *Staphylococcus aureus* and *Escherichia coli*. *Korean Journal of Food Science and Technology* 45 (4): 526–529.

Kim, B.S., Lee, H.O., Kim, J.Y. et al. (2011). An effect of light emitting diode (LED) irradiation treatment on the amplification of functional components of immature strawberry. *Horticulture, Environment, and Biotechnology* 52: 35–39.

Kim, M.J., Mikš-Krajnik, M., Kumar, A. et al. (2015). Antibacterial effect and mechanism of high-intensity 405 ± 5 nm light emitting diode on *Bacillus cereus*, *Listeria monocytogenes*, and *Staphylococcus aureus* under refrigerated condition. *Journal of Photochemistry and Photobiology B* 153: 33–39.

Kim, M.-J., Mikš-Krajnik, M., Kumar, A., and Yuk, H.-G. (2016). Inactivation by 405 ± 5 nm light emitting diode on *Escherichia coli* O157:H7, *Salmonella* Typhimurium, and *Shigella sonnei* under refrigerated condition might be due to the loss of membrane integrity. *Food Control* 59: 99–107.

Kim, M.-J., Bang, W.S., and Yuk, H.-G. (2017a). 405 ± 5 nm light emitting diode illumination causes photodynamic inactivation of *Salmonella* spp. on fresh-cut papaya without deterioration. *Food Microbiology* 62: 124–132.

Kim, M.-J., Tang, C.H., Bang, W.S., and Yuk, H.-G. (2017b). Antibacterial effect of 405 ± 5 nm light emitting diode illumination against *Escherichia coli* O157:H7, *Listeria monocytogenes*, and *Salmonella* on the surface of fresh-cut mango and its influence on fruit quality. *International Journal of Food Microbiology* 244: 82–89.

Kumar, A., Ghate, V., Kim, M.J. et al. (2016). Antibacterial efficacy of 405, 460 and 520 nm light emitting diodes on *Lactobacillus plantarum*, *Staphylococcus aureus* and *Vibrio parahaemolyticus*. *Journal of Applied Microbiology* 120: 49–56.

Kumar, A., Ghate, V., Kim, M.-J. et al. (2017). Inactivation and changes in metabolic profile of selected foodborne bacteria by 460 nm LED illumination. *Food Microbiology* 63: 12–21.

Lacombe, A., Niemira, B.A., Sites, J. et al. (2016). Reduction of bacterial pathogens and potential surrogates on the surface of almonds using high-intensity 405-nanometer light. *Journal of Food Protection* 79: 1840–1845.

Leite, D.P.V., Paolillo, F.R., Parmesano, T.N. et al. (2014). Effects of photodynamic therapy with blue light and curcumin as mouth rinse for oral disinfection: a randomized controlled trial. *Photomedicine and Laser Surgery* 32: 627–632.

Liu, S., Hu, L., Jiang, D., and Xi, W. (2019). Effect of post-harvest LED and UV light irradiation on the accumulation of flavonoids and limonoids in the segments of newhall navel oranges (*Citrus sinensis* Osbeck). *Molecules* 24 (9): 1755.

Loi, M., Liuzzi, V.C., Fanelli, F. et al. (2019). Effect of different light-emitting diode (LED) irradiation on the shelf life and phytonutrient content of broccoli (*Brassica oleracea* L. var. *italica*). *Food Chemistry* 283: 206–214.

Luksiene, Z. and Brovko, L. (2013). Antibacterial photosensitization-based treatment for food safety. *Food Engineering Reviews* 5: 185–199.

Luksiene, Z. and Paskeviciute, E. (2011). Novel approach to the microbial decontamination of strawberries: chlorophyllin-based photosensitization. *Journal of Applied Microbiology* 110: 1274–1283.

Maclean, M., Macgregor, S.J., Anderson, J.G., and Woolsey, G. (2009). Inactivation of bacterial pathogens following exposure to light from a 405-nanometer light-emitting diode array. *Applied and Environmental Microbiology* 75: 1932–1937.

Maclean, M., Murdoch, L.E., Macgregor, S.J., and Anderson, J.G. (2013). Sporicidal effects of high-intensity 405 nm visible light on endospore-forming bacteria. *Photochemistry and Photobiology* 89: 120–126.

Maroga, G.M., Soundy, P., and Sivakumar, D. (2019). Different postharvest responses of fresh-cut sweet peppers related to quality and antioxidant and phenylalanine ammonia lyase activities during exposure to light-emitting diode treatments. *Foods* 8 (9): 359.

Mitchell, C., Both, A.J., Bourget, C. et al. (2012). LEDs: the future of greenhouse lighting. *Chronica Horticulturae* 52: 6–12.

Moorhead, S., Maclean, M., Macgregor, S.J., and Anderson, J.G. (2016). Comparative sensitivity of Trichophyton and *Aspergillus conidia* to inactivation by violet-blue light exposure. *Photomedicine and Laser Surgery* 34: 36–41.

Murdoch, L.E., Maclean, M., Endarko, E. et al. (2012). Bactericidal effects of 405 nm light exposure demonstrated by inactivation of *Escherichia*, *Salmonella*, *Shigella*, *Listeria*, and *Mycobacterium* species in liquid suspensions and on exposed surfaces. *The Scientific World Journal*: 2012, 137805–137805.

Murdoch, L.E., Mckenzie, K., Maclean, M. et al. (2013). Lethal effects of high-intensity violet 405-nm light on *Saccharomyces cerevisiae*, *Candida albicans*, and on dormant and germinating spores of *Aspergillus niger*. *Fungal Biology* 117: 519–527.

Nájera, C., Guil-Guerrero, J.L., Enríquez, L.J. et al. (2018). LED-enhanced dietary and organoleptic qualities in postharvest tomato fruit. *Postharvest Biology and Technology* 145: 151–156.

Ngcobo, B.L., Bertling, I., and Clulow, A.D. (2020). Post-harvest alterations in quality and health-related parameters of cherry tomatoes at different maturity stages following irradiation with red and blue LED lights. *The Journal of Horticultural Science and Biotechnology* 96: 383–391.

de Oliveira, E.F., Tosati, J.V., Tikekar, R.V. et al. (2018). Antimicrobial activity of curcumin in combination with light against *Escherichia coli* O157:H7 and *Listeria innocua*: applications for fresh produce sanitation. *Postharvest Biology and Technology* 137: 86–94.

Penha, C.B., Bonin, E., Da Silva, A.F. et al. (2017). Photodynamic inactivation of foodborne and food spoilage bacteria by curcumin. *LWT – Food Science and Technology* 76: 198–202.

Pola, W., Sugaya, S., and Photchanachai, S. (2020). Color development and phytochemical changes in mature green chili (*Capsicum annuum* L.) exposed to red and blue light-emitting diodes. *Journal of Agricultural and Food Chemistry* 68: 59–66.

Prasad, A., Du, L., Zubair, M. et al. (2020). Applications of light-emitting diodes (LEDs) in food processing and water treatment. *Food Engineering Reviews* 12: 268–289.

Sliney, D.H. (2016). What is light? The visible spectrum and beyond. *Eye* 30: 222–229.

Srimagal, A., Ramesh, T., and Sahu, J.K. (2016). Effect of light emitting diode treatment on inactivation of *Escherichia coli* in milk. *LWT – Food Science and Technology* 71: 378–385.

Temba, B.A., Fletcher, M.T., Fox, G.P. et al. (2016). Inactivation of *Aspergillus flavus* spores by curcumin-mediated photosensitization. *Food Control* 59: 708–713.

Trzaska, W.J., Wrigley, H.E., Thwaite, J.E., and May, R.C. (2017). Species-specific antifungal activity of blue light. *Scientific Reports* 7: 4605.

United Nations, Department of Economic and Social Affairs, Population Division 2019. World Population Prospects 2019: Ten key findings.

Vogelmann, T.C. (1989). Penetration of light into plants. *Photochemistry and Photobiology* 50: 895–902.

Wu, J., Mou, H., Xue, C. et al. (2016). Photodynamic effect of curcumin on *Vibrio parahaemolyticus*. *Photodiagnosis and Photodynamic Therapy* 15: 34–39.

Xu, F., Cao, S., Shi, L. et al. (2014). Blue light irradiation affects anthocyanin content and enzyme activities involved in postharvest strawberry fruit. *Journal of Agricultural and Food Chemistry* 62: 4778–4783.

8

Pulsed Light

Vicente M. Gómez-López[1], Rajeev Bhat[2], and José A. Pellicer[3]

[1] *Cátedra Alimentos para la Salud, UCAM Universidad Católica San Antonio de Murcia, Murcia, Spain*
[2] *ERA-Chair Holder in Food By-products Valorization Technologies (VALORTECH), Estonian University of Life Sciences (EMÜ), Tartu, Estonia*
[3] *Molecular Recognition and Encapsulation Research Group (REM), Health Sciences Department, Universidad Católica de Murcia (UCAM), Campus de los Jerónimos, Guadalupe, Spain*

1 Introduction

Pulsed light is an emerging nonthermal technology based on the application of short and intense pulses of incoherent wide-spectrum light, in which the UV-C component of the spectrum plays the main role in its effects (Gómez-López et al. 2007). The main feature of pulsed light technology is its rapidity; it can promote effects in a very short time. They are also mercury-free and do not require warming up. It is more efficient to inactivate microorganisms than low or medium-pressure lamps as per time and fluence basis (McDonald et al. 2000; Bohrerova et al. 2008).

A survey published in 2015 among food professionals from industry, academia, and government worldwide, but with respondents mainly from the United States and Europe, concluded that pulsed light could be ready for commercial utilization in the next 10 years (Jermann et al. 2015). However, pulsed light technology has already been adopted by the food industry. Even though the bulk of the research about its application in food technology has been focused on the inactivation of microorganisms in foods (John and Ramaswamy 2020), it has found application in the industry in two specific cases: the inactivation of microorganisms in food packages such as cups (Figure 1) and in the generation of mushrooms with high vitamin D content. The main limiting factor of this technology is that it can only be effective in surfaces and liquids since UV light is readily absorbed at surfaces; hence, it has poor penetration ability.

Electromagnetic Technologies in Food Science, First Edition. Edited by Vicente M. Gómez-López and Rajeev Bhat.

Figure 1 Pulsed light cup sterilization system. *Source:* Kindly provided by Claranor.

2 Pulsed Light as a Technology Based on the Electromagnetic Spectrum

The emission of the xenon lamp used in pulsed light technology span approximately from 190 to 1100 nm. This includes three portions: infrared, visible, and UV light. Its microbicide action is caused by UV-C and part of the UV-B light (200–300 nm). The rest of the radiation is essentially useless and can have even deleterious effects, such as heating. This means that only part of the emission spectrum of the lamp is useful, which is likely because it is a technology adapted from other application fields.

The shape of the emission spectrum depends on the manufacturer of the pulsed light system and on its operating conditions (when the equipment model allows choosing the discharge voltage). The higher the discharge voltage, the higher the lamp temperature, which blue-shifts the emission spectrum according to the Wein's displacement law (Upadhyaya et al. 2004; Schaefer et al. 2007). Therefore, a higher voltage not only generates a more intense light but also the relative amount of UV light increases.

Since the effect of the lamp emission depends on its wavelength composition, one can think that treatment conditions are hardly equivalent among pieces of equipment made by different manufacturers and within the piece of equipment under different charging voltages. There are cases in which the spectrum lacks of UV-C light emission, common in researcher's own-made devices, which can even be questioned as being part of pulsed light technology. The shape of the spectrum is also important because pulsed light results are commonly reported in terms of fluence, which is a general measurement that does not take into account spectrum composition. Some researchers do a better approximation reporting their results in terms of UV fluence. A more accurate method based in the determination of photon fluence has been developed (Gómez-López and Bolton 2016), but this is not recommendable when the specific targets under study are many, such as effects on natural microbiota, hormesis, etc. and require an equipment with a collimated beam that does not exist in the market.

The part of the emission spectrum between 190 and 200 nm corresponds to the vacuum UV light. Since there is generally air between the lamp and the sample, this part of the spectrum does not reach the sample because it is absorbed by the oxygen present in the air. This fact has, however, consequences in the equipment design. This part of the spectrum generates ozone upon contact with oxygen, and this is harmful to workers if it accumulates at high concentrations. System designers avoid this issue by two possible ways. One is adding the lamp envelope with a dopant (a trace of a specific compound added to the quartz envelope) in order to intercept that wavelength before exiting the lamp. The other is exhausting it by an airflow or dissolving it in water circulating between the lamp and a quartz jacket, both alternatives are also useful to cool down the lamp because lamp heating reduces its lifetime and can even cause its catastrophic failure (Upadhyaya et al. 2004). In addition, keeping the lamp temperature relatively low allows the use of high pulse repetition rates; otherwise, the lamp would overheat and explode.

The microbicide effect of the wavelengths between 200 and 300 nm has been demonstrated by the action spectra reported by Wang et al. (2005) and Bohrerova et al. (2008). A small effect of the visible portion has been observed by Bohrerova et al. (2008) in phages but not in *Escherichia coli*.

The infrared part of the spectrum is essentially useless and can be harmful to food quality since it heats the samples. As any part of the lamp spectrum, infrared rays can be selectively screened by using appropriate filters although this possibility is not commonly exploited nor the possibility of coupling the thermal and photochemical action.

3 Photochemistry and Photophysics Laws

Since the major target for microbial inactivation is the DNA molecule due to photon absorption, the process is essentially photochemical. The fundamental parameter to characterize a photochemical process is fluence, which is the amount of light impinging per target area and has unit of $\mathrm{J\,cm^{-2}}$. A formal definition can be found at the International Union of Pure and Applied Chemistry (IUPAC) glossary (Braslavsky 2007). In pulsed light tests, the fluence of the pulse is generally know. Therefore, the fluence of the treatment can be found according to Eq. (1) or alternatively to Eq. (2).

$$F_0 = F_{0,\pi} * n \tag{1}$$

where $F_{0,\pi}$ is the light pulse fluence, and n is the number of pulses.

$$F_0 = F_{0,\pi} * f * t \tag{2}$$

where f is the frequency (Hz), and t is time (s) required for each treatment.

It is of paramount importance to be aware of the laws that govern the photochemistry and photophysic of light-based technologies, which must be taken into account when designing experiments and interpreting results.

Two are the laws that govern photochemical process. The first law of photochemistry, also known as the Grotthus–Draper law, states that light must be absorbed by a compound in order for a photochemical reaction to take place. The second law of photochemistry or Stark–Einstein law states that the extent of any photochemical process must be proportional to the total number of photons absorbed (Rohatgi-Mukherjee 1986).

Another important law for pulsed light technology is the inverse square law, which states that the fluence varies in inverse proportion to the square of the distance. This law is only valid at high lamp–target distances (Ryer 1997).

4 Factors Affecting Efficacy

Since fluence is the main factor affecting the efficacy of pulsed light treatments, the variables that determine fluence will govern pulsed light efficacy. These variables are the distance between lamp and target, the number of pulses, and the discharge voltage. The fluence is higher when the distance between the lamp and the sample is lower, when increasing the number of light pulses and at higher discharge voltages. The latter occurs because when the voltage supplied to the lamp increases, the temperature of the plasma of the lamp rises and the light that dissipates gets higher.

The effect of the distance on the microbicide efficacy of this technology can be observed, among many others, in the works by Jun et al. (2003), Gómez-López et al. (2005a), Luksiene et al. (2007), and Farrell et al. (2009a). Similarly, the geometry of the illumination is also important since the fluence distribution is usually not homogeneous within the reactor (Hsu and Moraru 2011).

The effect of discharge voltage can be appreciated in the works by Jun et al. (2003) and Luksiene et al. (2007).

The composition of the emission spectrum also determines pulsed light efficacy because different wavelengths have different microbicide efficacy as discussed before.

Other factors affecting pulsed light efficacy are the type of microorganism as commented before, its physiological state, substrate composition, and shadow effect.

The shadow effect is the most limiting factor in the application of pulsed light and restricts its use to surfaces and translucid liquids Any hurdle between the light and the target will decrease or annul its efficacy. This can be surface irregularities that can be hidden microorganisms, dense microbial populations or thick and/or opaque liquid columns. Cudemos et al. (2013) have shown that the level of inactivation is lower at high microbial densities, likely due to light attenuation; the microorganisms in the upper layers can be inactivated but protect the underlying ones. Similarly, the effect of light is attenuated in liquids according to the Lambert–Beer law, and light-absorbing liquids are associated with lower inactivation. For example, the inactivation of *E. coli* as function of food matrix is in decreasing order: maximum recovery diluent > apple juice > milk (Palgan et al. 2011). In light-absorbing liquids, the efficacy of pulsed light is increased by turbulence (Sauer and Moraru 2009).

The physiological state also affects the susceptibility of microorganisms, which are more vulnerable at the exponential phase (Cudemos et al. 2013). Regarding substrate composition, food with high levels of proteins or lipids should be more difficult to decontaminate than food with high humidity or with a high proportion of carbohydrates (Gómez-López et al. 2005b).

5 Pulsed Light Systems

The intense, UV-rich emission generated by pulsed light systems is generated by storing electric energy in capacitors for relatively long time and releasing it to a xenon lamp in shorter time by means of a fast switch. The electricity generates a plasma inside the lamp that releases light. There are pulsed light systems designed for laboratory and pilot-plant use and for the industry. Laboratory systems can be discontinuous and continuous. The former is the most common and consists of a treatment chamber with one or more lamps and a refrigeration system such as a fan or water jacket to avoid lamp overheating. They are used in laboratory experiments, and the most frequently used have been described in Fernández et al. (2009), Murugesan et al. (2012), and Aguiló-Aguayo et al. (2013). Experiments under continuous regime carried out on fluids such as milk and juices have been performed adapting quartz pipes to discontinuous systems (Krishnamurthy et al. 2007) or using continuous systems built as such. They can be falling film (Artíguez et al. 2011) or horizontal reactors (Chaine et al. 2012; Uslu et al. 2016). Solids can also be treated in continuous regime using a conveyor belt (Janve et al. 2014).

At industrial level, there are systems designed to treat food-contact surfaces such as bottle caps or cups such as that shown in Figure 2.

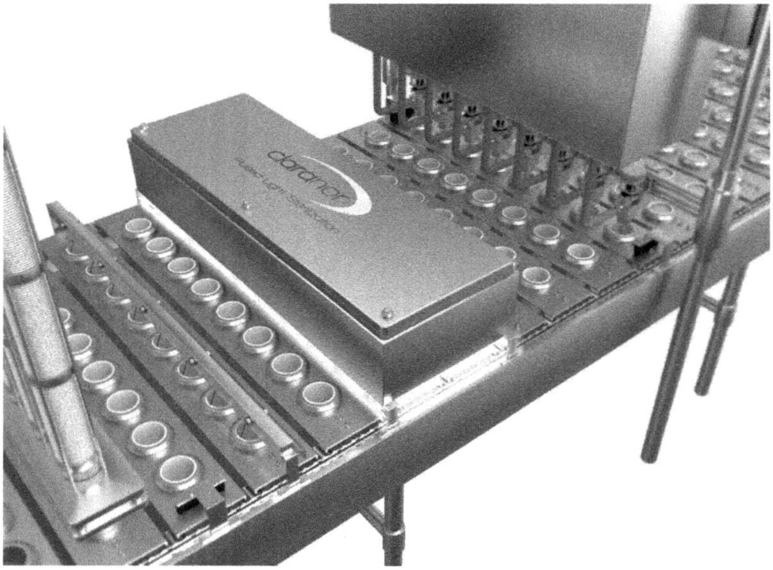

Figure 2 Artistic representation of an industrial pulsed light system to decontaminate cups. *Source:* Kindly provided by Claranor.

6 Effect on Microorganisms

Pulsed light has a broad action spectrum that includes viruses, Gram-positive and Gram-negative bacteria, yeasts, molds, fungal and bacterial spores, and protozoa. It is classic that the different kind of microorganisms are classified according to certain order of susceptibility depending on the inactivation technology or type of antimicrobial substance. However, this is not the case for pulsed light. Gómez-López et al. (2005a) could not find a susceptibility order after studying 25 bacteria and fungi, likely because the susceptibility depends fundamentally on the nucleotide sequence. The study showed indeed that there is a high variability in the susceptibility of microorganisms to pulsed light.

6.1 Action Spectrum

The action spectrum of a microorganism is the plot of the lethality as function of wavelength. The most microbicide wavelengths of the pulsed light spectrum are in the UV-C range and part of the UV-B range up to 300 nm. The peak germicidal wavelength for *E. coli* is 270 nm in the 230–300 nm range with no measurable inactivation at 300 nm and above (Wang et al. 2005). In phages, a minor effect of visible light wavelengths has been observed by Bohrerova et al. (2008), but not in *E. coli.*

It is worth studying the effect of very low UV-C wavelengths during pulsed light treatments. The determination of action spectra in light technologies has been dominated by the idea that the peak lethality matches the DNA absorption spectrum. However, many action spectra reported for conventional UV light do not confirm that idea. While that can be true for some microorganism, other action spectra differ from the DNA absorption spectrum with sharp increases below 240 nm and an increase around 280 nm according to more than a dozen of action spectra reviewed (Bolton 2017). According to Bolton (2017), the deviations seem to be correlated with the absorption spectra of aromatic amino acids, which would absorb UV light and transmit the excitation energy to nucleotides by excitation energy transfer.

6.2 Inactivation Mechanism

Early works about the microbicide efficacy of pulsed light frequently explained its inactivation mechanism based on the known effects of conventional UV-C light. There is currently specific research that has progressively unveiled the particular features of the effect of pulsed light on microorganisms and some differences have been discovered. Its effect has been classified into three types: photochemical, photothermal, and photophysical effects as proposed by Krishnamurthy et al. (2010). Depending on the analytical method used to measure or observe the inactivation, it is sometimes unclear if the results observed in some experiments are consequence of the photothermal or the photophysical effect. Anyway, photophysical effects should be consequence of one or both of the other effects.

The photochemical effect refers to the action of pulsed light on microbial DNA. Upon excitement of nucleotides, there is a formation of cyclobutane pyrimidine dimmers (Bohrerova et al. 2008; Aguirre et al. 2018). The dimmers form a notch in DNA structure that blocks the action of DNA polymerase, making DNA replication impossible and in consequence, the microorganisms cannot replicate, which is known as chlonogenic death. The formation of these dimmers is closely related to the action of wavelengths below 295 nm (Bohrerova et al. 2008). There are other photochemical effects such as single-strand breaks (Takeshita et al. 2003) and double-strand breaks (Cheigh et al. 2012), RNA breakage (Vimont et al. 2015), and pyrimidine (6–4) pyrimidine photoproduct (Aguirre et al. 2018). The accumulation of reactive oxygen species during pulsed light treatment, which can provoke damages in DNA, lipids, and proteins, has been observed in *Candida albicans* (Farrell et al. 2011), *Listeria innocua* (Kramer and Muranyi 2014), and *E. coli* O157:H7 (Zhu et al. 2019).

The photothermal effect refers to the disintegration of the microorganism as consequence of the fast heating of the cytoplasm (Wekhof et al. 2001). The explosion of microbial cells by pulsed light has been reported for *Aspergillus niger* and *Bacillus subtilis* spores (Wekhof et al. 2001) and vegetative cells (Nicorescu et al. 2013). No relationship has been established experimentally to predict when this effect occurs. Theoretically, the effect depends on the physical parameters of the microorganism (mass, size, thermal conductivity, and UV absorption) and on the thermal conductivity and UV absorption of the media supporting the microorganism (Wekhof et al. 2001), and it only takes place at high fluences per pulse (c. $0.5\,\mathrm{J\,cm^{-2}}$) (Wekhof 2000). The heating is localized on the microorganism since its heating rate and that of the surrounding media are different.

The photophysical effect refers to structural changes to bacterial cells as consequence of the constant disturbance caused by the high-energy pulses (Krishnamurthy et al. 2010). The changes are different from those characteristics of the photothermal effect. Perhaps the most notorious photophysical effect is the damage to cell membrane, which has been observed in cells of *Saccharomyces cerevisiae* (Takeshita et al. 2003; Ferrario et al. 2014), *C. albicans* (Farrell et al. 2009b), *Staphylococcus aureus* (Krishnamurthy et al. 2010), *L. innocua* (Kramer and Muranyi 2014), *E. coli* (Kramer and Muranyi 2014; Garvey et al. 2016), and *Bacillus cereus* and *Bacillus megaterium* (Garvey et al. 2016). Damage of viral capside has been observed in Phage MS2, hepatis A virus (Belliot et al. 2013), and murine norovirus 1 (Vimont et al. 2015). The occurrence of filamentous forms of *Listeria monocytogenes* after pulsed light treatment has also been observed (Bradley et al. 2012; Heinrich et al. 2016). The filamentous forms occur when the cells grow but cannot divide. This can have implications in food safety because filamentous forms can eventually divide (Heinrich et al. 2016).

6.3 Photoreactivation

It is known that bacteria inactivated by conventional UV light can repair cyclobutane thymine dimmers by several mechanisms. The most efficient repairing mechanism is called photoreactivation and requires visible light. The enzyme photolyase harvests visible light and excites the dimmer by electron transfer causing its split (Kao et al. 2005). However, when MacLean et al. (2008) studied the effect of different wavelengths in the range 320–450 nm on the photoreactivation of *S. aureus* and *L. monocytogenes*, they found that the optimal wavelengths for recovery of both bacteria were in the region of 360–380 nm. Otaki et al. (2003) demonstrated that bacteria inactivated by pulsed light can be photoreactivated and that this technology can decrease the capacity of photoreactivation of bacteria. The photoreparation rises with increasing visible fluence (Lamont et al. 2004). The photorepairing capacity of yeasts (Farrell et al. 2009b) and bacteria (Lasagabaster and Martínez de Marañón 2014) decreases with fluence until reaching a threshold beyond which microorganisms lose their capacity of photoreparation. This capacity is also lost if after pulsed light inactivation, the bacteria are kept in darkness before exposing them to daylight illumination; this effect is independent of treatment fluence (Lasagabaster and Martínez de Marañón 2014). From the practical point of view, Lamont et al. (2004) considered the consequences of photoreactivation irrelevant unless the food or water is subjected to very high visible light exposure because the amount of light required to cause photoreactivation is high.

6.4 Sublethal Injury

Microbial cells that are not killed by pulsed light treatment can become sublethally injured, which can be detected because these can grow faster in a nonselective than in a selective medium (Wuytack et al. 2003). Sublethal injury yields cells with difficulties to grow such as sublethally injured *L. innocua* cells, in which lag phase increases and growth rate decreases linearly with fluence (Aguirre et al. 2018). The lag phase of *L. monocytogenes* is also increased by pulsed light and is further higher at increasing concentrations of CO_2 (Van Houteghem et al. 2008).

6.5 Viable but Non-culturable State

Most of the research about the inactivation of microorganisms by pulsed light has used plate counts to enumerate survivors. However, pulsed light can promote some survivors that enter in a viable but non-culturable state (VBNC), which would underestimate the risks associated with pathogenic microorganisms. The occurrence of VBNC in pulsed-light-treated bacteria was identified in *Salmonella typhi* by Ben Said et al. (2012). The loss of culturability can occur at lower fluences than the loss of metabolic functions and the permeabilization of cell membranes in *L. innocua* and *E. coli* (Kramer and Muranyi 2014) and *S. cerevisiae* (Ferrario et al. 2014). A comprehensive review discussing the occurrence of the VBNC state after pulsed light treatment with reference to the inactivation kinetics has been recently published (Rowan et al. 2015).

7 Inactivation of Enzymes

Recently, the interest of light pulses has focused the attention on the inactivation of enzymes, especially those related to the undesirable properties that they exert in different food matrices.

One of the most common enzymes is peroxidase (POD). This enzyme is responsible for the browning effect in fruits and vegetables. Typical treatments to inactive this enzyme depend on thermal technologies; however, pulsed light is a green and nonthermal technology, and this is very useful to preserve the organoleptic properties. Pellicer and Gómez-López (2017) analyzed the inactivation of POD by using light pulses. More than 95% of this enzyme was inactivated at $128\,J\,cm^{-2}$. The inactivation followed a first-order kinetics; no crowding effect was observed, and the treatment provoked the ejection of the prostetic group. The application of light pulses gave rise to a decrease in the α-helices and an increase in the β-sheets compared with the native enzyme as measured by circular dichroism; however, Wang et al. (2017) reported a decrease in β-sheet composition as monitored by Raman spectroscopy.

A similar approach was used to study the inactivation of polyphenol oxidase (PPO) by using light pulses (Pellicer et al. 2018). More than 90% of the total enzymatic content was inactivated after applying $128\,J\,cm^{-2}$. This lack of activity was correlated by measuring the intrinsic fluorescence. Authors concluded that the inaction was an all or none process with no enzyme aggregation under the experimental conditions assessed. Previously, Manzocco et al. (2013) had reported the inactivation of PPO by pulsed light and found a dependence of the inactivation with the PPO concentration, likely due to the crowding effect. In this, high concentrations of enzyme would decrease its inactivation because their proximity would make the unfolding difficult.

Pellicer et al. (2019a) also studied the effect of light pulses in the structure of polygalacturonase (PG). The structure of the enzyme was analyzed by means of fluorescence emission spectra, free sulphydryl detection, and analysis of changes in parameter A and phase diagram. At $128\,J\,cm^{-2}$, the enzyme was inactivated more than 90%. According to the main results observed the inactivation of this enzyme, there is an all-or-none process where disulfide bridges are broken and the enzyme unfolds.

Lipoxygenase (LOX) is a nonheme iron-containing enzyme that catalyzes the deoxygenation of fatty acids. LOX is ubiquitous in plants, and its activity affects the stability of some foods. The inactivation of LOX in foods is very important because it catalyzes the oxidation of unsaturated fatty acids, causing undesirable by-products. The inactivation of this enzyme was reported by Pellicer et al. (2019b). LOX was fully inactivated after applying $96\,J\,cm^{-2}$, this inactivation was photochemical and all-or-none process. The treatment gave rise to the oxidation of aminoacids and the reduction of the α-helix content. The protein aggregates during treatment, likely due to hydrophobic interactions.

The final appearance of fruits juices depends on the activity of pectinmethylesterase (PME). The inactivation of PME is necessary to preserve cloudy juices since cloudiness is a determinant of the purchase decision of consumers in this type of drinks. The inactivation followed a first-order kinetic and more than 90% of the total enzymatic content was inactivated at $128\,J\,cm^{-2}$. The inactivation of PME by using this green technology was in accordance with the formation of a molten-globule state in all-or-none process where PME losses tertiary structure and aggregates, while the secondary structure was conserved (Pellicer et al. 2020).

8 Inactivation of Allergens

Food allergies are caused by adverse immunological responses to proteins, or antigens, which are components of food matrices. In the case of foods, these antigens are referred to as food allergens. The inactivation of allergens by using nonthermal treatments is a promising technology. Table 1 summarizes the main allergens inactivated by using light pulses. While pulsed light has been shown to be effective for allergen inactivation, the fluence necessary to achieve it is very high, many requiring treatments up to 4 minutes at three pulses per second and with a considerable evaporation of the extracts.

Table 1 Studies on the inactivation of allergens by pulsed light.

Food	Allergen	Maximal fluence (J cm^{-2})	References
Peanut butter	Peanut extract	221	Chung et al. (2008)
Milk	α-casein	NR	Anugu et al. (2009)
	β-lactoglogulin		
	α-lactalbumin		
Egg	Egg proteins	NR	Anugu et al. (2010)
Soybean	Glycinin	353	Yang et al. (2010)
	β-conglycinin		
Shrimp	Tropomyosin	194	Shriver et al. (2011)
Peanuts	Ara h2	223	Yang et al. (2012)
Almond	Almond proteins	NR	Li et al. (2013)
Peanuts	Peanuts kernels	NR	Zhao et al. (2014)
Milk	β-lactoglobulin	4	del Castillo-Santaella et al. (2014)

NR, not reported.

9 Effect on Lipids

It is widely known that lipid oxidation is favored by light. Therefore, it can be thought that there is a risk of inducing rancidity when treating fatty foods with pulsed light in the presence of oxygen, especially when the treatment is accompanied of temperature rise. However, the increase of rancidity to rejection levels would depend on the fluence applied. Furthermore, the short pulse duration can offer to pulsed light a benefit with respect to continuous UV light. Pulsed light may limit oxidation reactions because of the half-life of p-bonds and the short pulse duration, which prevent efficient coupling with oxygen (Fine and Gervais 2004). For example, the level of lipid peroxidation in skinless breast chicken increased with fluence; however, the same fluence considered satisfactory to inactivate *Salmonella* Typhimurium and *L. monocytogenes* did not cause changes detectable by a sensory panel (Paskeviciute et al. 2011). Similar results have been reported for cooked ham (Hierro et al. 2011) and chicken frankfurters (Keklik et al. 2009); however, bologna did not develop rancidity (Hierro et al. 2011). No fatty acid oxidation has been detected in pulsed-light-treated milk but vitamin A was decomposed (Elmnasser et al. 2008). No lipid oxidation was detected in halved walnuts (Izmirlioglu et al. 2020).

10 Effect on Health-Related Compounds

Mild stresses are used in postharvest technology to generate positive physiological responses such as an increased resistance to mold attach or generation of health-related compounds. Pulsed light can be used for these purposes. Numerous studies avail the application of pulsed light for the increase general markers such as antioxidant capacity and total phenolic compounds or enhancing concentrations of specific compounds. The induced changes are generally not instantaneous but produced during storage. Pulsed light increases the concentration of carotenoids and total antioxidant capacity of figs (Rodov et al. 2012), total phenol content of elderberry fruit (Murugesan et al. 2012) and Annurca apple (Pataro et al. 2015a). In tomatoes, it increases the content of total carotenoid, lycopene, phenolic compounds, and antioxidant activity (Aguiló-Aguayo et al. 2013; Pataro et al. 2015b).

11 Effect on Vitamin D

Vitamin D can be synthetized by exposing mushrooms to pulsed light. Mushrooms contain ergosterol, which can be converted to ergocalciferol (vitamin D_2) by UV light by means of a simple photochemical reaction. The dramatic increase of vitamin D in mushrooms can be achieved by pulsed light in few seconds, and it is a process that is currently in use at industrial level. The process is much more efficient than the application of conventional UV light for the same purpose, which needs 5–20 minutes of exposure versus just 1 second using pulsed light treatment (Koyyalamudi et al. 2011). The vitamin D content of white button (*Agaricus bisporus*) mushrooms increases with the number of pulses, becomes higher in sliced versus whole mushrooms and in white versus brown mushrooms. During refrigerated storage, the vitamin D_2 content of sliced white button mushrooms decreases after three days and then remains stable up to 11 days (Kalaras et al. 2012). Similar results have been obtained using the same approach to treat mushrooms from the *Pleurotus* genus with enhanced, with losses of vitamin D_2 of 28.9% and 37.7% after 60 days of storage at 4 and 25 °C, respectively (Chen et al. 2015).

12 Effect on Pesticides

Pulsed light can be used to degrade pesticides. This is important because fruits can contain pesticide residues that should be eliminated to protect the consumer. Moreover, wastewater produced by fruit and vegetable washing in the food industry can contain considerable amounts of pesticides that can enter the environment and polluting it. An environmentally friendly food industry should take into consideration the amount of pesticides that are discharged to the environment. Pulsed light is able to degrade atrazine (Baranda et al. 2012) and other herbicides such as phosmet, azinphos-methyl, simazine, pirimiphos-methyl, methyl-parathion, chlorpyrifos-ethyl (Baranda et al. 2014), malathion, and bromopropylate (Baranda et al. 2017). These compounds are modified by pulsed light giving place to products that are further degraded following a degradation pathway specific of each compound. However, it is well known in the research field of environmental pollution that light-based methods to degrade pollutants can give place to by-products with higher toxicity than their parent compounds. This is the case observed when pulsed light has been used together with H_2O_2 to degrade another toxic compound, the azo dye Direct red 83 : 1 (Gómez-Morte et al. 2021). Even more, some by-products can be refractory to pulsed light and remain in water. Therefore, the use of pulsed light for this application requires eco-toxicological

Figure 3 Continuous pulsed light system for liquids treatment. *Source:* Uslu et al. (2016).

studies in order to better assess the efficacy of this technology from the environmental point of view. Alternatively, mineralization indicators such as total organic carbon or chemical oxygen demand can be used to monitor pesticide degradation. The mineralization of organic compounds means to transform them into inorganic compounds such as CO_2, which implies the reduction of the concentration of by-products potentially more toxic than their parent compounds. A flow-through pulsed light reactor has been shown to be useful for mineralization purposes (Figure 3).

13 Energy Efficiency

Very few studies report the energy consumption required by pulsed light to achieve specific goals, which may have slowed down its adoption by the food industry. The adoption of the figure-of-merit electric energy per order, which is the electric energy in kilowatt hours required to degrade a contaminant by one order of magnitude in a unit volume (Bolton et al. 2001), may be beneficial for this aim. The main problem precluding its use is that in many pulsed light systems used in research laboratories, a hard-to-quantify amount of lamp emission does not reach the target (Navarro et al. 2017). The use of this figure of merit in the pulsed light context can be appreciated in Ferrario and Guerrero (2016, 2017).

14 Legislations (Regulations and Safety) of Pulsed Light

Consumers' safety is one of the paramount criteria that needs to be considered when a novel technology is applied for food products. U. S. Food and Drug Administration (FDA) has already approved use of pulsed light technology for decontamination of food and/or food contact surfaces (FDA Code 21CFR179.41). As per the "Federal Register Rules and Regulations," application of pulsed light for safe treatment of foods is cleared as afar as radiation source used is xenon flash lamps, which are designed to emit broadband radiation (wavelength range of 200–1100 nm) with pulse duration not exceeding 2 ms and 12.0 J cm^{-2}. Further, the technology needs to be used for surface microbial control and with minimal treatment to achieve/meet the specified technical effects (Shank 1996). In addition, in the European Union, as of now there are no distinct legislations set aside on use or application of pulse light technology in food applications. However, there is a mention on the specifications of pulse light that can be used (EU Commission Implementing Regulation 2011).

Significant concerns over food safety on the presence of viable microorganisms after pulse light treatments are reported (Schottroff et al. 2018). Nevertheless, when applying pulse light on a commercial scale, it is worth to consider a significant log reduction of pathogenic microbes in accordance with the essential hazard analysis and critical control points (HACCP) regulations of a food industry. Further, from physical safety point of view, further research activities are still warranted to design suitable and user-friendly treatment chamber, which are capable of meeting the food industrial demands.

15 Conclusions and Future Outlook

Pulsed light has demonstrated to be a technology that produces very fast results using short treatment times and can be used to reduce microbial counts, inactivate enzymes and allergens, enhance the concentration of health-related compounds and vitamin D, and reduce the concentration of pesticides and mycotoxins. While initially devoted for microbial decontamination of foods, its current industrial use is in the decontamination of food-contact surfaces and to increase the vitamin D content of mushrooms. A wider adoption of this technology by the food industry may be encouraged by better dosimetry that should allow an easier scaling-up and by reporting the energy consumption required by this technology to achieve specific goals.

Conflict of Interest Statement

The authors declare no conflicts of interest.

References

Aguiló-Aguayo, I., Charles, F., Renard, C.M.G.C. et al. (2013). Pulsed light effects on surface decontamination, physical qualities and nutritional composition of tomato fruit. *Postharvest Biology and Technology* 86: 29–36. https://doi.org/10.1016/j.postharvbio.2013.06.011.

Aguirre, J.S., García de Fernando, G., Hierro, E. et al. (2018). Characterization of damage on *Listeria innocua* surviving to pulsed light: effect on growth, DNA and proteome. *International Journal of Food Microbiology* 284: 63–72. https://doi.org/10.1016/j.ijfoodmicro.2018.07.002.

Anugu, A., Yang, W., and Krishnamurthy, K. (2009). Efficacy of pulsed ultraviolet light for reduction of allergenicity in isolated milk proteins. Abstract presented at the IFT, Anaheim, CA (2009).

Anugu, A.K., Yang, W., Shriver, S. et al. (2010). Efficacy of pulsed ultraviolet light on reducing the allergenicity of isolated egg proteins. IFT Annual Meeting, Chicago, IL, USA. pp. 17–20.

Artíguez, M.L., Lasagabaster, A., and Martínez de Marañón, I. (2011). Factors affecting microbial inactivation by pulsed light in a continuous flow-through unit for liquid products treatment. *Procedia Food Science* 1: 786–791. https://doi.org/10.1016/j.profoo.2011.09.119.

Baranda, A.B., Barranco, A., and Martínez de Marañón, I. (2012). Fast atrazine photodegradation in water by pulsed light technology. *Water Research* 46: 669–678. https://doi.org/10.1016/j.watres.2011.11.034.

Baranda, A.B., Fundazuri, O., and Martínez de Marañón, I. (2014). Photodegradation of several triazidic and organophosphorus pesticides in water by pulsed light technology. *Journal of Photochemistry and Photobiology A: Chemistry* 286: 29–39. https://doi.org/10.1016/j.jphotochem.2014.03.015.

Baranda, A.B., Lasagabaster, A., and Martínez de Marañón, I. (2017). Static and continuous flow-through pulsed light technology for pesticide abatement in water. *Journal of Hazardous Materials* 340: 140–151. https://doi.org/10.1016/j.jhazmat.2017.07.012.

Belliot, G., Loutreul, J., Estienney, M. et al. (2013). Potential of pulsed light to inactivate bacteriophage MS2 in simple liquid medium and on complex foodstuffs. *Food and Environmental Virology* 5: 176–179. https://doi.org/10.1007/s12560-013-9110-8.

Ben Said, M., Otaki, M., and Hassen, A. (2012). Use of lytic phage to control *Salmonella typhi*'s viability after irradiation by pulsed UV light. *Annales de Microbiologie* 62: 107–111. https://doi.org/10.1007/s13213-011-0234-5.

Bohrerova, Z., Shemer, H., Lantis, R. et al. (2008). Comparative disinfection efficiency of pulsed and continuous-wave UV irradiation technologies. *Water Research* 42: 2975–2982. https://doi.org/10.1016/10.1016/j.watres.2008.04.001.

Bolton, J.R. (2017). Action spectra: a review. *IUVA News* 19 (2): 10–12.

Bolton, J.R., Bircher, K.G., Tumas, W., and Tolman, C.A. (2001). Figures-of-merit for the technical development and application of advanced oxidation technologies for both electric- and solar-driven systems. *Pure and Applied Chemistry* 73: 627–637. https://doi.org/10.1351/pac200173040627.

Bradley, D., McNeil, B., Laffey, J.G., and Rowan, N.J. (2012). Studies on the pathogenesis and survival of different culture forms of *Listeria monocytogenes* to pulsed UV-light irradiation after exposure to mild-food processing stresses. *Food Microbiology* 30: 330–339. https://doi.org/10.1016/j.fm.2011.12.024.

Braslavsky, S.E. (2007). Glossary of terms used in photochemistry. *Pure and Applied Chemistry* 79: 293–465. https://doi.org/10.1351/pac200779030293.

del Castillo-Santaella, T., Sanmartín, E., Cabrerizo-Vílchez, M.A. et al. (2014). Improved digestibility of β-lactoglobulin by pulsed light processing: a dilatational and shear study. *Soft Matter* 10: 9702–9714. https://doi.org/10.1039/C4SM01667J.

Chaine, A., Levy, C., Lacour, B. et al. (2012). Decontamination of sugar syrup by pulsed light. *Journal of Food Protection* 75: 913–917. https://doi.org/10.4315/0362-028X.JFP-11-342.

Cheigh, C.-I., Park, M.-H., Chung, M.-S. et al. (2012). Comparison of intense pulsed light- and ultraviolet (UVC)-induced cell damage in *Listeria monocytogenes* and *Escherichia coli* O157:H7. *Food Control* 25: 654–659. https://doi.org/10.1016/j.foodcont.2011.11.032.

Chen, S.Y., Huang, S.J., Cheng, M.C. et al. (2015). Enhancement of vitamin D_2 content in *Pleurotus* mushrooms using pulsed light. *Journal of Food Processing and Preservation* https://doi.org/10.1111/jfpp.12443.

Chung, S.Y., Yang, W., and Krishnamurthy, K. (2008). Effects of pulsed UV-light on peanut allergens in extracts and liquid peanut butter. *Journal of Food Science* 73: C400–C404. https://doi.org/10.1111/j.1750-3841.2008.00784.x.

Cudemos, E., Izquier, A., Medina-Martínez, M.S., and Gómez-López, V.M. (2013). Effects of shading and growth phase on the microbial inactivation by pulsed light. *Czech Journal of Food Science* 31: 189–193. https://doi.org/10.17221/145/2012-CJFS.

Elmnasser, N., Dagalarrondo, M., Orange, N. et al. (2008). Effect of pulsed-light treatment on milk proteins and lipids. *Journal of Agricultural and Food Chemistry* 56: 1984–1991. https://doi.org/10.1021/jf0729964.

EU Commission Implementing Regulation (2011). EU No 1204/2011 of 18 November 2011 concerning the classification of certain goods in the combined nomenclature. https://eur-lex.europa.eu/eli/reg_impl/2011/1204/oj (accessed 8th September 2020).

Farrell, H.P., Garvey, M., Cormican, M. et al. (2009a). Investigation of critical inter-related factors affecting the efficacy of pulsed light for inactivating clinically relevant bacterial pathogens. *Journal of Applied Microbiology* 108: 1494–1508. https://doi.org/10.1111/j.1365-2672.2009.04545.x.

Farrell, H., Garvey, M., and Rowan, N. (2009b). Studies on the inactivation of medically important *Candida* species on agar surfaces using pulsed light. *FEMS Yeast Research* 9: 956–966. https://doi.org/10.1016/j.mimet.2010.12.021.

Farrell, H., Hayes, J., Laffey, J., and Rowan, N. (2011). Studies on the relationship between pulsed UV light irradiation and the simultaneous occurrence of molecular and cellular damage in clinically-relevant *Candida albicans*. *Journal of Microbiological Methods* 84: 317–326. https://doi.org/10.1016/j.mimet.2010.12.021.

Fernández, M., Manzano, S., de la Hoz, L. et al. (2009). Pulsed light inactivation of *Listeria monocytogenes* through different plastic films. *Foodborne Pathogens and Disease* 6: 1265–1267. https://doi.org/10.1089/fpd.2009.0328.

Ferrario, M. and Guerrero, S. (2016). Effect of a continuous flow-through pulsed light system combined with ultrasound on microbial survivability, color and sensory shelf life of apple juice. *Innovative Food Science and Emerging Technologies* 34: 214–224. https://doi.org/10.1016/j.ifset.2016.02.002.

Ferrario, M. and Guerrero, S. (2017). Impact of a combined processing technology involving ultrasound and pulsed light on structural and physiological changes of *Saccharomyces cerevisiae* KE 162 in apple juice. *Food Microbiology* 65: 83–94. https://doi.org/10.1016/j.fm.2017.01.012.

Ferrario, M., Guerrero, S., and Alzamora, S.M. (2014). Study of pulsed light-induced damage on *Saccharomyces cerevisiae* in apple juice by flow cytometry and transmission electron microscopy. *Food and Bioprocess Technology* 7: 1001–1011. https://doi.org/10.1007/s11947-013-1121-9.

Fine, F. and Gervais, P. (2004). Efficiency of pulsed UV light for microbial decontamination of food powders. *Journal of Food Protection* 67: 787–792.

Garvey, M., Stocca, A., and Rowan, N. (2016). Use of a real time PCR assay to assess the effect of pulsed light inactivation on bacterial cell membranes and associated cell viability. *Water Environment Research* 88: 168–174. https://doi.org/10.2175/106143016X14504669767210.

Gómez-López, V.M. and Bolton, J.R. (2016). An approach to standardize methods for fluence determination in bench-scale pulsed light experiments. *Food and Bioprocess Technology* 9: 1040–1048. https://doi.org/10.1007/s11947-016-1696-z.

Gómez-López, V.M., Devlieghere, F., Bonduelle, V., and Debevere, J. (2005a). Factors affecting the inactivation of micro-organisms by intense light pulses. *Journal of Applied Microbiology* 99: 460–470. https://doi.org/10.1111/j.1365-2672.2005.02641.x.

Gómez-López, V.M., Devlieghere, F., Bonduelle, V., and Debevere, J. (2005b). Intense light pulses decontamination of minimally processed vegetables and their shelf-life. *International Journal of Food Microbiology* 103: 79–89. https://doi.org/10.1016/j.ijfoodmicro.2004.11.028.

Gómez-López, V.M., Ragaert, P., Debevere, J., and Devlieghere, F. (2007). Pulsed light for food decontamination: a review. *Trends in Food Science and Technology* 18: 464–473. https://doi.org/10.1080/10408390701638878.

Gómez-Morte, T., Gómez-López, V.M., Lucas-Abellán, C. et al. (2021). Removal and toxicity evaluation of a diverse group of drugs from water by a cyclodextrin polymer/pulsed light system. *Journal of Hazardous Materials* 402: 123504. https://doi.org/10.1016/j.jhazmat.2020.123504.

Heinrich, V., Zunabovic, M., Petschning, A. et al. (2016). Previous homologous and heterologous stress exposure induces tolerance development to pulsed light in *Listeria monocytogenes*. *Frontiers in Microbiology* 7: 490. https://doi.org/10.3389/fmicb.2016.00490.

Hierro, E., Barroso, E., de la Hoz, L. et al. (2011). Efficacy of pulsed light for shelf-life extension and inactivation of *Listeria monocytogenes* on ready-to-eat cooked meat products. *Innovative Food Science and Emerging Technologies* 12: 275–281. https://doi.org/10.1016/j.ifset.2011.04.006.

Hsu, L. and Moraru, C.I. (2011). Quantifying and mapping the spatial distribution of fluence inside a pulsed light treatment chamber and various liquid substrates. *Journal of Food Engineering* 103: 84–91. https://doi.org/10.1016/j.jfoodeng.2010.10.002.

Izmirlioglu, G., Ouyang, B., and Demirci, A. (2020). Utilization of pulsed UV light for inactivation of *Salmonella* Enteritidis on shelled walnuts. *LWT - Food Science and Technology* 134: 110023. https://doi.org/10.1016/j.lwt.2020.110023.

Janve, B.A., Yang, W., Marshall, M.R. et al. (2014). Nonthermal inactivation of soy (*Glycine max* sp.) lipoxygenase by pulsed ultraviolet light. *Journal of Food Science* 79: C8–C18. https://doi.org/10.1111/1750-3841.12317.

Jermann, C., Koutchma, T., Margas, E. et al. (2015). Mapping trends in novel and emerging food processing technologies around the world. *Innovative Food Science and Emerging Technologies* 31: 14–27. https://doi.org/10.1016/j.ifset.2015.06.007.

John, D. and Ramaswamy, H.S. (2020). Comparison of pulsed light inactivation kinetics and modeling of *Escherichia coli* (ATCC-29055), *Clostridium sporogenes* (ATCC-7955) and *Geobacillus stearothermophilus* (ATCC-10149). *Current Research in Food Science* 3: 82–91. https://doi.org/10.1016/j.crfs.2020.03.005.

Jun, S., Irudayaraj, J., Demirci, A., and Geiser, D. (2003). Pulsed UV-light treatment of corn meal for inactivation of *Aspergillus niger* spores. *International Journal of Food Science and Technology* 38: 883–888. https://doi.org/10.1046/j.0950-5423.2003.00752.x.

Kalaras, M.D., Beelman, R.B., and Elias, R.J. (2012). Effects of postharvest pulsed UV light treatment of white button mushrooms (*Agaricus bisporus*) on vitamin D2 content and quality attributes. *Journal of Agricultural and Food Chemistry* 60: 220–225. https://doi.org/10.1021/jf203825e.

Kao, Y.-T., Saxena, C., Wang, L. et al. (2005). Direct observation of thymine dimer repair in DNA by photolyase. *Proceedings of the National Academy of Sciences* 102: 16128–16132. https://doi.org/10.1073/pnas.0506586102.

Keklik, N.M., Demirci, A., and Puri, V.M. (2009). Inactivation of *Listeria monocytogenes* on unpackaged and vacuum-packaged chicken frankfurters using pulsed UV-light. *Journal of Food Science* 74: M431–M439. https://doi.org/10.1111/j.1750-3841.2009.01319.x.

Koyyalamudi, S.R., Jeon, S.C., Pang, G. et al. (2011). Concentration of vitamin D_2 in white button mushroom (*Agaricus bisporus*) exposed to pulsed UV light. *Journal of Food Composition and Analysis* 24: 976–979. https://doi.org/10.1016/j.jfca.2011.02.007.

Kramer, B. and Muranyi, P. (2014). Effect of pulsed light on structural and physiological properties of *Listeria innocua* and *Escherichia coli*. *Journal of Applied Microbiology* 116: 596–611. https://doi.org/10.1111/jam.12394.

Krishnamurthy, K., Demirci, A., and Irudayaraj, J. (2007). Inactivation of *Staphylococcus aureus* in milk using flow-through pulsed UV-light treatment system. *Journal of Food Science* 72: M233–M239. https://doi.org/10.1111/j.1750-3841.2007.00438.x.

Krishnamurthy, K., Tewari, J.C., Irudayaraj, J., and Demirci, A. (2010). Microscopic and spectroscopic evaluation of inactivation of *Staphylococcus aureus* by pulsed UV light and infrared heating. *Food and Bioprocess Technology* 3: 93–104. https://doi.org/10.1007/s11947-008-0084-8.

Lamont, Y., MacGregor, S.J., Anderson, J.G., and Fouracre, R.A. (2004). Effect of visible light exposure on *E. coli* treated with pulsed UV-rich light. IEEE International Power Modulator Conference, San Francisco, CA, USA. https://doi.org/10.1109/MODSYM.2004.1433653.

Lasagabaster, A. and Martínez de Marañón, I. (2014). Survival and growth of *Listeria innocua* treated by pulsed light technology: impact of post-treatment temperature and illumination conditions. *Food Microbiology* 41: 76–81. https://doi.org/10.1016/j.fm.2014.02.001.

Li, Y., Yang, W., Chung, S.Y. et al. (2013). Effect of pulsed ultraviolet light and high hydrostatic pressure on the antigenicity of almond protein extracts. *Food and Bioprocess Technology* 6: 431–440. https://doi.org/10.1007/s11947-011-0666-8.

Luksiene, Z., Gudelis, V., Buchovec, I., and Raudeliuniene, J. (2007). Advanced high-power pulsed light device to decontaminate food from pathogens: effects on *Salmonella* Typhimurium viability *in vitro*. *Journal of Applied Microbiology* 103: 1545–1552. https://doi.org/10.1111/j.1365-2672.2007.03403.x.

Maclean, M., Murdoch, L.E., Lani, M.N. et al. (2008). Photoinactivation and photoreactivation responses by bacterial pathogens after exposure to pulsed UV-light. Proceedings of the 2008 IEEE International Power Modulators and High Voltage Conference, Las Vegas, USA (2008).

Manzocco, L., Panozzo, A., and Nicoli, M.C. (2013). Inactivation of polyphenol oxidase by pulsed light. *Journal of Food Science* 78: E1183–E1187. https://doi.org/10.1111/1750-3841.12216.

McDonald, K.F., Curry, R.D., Clevenger, T.E. et al. (2000). A comparison of pulsed and continuous ultraviolet light sources for the decontamination of surfaces. *IEEE Transactions on Plasma Science* 28: 1581–1587. https://doi.org/10.1109/27.901237.

Murugesan, R., Orsat, V., and Lefsrud, M. (2012). Effect of pulsed ultraviolet light on the total phenol content of elderberry (*Sambucus nigra*) fruit. *Food and Nutrition Sciences* 3: 774–783. https://doi.org/10.4236/fns.2012.36104.

Navarro, P., Gabaldón, J.A., and Gómez-López, V.M. (2017). Degradation of an azo dye by a fast and innovative pulsed light/H_2O_2 advanced oxidation process. *Dyes and Pigments* 136: 887–892. https://doi.org/10.1016/j.dyepig.2016.09.053.

Nicorescu, I., Nguyen, B., Moreu-Ferret, M. et al. (2013). Pulsed light inactivation of *Bacillus subtilis* vegetative cells in suspensions and spices. *Food Control* 31: 151–157. https://doi.org/10.1016/j.foodcont.2012.09.047.

Otaki, M., Okuda, A., Tajima, K. et al. (2003). Inactivation differences of microorganisms by low pressure UV and pulsed xenon lamps. *Water Science and Technology* 47: 185–190.

Palgan, I., Caminiti, I.M., Muñoz, A. et al. (2011). Effectiveness of high intensity light pulses (HILP) treatments for the control of *Escherichia coli* and *Listeria innocua* in apple juice, orange juice and milk. *Food Microbiology* 28: 14–20. https://doi.org/10.1016/j.fm.2010.07.023.

Paskeviciute, E., Buchovec, I., and Luksiene, Z. (2011). High-power pulsed light for decontamination of chicken from food pathogens: a study on antimicrobial efficiency and organoleptic properties. *Journal of Food Safety* 31: 61–68. https://doi.org/10.1111/j.1745-4565.2010.00267.x.

Pataro, G., Donsi, G., and Ferrari, G. (2015a). Post-harvest UV-C and PL irradiation of fruits and vegetables. *Chemical Engineering Transactions* 44: 31–36. https://doi.org/10.3303/CET1544006.

Pataro, G., Sinik, M., Capitoli, M.M. et al. (2015b). The influence of post-harvest UV-C and pulsed light treatments on quality and antioxidant properties of tomato fruits during storage. *Innovative Food Science and Emerging Technologies* 30: 103–111. https://doi.org/10.1016/j.ifset.2015.06.003.

Pellicer, J.A. and Gómez-López, V.M. (2017). Pulsed light inactivation of horseradish peroxidase and associated structural changes. *Food Chemistry* 237: 632–637. https://doi.org/10.1016/j.foodchem.2017.05.151.

Pellicer, J.A., Navarro, P., and Gómez-López, V.M. (2018). Pulsed light inactivation of mushroom polyphenol oxidase: a fluorometric and spectrophotometric study. *Food and Bioprocess Technology* 11: 603–609. https://doi.org/10.1007/s11947-017-2033-x.

Pellicer, J.A., Navarro, P., and Gómez-López, V.M. (2019a). Pulsed light inactivation of polygalacturonase. *Food Chemistry* 271: 109–113. https://doi.org/10.1016/j.foodchem.2018.07.194.

Pellicer, J.A., Navarro, P., Hernández-Sánchez, P., and Gómez-López, V.M. (2019b). Structural changes associated with the inactivation of lipoxygenase by pulsed light. *LWT - Food Science and Technology* 113: 108332. https://doi.org/10.1016/j.lwt.2019.108332.

Pellicer, J.A., Navarro, P., and Gómez-López, V.M. (2020). Pectin methylesterase inactivation by pulsed light. *Innovative Food Science and Emerging Technologies* 62: 102366. https://doi.org/10.1016/j.ifset.2020.102366.

Rodov, V., Vinokur, Y., and Horev, B. (2012). Brief postharvest exposure to pulsed light stimulates coloration and anthocyanin accumulation in fig fruit (*Ficus carica* L.). *Postharvest Biology and Technology* 68: 43–46. https://doi.org/10.1016/j.postharvbio.2012.02.001.

Rohatgi-Mukherjee, K.K. (1986). *Fundamental of Photochemistry*. New Delhi: New Age International.

Rowan, N.J., Valdramidis, V.P., and Gómez-López, V.M. (2015). A review of quantitative methods to describe the efficacy of pulsed light generated data that embraces the occurrence of viable but non culturable stated microorganisms. *Trends in Food Science and Technology* 44: 79–92. https://doi.org/10.1016/j.tifs.2015.03.006.

Ryer, A.D. (1997). *The Light Measurement Handbook*. Newburyport, MA, USA: International Light.

Sauer, A. and Moraru, C.I. (2009). Inactivation of *Escherichia coli* ATCC 25922 and *Escherichia coli* O157:H7 in apple juice and apple cider, using pulsed light treatment. *Journal of Food Protection* 72: 937–944. https://doi.org/10.4315/0362-028x-72.5.937.

Schaefer, R., Grapperhaus, M., Schaefer, I., and Linden, K. (2007). Pulsed UV lamp performance and comparison with UV mercury lamps. *Journal of Environmental Engineering and Science* 6: 303–310. https://doi.org/10.1139/s06-064.

Schottroff, F., Fröhling, A., Zunabovic-Pichler, M. et al. (2018). Sublethal injury and viable but non-culturable (VBNC) state in microorganisms during preservation of food and biological materials by non-thermal processes. *Frontiers in Microbiology* 9: 2773. https://doi.org/10.3389/fmicb.2018.02773.

Shank, F.R. [FR Doc. 96–20853 Filed 8–14–96; 8:45 am] BILLING CODE 4160–01–F. Federal Register / Vol. 61, No. 159 / Thursday, August 15, 1996 / Rules and Regulations 42383. https://www.accessdata.fda.gov/scripts/cdrh/cfdocs/cfcfr/CFRSearch.cfm?fr=179.41 (accessed 8 September 2020).

Shriver, S., Yang, W., Chung, S.Y., and Percival, S. (2011). Pulsed ultraviolet light reduces immunoglobulin E binding to Atlantic white shrimp (*Litopenaeus setiferus*) extract. *International Journal of Environmental Research and Public Health* 8: 2569–2583. https://doi.org/10.3390/ijerph8072569.

Takeshita, K., Shibato, J., Sameshima, T. et al. (2003). Damage of yeast cells induced by pulsed light irradiation. *International Journal of Food Microbiology* 85: 151–158. https://doi.org/10.1016/S0168-1605(02)00509-3.

Upadhyaya, G.S., Curry, R.D., Nichols, L. et al. (2004). The design and comparison of continuous and pulsed ultraviolet reactors for microbial inactivation in water. *IEEE Transactions on Plasma Science* 32: 2032–2037.

Uslu, G., Demirci, A., and Regan, J.M. (2016). Disinfection of synthetic and real municipal wastewater effluent by flow-through pulsed UV-light treatment system. *Journal of Water Process Engineering* 10: 89–97. https://doi.org/10.1016/j.jwpe.2016.02.004.

Van Houteghem, N., Devlieghere, F., Rajkovic, A. et al. (2008). Effects of CO_2 on the resuscitation of *Listeria monocytogenes* injured by various bactericidal treatments. *International Journal of Food Microbiology* 123: 67–73. https://doi.org/10.1016/j.ijfoodmicro.2007.12.002.

Vimont, A., Fliss, I., and Jean, J. (2015). Efficacy and mechanisms of murine norovirus inhibition by pulsed-light technology. *Applied and Environmental Microbiology* 81: 2950–2957. https://doi.org/10.1128/AEM.03840-14.

Wang, T., MacGregor, S.J., Anderson, J.G., and Woolsey, G.A. (2005). Pulsed ultra-violet inactivation spectrum of *Escherichia coli*. *Water Research* 39: 2921–2925. https://doi.org/10.1016/j.watres.2005.04.067.

Wang, B., Zhang, Y., Venkitasamy, C. et al. (2017). Effect of pulsed light on activity and structural changes of horseradish peroxidase. *Food Chemistry* 234: 20–25. https://doi.org/10.1016/j.foodchem.2017.04.149.

Wekhof, A. (2000). Disinfection with flash lamps. *PDA Journal of Pharmaceutical Science and Technology* 54 (3): 264–276.

Wekhof, A., Trompeter, F.J., and Franken, O. (2001). Pulse UV disintegration (PUVD): a new sterilization mechanism for packaging and broad medical-hospital application. *First International Conference on Ultraviolet Technologies*, Washington, DC, USA (2001).

Wuytack, E.Y., Thi, L.D., Aertsen, A. et al. (2003). Comparison of sublethal injury induced in *Salmonella enterica* serovar Typhimurium by heat and by different nonthermal treatments. *Journal of Food Protection* 66: 31–37. https://doi.org/10.4315/0362-028x-66.1.31.

Yang, W.W., Chung, S.Y., Ajayi, O. et al. (2010). Use of pulsed ultraviolet light to reduce the allergenic potency of soybean extracts. *International Journal of Food Engineering* 6 https://doi.org/10.2202/1556-3758.1876.

Yang, W.W., Mwakatage, N.R., Goodrich-Schneider, R. et al. (2012). Mitigation of major peanut allergens by pulsed ultraviolet light. *Food and Bioprocess Technology* 5: 2728–2738. https://doi.org/10.1007/s11947-011-0615-6.

Zhao, X., Yang, W., Chung, S.Y. et al. (2014). Reduction of IgE immunoreactivity of whole peanut (*Arachis hypogaea* L.) after pulsed light illumination. *Food and Bioprocess Technology* 7: 2637–2645. https://doi.org/10.1007/s11947-014-1260-7.

Zhu, Y., Li, C., Cui, H., and Lin, L. (2019). Antimicrobial mechanism of pulsed light for the control of *Escherichia coli* O157:H7 and its application in carrot juice. *Food Control* 106: 106751. https://doi.org/10.1016/j.foodcont.2019.106751.

9

Infrared Radiation

Yvan Llave and Noboru Sakai

Department of Food Science and Technology, Tokyo University of Marine Science and Technology, Tokyo, Japan

1 Introduction

Infrared radiation (IR) is energy in the electromagnetic (EM) waveform that impinges on IR-exposed material, penetrates it, and is converted to sensible heat. In general terms, the penetration capability of IR depends on the properties of the treated material and the temperature of the radiator. IR heating has been widely used for many years in industrial applications, e.g. home heaters, furnaces in forging processes, the hot forming of sheet metal, drying car paint, and soldering systems (Lee 2021). In contrast, IR technologies for processing foods and agricultural products have not developed to a similar extent, due to the slower progress in research addressing the heterogeneous agricultural landscape. The major obstacle to the adoption of IR technology in the agriculture and food industries was proposed to be a lack of understanding of this technology.

However, over the last two decades, IR technology has resurfaced as an alternative to most processing technologies in the food sector (Sakare et al. 2020; Aboud et al. 2019; Onwude et al. 2016; Hassoun et al. 2020). IR has been very useful in food processing in the following three areas: (i) food processes involving heating; (ii) spectroscopic measurements of the chemical composition (analytical application) of foods, e.g. the measurement of the water content in food products; and (iii) noncontact temperature measurements of foods. This chapter is focuses on the first area (i.e. food processes involving heating) for the food processing sector. For detailed information about the other two applications, the publications of Arana (2012), Atungulu and Pan (2011), and Cortés et al. (2019) are recommended.

IR technology has been commonly used in the food industry for the dehydration of vegetables, fish, pasta, and rice; heating flour; frying meat; roasting cereals, coffee, and cocoa; baking biscuits and bread, cheese, eggs, frankfurters, and hotdogs; thawing of frozen foods; and disinfection of packaging materials. However, due to the limited penetration of IR heating into the food processing industry, combinations of IR heating with microwave and other common conductive and convective modes of heating have been used in the past with great potential to

Electromagnetic Technologies in Food Science, First Edition. Edited by Vicente M. Gómez-López and Rajeev Bhat.

improve temperature uniformity. In addition, the selective IR absorption of foods has been reported to provide the basis for a convenient approach now practiced in the food industry; "selective heating" is the use of the IR spectrum (after it is passed through suitable filters) to allow radiation within a specific spectral range for a specific application.

In this chapter, the basic principles of the IR heating of foods and the most relevant applications of IR heating in food processing operations – taking into consideration both their physical properties and their effect on quality retention – are provided. The recent combinations of IR heating with other technologies such as convection heating, microwave heating, freeze-drying (FD), and others in terms of heating performance are evaluated. In addition, the recent mathematical models developed in this area for the virtualization of the heating processes using computer simulation approaches are discussed in detail.

2 Fundamentals and Theory of Infrared Radiation

2.1 Principles of Infrared Radiation Heating

2.1.1 Infrared Wavelength

As do microwave, radiofrequency, and induction heating, IR transfers thermal energy in the form of EM waves and encompasses that portion of the EM spectrum that borders on visible light and microwaves. The exposure of food to EM radiation results in changes in the electronic, vibrational, and rotational states of atoms and molecules. The types of mechanisms underlying this energy's absorption are dependent on the wavelength of the radiation. The changes in the electronic, vibrational, and rotational states correspond to wavelengths in the range of 0.2–0.7 μm (ultraviolet and visible rays), 0.75–1.4 μm (near-infrared radiation/NIR), and 3–100 μm (far-infrared radiation/FIR), and > 100 μm (microwaves), respectively (Decareau 1985) (see Figure 1).

With the exposure of food material to IR (wavelength 0.78–1000 μm), the heat energy generated can be absorbed by the food. In general, the food substances absorb FIR energy

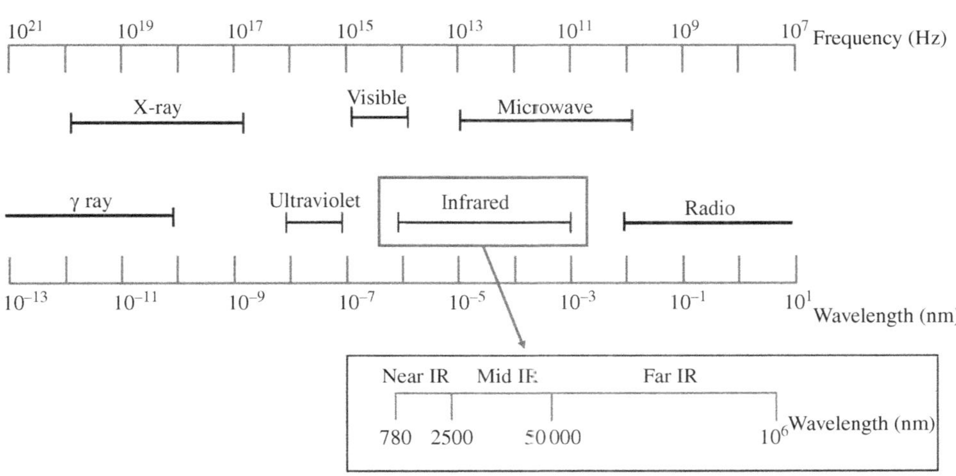

Figure 1 Various wavelengths of electromagnetic waves. *Source:* Decareau (1985).

most efficiently through the mechanism of changes in the molecular vibrational state, which can lead to radiative heating. Water and organic compounds (such as proteins and starches) are the main components of food, and they absorb FIR energy at wavelengths >2.5 μm (Sakai and Hanzawa 1994).

2.1.2 Basics Laws of Infrared Radiation

The wavelength at which the maximum radiation occurs is determined by the temperature of the IR heating element(s). This relationship is described by the basic laws for blackbody radiation including Planck's law, Wien's displacement law, and Stefan–Boltzman's law, which are thoroughly described in previous chapters and reviews (Sakai and Hanzawa 1994; Jun et al. 2011; Ganapathy et al. 2018). A summarized explanation of these basic laws is as follows.

Planck's Law Planck's law presents the spectral distribution of radiation from a blackbody source that emits 100% IR at a given single temperature. IR sources are made up of thousands of point sources at different temperatures. By combining the point sources, an entire spectral distribution for specific regions can be obtained. An approximation of the spectral distribution obtained with the average surface temperature and emissivity values can thus be used to characterize the IR.

Max Planck (1901) reported the spectral blackbody emissive power distribution, now commonly known as Planck's law (Eq. (1)), for a black surface bounded by a transparent medium with refractive index n, as:

$$E_{h\lambda}(T,\lambda) = \frac{2\pi h c_0^2}{n^2\lambda^5\left(e^{hc_0/n\lambda kT} - 1\right)} \tag{1}$$

where k is known as Boltzmann's constant (1.3806×10^{-23} J K^{-1}) and n is the refractive index of the medium. By definition, the refractive index of a vacuum is $n = 1$. For most gases, the refractive index is very close to unity. The variable λ is the wavelength (μm), T is the source temperature (K), c_0 is the speed of light (km s^{-1}), and h is Planck's constant (6.626×10^{-34} J s). Overall, the level of emissive power rises with an increase in the temperature, while the wavelength of the corresponding maximum emissive power shifts toward shorter wavelengths. The total amount of IR emissive power within a specific region considered can be estimated by the integration of Planck's law at a given temperature with respect to the wavelength. The amount of radiant infrared energy delivered to the food in an IR oven increases and the peak wavelength decreases as the temperature of the heater is raised (Figure 2). Thus, Planck's law can be applied to estimate the total amount of radiative heat flux when a specific surface temperature of the heating element is known.

Wien's Displacement Law Wien's displacement law gives the wavelength (denoted as the peak wavelength) at which the spectral distribution of radiation emitted by a blackbody reaches a maximum emissive power. The maximum of the curves can be determined by differentiating Eq. (2):

$$\frac{d}{d(n\lambda T)}\left(\frac{E_{h\lambda}}{n^3\lambda^5}\right) = 0 \tag{2}$$

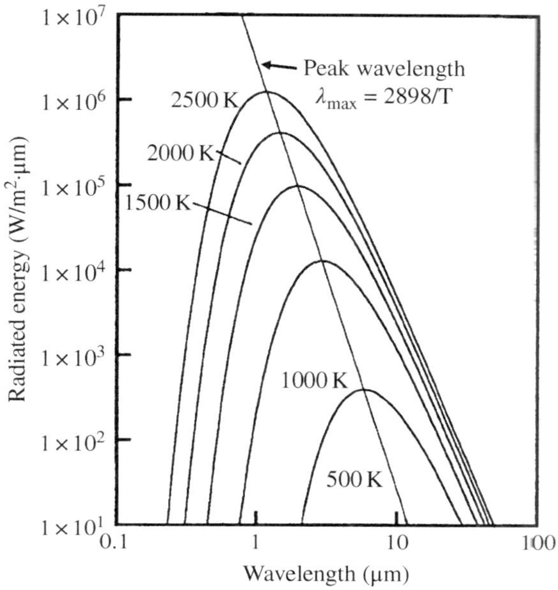

Figure 2 Spectral characteristics of blackbody radiation at different temperatures.

The source temperatures of IR emitters needed for a desired spectral distribution can be estimated by:

$$\lambda_{max} = \frac{2898}{T} \tag{3}$$

where T is the source temperature and λ_{max} is the peak wavelength. If the source temperature is known, the peak wavelength can be derived from Eq. (3).

Stefan–Boltzman's Law Stefan–Boltzmann's law gives the total power radiated at a specific temperature from an IR source. The entire amount of heat flux estimated using this law should be consistent with the integration of the spectral amount of heat flux estimated using Planck's:

$$E_b(T) = \int_0^{\infty} \frac{C_1 d(n\lambda T)}{(n\lambda T)^5 \left(e^{\frac{C_2}{n\lambda T}} - 1 \right)} = n^2 \sigma T^4 \tag{4}$$

where $C_1 = 2\pi hc_0{}^2 = 3.7419 \times 10^{-16}\,\mathrm{W\,m^2}$, $C_2 = hc_0/k = 14\,388\,\mu\mathrm{m\,K}$, and σ is known as the Stefan–Boltzmann constant ($5.670 \times 10^{-8}\,\mathrm{W\,m^{-2}\,K^{-4}}$). Stefan–Boltzmann's law is useful for a prompt estimation of the total amount of heat flux at a given source temperature.

Equation (4) is the equation for an ideal object (blackbody); however, the actual object (gray body) is governed by the following equation:

$$E = \varepsilon\sigma T^4 \tag{5}$$

where ε $(0 < \varepsilon < 1)$ is the emissivity.

2.2 Characteristics of Thermal Radiation

2.2.1 Types of Infrared Radiation

IR can be divided into three different categories: NIR (0.75–1.4 μm), mid-infrared radiation (MIR; 1.4–3 μm), and FIR (3–1000 μm) (Sakai and Hanzawa 1994). In general, FIR radiation is advantageous for food processing because most food components absorb radiative energy in the FIR region (Sandu 1986). The drying of thin layers seems to be more efficient at the FIR region, whereas the NIR region should give better results for the drying of thicker bodies. Several studies investigating the superiority of FIR versus NIR were summarized by Sakai and Hanzawa (1994).

2.2.2 Heat Generation

Atoms have electrically charged particles such as protons and electrons, and an electric field is present around these charged particles. When these charged particles move, a magnetic field is generated. Atoms that are present in hot objects will vibrate violently, disturbing the electric and magnetic fields. This creates EM waves, and thus hot substances radiate EM waves. When these EM waves fall on a cold body, the molecules are excited and the atoms undergo vibrational motion. As the vibration increases, the cold body is heated up. Heat energy can thus be transmitted without any physical contact and without a medium. However, even a thin layer of food material on the surface can affect the IR process: the IR cannot penetrate deeply and heats the object to a depth of only a few millimeters below the surface. The IR is then absorbed by the object's surface and transferred through the thickness of the object by means of conduction.

2.2.3 Sources of Infrared Heating

Two types of IR equipment are generally used: the cabinet type and the conveyor belt type. Cabinet equipment is simple in structure and easy to operate, but it cannot provide continuous-mode operation. For this reason, methods using a conveyor belt or moving bed (which is suitable for continuous-mode operation) have been gradually developed (Figure 3). IR emitters are the most critical part of IR equipment, and they are classified as described below.

Types of Infrared Emitters by Wavelengths Based on their wavelengths, IR emitters can be classified into short-wave, long-wave, and medium-wave emitters depending on the temperature applied to the emitter. Medium-wave IR emitters radiate in the range 1.4–3.0 μm, which can be used for the drying and curing of food products. The heat transfer rate, thermal efficiency, and effects of the radiation on heated materials are closely related to the range of wavelengths emitted from the heating source.

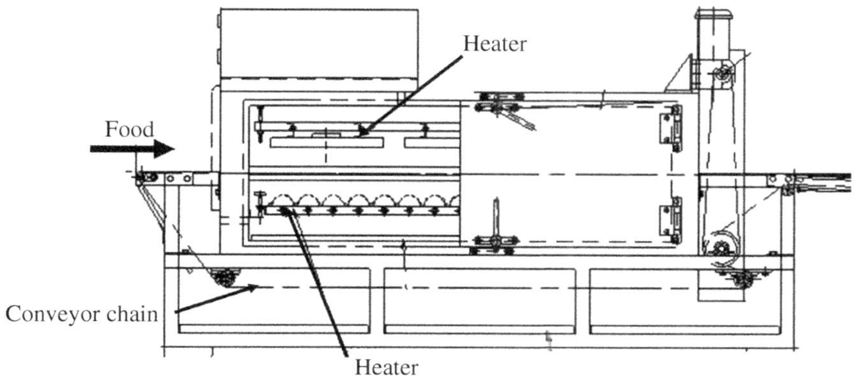

Figure 3 Schematic diagram of a conveyor chain FIR oven.

Types of IR Emitters Based on Heat Generation IR emitters can also be classified based on their design as an electrical- or gaseous-type emitter. The cost of gas heaters is higher, but their operating cost is relatively lower compared to the systems that produce IR electrically (electric heaters). IR electric heaters are more widespread than gaseous heaters because of their ease of control, fast heating rate, and clean energy. Electric IR emitters are more flexible in producing the required wavelength for a particular application. In general, the operating efficiency of an electric IR heater ranges from 40% to 70%, while that of gas-fired IR heaters ranges from 30% to 50% (Hung et al. 1995). Electrical emitters can reach a radiant intensity of up to 400 kW m^{-2}, while gas-fired emitters generally do not exceed the radiant intensity of 22 kW m^{-2} (Nindo and Mwithiga 2010).

The forms of the electrical emitters vary, including plate-shaped, pipe-shaped, and lamp-shaped forms. Metals, ceramics, quartz, and halogen are often used as thermal substrates, as they can emit IR radiation at different wavelengths. Curved ceramic emitters which emit a wide range of the MIR and FIR spectra have higher emissive power, more uniform heating, and a broader operating range than some other types of emitters. Electrical IR emitters consist of a metal filament placed inside a sealed container that may be empty or filled with an inert gas; the IR is produced by heating the filament to a particular temperature with an electric heater by passing an electric current through a high-resistance wire such as nichrome (NiCr) wire, iron-chromium wire, or tungsten filament. However, NIR heats the emitter by directly passing electricity, while FIR passes electricity through the nichrome wire to indirectly heat the emitter (ceramic). The types of electric IR emitters include reflector-type emitters, incandescent lamps, quartz tubes, and resistance elements such as metallic tubes, ceramic tubes, and nonmetallic rods (Vidyarthi et al. 2019).

Gaseous IR emitters are powered by propane or natural gas. There are two types of gaseous IR emitters: direct-flame or open-flame IR emitters, and catalytic flameless infrared (CIR) emitters. The open-flame IR emitters release IR radiation via normal flame combustion. Direct-flame IR emitters have restricted usage in food processes because of safety and environmental concerns. CIR heating uses natural gas or propane, which is passed over a mesh catalyst pad to produce thermal radiant energy through a catalytic reaction. This reaction occurs below the ignition temperature of gas so that no flame is produced. The EM radiant energy from CIR has peak wavelengths in the range of medium- to far-infrared. The peak wavelengths match reasonably well with the three absorption peaks of liquid water, which could result in rapid moisture removal. Since CIR directly converts natural gas to radiant energy, it is more energy efficient than typical IR emitters using electricity.

2.3 Special Features of Infrared Radiation

2.3.1 Factors Related to the Penetration of IR

The depth of the penetration of IR into food products depends on the composition and structure of the foods and on the wavelengths of IR. The basic factors are follows:

- IR can penetrate products and directly transfer thermal energy to a certain depth. The depth to which IR can penetrate (the penetration depth) is generally dependent on the absorbance of the product. FIR energy scarcely penetrates, and almost all of its energy is converted to heat at the surface of a food product.
- Foods of various sizes absorb different amounts of thermal energy under the same heating time and heating configuration, and they consequently have different heating performances and product qualities.
- In the double-sided heating scenario, manipulating the emitter gap (i.e. the distance between two parallel IR emitters) can be an effective way to adjust the IR intensity and thus ensure a sufficient radiation heat exchange between the IR emitters and the food surface, which is fundamentally reflected by changes in the radiation view factor.
- A longer exposure to IR heating may provide sufficient thermal energy but leads to a deterioration of food quality and nutritional loss due to overheating, whereas shorter heating times may not be able to achieve the desired heating requirement, such as peel separation in fruit peeling applications, thus reducing the peelability.

2.3.2 Advantages of IR Processing

The basic advantages and limitations of IR processing have been extensively reviewed (Krishnamurthy et al. 2008a; Sakai and Hanzawa 1994; Datta and Almeida 2014). Examples of advantages include heating under controlled environments, geometry simplicity, easy load feeding, space saving, a high degree of process control parameters, uniform heating, high heat transfer rates, reduced processing times and energy consumption, and improved product quality and safety. IR processing is also environmentally friendly; since air is transparent to IR radiation, and the process can be done at ambient air temperatures (Nowak and Lewicki 2004). Some additional reported advantages of IR processing are as follows:

- During IR heating, surface irregularities have a smaller effect on the rate of heat transfer, resulting in more uniform heating.
- The slight degree of thermal inertia of the radiation by IR removes the necessity for a long preheating cycle. This means that the wavelengths used in IR are such that it is possible to start up and reach the working temperature in seconds, while also offering excellent process control.
- IR heating is useful in noncontact heating, providing high efficiency as it does not heat the surrounding air. Since there is no medium required for transferring heat, IR heating is much faster than other types of heating. This is very advantageous for heating moving

products and for heating products for which physical contact would affect their surface finish temperature.
- In IR heating, a certain amount of radiation which is not absorbed can also be reflected back to the product for heating by using reflectors, increasing the heating efficiency.

2.3.3 Limitations of Infrared Radiation Processing

The main reported limitations of IR processing are

- Although the absorbed energy at the sample's surface during IR heating is transferred by conduction to other areas within the food material, if the sample's volume increases, this conduction is limited, and thus the total energy absorbed is limited. As a result, penetrative radiation energy does not make a significant contribution to internal heating. Thus, in order to achieve the optimum use of energy, the combination of IR heating with microwave or other common conductive and convective modes of heating holds great potential.
- As the IR energy is absorbed on the surface, it allows only a shallow layer to be dried. Rastogi (2012) observed that with increased bed depths, a wide variation in moisture distribution was obtained. The layer close to the IR source dried more rapidly compared to the layer that was deep inside. Hence, in a deep bed, external agitation such as that provided by vibration would be helpful to turn the bed, so that each part of a sample can receive uniform radiation.
- IR is not sensitive to reflective properties of coatings.
- It is quite likely that the use of IR heating for prolonged exposures of biological materials will cause swelling and ultimately the fracturing of foods.

2.4 Interaction of Infrared Radiation with Food

2.4.1 Fundamentals of Interaction with Foods

Generally, foods are complex mixtures of different biochemical macromolecules, biological polymers, inorganic salts, and water. The IR spectra of such mixtures originate with the mechanical vibrations of molecules or particular molecular aggregates within a very complex phenomenon of reciprocal overlapping (Halford 1957). The strongest absorption is often localizable within the so-called group frequencies, which are generated by the vibrations of these molecular aggregates of molecular structural groups. Analyses of the influence of the molecular environment on the IR spectra of complex biochemicals indicated that there are two groups of parameters acting on the group frequencies: intramolecular and extramolecular environmental parameters. The first group has interactions due to the chemical bonds characterizing the given biochemical molecule itself, and the second group has well-known hydrogen bonds (as important forms of interactions with the extramolecular environment).

Absorption intensities at different wavelengths differ by food components. Water and organic compounds such as proteins and starches, which are the main components of foods, absorb FIR energy at wavelengths >2.5 μm (Sakai and Hanzawa 1994). Due to a lack of detailed information, the data shown in Table 1 on the absorption of IR by the principal food constituents can be regarded as only approximate values (Sakai and Hanzawa 1994) for the IR absorption bands associated with food heating.

Table 1 IR absorption band characteristics of chemical groups relevant to the heating of food.

Chemical group	Absorption wavelength (μm)	Relevant food components
Hydroxyl group (O–H)	2.7–3.3	Water, carbohydrates
Aliphatic carbon–hydrogen	3.25–3.7	Fats, carbohydrates, protein
Carbonyl group (C=O) (ester)	5.71–5.76	Fats
Carbonyl group (C=O) (amyl)	5.92	Proteins
Nitrogen–hydrogen group (-NH-)	2.83–3.33	Proteins
Carbon–carbon double bond (C=C)	4.44–4.76	Unsaturated fats

Source: Based on Sakai and Hanzawa (1994).

The data provided in Table 1 demonstrate that there is a strong absorption of foods due to longitudinal vibrations. A material's absorption of radiation does not make it saturated with IR, because the molecules excited by the vibratory movement continuously lose energy in random directions as a result of collisions between the molecules, which transfer energy to the surrounding environment in the form of heat (Yadav et al. 2020).

Food materials also interact with light, and crucial optical principles such as regular reflection, body reflection, and light scattering must also be considered. Regular reflection takes place at the surface of a material. For body reflection, the light enters the material, becomes diffuse due to light scattering, and undergoes some absorption; the remaining light leaves the material close to where it enters. Regular reflection produces only the gloss or shine of polished surfaces, whereas body reflection produces the colors and patterns that constitute most of the information obtained visually (Krishnamurthy et al. 2008a). For materials with a rough surface, both regular and body reflections can be observed. For example, at the NIR wavelength region ($\lambda < 1.25\,\mu m$), approx. 50% of the radiation is reflected back, and <10% of the radiation is reflected back at the FIR wavelength region (Skjoldebrand 2001). Most organic materials reflect 4% of the total reflection, producing a shine of polished surfaces. The rest of the reflection occurs where radiation enters the food material and scatters, producing different colors and patterns (Dagerskog 1979). From the food engineering standpoint, it is very important to fully understand the above optic-thermal phenomena associated with IR and food products.

2.4.2 Selective Infrared Radiation Absorption of Foods

An ideal IR heating system optimally raises the temperature of a target with the least energy consumption. Such a system may be comprised of a device that can directly convert its electrical power input into a radiant EM energy output, with the chosen single-band or narrowband wavelengths that are aimed at a target, so that the energy comprising the irradiation is partially or fully absorbed by the target and converted to heat. This can be achieved because IR can be controlled or filtered to allow radiation within a specific spectral range to pass through, with the use of suitable optical band-pass filters. Such controlled radiation can stimulate the maximum response of the target object when the emission band of the IR and the peak absorbance band of the target object are identical.

The higher the efficiency of the conversion of the electrical input into a radiant EM output, the higher the overall system efficiency. In addition, the more efficiently the radiant EM waves are aimed to expose only the desired areas on the target, the more efficiently the system will accomplish its work. Some research advances have been made in efforts to directly output substantial quantities of IR at selected wavelengths for the purpose of replacing broadband-type devices with narrowband irradiation sources. Recent advances in semiconductor processing technology have resulted in the availability of direct electron-to-photon solid-state emitters that operate in the general MIR range > 1 μm. These solid-state devices operate analogously to common light-emitting diodes (LEDs), but they do not emit visible light; rather, they emit thermal IR energy at the longer MIR wavelengths. To distinguish this new class of devices from the conventional shorter-wavelength devices (LEDs), these devices are more appropriately described as radiance or radiation-emitting diodes (REDs). These devices have the property of emitting radiant EM energy in a tightly limited wavelength range. With the proper semiconductor processing operations, REDs can be tuned to emit at specific wavelengths that are most advantageous to a particular radiant treatment application.

3 Infrared Radiative Properties of Food Materials

The effect of IR on optical and physical properties of food materials is crucial for the design of IR heating systems and the optimization of the thermal processing of food components. Two key radiative aspects of interest for designing an IR heater are its spectral distribution and energy intensity.

3.1 Attenuation of Radiation

The two underlying mechanisms that explain the attenuation of EM radiation as it propagates through a medium are absorption and scattering. Absorption phenomena describe the conversion of radiation to some other form of energy (or some other spectral distribution). Scattering mechanisms redirect the radiant energy from its original direction of propagation based on the combined effects of reflection, refraction, and diffraction. The sum of the mechanisms of the attenuation of EM radiation as it passes through a medium (absorption plus scattering) is generally called the attenuation of radiation.

The three fundamental radiative properties are the reflectivity (ρ) as the ratio of the reflected part of the incoming radiation to the total incoming radiation, the absorptivity (α) as the ratio of the absorbed part of the incoming radiation to the total incoming radiation, and the transmissivity (τ) as the ratio of the transmitted part of the incoming radiation to the total incoming radiation (Figure 4). Under these terms, the energy balance leads to the well-known relation given by:

$$\rho + \alpha + \tau = 1 \tag{6}$$

However, if the food is thick it will be hardly penetrated (so, $\rho + \alpha = 1$), and it is accepted that the absorption rate will be high since the emissivity of foods (organic matters) is also high. The energy transmitted into the food is attenuated exponentially with the penetration distance. The attenuation factor determines the energy absorption within the food as a function of the depth from the surface of the food, as described by Lambert–Beer's law:

$$I_\lambda = I_{\lambda 0} \exp\left(-\alpha_\lambda x\right) \tag{7}$$

where I_λ is the energy flux at the wavelength of λ and α_λ is the spectral attenuation factor. The attenuation factor of water changes remarkably with the wavelength. For details, readers are referred to the work of Sakai and Mao (2006). Understanding the attenuation of radiation is crucial, because most IR heat transfer models are based on the amount of local heat flux imparted to the food material in relation to the penetration depth.

3.2 Properties Related to the Radiative Heat Transfer of Foods

Thermal radiation involves the absorption, reflection, and transmission of incident radiation, as well as the emission of radiation. The first three characteristics depend on the spectral composition of the radiation source and the degree of polarization and spatial properties of incident radiation i.e., on the irradiation conditions and the state and properties of the irradiated substance itself. The emission of radiation by a substance is determined solely by the state and the emissivity of the substance.

4 Applications of Infrared Radiation in Food Processing

4.1 Traditional Applications for Foods

4.1.1 Infrared Radiation Drying

Drying is defined as a process of moisture removal achieved by simultaneous heat and mass transfer. The moisture can be either transported to the surface of the product and then evaporated or evaporated internally at a liquid–vapor interface and then transported as vapor to the surface. In IR drying, the radiation is generated by heating the product to a high temperature, resulting in a direct penetration of the surface by the radiation. The IR

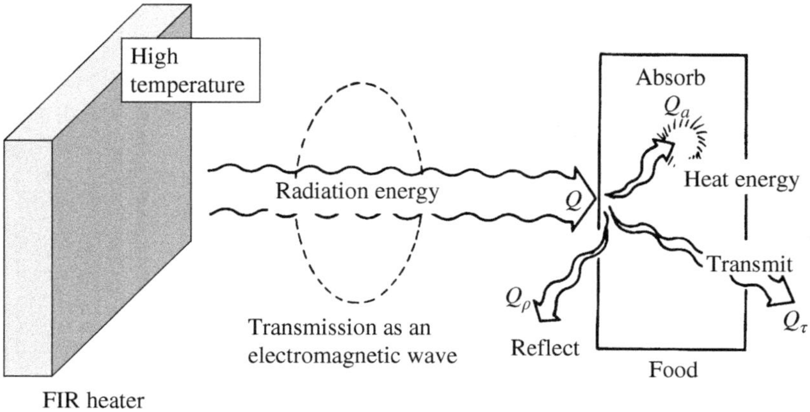

Figure 4 Schematic diagram of the IR radiation applied to food.

drying of foods is without a doubt the most common application of IR heating in the food industry. IR drying improves food stability since it considerably reduces the water and microbiological activity of the material and minimizes physical and chemical changes during its storage.

One of the most important physical changes that food products experience during drying is the reduction of their external volume. The loss of water and heating cause stress in the cellular structure of the food, leading to changes in shape and decreases in dimensions. The shrinkage of food materials has a negative consequence on the quality of the dehydrated product. Changes in shape and increased hardness usually cause a negative impression among consumers, although there are some dried products that traditionally have a shrunken (and expected) aspect, such as raisins, dried plums (prunes), dried peaches, and dates. IR drying has also been investigated as a potential method for obtaining high-quality dried foodstuffs, as was reported by Togrul (2005) and Pan et al. (2008b) for the case of apples and bananas, respectively, and the works can be observed in Table 2.

4.1.2 Infrared Radiation Pasteurization

During an IR heating process, intense heat might accumulate on the surface of the heated object, causing the surface temperature to increase rapidly. If the IR exposure time is properly controlled, the surface temperature can be preferentially raised to a degree that is sufficient to inactivate targeted pathogenic microorganisms without substantially increasing the interior temperature. IR heating inactivates pathogenic microorganisms by damaging their intracellular components such as DNA, RNA, ribosomes, the cell envelope, and cell proteins (Krishnamurthy et al. 2008a). The absorption of IR by water molecules in a microorganism is a significant factor in the microorganism's inactivation since the water absorbs the IR radiation, resulting in a rapid increase in temperature.

Selected applications of IR heating for surface pasteurization have been described in previous chapters and reviews (Sakai and Hanzawa 1994; Krishnamurthy et al. 2008a). For details, refer to these works.

4.1.3 Infrared Radiation Grilling, Broiling, and Roasting

The rapid surface heating provided by IR can be used to seal in a food product's moisture and flavor or aroma compounds without burning the surface of the food black, resulting in a highly acceptable sensory quality of the product. Several studies have been conducted in the field of IR grilling, broiling, and roasting, the most relevant of which are summarized in Table 3. In addition, in Figure 5, an example of IR grilling application of eggplants slices and their changes in color, temperature, and shrinkage are showed.

4.1.4 Infrared Radiation Blanching

Blanching is an essential process to inactivate the enzymes responsible for the quality deterioration of fruits and vegetables in storage. Blanching also serves the purposes of microbial population reduction, color stabilization, and facilitation for further processing and handling. A newer method of simultaneous IR dry blanching and dehydration uses efficient IR heating to combine blanching and dehydration into a one-step process that is simpler and more energy efficient than conventional methods.

Yi et al. (2007) investigated the effectiveness of various combinations of ascorbic acid, citric acid, and calcium chloride dipping treatments for reducing the enzymic browning of

Table 2 Selected infrared drying applications of foodstuffs.

Food product	Processing method	Effect of treatment	References
Parbaked baguettes	IR and jet impingement compared to the conventional oven	Increased the rate of color development of the crust and shortened the time	Olsson et al. (2005)
Soybean grains	Combined NIR with fluidized-bed drying (using 4, 6, and 8 kW) under constant velocity and temperature	The NIRs of 4 and 6 kW were found to significantly reduce the cracking of soybean grains and obtain acceptable final products	Dondee et al. (2011)
Beef jerky	Mid- and far-infrared drying (MIR 2.9 ~ 3.1 μm, FIR 5.8 ~ 6.2 μm, 70 °C, 1 $m\,s^{-1}$, MFIR) was compared with hot air drying (70 °C, 1 $m\,s^{-1}$, HA)	MFIR could improve dehydration efficiency of beef jerky by promoting protein denaturation, inducing the shift in immobilized water to free water and reducing the activation energy. It was suggested that MFIR has high application potential on the dehydration of jerky	Li et al. (2018)
Sunflower seeds	Vibration IR monolayer drier	With an increase in specific power of up to 2.5 times, the drying process decreases in proportion. The drying time to the relative humidity of the product at 6–7% takes 30–60 min. There was no cracking of the husk	Bandura et al. (2019)
Ginger slices	A dehydration process by approaching the constant period of dehydration to the theory of mass and heat transfer process to the wet bulb thermometer and the decreasing period of dehydration to liquid diffusion theory was investigated. 5.0 mm thickness and 2.0 cm diameter slices were used to a dryer with infrared radiation at 50, 60, 70, 80, 90, and 100 °C until constant mass	Heat and mass transfer coefficients, and effective diffusion coefficient increased linearly with temperature increasing, resulting in values ranging from 69.40 to 92.23 $W\,m^{-2}\,°C^{-1}$, 0.062 to 0.089 $m\,s^{-1}$, and 3.81×10^{-9} to $1.13 \times 10^{-8}\,m^2\,s^{-1}$. Variation of heat and mass transfer coefficients was described by a linear model and the variation of effective diffusion coefficient with the temperature was described with the Arrhenius relation, whose activation energy was 22.07 $kJ\,mol^{-1}$	Correa et al. (2019)
Linden (*Tilia platyphyllos Scop.*) leaves	Thin layer drying kinetics of linden leave samples were experimentally investigated in an infrared (IR) dryer	The results showed that the drying time decreased from 50 to 20 min with increased IR temperature from 50 to 70 °C. The lightness and greenness of the dried linden leaves were significantly changed compared with fresh samples. The projected area of dried sample decreased similar to the drying time	Selvi (2020)

Source: Adapted from Togrul (2005) and Pan et al. (2008a).

Table 3 Selected infrared grilling, broiling, and roasting applications of foodstuffs.

Food product	Processing method	Effect of treatment	References
Deli turkey	NIR roasting (1.1–1.3 μm, 6 kW)	The chemical properties were significantly affected	Muriana et al. (2004)
Coffee beans	IR roasting process might take up to 20 min at 180–230 °C	IR energy can be used as an alternative to roasting	Sakai and Mao (2006)
Green tea leaves	FIR irradiation	The total phenols, flavanols, epigallocatechin, epigallocatechin gallate contents, ascorbic acid, caffeine contents, and nitrite scavenging activities were increased	Kim et al. (2006)
Green coffee beans	A combination of IR, microwave, conduction, convection, and latent steam heating	The roasted coffee was more flavorful and the process was energy efficient	Poss (2007)
Coffee bean	Developed an oven with an IR ceramic heating plate with a gate attached with a rotary sliding unit	A more appealing brown color and roasted appearance was obtained in addition to effective pasteurizing of the surface	Chung (2008)
Sesame seeds	IR heating	Lignan sesamolin was degraded to sesamol, which increases the oxidative stability of sesame oil synergistically with tocopherols. The γ-tocopherol content was also decreased by 17% and 25% in oils treated at 200 and 220 °C for 30 min, respectively	Kumar et al. (2009)
Red sea bream muscles blocks	FIR grilling and convective (superheated steam, SHS) heating were compared using air, dry air, and N_2 as heating media	The temperature and color of the sample surfaces were monitored over time. The absence of O_2 in the heating medium could be the reason for the delay in the browning reaction during heating using N_2	Matsuda et al. (2013)
Japanese eggplant	Evaluation of color changes during FIR broiling using computer vision system (CVS)	A nonhomogeneous browning distribution of eggplant can be explained by the structural changes induced during broiling	Llave et al. (2017)

apple cubes using the FIR dry-blanching process. Dipping treatments with calcium chloride and ascorbic acid (0.5% each for 5 minutes) resulted in firmer samples and was recommended as a best pretreatment for apple cubes. Yi and Zhongli (2009) described the IR drying of apples following Page's model. Heating under high radiation intensity (5000 W m^{-2}) inactivated polyphenol oxidase (PPO) and peroxidase (POD) much faster than that under low radiation intensity (3000 W m^{-2}). Under the same radiation intensity, thinner slices were blanched more quickly than thicker slices. Compared to PPO, POD was observed to be more heat-resistant. The inactivation of PPO and POD followed first-order

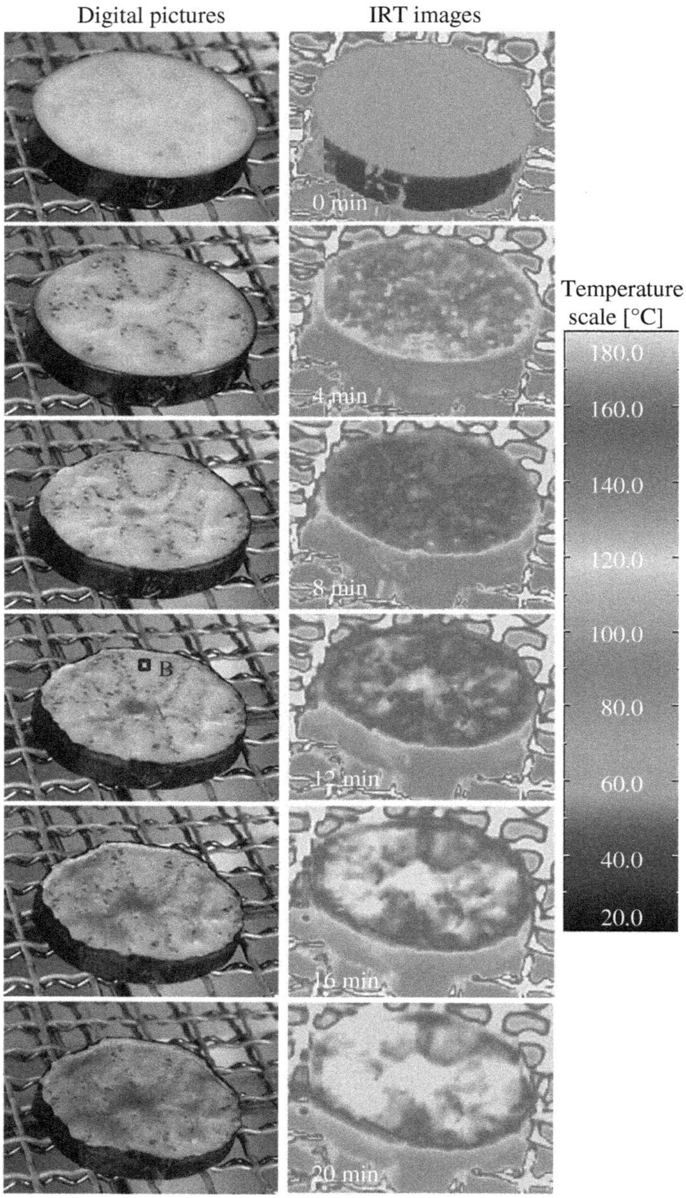

Figure 5 Temperature distribution on the upper surface of eggplant during FIR broiling, and digital pictures captured at several time intervals (infrared thermography images in °C). (*See insert for color representation of this figure*). *Source:* Llave et al. (2017).

kinetics and fractional conversion models, respectively. The surface color changes of the apple slices were due mainly to enzymatic browning, which occurred during the process.

Boudhrioua et al. (2009) demonstrated that the application of IR for blanching and/or drying of olive leaves resulted in a significant increase in the content of total phenols compared to that of fresh leaves, and that this use of IR preserved the green color of fresh olive leaves, enhancing their luminosity. This method may be used for preserving olive leaves before their use in food or cosmetic applications.

4.1.5 Infrared Radiation Baking

Compared to conventional baking, IR baking is more efficient at both the surface and the central sections. The baking time reduction afforded by IR is due to the more effective heat transfer to the surface than that provided by convection or conduction heating. Rastogi (2012) reported that compared to the use of conventional baking, the weight loss of IR-baked materials was lower, leading to higher water content in the center during baking and a better and longer shelf life. The energy consumption was also comparable. The interest in FIR ovens is driving the demands of baking industries for less time needed, less energy required, and better quality obtained.

Yung et al. (2008) reported that sponge cake baked in an FIR oven after seven days of storage was softer than the same type of cake baked in an electric oven. No significant differences in the volume, water activity, staling rate, or sensory scores were observed.

4.1.6 Infrared Radiation Cooking

IR cooking is of particular interest to the processed meat sector, because conventional cooking ovens require high velocity; however, hot air convection can cause surface deterioration, overheating, oxidation, charring, impingement damage, low yields, difficult emissions, and high energy costs. IR cooking has intrinsic advantages in achieving these goals because there is no direct intention or necessity to heat the air, keeping the oven temperatures and humidity down. This results in energy savings and the conservation of cooked material. A further advantage of this method is the ease with which heat can be applied evenly over a broad surface area (Rastogi 2012).

Sheridan and Shilton (1999) evaluated the efficacy of cooking hamburger patties by IR heating at 2.7 and 4.0 μm. With a higher-temperature (lower-wavelength) energy source, the change in the core temperature of the patties was closer to the change in their surface temperature, and the necessary cooking time was shorter.

4.2 Rough Rice Drying

Rough rice drying is a critical postharvest handling process, and it has direct effects on the milling rice quality, subsequent handling processes, and the commercial value of a rice crop. Rough rice is normally harvested at moisture contents higher than the 14% (wet basis) required for safe storage. The rough rice is typically dried using heated air, which is a slow process because only relatively low air temperatures can be used, in order to avoid reducing the milling quality of the rice. The heated air warms the outer layer of the rice kernel first and causes the moisture to evaporate from the kernel surface into the drying air. As the moisture is removed from the outer layers of the grain, a moisture gradient is established within the kernel (Pan et al. 2011). This gradient causes stresses in the grain, causing the rice kernel to fissure after drying. Therefore, to minimize the moisture gradient generated during conventional rice drying, multiple drying passes are typically used. The multiple drying normally removes a relatively small amount of moisture (2–3%) in each drying pass by exposing the rough rice to relatively low heated air temperature (Kunze and Calderwood 2004).

With the increase in rice production due to high-yielding varieties and rapid harvesting and transport capabilities, it is important to develop methods for drying that have high energy efficiency and result in high-quality rice, such as IR heating applications. Many studies have investigated the optimization of rice drying by IR. The most relevant applications are compiled in Table 4.

4.3 Fruit and Vegetable Peeling

The current industrialized wet-lye method for peeling fruits is a very energy- and water-intensive process. In addition, after lye peeling, the wastewater contains very high salinity (100–200 $g l^{-1}$) and organic loads, resulting in negative environmental footprints and high costs of wastewater management (Garcia and Barrett 2006). IR has a rapid surface heating characteristic that allows the effective heating of only a shallow layer of a fruit or vegetable's surface, achieving peel separation while preserving the nutrients and quality in the edible portion of the products. Because IR does not require any heating medium such as lye, water, or steam, the IR heating technique is being used as a novel dry-peeling method for peaches, tomatoes, and other vegetables. The intensity of the IR affects the heat fluxes impinged on the fruit surface. The higher the IR heat flux that irradiates onto the fruit surface, the more effective the IR radiation will be for peeling.

In order to ensure the good peelability and high quality of peeled end products, an important aspect of the IR dry-peeling method is to heat the fruit surface rapidly to a proper temperature while maintaining the fruit's interior temperature as low as possible. It is thus necessary to identify the optimal heating conditions for fruit peeling and to monitor the temperature changes on the fruit surface and in the interior flesh during IR heating. Some of the recent reported works in this field are compiled in Table 4.

As the contents of Table 4 indicate, the emerging IR dry-peeling technique offers a novel approach to eliminate the uses of chemicals and water in the peeling process while maintaining high-quality peeled products.

4.4 Disinfestation and Pest Management

Cereals are susceptible to fungal infection during their growth, harvesting, handling, and storage. The formation of aflatoxins can be prevented by disinfecting the cereals and then storing the dried cereal under proper conditions. Effective disinfection processes must be used to prevent mold growth, such as on rough rice. Drying rice to below the critical moisture content that is necessary for the growth of fungi is the most effective and widely used method to control microbial growth on rice. Convective heated air drying is the most common drying process used in the rice industry, by which rice grains are exposed to relatively low air temperatures (~43 °C) to avoid lowering the rice milling quality. However, these drying temperatures are below the temperature needed to inactivate *Aspergillus flavus* spores. Thus, convective drying is not effective for the disinfection of rough rice. Alternative IR radiation processing methods that can simultaneously dry and disinfect rough rice are needed.

Wang et al. (2014) revealed that high drying rates, good milling and sensory qualities, and the effective removal of insects were achieved using IR radiation to heat rough rice to a surface temperature of 60 °C followed by a tempering treatment (at 60 °C) for 4 hours and natural cooling. IR heating followed by tempering thus has the potential to

Table 4 Selected applications of infrared rice drying and infrared peeling of fruit and vegetables.

Food product	Processing method	Effect of treatment	References
Three varieties of high-volatile rice (shankar, basmati, and slenderness)	Serial vibration IR dryer with a radiation intensity of 3100 and 4290 $W\,m^{-2}$ and a depth of 12 and 16 mm grain layer	Drying rate depends on the intensity of the radiation. At a given temperature of air for drying (40 °C), the increase in the intensity of IR reduced the drying time in both fixed and vibration modes	Das et al. (2009)
Tomatoes	IR dry-peeling	IR treatment achieved similar peelability but yielded lower peeling loss and firmer peeled product for the same or slightly longer peeling time compared to the regular lye-peeling method	Li (2012)
Peaches	IR with three emitter gaps for a range of heating times from 90 to 180 s was used as well as wet-lye peeling as a control	180-s IR heating for medium-sized peaches under an emitter gap of 90 mm yielded 84/100 mm^2 peelability and 90/100 g peeling yield produced peeled products with comparable firmness and color to wet-lye-peeled peaches. Thermal expansion of cell walls and collapse of cellular layers adjacent to skins were found in IR-heated peaches and differed from the microstructural changes observed in lye-heated samples, indicating their mechanistic difference	Li et al. (2014)
Rough rice	IR drying characteristics, moisture diffusivity, milling quality, and disinfestation effect of rough rice were investigated	High drying rate, good milling quality, effective disinfestation and disinfection of rough rice, and effective stabilization can be achieved using IR heating with high processing efficiency and low energy consumption	Khir et al. (2014)
Rough and brown rice	Tempering treatments and IR heating (4685 $W\,m^{-2}$) were compared to hot air at 43 °C and ambient air	A promising potential to be used to inactivate the lipase enzyme and improve shelf lives was reported	Ding et al. (2015a)
White rice	Effect of IR drying followed by tempering and natural cooling on the change of physicochemical characteristics	IR drying slightly decreased the gelatinization temperature, enthalpy, and viscosities, reduced the changes in microstructure, and maintained cooking characteristics during storage	Ding et al. (2015b)
Jujube	IR peeling using two electric IR emitters with an IR radiation intensity (5.25–6.07 $W\,cm^{-2}$), emitter distance (75–85 mm), and heating time (40–60 s)	The results indicated that the best peeling with minimal thermal damage conditions was low IR radiation intensity (5.25 $W\,cm^{-2}$), short distance from the emitter (75 mm), and moderate heating time (56 s), which yielded lower peeling loss and less color change compared to lye-peeling	Wang et al. (2016)

simultaneously dry, disinfect, disinfest, and improve the milling quality of rough rice. The Wang laboratory investigated the effect of IR heating and tempering treatments on the disinfection of *A. flavus* in freshly harvested rough rice and storage rice. Rice samples with initial moisture contents of 14.1–27.0% (wet basis) were infected with *A. flavus* spores before the tests. The infected samples were heated by IR radiation to 60 °C in <1 minute, and then samples were tempered at 60 °C for 5, 10, 20, 30, 60, or 120 minutes. High heating rates and corresponding high levels of moisture removal were achieved using IR heating. The highest total moisture removal was 5.3% for the fresh rice with an initial moisture content of 27.0% after IR heating and then 120 minutes of tempering. IR heating followed by tempering for 120 minutes resulted in 2.5- and 8.3-log reductions of *A. flavus* spores in rough rice with the lowest and highest initial moisture contents, respectively.

4.5 Surface Disinfection in the Food Industry

IR can be effectively used for enzyme inactivation and to inactivate bacteria, spores, yeast, and mold in both liquid and solid foods (Krishnamurthy et al. 2008b). Several reviews have summarized these chemical and microbiological applications. IR heating was investigated by Bingol et al. (2011) for the inactivation of Salmonella on raw almonds. When almonds were heated to 100°, 110°, and 120 °C with IR and then quickly cooled to 90 °C and held for 10–15 minutes, the Salmonella population reduction was 5 log CFU g^{-1}. However, when almonds were cooled to 80 °C, a 4-log reduction was achieved after a holding time of 22 minutes. Hamanaka et al. (2011) investigated the surface decontamination of figs using IR and UV radiation and obtained a 1-log reduction in total fungi after a single 30-second IR heat treatment. For additional applications of this technology, see the review by Krishnamurthy et al. (2008a).

5 Integrated Heating Technologies

Efficient IR processing requires different means to maximize the advantages of the ordinary IR heating and overcome its inherent shortcomings. To address this challenge, IR heating is recently being applied together with conventional heating methods. However, the conventional processing technologies present problems that are contrary to the goal of establishing environmentally friendly and resource-saving methods, as they are time-consuming and energy-wasting and result in environmental pollution. In this sense, efficient hybrid IR processing should not only maintain the nutrients, texture, flavor, and other qualities of food during processing, but also improve the processing efficiency, energy, and resource utilization as much as possible. Lao et al. (2019) reported that in order to make IR processing a better alternative to conventional thermal processing techniques, the most important aspect is to control the interaction between heated foods and the IR. Therefore, when the advantages of two heating methods are combined, the energy efficiency of the heating process can be enhanced, and the degradation of heated product quality can be reduced. Some of these hybrid IR applications are described as follows.

5.1 Infrared Radiation and Convective Heating

Considering the combinations of IR with other heating technologies, IR and convection are the most attractive options in terms of versatility and fuel economy. The application of combined EM radiation and conventional convective heating is considered to be more efficient than radiation or convective heating alone, as it provides a synergistic effect.

In Table 5, the most relevant applications of IR and convection heating are summarized. In general terms, it can be concluded that in almost all cases, the combination of these two

Table 5 Selected applications of infrared and convective hybrid systems.

Food product	Processing method	Effect of treatment	References
Bread	IR heating and jet impingement treatment compared to conventional heat treatment	IR heating and jet impingement resulted in rapid drying and enhanced color development, compared to conventional heat treatment. Though the thickness of the bread crust increased faster, a short IR treatment time enabled the formation of a thinner crust	Olsson et al. (2005)
Sliced high-solids onions	Comparison of catalytic infrared (CIR) drying with forced air convection (FAC) heating and the quality characteristics of the dried foods	CIR both with and without air recirculation had higher maximum drying rates, shorter drying times, and greater drying constants than FAC at moisture contents greater than 50% (d.b.)	Gabel et al. (2006)
Meat patties (hamburgers)	Influence of different IR grilling/ hot air cooking conditions on moisture and fat content and on texture and color attributes	It resulted in less significant differences in moisture and total weight loss of the processed meat patties under the combined heating systems than using hot air cooking conditions only	Braeckman et al. (2009)
Strawberry	The effects of different drying conditions, such as infrared power, drying air temperature, and velocity, on quality of strawberry were evaluated	Drying time decreased with increased infrared power, air temperature, and velocity. An increase in power from 100 to 300 W, temperature from 60 to 80 °C, and velocity from 1.0 to 2.0 $m\,s^{-1}$ decreased fruit color quality index. For total phenol and anthocyanin content, 300 W, 60 °C, and 1.0 $m\,s^{-1}$ were superior to the other experimental conditions	Adak et al. (2017)
Kaki (*Diospyros kaki*) chips	A prototype dryer that combines an air convection oven and a pair of new FIR heaters sets: a set of carbon nanotube (CNT)/ carbon nanofiber/carbon nanocellulose (GTF) heaters and a set of CNT heaters	During low-temperature drying (<40 °C), higher-quality retention was observed when GTF heaters combined with air convection ovens were used, in addition to a reduction in drying time	Llave and Hayakawa (2019)

(*Continued*)

Table 5 (Continued)

Food product	Processing method	Effect of treatment	References
Sweet potato (*Ipomoea batatas L.*)	Simultaneous IR and hot air drying (IR-HAD), two-stage sequential hot air and IR drying (HAD+IR), two-stage sequential IR and hot air drying (IR + HAD), and intermittent infrared radiation and hot air drying (IIR + HAD)	Overall, considering the drying characteristics, drying time, specific energy consumption, and color of sweet potato, the IIR + HAD was found to be the most suitable for the drying of sweet potato	Onwude et al. (2019)
Paddy rice grains	A prototype dryer that combines an air convection oven and a pair of new FIR heaters sets: a set of CNT/carbon nanofiber/carbon nanocellulose (GTF) heaters and a set of CNT heaters	By using a pair of FIR heaters and the air convection oven together, an improvement of the drying speed (reduction rate of water content from 0.5% to 2% h^{-1}) compared to conventional methods was confirmed	Llave et al. (2020)

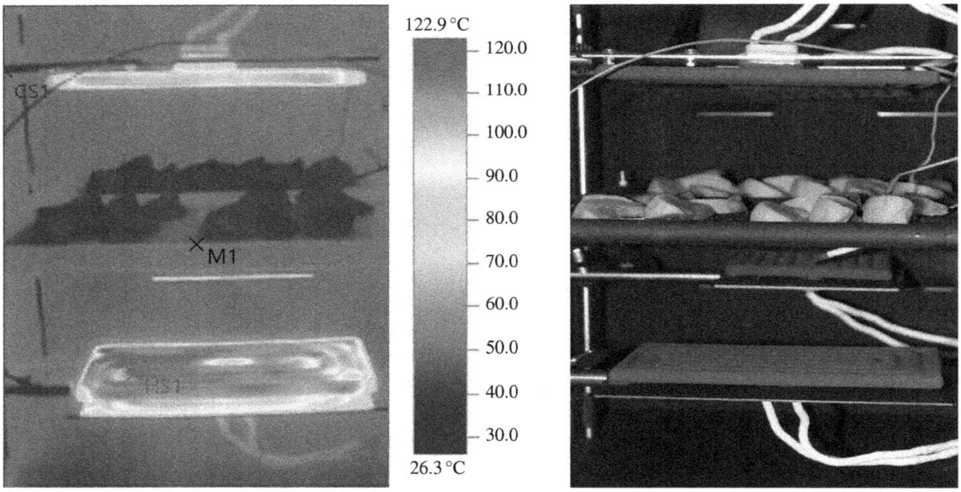

Figure 6 Infrared radiation temperature (IRT) image showing the final temperature distribution in the samples and heater surface (*left*), and an internal view of the oven with samples (*right*) of kaki chips drying in a hybrid convection oven with novel IR heaters. (*See insert for color representation of this figure*).

heating technologies brings advantages in heating performance and/or quality attributes retention of the treated samples (e.g. kaki chips application as in Figure 6).

5.2 Infrared Radiation and Microwave Heating

In the combined IR–microwave heating method, microwave heating is used to accelerate the entire baking process, and the IR heating is used to promote browning and crisping reactions on the product surfaces. The effects of IR-microwave baking on quality control and improvement texture, volume, porosity, and color of breads and cakes have been described (Sakiyan et al. 2011). Turabi et al. (2008) indicated that gluten-free cakes made from rice flour baked in a microwave–IR combination oven formulated with xanthan gum had better quality characteristics than cakes containing the xanthan–guar gum blend. Ozkoc et al. (2009) indicated that the use of IR–microwave combination heating and the addition of a xanthan–guar gum blend decreased the hardness, retrogradation enthalpy, and total mass crystallinity values of bread samples, also showing a delay in staling.

5.3 Infrared Radiation and Freeze-Drying

Although many advantages of the use of FD have been described (Krishnamurthy et al. 2008a), some problems with this technology are also evident (Bae et al. 2010); for example, FD processes involve high energy costs for operating the refrigerators and vacuum pumps to freeze and dehydrate products, and the limited heat and mass transfer rate in the FD conditions generally leads to long drying times and low process productivity. For these reasons, IR heating is often used in conjunction with FD (in order to accelerate the sublimation process).

Lin et al. (2005) evaluated the combination of IR heating with FD for sweet potatoes. It was reported to reduce the processing time by less than half. Compared to regular FD, IR-assisted FD brought lower color differences and faster dehydration. IR heating leading to a higher dehydration ratio implies that IR heating reduces serious product shrinkage. Pan et al. (2008b) employed a sequential IR and FD to dry fruits. By IR FD, better color, higher shrinkage, higher crispiness but poor rehydration capacity compared to those produced by using single-stage FD were obtained. Antal (2015) evaluated a combination of FD with IR drying and hot air drying (IR-FD and HAD-FD, respectively) to dry Idared apple cubes. A $5\,\text{kW}\,\text{m}^{-2}$ IR power intensity was used. The IR heating had a higher drying rate than hot air during the pre-dehydration. The dried material produced with IR-FD had desirable color, higher rehydration rate, and lower firmness compared to that dried by HAD-FD. The quality of single-stage FD samples was close to the of IR-FD materials. It was observed that the IR-FD method greatly decreased the energy consumption compared to FD and HAD-FD drying treatments.

5.4 Infrared Radiation and Vacuum Drying

With the vacuum drying method, the quality of dried food is higher than that obtained with other methods because of the absence of oxygen in the internal dryer cabinet and the reduction of unwanted reactions in the food. The application of a vacuum in food drying causes the expansion of air and vapor and creates a puff state in the matter. Due to the high energy consumption in this method, vacuum drying is recommended for highly sensitive and high value-added products. For vacuum drying, the moisture within the product being dried evaporates at lower temperatures ($<100\,°\text{C}$), giving better product quality – especially in the cases of foods or agricultural products, which are heat-sensitive in nature (Hafezi et al. 2015).

In vacuum drying, the air surrounding wet material flows through a suction device (a vacuum pump) to remove the humidity released by the material from its vicinity, so that the material faces less resistance while avoiding the saturation of the material's surface with moisture. IR drying, in contrast, has the unique characteristics of an energy transfer mechanism. Hafezi et al. (2015) studied the effect of drying behavior on the drying rate and quality characteristics of potato slices in a vacuum-IR drying system. They investigated the effects of the IR powers (100, 150, and 200 W) and vacuum levels (20, 80, 140 mmHg and the atmosphere pressure) at different slice thicknesses (1, 2, and 3 mm) on the drying rate, shrinkage percentage, and rehydration capacity. They concluded that the IR power level has a significant effect on the processing time and the drying rate. With increasing radiation power, the drying time is reduced and consequently the drying rate is increased. The shrinkage percentage increased with the increase in the sample thickness. In other words, shrinkage was decreased at different thicknesses with an increase in the IR power and vacuum level. However, with increasing shrinkage, the rehydration capacity was decreased.

6 Mathematical Modeling and Simulations

Mathematical modeling is one of the most useful tools in the design and control of IR heating systems. For the improvement of the functionality of IR radiation systems in terms of the heating rate and temperature uniformity, several factors can be considered: system rotation, optimized heating cycles, and different heater geometries. Computer simulation approaches can consider these factors in a comprehensive manner.

Estimations of the amount of IR energy needed for food products requires the knowledge of the emissive power from IR emitters with different surface temperatures, the travel pathways of photo particles, and view factors. In addition, radiative properties of food products such as reflectivity and transmissivity at the interface, absorption and emission within the medium, and heat transfer incorporating the thermal properties of food products are needed (Jun et al. 2011). Models coupling heat transfer and microbial inactivation are also needed to predict the population of microorganisms during non-isothermal processes. Since IR heating is a complex phenomenon for achieving the inactivation of different groups of microorganisms, a proper model simulating simultaneous kinetics in the combined mode of heating is highly desirable.

6.1 Basics of Computer Simulations of Infrared Radiation Processes

Sakai and Hanzawa (1994) assumed that most FIR radiation energy would be absorbed at the surface of a food product due to the predominant energy absorption of water. Energy would thereafter be transported by heat conduction in the food. Based on this assumption, the governing equations and boundary conditions to explain the heat transfer derived from energy balance in a food product can be solved using Galerkin's finite element method. The selection of the appropriate boundary conditions depends on several factors which include the type and geometry of the sample, the selected solution method, the characteristics of the oven, the assumptions, and others.

As an example of the mathematics included in a computer simulation model of IR heating of foods, the study by Llave et al. (2016) related to the mathematical modeling of eggplant undergoing simultaneous heat and mass transfer during oven roasting is shown. The present model assumed:

- Cylindrical symmetry, which allowed restriction of the analysis to a 2D geometry.
- Negligibility of any interaction at the contact points between the eggplant and the sample holder.
- Eggplant is a fictitious continuum, although it is a hygroscopic porous medium.
- Negligibility of the convective heat transfer and the gas pressure gradients in the food sample.
- Negligibility of hardening and surface cracking effects, because the process was conducted until a point before the occurrence of fractures, avoiding prolonged roasting processes.

6.1.1 Moisture Transfer

The mathematical model of moisture transfer in 2D according to Fick's equation plus water evaporation is as follows:

$$\rho_s \frac{\partial W_w}{\partial \tau} = \frac{1}{r}\left\{\frac{\partial}{\partial r}\left(D_w r \rho_s \frac{\partial W_w}{\partial r}\right) + \frac{\partial}{\partial z}\left(D_w r \rho_s \frac{\partial W_w}{\partial z}\right)\right\} - V_e \tag{8}$$

where ρ_s is the density of a porous solid (kg-solid m^{-3}), W_w is the moisture content (kg-water kg-solid^{-1}), τ is time (s), r is the radial distance in cylindrical coordinates (m), D_w is the moisture diffusivity ($m^2\ s^{-1}$), z is the thickness distance in cylindrical coordinates (m), and V_e is the internal evaporation rate (kg-water $s^{-1}\ m^{-3}$).

6.1.2 Heat Transfer

For a cylindrical eggplant sample, the following governing equation was used to model 2D axisymmetric transient heat conduction:

$$\rho_b C_p \frac{\partial T}{\partial \tau} = \frac{1}{r}\left\{\frac{\partial}{\partial r}\left(kr\frac{\partial T}{\partial r}\right) + \frac{\partial}{\partial z}\left(kr\frac{\partial T}{\partial z}\right)\right\} - H_L V_e \tag{9}$$

where ρ_b is the density ($kg\,m^{-3}$), C_p is the specific heat ($kJ\,kg^{-1}\,K^{-1}$), T is the temperature (K), k is the thermal conductivity ($W\,m^{-1}\,K^{-1}$), and H_L is the latent heat of vaporization ($J\,g^{-1}$).

6.1.3 Boundary Conditions

The initial and boundary conditions used in the moisture and heat transfer equations are as follows:

For Moisture Transfer

$$\text{Initial C.}: \ W_w = W_{w0} \ \text{ when } \ t = 0 \tag{10}$$

$$\text{Boundary C.}: -D_w \text{grad}\left(W_w\right)\cdot n = h_m\left(W_w - W_{we}\right) \tag{11}$$

where h_m is the convective surface mass transfer coefficient ($m\,s^{-1}$), W_{we} is the equilibrium moisture content (kg-water kg-solid^{-1}) which was assumed as 0 from the beginning of the process for simplicity, and n is the outward normal unit vector (−).

For Heat Transfer

$$\text{Initial C.}: \ T = T_0 \ \text{when} \ t = 0 \tag{12}$$

$$\text{Boundary}: -k\text{grad}(T) \cdot n = -H_L h_m \rho_S (W_W - W_{We}) + h_t (T_{\text{air}} - T) + \phi\sigma_c (T_h^4 - T^4) \tag{13}$$

where h_t is the convective surface heat transfer coefficient (W m^{-2} K^{-1}), T_h is the wall temperature (K), φ is the summary radiant heat transfer coefficient (–), and σ_c is the Stefan–Boltzmann constant (W m^{-2} K^{-4}).

6.2 Heat and Mass Transfer Modeling of the Infrared Radiation Heating of Foods

Table 6 summarizes some of the pioneering models that have been developed in the food processing sector.

According to Tanaka et al. (2006), most of the pioneering mathematical modeling works in the food processing field were carried out by solving the heat transfer equation as a deterministic model with appropriate boundary and initial conditions. Although the random nature of photons plays an important role in radiation heat transfer, it was neglected or simplified in the process of model development in the previous works. The thermal radiation phenomenon should be regarded as a random process because it is impossible to predict the birth, traveling, and death of photons using deterministic models. The Monte Carlo (MC) simulation is useful in developing a radiation heat transfer model because it can describe the aforementioned random nature of photons. Tanaka et al. (2006) thus claimed that by combining a MC FIR model with convection–diffusion air flow and heat transfer models, an FIR heating process could begin to be simulated appropriately. Tanaka et al. (2007) combined MC FIR radiation simulations with convection-diffusion air flow and heat transfer simulations to investigate the suitability of the method for the surface decontamination of strawberries. The model was a powerful tool to quickly and comprehensively address complex heating configurations that include radiation, convection, and conduction. Computations were validated against measurements with a thermographic camera. Compared to the air convection heating, the FIR heating obtained more uniform surface heating, with a maximum temperature well below the critical limit of ~50 °C.

However, for a more accurate and integral computer simulation of the IR heating process, the developed models should be extended to consider mass transfer and the shrinkage of the treated food as well as the volumetric dissipation of the radiation power. Some of these approaches are shown in the next section.

6.3 Computer Simulations of Novel IR Heating Applications of Foods

Studies describing computer simulation models of recent applications of IR heating alone or in combination with other heating systems are increasing in number. These advanced studies include computational fluid dynamics, several multiphysics modeling methods (e.g. conjugate modeling), multiscale modeling and modeling of material properties and the associated propagation of material property variability (Defraeye 2014). Some of them are summarized in Table 7.

Table 6 Selected heat and mass transfer modeling approaches of IR heating.

Food product	Evaluated method/applied model	Relevant findings	References
Frozen tuna	An apparent specific model to predict 2-dimensional heat transfer during FIR heating	To prevent overheating of the tuna, the surface temperature was controlled by intermittent irradiation treatment. The experimentally determined temperature distributions in samples agreed well with theoretical predictions	Sakai et al. (1995)
Apple	Developed 66 different model equations assessing the temperature, and time dependence of IR drying was evaluated	Wherein the model derived from modified Page II had the lowest root mean squared error, mean bias error, and chi-square along with the highest modeling efficiency and regression coefficient for the variation of moisture ratio with time	Midilli et al. (2002)
Apple	To create new suitable models, including combined effects of drying time and temperature, during IR drying	In order to explain the drying behavior of apple, 10 different drying models (Newton, Page, modified Page, Wang and Singh, Henderson and Pabis, logarithmic, diffusion approach, simplified Fick's diffusion (SFFD) equation, modified Page equation-II, and Midilli equation) were developed and validated	Togrul (2005)
Paddy grain	To predict the effect of FIR irradiation in a series paddy drying process comprising of fluidized bed drying, transport of paddy, FIR irradiation, tempering, and ambient air ventilation by a set of coupled heat and mass transfer equations	The model was capable of reasonably predicting the temperature and moisture distributions inside a paddy grain	Meeso et al. (2007)
Rainbow trout (fish)	Three different infrared power levels of 83, 104, and 125 W were used. For determination of drying kinetics of fish, the experimental data for moisture content changes with time were evaluated using various empirical drying models, while the effective diffusion coefficient was evaluated using Fick's law of diffusion	The logarithmic model was found to be the most suitable in describing the drying characteristics of rainbow trout fillets. Kinetic parameters for the color change were determined using Hunter "L," "a," "b" values and total color difference values. "L" and "b" values increased and "a" value decreased after drying	Ismail and Kocabay (2018)
Tomato	A predictive 3D numerical model was developed to simulate heat transfer in a rotating mode undergoing IR heating	The net radiative heat flux on the tomato surface was approximated by a suitable periodic function: the related angular velocity accounted for tomato rotation, while its amplitude was linked with view factors between the tomato and the heating source. It was observed that temperatures predicted by the simplified model (1D axisymmetric model involving IR heating) agreed well with the 3D model output	Cuccurullo and Giordano (2019)

Table 7 Novel heat and mass transfer modeling approaches of IR heating.

Food product	Evaluated method/ applied model	Relevant findings	References
Milk	The efficacy of the IR heating process for inactivation of *S. aureus* was ensured by simulating the heating patterns using computational fluid dynamics	The predicted temperature values were found to be in good agreement with the experimental data. Transmission electron microscopic observation and IR spectroscopy of IR-treated *S. aureus* cells clearly verified cell wall damage, cytoplasmic membrane shrinkage, cellular content leakage, and mesosome disintegration	Krishnamurthy et al. (2008b)
Banana	A mathematical model, which allowed prediction of the moisture content and temperature, undergoing combined FIR and vacuum drying was developed. Various sample thicknesses for drying temperature of 50 °C and pressure of 5 kPa were evaluated	It was demonstrated that combined FIR heating and vacuum drying can be effectively used to produce fat-free banana-based snacks. Experimental and calculated moisture content and temperature agreed well, especially when the penetration depth was assumed to be 2 mm	Swasdisevi et al. (2009)
Frozen potato puree	Microwave tempering and NIR-assisted microwave tempering simulation using the finite difference method	The increase in microwave or IR power levels decreased tempering times. The temperature distribution inside the sample was modeled and predicted values showed good agreement with the experimental results	Seyhun et al. (2009)
Skim milk solution	A numerical model to accurately predict the FD processes with IR radiation heating was developed. In addition, a spectral two-flux radiation model was combined with the fixed grid numerical framework for detailed consideration of the spectral absorbing and emitting characteristics of moisture-rich products, as well as the spectral composition of light from IR heaters	Steady and transient conduction-radiation problems were successfully solved using the two-flux model, which validated the present calculation for radiative transfer and energy conservation equations. It was reported that the numerical model proposed in this study can be a useful tool for predicting and optimizing FD processes with IR heating	Bae et al. (2010)
Tomato	A previously developed predictive mathematical model to simulate heat transfer during IR heating in a dry-peeling process was validated with experimental data	Uniformity of surface temperature distribution was quantified by surface-averaged temperatures and a derived temperature uniformity index. IR heating induced a dramatic temperature increase on the tomato surface, which extended to 0.6 mm beneath (>90 °C) during a 60-s heating period, whereas interior temperature at the tomato center remained low (<30 °C)	Li and Pan (2014)

7 Future Research to Enhance Practical Applications of Infrared Heating

For the design and operation of IR heaters for food applications, radiation theory provides an excellent tool to augment rough experimental data with Planck's law, Wien's displacement law, Stefan–Boltzmann's law, exponential decay, view factors, and the interaction of IR with food. However, since foods are rarely smooth and diffuse and since most of the cases are heterogeneous food materials, further studies are needed to obtain a complete set of experimental data for radiative properties such as surface roughness, extinction coefficients, and the radiation penetration depth versus frequencies. In addition, the determination of the effects of irradiation on nutrition and sensory characteristics and physicochemical properties as well as interactions of food components under IR radiation could further support the uses of IR for food processing. Therefore, more knowledge about the interactions between processes and products is needed.

The applications of novel hybrid heating systems are growing as food equipment manufacturers begin to realize their full potential. All of the different heating techniques used in the integrated heating systems have their own limitations concerning application areas and possibilities. Good background knowledge may enable the combinations of techniques in the most optimal way. More experimental and practical data must be accumulated to ensure the success of the combinations of IR heating techniques with other heating technologies.

Foods' selective absorption of IR as provided by novel IR devices that offer the ability to control and generate only wavelength-specific radiant energy output might greatly improve the efficiency of various process heating applications. Such controlled radiation can stimulate the maximum optical response of the target object when the emission band of infrared and the peak absorbance band of the target object are identical. Although such manipulations of IR for the selective heating of foods could be very useful, there have been few investigations of selective heating using IR in foods.

The 3D modeling of food products is recognized to be a necessary tool in process development and product design. However, there are limited works exploring the whole multiphysics occurring during IR heating of foods. Most crucially, integrating microbial death kinetics with changed in chemical kinetics due to IR heating will provide a holistic approach to the understanding of complex microbial and chemical process kinetics and interactions as well as to the design of new and more effective heating systems.

8 Conclusions and Future Outlook

Although IR heating is attractive primarily for surface heating applications, it is a widely used technique in the food processing field. Novel applications of IR heating are being pursued, and efforts to better understand this unique process are increasing.

The selection of appropriate processing parameters has a great influence on the efficiency of IR processing and food quality. It is therefore necessary to improve the processing parameters in accord with the specific processing requirements and the various material parameters such as moisture content, thickness, and the physical and chemical properties of food materials. In other words, the choice of the optimal IR wavelength, radiation power, and radiation intensity for the specific food under certain processing requirements is necessary to make the heated materials absorb the IR radiation more accurately. In addition, based on some of the research findings presented in this chapter, it is apparent that higher IR power and intensity are not always the better choice. Both the processing rate and the quality of products should be taken into consideration while choosing processing parameters.

Combinations of IR heating with other heating technologies hold great potential to achieve the optimal energy and practical applicability of IR heating in the food processing industry. It is quite likely that the use of IR heating in the food processing sector will increase in the near future, especially for the tasks of drying and minimal processing such as fruit and vegetable peeling, disinfestation, and disinfection of food surfaces. The further development of these novel applications and process designs can be facilitated and improved using the increasingly accurate computer simulation models.

References

Aboud, S.A., Altemini, A.B., Al-Hilphy, A.R. et al. (2019). A comprehensive review on infrared heating applications in food processing. *Molecules* 24: 4125. https://doi.org/103390/molecules24224125.

Adak, N., Heybeli, N., and Ertekin, C. (2017). Infrared drying of strawberry. *Food Chemistry* 219: 109–116.

Antal, T. (2015). Comparative study of three drying methods: freeze, hot air assisted freeze and infrared-assisted freeze modes. *Agronomy Research* 13 (4): 863–878.

Arana, I. (2012). *Physical Properties of Foods – Novel Measurements Techniques and Applications*. Boca Raton, FL: CRC Press Taylor & Francis Group.

Atungulu, G.G. and Pan, Z. (2011). Infrared radiative properties of food materials. In: *Infrared Heating for Foods and Agricultural Processing* (eds. Z. Pan and G.G. Atungulu), 19–40. Boca Raton, FL: CRC Press Taylor & Francis Group.

Bae, S.H., Nam, J.H., Song, C.S. et al. (2010). A numerical model for freeze drying processes with infrared radiation heating. *Numerical Heat Transfer Part A* A58: 333–355.

Bandura, V., Mazur, V., Yaroshenko, L. et al. (2019). Research on sunflower seeds drying process in a monolayer tray vibration dryer based on infrared radiation. *INMATEH Agricultural Engineering* 57 (1): 233–242.

Bingol, G., Yang, J., Brandl, M.T. et al. (2011). Infrared pasteurization of raw almonds. *Journal of Food Engineering* 104: 387–393.

Boudhrioua, N., Bahloul, N., Slimen, I.B. et al. (2009). Comparison on the total phenol contents and the color of fresh and infrared dried olive leaves. *Industrial Crops and Products* 29: 412–419.

Braeckman, L., Ronsse, F., Cueva-Hidalgo, P. et al. (2009). Influence of combined IR grilling and hot air cooking conditions on moisture and fat content, texture and color attributes of meat patties. *Journal of Food Engineering* 93: 437–443.

Chung, D. (2008). Coffee bean roaster and method for roasting coffee beans using the same. US Patent 2008/0268119 A1, filed 30 June 2008 and issued 30 October 2008.

Correa, P.C., Baptestini, F.M., Zeymer, J.S. et al. (2019). Dehydration of infrared ginger slices: heat and mass transfer coefficient and modeling. *Ciencia e Agrotecnologia* 43: 1–11.

Cortés, V., Blasco, J., Aleixos, N. et al. (2019). Monitoring strategies for quality control of agricultural products using visible and near-infrared spectroscopy: a review. *Trends in Food Science & Technology* 85: 138–148.

Cuccurullo, G. and Giordano, L. (2019). Simplified numerical modelling of infrared radiation effects in tomato dry peeling. *Journal of Physics: Conference Series* 1224: 012017.

Dagerskog, M. (1979). Infrared radiation for food processing II. Calculation of heat penetration during infrared frying of meat products. *Lebens Wissen Technology* 12 (5): 252–257.

Das, I., Das, S.K., and Bal, S. (2009). Drying kinetics of high moisture paddy undergoing vibration-assisted infrared (IR) drying. *Journal of Food Engineering* 95: 166–171.

Datta, A.K. and Almeida, M. (2014). Properties relevant to infrared heating of food. In: *Engineering Properties of Foods*, 4e (eds. M.A. Rao, S.S.H. Rizvi, A.K. Datta and J. Ahmed), 281–310. Boca Raton, FL: CRC Press Taylor & Francis Group.

Decareau, R.V. (1985). *Microwaves in the Food Processing Industry*. Orlando, FL: Academic Press.

Defraeye, T. (2014). Advanced computational modelling for drying processes – a review. *Applied Energy* 131: 323–344.

Ding, C., Khir, R., Pan, Z. et al. (2015a). Improvement in shelf life of rough and brown rice using infrared radiation heating. *Food and Bioprocess Technology* https://doi.org/10.1007/s11947-015-1480-5.

Ding, C., Khir, R., Pan, Z. et al. (2015b). Effect of infrared and conventional drying methods on physicochemical characteristics of stored white rice. *Cereal Chemistry* 92 (5): 441–448.

Dondee, S., Meeso, N., Soponronnarit, S. et al. (2011). Reducing cracking and breakage of soybean grains under combined near-infrared radiation and fluidized-bed drying. *Journal of Food Engineering* 104 (1): 6–13.

Gabel, M.M., Pan, Z., Amaratunga, K.S.P. et al. (2006) Catalytic infrared dehydration of onions. *Journal of Food Science* 71: E351–E357.

Ganapathy, S.R., Palani, S., Saran, R. et al. (2018). Design of radiation heat source for paint curing ovens. *International Journal of Pure and Applied Mathematics* 119 (5): 1039–1046.

Garcia, E. and Barrett, D.M. (2006). Peelability and yield of processing tomatoes by steam or lye. *Journal of Food Processing and Preservation* 30 (1): 3–14.

Hafezi, N., Sheikhdavoodi, M.J., and Sajadiye, S.M. (2015). Evaluation of quality characteristics of potato slices during drying by infrared radiation heating method under vacuum. *International Journal of Agricultural and Food Research* 4 (3): 1–8.

Halford, R.S. (1957). The influence of molecular environment on infrared spectra. *Annals of the New York Academy of Sciences* 69: 63–69.

Hamanaka, D., Norimura, N., Baba, N. et al. (2011). Surface decontamination of fig by combination of infrared radiation heating with ultraviolet irradiation. *Food Control* 22: 375–380.

Hassoun, A., Ojha, S., Tiwari, B. et al. (2020). Monitoring thermal and non-thermal treatments during processing of muscle foods: a comprehensive review of recent technological advances–review. *Applied Sciences* 10: 6802. https://doi.org/10.3390/app10196802.

Hung, J.Y., Wimberger, R.J., and Mujumdar, A.S. (1995). Drying of coated webs. In: *Handbook of Industrial Drying*, 2e (ed. A.S. Mujumdar), 1007–1038. New York: Marcel Dekker Inc.

Ismail, O. and Kocabay, O.G. (2018). Infrared and microwave drying of rainbow trout: drying kinetics and modelling. *Turkish Journal of Fisheries and Aquatic Sciences* 18: 259–266.

Jun, S., Krishnamurthy, K., Irudayaraj, J. et al. (2011). Fundamentals and theory of infrared radiation. In: *Infrared Heating for Foods and Agricultural Processing* (eds. Z. Pan and G.G. Atungulu), 1–18. Boca Raton, Fl: CRC Press Taylor & Francis Group.

Khir, R., Pan, Z., Thompson, J.F. et al. (2014). Moisture removal characteristics of thin layer rough rice under sequenced infrared radiation heating and cooling. *Journal of Food Processing and Preservation* 38: 430–440.

Kim, S.Y., Jeong, S.-M., Jo, S.-C. et al. (2006). Application of far infrared irradiation in the manufacturing process of green tea. *Journal of Agricultural and Food Chemistry* 54: 9943–9947.

Krishnamurthy, K., Khurana, H.K., Soojin, J. et al. (2008a). Infrared heating in food processing: an overview. *Comprehensive Reviews in Food Science and Food Safety* 7: 2–13.

Krishnamurthy, K., Jun, S., Irudayaraj, J. et al. (2008b). Efficacy of infrared heat treatment for inactivation of *Staphylococcus aureus* in milk. *Journal of Food Process Engineering* 31: 798–816.

Kumar, C.M., Rao, A.G.A., and Singh, A.S. (2009). Effect of infrared heating on the formation of sesamol and quality of defatted flours from *Sesamum indicum* L. *Journal of Food Science* 74: H105–H111.

Kunze, O.R. and Calderwood, D.L. (2004). Rough rice drying: moisture adsorption and desorption. In: *Rice: Chemistry and Technology* (ed. E.T. Champagne), 223–268. St. Paul, MN: American Association of Cereal Chemists.

Lao, Y., Zhang, M., Chitrakar, B. et al. (2019). Efficient plant foods processing based on infrared heating. *Food Reviews International* https://doi.org/10.1080/87559129.2019.1600537.

Lee, E.H. (2021). A review on applications of infrared heating for food processing in comparison to other industries. *Innovative Food Processing Technologies*: 431–455. https://doi.org/10.1016/B978-0-08-100596-5.22670-X.

Li, X. (2012). A study of infrared heating technology for tomato peeling: process characterization and modeling. 9781267968708.

Li, X. and Pan, Z. (2014). Dry peeling of tomato by infrared radiative heating: part II. Model validation and sensitivity analysis. *Food and Bioprocess Technology* 7: 2005–2013.

Li, X., Zhang, A., Atungulu, G.G. et al. (2014). Effects of infrared radiation heating on peeling performance and quality attributes of clingstone peaches. *LWT- Food Science and Technology* 55: 34–42.

Li, X., Xie, X., Zhang, C.H. et al. (2018). Role of mid- and far-infrared for improving dehydration efficiency in beef jerky drying. *Drying Technology* 36: 283–293.

Lin, Y.P., Tsen, J.H., and King An-Erl, V. (2005). Effects of far-infrared radiation on the freeze-drying of sweet potato. *Journal of Food Engineering* 68: 249–255.

Llave, Y. and Hayakawa, K. (2019). Analysis of color and moisture content changes during drying of kaki (*Diospyros kaki*) chips. In: *Proceedings of the 20th Japan Food Engineering Conference (JSFE)*, 99. Japan Society for Food Engineering.

Llave, Y., Takemori, K., Fukuoka, M. et al. (2016). Mathematical modeling of shrinkage deformation in eggplant undergoing simultaneous heat and mass transfer during convection-oven roasting. *Journal of Food Engineering* 178: 124–136.

Llave, Y., Takemori, K., Fukuoka, M. et al. (2017). Analysis of browning of broiled foods by noncontact techniques: a case study for Japanese eggplant (*Solanum melongena*). *Journal of Food Process Engineering* 40 (1): e12347. https://doi.org/10.1111/jfpe.12347.

Llave, Y., Satone, H., and Hayakawa, K. (2020). Application of new far infrared radiation (FIR) heaters in paddy rice grain drying. In: *Proceedings of the 21th Japan Food Engineering Conference (JSFE)*, 23. Japan Society for Food Engineering.

Matsuda, H., Llave, Y., Fukuoka, M. et al. (2013). Color changes in fish during grilling – influences of the heat transfer and heating medium on browning color. *Journal of Food Engineering* 116: 130–137.

Meeso, N., Nathakaranakule, A., Madhiyanon, T. et al. (2007). Modelling of far infrared irradiation in paddy drying process. *Journal of Food Engineering* 78: 1248–1258.

Midilli, A., Kucuk, H., and Yapar, Z. (2002). A new model for single-layer drying. *Drying Technology* 20 (7): 1503–1513.

Muriana, P., Gande, N., Robertson, W. et al. (2004). Effect of prepackage and post package pasteurization on postprocess elimination of *Listeria monocytogenes* on deli Turkey products. *Journal of Food Protection* 67 (11): 2472–2479.

Nindo, C.I. and Mwithiga, G. (2010). Infrared drying. In: *Infrared Heating for Food and Agricultural Processing* (eds. Z. Pan and G.G. Atungulu), 89–100. Boca Raton, FL: CRC Press Taylor & Francis Group.

Nowak, D. and Lewicki, P.P. (2004). Infrared drying of apple slices. *Innovative Food Science and Emerging Technologies* 5 (3): 353–360.

Olsson, E.E.M., Tragardh, A.C., and Ahrn, L.M. (2005). Effect of near-infrared radiation and jet impingement heat transfer on crust formation of bread. *Journal of Food Science* 70 (8): E484–E491.

Onwude, D.I., Hashim, N., and Chen, G. (2016). Recent advances of novel thermal combined hot air drying of agricultural crops–review. *Trends in Food Science & Technology* 57: 132–145.

Onwude, D.I., Hashim, N., Abdan, K. et al. (2019). The effectiveness of combined infrared and hot-air drying strategies for sweet potato. *Journal of Food Engineering* 241: 75–87.

Ozkoc, S.O., Sumnu, G., Sahin, S. et al. (2009). Investigation of physicochemical properties of breads baked in microwave and infrared microwave combination ovens during storage. *European Food Research and Technology* 228: 883–893.

Pan, Z., Khir, R., Godfrey, L.D. et al. (2008a). Feasibility of simultaneous rough rice drying and disinfestations by infrared radiation heating and rice milling quality. *Journal of Food Engineering* 84: 469–479.

Pan, Z., Shih, C., McHugh, T.H. et al. (2008b). Study of banana dehydration using sequential infrared radiation heating and freeze-drying. *LWT - Food Science and Technology* 41 (10): 1944–1951.

Pan, Z., Khir, R., Bett-Garber, K.L. et al. (2011). Drying characteristics and quality of rough rice under infrared radiation heating. *Transactions of the ASABE* 54 (1): 203–210.

Plank, M. (1901). Distribution of energy in the spectrum. *Annalen der Physik* 4 (3): 553–563.

Poss, G.T. (2007). Roasting coffee beans. US Patent 7 235 764 B2.

Rastogi, N.K. (2012). Recent trends and developments in infrared heating in food processing. *Critical Reviews in Food Science and Nutrition* 52: 737–760.

Sakai, N. and Hanzawa, T. (1994). Applications and advances in far-infrared heating in Japan. *Trends in Food Science & Technology* 5: 357–362.

Sakai, N. and Mao, W. (2006). Infrared heating. In: *Thermal Food Processing: New Technologies and Quality Issues* (ed. D.W. Sun), 493–544. Boca Raton, FL: CRC Press.

Sakai, N., Morita, N., Qiu, P. et al. (1995). Two dimensional heat transfer analysis of the thawing process of tuna by far infrared radiation. *Food Science and Technology Research* 42: 524–530.

Sakare, P., Prasad, N., Thombare, N. et al. (2020). Infrared drying of food materials: recent advances. *Food Engineering Reviews* 12: 381–398.

Sakiyan, O., Sumnu, G., Sahin, S. et al. (2011). A study on degree of starch gelatinization in cakes baked in three different ovens. *Food and Bioprocess Technology* 4 (7): 1237–1244.

Sandu, C. (1986). Lnfrared radiative drying in food engineering: a process analysis. *Biotechnology Progress* 2 (3): 109–119.

Selvi, K.C. (2020). Investigating the influence of infrared drying method on linden (*Tilia platyphyllos Scop.*) leaves: kinetics, color, projected area, modeling, total phenolic, and flavonoid content. *Plants* 9: 916. https://doi.org/10.3390/plants9070916.

Seyhun, N., Ramaswamy, H., Sumnu, G. et al. (2009). Comparison and modeling of microwave tempering and infrared assisted microwave tempering of frozen potato puree. *Journal of Food Engineering* 92: 339–344.

Sheridan, P. and Shilton, N. (1999). Application of far-infrared radiation to cooking of meat products. *Journal of Food Engineering* 41: 203–208.

Skjoldebrand, C. (2001). Infrared heating. In: *Thermal Technologies in Food Processing* (ed. P. Richardson), 208–228. England: Woodhead Publishing Limited, Abington.

Swasdisevi, T., Devahastin, S., Sa-Adchom, P. et al. (2009). Mathematical modeling of combined far infrared and vacuum drying banana slice. *Journal of Food Engineering* 92: 100–106.

Tanaka, F., Morita, K., Iwasaki, K. et al. (2006). Monte Carlo simulation of far infrared radiation heat transfer: theoretical approach. *Journal of Food Process Engineering* 29: 349–361.

Tanaka, F., Verboven, P., Scheerlinck, N. et al. (2007). Investigation of far infrared radiation heating as an alternative technique for surface decontamination of strawberry. *Journal of Food Engineering* 79: 445–452.

Togrul, H. (2005). Simple modeling of infrared drying of fresh apple slices. *Journal of Food Engineering* 71: 311–323.

Turabi, E., Sumn, G., and Sahin, S. (2008). Optimizationof bakingof rice cakes in infrared-microwave combination oven by response surface methodology. *Food and Bioprocess Technology* 1: 64–73.

Vidyarthi, S., Li, X., and Pan, Z. (2019). Peeling of tomatoes using infrared heating technology. In: *Tomato Chemistry, Industrial Processing and Product Development* (ed. S. Porreta), 180–200. Royal Society of Chemistry https://doi.org/10.1039/9781788016247.

Wang, B., Khir, R., Pan, Z. et al. (2014). Effective disinfection of rough rice using infrared radiation heating. *Journal of Food Protection* 77 (9): 1538–1545.

Wang, B., Venkitasamy, C., Zhang, F. et al. (2016). Feasibility of jujube peeling using novel infrared radiation heating technology. *LWT - Food Science and Technology* 69: 458–467.

Yadav, G., Gupta, N., Sood, M. et al. (2020). Infrared heating and its application in food processing. *The Pharma Innovation Journal* 9 (2): 142–151.

Yi, Z. and Zhongli, P. (2009). Processing and quality characteristics of apple slices under simultaneous infrared dry blanching and dehydration with continuous heating. *Journal of Food Engineering* 90: 441–452.

Yi, Z., Zhongli, P., and McHugh, T.H. (2007). Effect of dipping treatments on color stabilization and texture of apple cubes for infrared dry blanching process. *Journal of Food Processing and Preservation* 31: 632–648.

Yung, S.S., Wen, C.S., Ming, H.C. et al. (2008). Effect of far infrared oven on the qualities of bakery products. *Journal of Culinary Science and Technology* 6: 105–118.

10

Microwaves

Rifna E. Jerome and Madhuresh Dwivedi

Department of Food Process Engineering, National Institute of Technology Rourkela, Rourkela, Odisha, India

1 Introduction

Thermal processing is widely applied in food industry for shelf life extension and preservation of the food products. Application of heat continues to be most vital in food-processing unit operations, not only to accomplish acceptable changes in eating standard but also for preservative effect by destruction of microorganisms, pest, and inactivation of enzymes (Xanthakis et al. 2018). Heat used in diverse unit operations is emitted through various heat transfer modes such as conduction, convection, and radiation. Conduction is the heat transfer mode wherein the heat energy is produced with microscopic percussions of particles and flow of electrons within the body. Convection is the heat transfer mode in which groups of molecules flow in a fluid. Radiation is the heat transfer mode where electromagnetic wave moves from one body to another. Nevertheless, there are various uses, advantages, and disadvantages to each of the aforementioned mode of heat transfers. Likewise, there exist also various drawbacks related to conventional heating techniques applied for thermal processing of food, wherein dielectric heating acts as a potential alternative (Figure 1).

The progress of dielectric heating uses in food industry initiated in the radio frequency range during the 1930s (Püschner 1966). The preferred energy transfer rate augmentation steered to an amplified frequency: the microwaves (MWs). The microwave technique survived and still continues to be successful owing to the reason that microwave could heat more swiftly and uniformly the food products with direct interaction between microwaves and samples (Datta and Nelson 2000). Microwaves are electromagnetic waves falling within the frequency range between 3 kHz and 300 GHz. There are numerous applications of microwave radiation such as drying, heating, thawing, reheating, dehydration, blanching, microbial inactivation, and waste treatment (Ayappa et al. 1991). Drying is the unit operation where moisture is removed from the food material to a predetermined level to enhance the product stability and shelf life (Mujumdar 2000). Conventional drying technique though is extensively applied for drying of agricultural produce; mostly, it results in

Electromagnetic Technologies in Food Science, First Edition. Edited by Vicente M. Gómez-López and Rajeev Bhat.

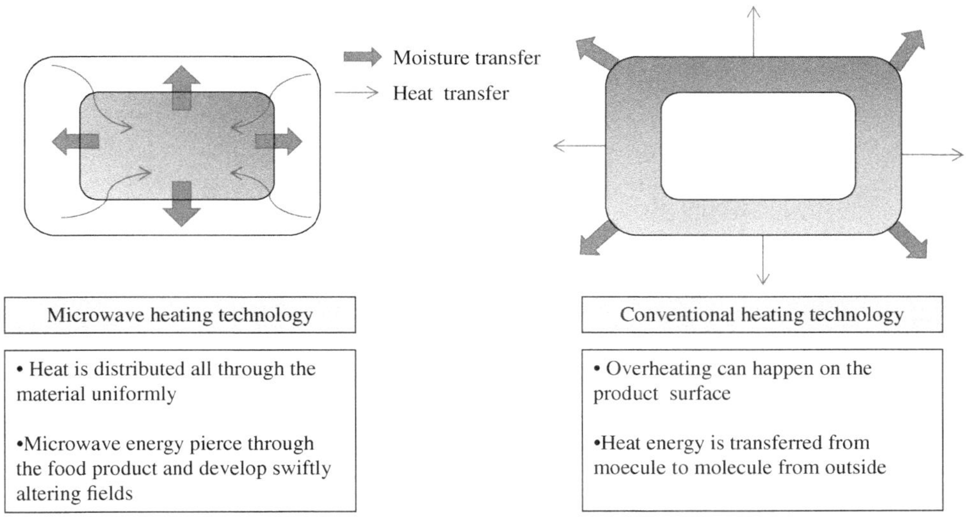

Figure 1 Microwave versus conventional heating technique. (*See insert for color representation of this figure*).

considerable variations in quality attributes of food products such as taste, color, nutritional profile, and final product quality. Furthermore, changing of inner solutes from interior to surface of the product owing to prolonged drying period, elevated temperature, and high initial installment cost of novel non-destructive technologies limits the use of conventional and process analytical technologies in food and pharmaceutical industries (Das et al. 2017; Jerome et al. 2019; Maskan 2001). Microwaves are a basic appliance in most of the households. In the United States, European countries, and most of Asian countries, microwave heating is the third most prominent heating technique for foods (Kalla and Devaraju 2017). Use of microwave energy in the drying process has been popularized because of bare minimum quality deterioration, quick drying, and effectual heat distribution (Diaz et al. 2003). Washington State University (WSU) has performed various research works that have highlighted commercial uses of 915 MHz microwave pasteurization methodology for sterilization of packed foods. Furthermore, microwave-sterilized processed foods including rice dishes and sauces are commercially available in European markets (Ekezie et al. 2017; Kumar and Karim 2019).

Blanching of horticultural produce is performed to soften vegetable tissues and inactivate the targeted enzymes. It has been reported that microwave-assisted blanching produces effluent-free process when compared to hot water treatment techniques (Lanqingzi 2005). Thawing is the technique to convert the frozen food to semi-liquid or liquid phase. The characteristic attribute of microwaves to pierce through and release the heat waves within the core of chosen sample makes them prospective in enhancing thawing technique (Li and Sun 2002). Sterilization and pasteurization methods are applied to destroy microbes and inactivate targeted enzymes in food products. Microwave-assisted processing of liquid products chiefly targets on the inactivation of microbes (Al-Hilphy and Ali 2013), enzyme inactivation (Salazar-González et al. 2012), and improving physicochemical properties (Iuliana et al. 2015).

MW-assisted thermal treatment is a suitable method as it has been proven to be rapid, clean, easy, and safe technology with volumetric heating of product. For thermal applications, use of microwave energy is the matter of study since many years. The purpose of this chapter is to confer the application of microwave energy in the food-processing sector and to spotlight the recent developments and its effect (Constantin and Csatlos 2010; Vadivambal and Jayas 2007) on enzymes, pathogen inactivation, and physicochemical attributes of food products. Furthermore, the chapter elucidates subsisting knowledge in the sector of microwave heating and evaluates insights for the correlation among product characteristics prior to microwave treatment, and engineering processes are discussed as well.

2 Microwave Heating Mechanism and Principle

All through microwave processing, heating of the food sample occurs owing to the incorporation of microwave energy by the food product which leads into internal heating of food product chosen (Arnoldi 2001). An important verity that needs to be kept in mind during microwave application is that microwaves and dielectrics are not a kind of heat but heat is produced through its interface with sample materials. There are numerous principles that are applied in this energy exchange such as dipole rotation, ionic conduction, interface polarization, dipole stretching, piezoelectricity, nuclear magnetic resonance, ferroelectric hysteresis, and ferromagnetic resonance (Mujumdar 2014). Microwave radiations generally penetrate into food samples and interrelate with water and solid components existing in a product which outcomes in volumetric heating owing to flow of dissolved ions and dipole action of solvent. The dipole action is initiated owing to deviations in electric and magnetic fields; furthermore, water is also another vital source for microwave interaction to occur. Heating of the food product when subjected to microwave energy occurs because of the to and fro motions of ionic particles produced with high-frequency oscillating electric fields (Bhatt et al. 2020). There are various factors that affect efficacy of microwave application in food processing. The important microwave parameters involve heat distribution, uniformity, penetration depth, and dielectric properties of the food product.

2.1 Dielectric Properties of Food Product

The dielectric properties are extremely necessary to propose and develop both microwave heating techniques and manufacturing of equipment. Techniques applied for the determination of dielectric properties of food products chiefly rely upon the degree of accuracy and frequency essential, according to food chosen, for microwave treatment (Curet et al. 2014). The method of analysis includes production of microwave energy at the frequency of regard, pointing it into product to be analysed, assessment of variations in the signal produced owing to product. The capability to transform microwave energy into heat can be implicit by dielectric properties of product. The most important property which governs microwave heating is the dielectric permittivity (ε^*).

The ability to transform microwave radiations into heat energy could be understood by dielectric attributes of product. The most vital attribute that rules microwave treatment is the dielectric permittivity (ε^*), expressed in Eq. (1). The actual part of the dielectric

Table 1 The microwave dielectric properties of water at indicated temperature and frequency.

	Dielectric constant (ε')		Dielectric loss factor (ε'')	
	Temperature (°C)		Temperature (°C)	
Frequency (GHz)	20	50	20	50
0.60	80.30	69.90	2.75	1.25
1.70	79.20	69.70	7.90	3.60
3.00	77.40	68.40	13.00	5.80
4.60	74.00	68.50	18.80	9.40
7.70	67.40	67.20	28.20	14.50
9.10	63.00	65.50	31.50	16.50
12.50	53.60	61.50	35.50	21.40
17.40	42.00	56.30	37.10	27.20
26.80	26.50	44.20	33.90	32.00
36.40	17.60	34.30	28.80	32.60

ε', dielectric constant; ε'', dielectric loss factor.
Source: Adapted from Hasted (1973).

attribute is acknowledged as dielectric constant, which describes the potential of material to accumulate electric energy and the imaginary phase of the dielectric factor is considered as dielectric loss factor that elucidates the capability to change electric energy into heat energy. The dielectric properties of water is listed in Table 1 for few selected frequencies at temperatures of 20 and 50 °C from data produced by (Hasted 1973).

$$\varepsilon^* = \varepsilon' - j\varepsilon'' \tag{1}$$

where ε^* is dielectric permittivity, ε' is dielectric constant, ε'' is dielectric loss factor, and $j = \sqrt{(-1)}$, where j in the Cartesian plane depicts an imaginary number.

The ratio of dielectric loss to dielectric constant is given by loss tangent and is expressed as

$$\tan\delta = \frac{K''}{K'} = \frac{\varepsilon''}{\varepsilon'} \tag{2}$$

where k'' is relative dielectric loss, $k'' = \varepsilon''/\varepsilon_0$ k' is relative dielectric constant, and $k' = \varepsilon'/\varepsilon_0$, where ε_0 is the permittivity of free space ($\varepsilon_0 = 8.854 \times 10^{-12}\,\mathrm{F\,m^{-1}}$).

The dielectric properties of the food products are chiefly governed by the microwave frequency and temperature applied. With regard to the degree of microwave absorption, the products are classified into three categories:

i) Opaque, that reflects the microwaves from its surface.
ii) Transparent, microwaves are allowed to pass within the materials with little lessening.
iii) Absorber, that absorbs microwave energy very robustly.

Table 2 Significant studies on microwave technology in food applications.

Application	Food product	Treatment condition	Quality attributes	Major results	References
Drying	Raspberry	600 W, 70 °C	Texture, retention of anthocyanin, rehydration property	17.5% enhanced anthocyanin recovery, 25.36% increased rehydration ratio	Si et al. (2016)
	Eggplant	630 W	Rehydration capacity	37% increase in rehydration value	Chayjan and Kaveh (2016)
	Banana	100–300 W	Color, drying time	Time saved by 96%	Omolola et al. (2014)
	Green pepper	100 W	Water activity and rehydration capacity	Significant increase in rehydration property and energy consumption decreased by 57%	Darvishi et al. (2014)
Baking	Gluten-free bread	20% microwave power	Volume and hardness	Increased volume and hardness value reduced	Therdthai et al. (2016)
	Cake	100–900 W	Volume, microstructure	Volume increased with large pores	Megahey et al. (2005)
	Soy cake	700 W	Volume	Enhancement in specific volume to 1.37 $m^3\ kg^{-1}$	Şakıyan (2015)
Microbial inactivation	Paprika powder	650 W	Water activity	Increased water activity	Eliasson et al. (2015)
	Black pepper	800 W	Volatile oil content, piperine	Decreased volatile oil content, piperine	Jeevitha et al. (2016)
	Peanut	360, 480, and 600 W	Moisture content, free fatty acid	Decreased moisture content, preserved free fatty acid	Patil et al. (2019)
Roasting	Pistachios	682 W	Texture and color	Significantly preserved color and texture	Hojjati et al. (2015)
Tempering	Potato	706 W	Dielectric properties	Loss factor and dielectric constant increased	Seyhun et al. (2009)

Consequently, understanding of dielectric factors is necessary to classify the food products on the basis of microwaves absorption and also to fabricate dielectric drying of equipment and process and procedure design (Table 2).

The penetration depth (D_p) could be described as the distance at which the power density falls to a value of 1/e from its actual value at the surface and is denoted as (Metaxas and Meredith 1983)

$$D_P = \frac{C}{\sqrt{2\pi f}\left[K'\left\{\sqrt{1+\left(K''/K'\right)^2}-1\right\}\right]^{1/2}} \tag{3}$$

where c is the velocity of light, $c = (\mu_o\varepsilon_o)^{-1/2}$, f is the frequency of microwave irradiation at desired power levels, and μ_o is the permeability of free space ($\mu_o = 4\pi \times 10^{-7}\,\text{H m}^{-1}$).

Equation (3) is applicable for food products with non-magnetic property. The microwave power that alters with the square of the electric field is expressed as

$$q = \frac{1}{2}\omega\varepsilon_0 K''|E|^2 \tag{4}$$

wherein E is the electric field intensity and ω is the angular frequency. Separately from dielectric properties of materials and penetration depth, few other properties also affect microwave-assisted food processing such as the microwave cavity design (geometry of cavity), product moisture content, microwave frequency, density of the product, composition of food, and shape and size of the product (Nelson and Datta 2001). The product moisture content or quantity of water accessible in the product is the vital attribute regarded in deciding its dielectric values as water is an excellent absorber of microwave radiations.

2.2 Factors Affecting Microwave Heating

A range of physico-thermal, electrical, and dielectric properties of the food products determine the heating performance of the food sample and absorption of microwave energy. Section below describes few such factors.

2.2.1 Moisture Content and Temperature Dependency

Percentage of free moisture there in the food product is the key factor that controls the heating of food sample within a microwave oven (Nelson and Kraszewski 1990). Efficient heating occurs within microwave equipment when the moisture content of the product is elevated as a result of increased dielectric loss factor (Schiffmann 1986). Likewise, the dipole rotation is related to free water existing within the product, and the chemical composition of food product also governs dielectric properties of the product. The dependency of dielectric constant on temperature is a bit complex, and it may differ as the product temperature increases or decreases.

2.2.2 Effect of Composition of Food Product

The dielectric values of food samples also significantly depend on the chemical composition of the product, which convolutes the study of dielectric attributes of the entire product with regard to a particular component or ingredient. The organic constituents of the food product are generally regarded as transparent to energy since they are dielectrically inert ($\varepsilon' < 3$ and $\varepsilon'' < 0.1$) (Mudgett 1985). Furthermore, salt fraction in selected food influences the dielectric properties of the product to greater extent (Ohlsson and Bengtsson 1975). It has also been observed that the dielectric factor value increases gradually as the temperature increases, fat content decreases, and moisture percentage increases (Teseme and Weldeselassie 2020; To et al. 1974; Zhu and Guo 2017).

2.2.3 Effect of Microwave Frequency

It has been studied that as the microwave frequency and water content decrease, the penetration depth of emitted MW increases (Coronel et al. 2003). Equation (5) demonstrates the relation among wavelength and penetration depth:

$$D = \frac{\lambda_0 (K')^{1/2}}{2\pi K''} \tag{5}$$

where D = penetration depth (at which 63% of incident energy is absorbed), λ_0 = wavelength in free space (cm), K' = relative dielectric constant, K'' = relative dielectric loss.

Penetration depth rises with wavelength and falls down as microwave radiation frequency increases. The dielectric loss factor is greatly influenced by frequency, for aqueous solutions, that is a low frequency produces a lesser dielectric loss.

2.2.4 Product Parameters

Product parameters, such as geometry, density, and its mass, also control the microwave-assisted heating profile of the food. Products with bigger geometry absorb a greater amount of microwave energy and results in longer heating period. The air entrapped within the product makes the chosen product to be a good insulator (Schiffmann 1990). The size and shape of the product govern the product heating profile during microwave treatment of food. Food products that possess edges and corners exhibit high tendency for localized heating owing to multidirectional circulation of microwave radiation (Campanone and Zaritzky 2005).

2.3 Non-uniformity in Temperature Distribution

Despite the fact that microwave-assisted heating is a rapid heating technique when compared to conventional techniques, the non-uniformity in temperature profile is one of the major significant trouble linked with microwave heating. Progression of hot spots all through microwave-related heating is the major issue coupled with this electromagnetic technique and is widely dependent upon the geometry of the food product (Campanone and Zaritzky 2005). Owing to lack of temperature uniformity, it was reported that few portion of food products get heated quickly and to greater extent, whereas the residual portion gets heated to a very small extent (Vadivambal and Jayas 2010). Few projected remedies to curtail temperature non-uniformity are as mentioned below:

i) Use of hurdle technique, such as microwave and conventional methods
ii) Proper microwave oven design
iii) Regulating the product geometry
iv) Managing heating cycle
v) Heating of food product at reduced microwave power for an increased period
vi) Limiting the product thickness to 25 mm (Kalla and Devaraju 2017; Ohlsson and Thorsell 1984).

3 Microwave Application in Food Industries

Owing to the high number of microwave ovens in domestic level, the food-associated industries not only use microwave energy for processing operation but also develop product properties and food products particularly suitable for microwave heating. This technique of product development is acknowledged as product formulation or engineering.

3.1 Microwave-Assisted Cooking and Baking

Within last 10 years, domestic microwave oven has become a chief component in most kitchens. Different components of domestic microwave oven are shown in Figure 2 such as power adjustor, air outlet, transformer that act as high-voltage power source, control panel, magnetron cavity that converts high-voltage electric energy into microwave energy, waveguide, turntable base plate, metal wave guide stirrer, and a limit switch for safety reason to switch off magnetron unit if door is half or fully opened (Chandrasekaran et al. 2013). Microwave-assisted cooking of food specimen is the major purpose of microwave radiation technique. During cooking and baking process, microwaves alter color, taste, quality, and texture of products under treatment. Quality variations that occur during conventional and microwave baking of bread were studied by Içöz et al. (2004). Authors observed that during baking operations, it is vital to acquire desired level of browning and good texture at fixed moisture level. During microwave-assisted baking process, crust formation and crust browning are limited owing to the cold air neighboring the product and moisture condensation occurring from the product, which results in products lacking crispness (Orsat et al. 2017). To overcome this limitation, susceptors were later included in the design at the bottom of the surface product. Susceptors are fabricated out of material that absorbs microwaves and transforms the absorbed microwave energy into heat energy to provide the heat to products by method of convection and conduction.

Through another study, energy usage, cooking efficiency of domestic appliances such as electric rice cooker (ERC), liquefied petroleum gas (LPG) pressure cooker, and microwave was compared (Lakshmi et al. 2007). Microwave-assisted cooking of rice was completed in

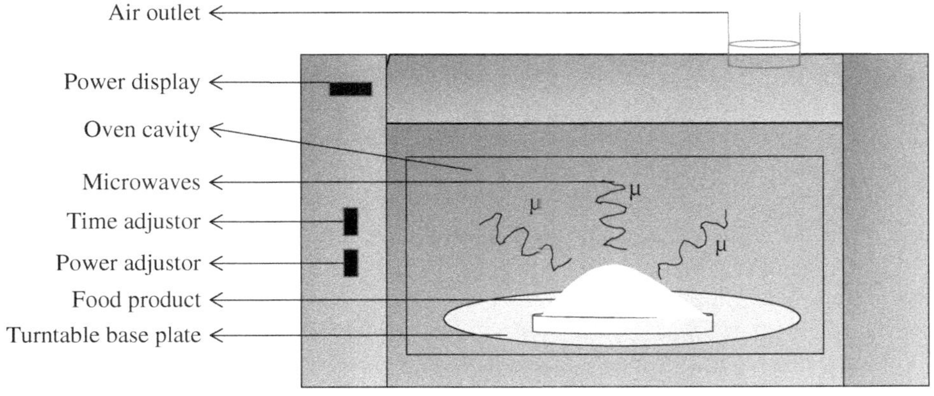

Figure 2 Domestic oven and important parts. (*See insert for color representation of this figure*).

shorter period as compared to other chosen cooking techniques, but energy efficiency was studied to be less significant in case of microwave cooking which warrants the need for future improvement.

Thorough references to the baking procedures of cakes, bread, pastry, etc., with the aid of microwaves on commercial scale can be studied. An improved throughput is attained by a quickening of the baking wherein further space requirements for microwave power generators are insignificant (Yolacaner et al. 2017). Microwaves in baking operations are generally used in amalgamation with infrared or conventional surface baking; this helps in eliminating the problem of surface browning and crust formation (Therdthai et al. 2016). In distinction to baking with conventional methods, microwave heating inactivates enzymes at a faster rate (this is owing to the rapid and homogeneous temperature elevation in the whole product) to avoid the starch from wide breakdown and develops adequate CO_2 and steam to create a extremely porous natured baked products (Decareau 1986). One intricacy to be conquered was development of microwavable baking pan that is adequately heat resistant and not too costly for commercial application. However, understanding its importance in 1982, patent was issued tacking this problem by incorporating metal baking pans in microwave ovens (Schiffmann et al. 1981).

The vital application of microwaves in the baking industry presently is for the microwave finishing, when reduced heat conductivity leads to significant elevated baking times in the conventional techniques. A diverse process that could also be geared by application of microwave heating is the process of pre-cooking. It has been well recognized for pre-cooking of bacon, meat patties and poultry (Helmar and Marc 2007). Microwaves are the chief energy source to coagulate the proteins and deliver the fat by an enhanced temperature. At the same time, the elimination of surface water is initiated by the convective flow of air. Another benefit of this technique is the release of important by-products, specifically rendered fat of eminent quality, which is applied as food flavoring (Schiffmann 1986).

3.2 Microwave-assisted Drying

Drying is the removal of moisture content from food products to predetermined level by creating vapor pressure difference between product and surrounding using different methods (Orsat et al. 2007). The main goal of drying is to enhance shelf life and preservation over microwave treatment by maintaining the product nutritional characteristics to significant level and physical attributes such as color and texture to appreciable levels. There are different drying methods used such as sun drying, solar drying, fluidized bed drying, hot air drying, spray drying, recirculatory air drying, freeze drying, and osmotic drying (Mujumdar 2000). Drying methods are applied according to product form, operational and industrial need, resources available, climatic condition and targeted outputs, and so on. Also, there are different technical issues associated with conventional drying operation such as extended drying time, quality alterations in the product, energy efficiency, temperature uniformity within product, and textural changes. Nowadays, microwave-assisted drying of food products is gaining popularity because of many advantages such as quick heating, volumetric heating, less quality alteration of product, and improvement of quality of some food products and very effective during falling rate period (Rattanadecho and Makul 2016). Volumetric heating leads to vapor transmission from inside of product to surface as pressure gradient is developed. Thus, shrinkage of product sample can be prevented in microwave-assisted drying. The synergistic effect of microwave and other conventional or non-conventional drying techniques can augment product superiority and drying efficiency, which was proven to be far superior when compared to conventional techniques alone or microwave only (Rifna and Dwivedi 2021; Zhang et al. 2006).

3.3 Microwave-Assisted Blanching

Blanching is a pretreatment process applied to fruit and vegetables at temperature enough to inactivate enzymes present in the tissue. Blanching inhibits the enzyme activity, sets the color of the product, and also shortens the time required for drying. Some food industries are using microwave blanching to blanch fruits and vegetables as it is effluent-free technology. Microwave-assisted blanching permits efficient heat transfer with minimum or no water requirement, which decreases nutrients lost to leaching as compared to immersed water blanching (Lin and Brewer 2005). Microwave-assisted blanching retains nutrients to significant level when compared to conventional hot water or steam blanching techniques owing to diminution in leaching losses in the process of microwave-assisted blanching (Ramesh et al. 2002). Color preservation and quality retention of frozen peas were found to be greater in microwave-assisted blanching when compared to water and steam blanching (Lin and Brewer 2005).

Through research works conducted till date, it can be concluded that microwave blanching of fresh food samples is far superior method than various other blanching techniques as it results in minimal degradation of sample nutrient composition unlike wet blanching and steam blanching techniques (Ranjan et al. 2017; Ruiz-Ojeda and Peñas 2013). This could be assumed with respect to the information that microwave blanching does not entail water as requisite in the method of wet blanching and, henceforth, decreases the probability of microbial spoilage. Some of the advantages contrasted with conventional heating techniques involve rapidity of operation. Hence, there are lesser leaching of soluble nutrients and vitamins, and the production of waste water is eradicated or greatly reduced. However, Microwave blanching of horticultural produce is still inadequate, as it causes slight blackening of product and minimal product structural changes (Trilokia et al. 2019). Henceforth, future works are desirable to optimize the microwave blanching technique with the aim to diminish the maintenance of enzymes which aids in deprivation and to enhance the preservation of product color and structure.

3.4 Microwave-Assisted Microbial Inactivation

Pasteurization or sterilization of food products refers to eradication or complete elimination of all microbial species or targeted or harmful microorganisms in foodstuffs to increase shelf life, food quality, and safety (Nott and Hall 1999). There are various techniques used to kill microorganisms such as ethylene dioxide, irradiation method, fumigation, steam sterilization, high-pressure processing, etc., and its application changes as the product varies. To assure complete eradication of harmful microorganisms, food products are treated for a specific time and temperature. Pasteurization also comprises inactivation of redundant enzymes, which results in cloud loss in certain fruit juices. Microbes or enzymes are inactivated using microwave energy at sub-lethal temperature. This has been explained by various techniques such as cell membrane rupture, selective heating, electroporation, and magnetic field coupling (Kozempel et al. 1998). The discerning heating states that microwaves heat up the microorganisms particularly and attain the temperature higher than the adjacent fluid that results in inactivation of microorganisms more rapidly. The selective heating proposes that microwaves heat the microorganisms selectively and reach the temperature greater than of the surrounding fluid which results in destruction of microorganisms more quickly. The theory of electroporation suggests that the electrical potential through the cell membranes is affected by pores, which results in the leakage of cellular materials. Cell membrane breaks down due to stress on cell membrane, according to cell membrane theory. As per magnetic coupling theory, the coupling of critical molecules like DNA or protein with electromagnetic energy disruptures the internal components of the cell. With the proper conditions of applying microwave and knowledge of dielectric properties of food products, desired process lethality can be achieved (Salazar-González et al. 2012). The potential of microwave radiation and its mechanism of action for inactivation of microbial spore from low moisture foods have been critically reviewed by Rifna and Dwivedi (2021) and Rifna et al. (2019). The impact of MW treatment at process condition of 650 W and 98 °C on mesophilic bacteria in the paprika powder was studied by Eliasson et al. (2015). The microbial count post- and pretreatment was demonstrated with thermal imaging technique. From the earlier study, it was reported that the microwave treatment at 650 W for holding time of 20 minutes attained 4.8-log reduction of targeted organism. Likewise Jeevitha et al. (2016) studied the log reduction in black pepper with regard to *Salmonella* and *E. coli*. Microwave condition of 800 W power achieved absolute significant sterilization of 6-log reduction. Pina-Pérez et al. (2014) studied the impact of MW technique on powdered food products that produced a microbial diminution level higher or equivalent to 5 $\log_{10}$ cycles at microwave power of 900 W. Nevertheless, it is essential to perform more experiments to verify its consequence on microbial inactivation on various solid and liquid food products whose request has been established to enhance principally in the market in recent times.

3.5 Microwave-Assisted Extraction

There are different methods used for the extraction of various components from food matrix such as Soxhlet extraction, hydro distillation, steam distillation, and simultaneous distillation–extraction. Recently, there has been extensive curiosity in utilization of microwave for extraction of different components such as essential oil from herbs (Lucchesi et al. 2004), aromas (Chee et al. 1996), phenols (Proestos and Komaitis 2008), pesticides (Onuska and Terry 1993), and dioxins (Kodba and Marsel 1999) extracted efficiently from different matrices such as food, plant materials, animal tissues, soils, and sediments. Ciriminna et al. (2019) used microwave-assisted extraction (MWAE) (one hour at 70 °C) to get pure extracts of betalain (a polyphenolic pigment) from the peel of *Opuntia ficus-indica* fruits gathered. Authors demonstrated that the preservation of identified polyphenolic molecules after MWAE possessed significant stabilizing effect, preserving the stability (with regard to both extract color and purity) of targeted compound. Thus, it can be concluded that this research opened up an effective solution to tackle the betalain pigment chemical instability issue which till date was the cause restricting its broad-scale industrial application. Similarly, Liazid et al. (2011) applied MWAE to extract anthocyanin pigment from grape peels and observed that anthocyanin extract was more stable at temperature 100 °C, while above 100 °C the extract stability and yield decreased indicating the compound degradation. The main benefits of MWAE are minimum requirement of solvent and energy and reduction of extraction time (Lucchesi et al. 2004). All findings of the aforementioned works show that microwave-assisted solvent extraction is a feasible novel extraction technique when compared to conventional techniques.

4 Safety of Food Processed in Microwave for Consumers

The food products prepared and processed through this novel electromagnetic technology is safe and sound for consumption. For the reason that the energy from microwave is converted to heat energy as quickly as it is engrossed by the food component, it cannot transform the food products contaminated or radioactive (Tang et al. 2018). As the microwave radiation is switched off and the foodstuff is taken out from the oven, there will be no leftover radiation lasting in the food. On this aspect, an microwave oven is more likely to an electric light that stops shining when the switch is turned off (Gallawa 2005). A number of countries, with the International Committee on Electromagnetic Safety (ICES), Food and Drug Administration (FDA), and the Institute of Electrical and Electronics Engineers (IEEE), have formulated a product emission range of 50 W m^{-2} within a region 5 cm away from oven surface. Practically, radiations from domestic ovens are significantly underneath the above-said international range and have interlocks which avoid human beings subjected to microwaves while the microwave oven is on. The International Commission on Non-Ionizing Radiation Protection (ICNIRP) has produced guiding principle on exposure ranges for complete electromagnetic fields (EMF) region of spectrum. Exposure ranges for human resources and respect to the public have been well established at below ranges where any dangerous heating happens from microwave exposure.

5 Merits and De-merits of Microwave Heating Applications

Today's application of microwave energy varies from these well-acknowledged uses such as sterilization and pasteurization to pooled techniques, for instance, microwave vacuum drying. The moderately slow reach of food business microwave uses has an array of reasons: primarily it's the conservatism of the food-processing units (Decareau 1985) and its comparatively reduced research financial plan. Associated to this, there are problems in solving the issues related to microwave heating applications. The chief problem is that, to produce good outcomes, they require a high effort of engineering aptitude.

Unlike from traditional heating methods, where agreeable outcomes could be attained effortlessly by perception, high-quality microwave application outcome frequently do demand an array of experience to recognize and modest effects such as non-uniform heating, etc. (Tang et al. 2018). A further demerit of microwave radiation application as compared to conventional technique is the requisite for electrical energy which is accounted to be its most costly form. Cavity effects are due to the design of cavity, the shape of the cavity, and the location of inlet point for the microwave. Product interaction includes dielectric loss factor, product thickness, penetration depth, and size and shape of the product (Kelen et al. 2006).

On the other hand, microwave radiation energy has a wide array of qualitative and quantitative merits over conventional techniques that formulate its acceptance a severe suggestion (Manickavasagan et al. 2006). One main benefit is the position where the heat is produced, chiefly the product itself. Owing to this, the consequence of tiny heat transfer coefficients does not occupy such a significant position. Henceforth, products with greater geometry could be heated in a reduced period of time and with a better uniform temperature profile. These merits often capitulate an enlarged production.

6 Conclusion and Outlook

The mechanism of microwave heating, factors affecting microwave heating, different microwave-assisted unit operations, and drawbacks associated with the processing of food products are discussed all through this chapter. Microwave-assisted heating has several advantages as compared to conventional methods, but microwave heating is still not applied widely for commercial purposes because of some technical as well as cost factors. It wants noteworthy research targeted at enhancements in certain areas. Microwave-assisted processing of food products are required to conduct at large scale or pilot scale rather than laboratory scale so that outcomes might be applicable for industry level. Further, more investigation on intricate nature of microwave-product interaction needs to be done for better understanding of microwave-assisted processes. The come through of microwave energy in the food industry owing to its significant prospective has been reported many periods before, however it has been deferred each single time till now. That is the reason why we are precautious in foretelling the prospect of microwave technology in industrial applications. Nevertheless, we consider that the possible of microwave technology in the food applications is far from being exhausted. Thus, microwave assisted processing of food products to appear to have great potential in the near future. Novel dielectric characteristic data on food products and more robust mathematical tools are significant to novel applications. Proceeding works on microwave method design and optimization are to be performed to implement further FDA approvals for microwave treated foods. The applicability of microwaves to enhance nutrient retention in end product and decrease the requisite for added salt component is also being studied. Henceforth, microwave technology is anticipated to attain a robust space in the food sector during the time period ahead.

References

Al-Hilphy, A.R.S. and Ali, H.I. (2013). Milk flash pasteurization by the microwave and study its chemical, microbiological and thermo physical characteristics. *Journal of Food Processing & Technology* 4 (7): 1–5.

Arnoldi, A. (2001). Thermal processing and food quality: analysis and control. In: *Thermal Technologies in Food Processing* (ed. P. Richardson), 138–159. Taylor & Francis.

Ayappa, K., Davis, H., Davis, E., and Gordon, J. (1991). Analysis of microwave heating of materials with temperature-dependent properties. *AIChE Journal* 37 (3): 313–322.

Bhatt, K., Vaidya, D., Kaushal, M. et al. (2020). Microwaves and radiowaves: in food processing and preservation. *International Journal of Current Microbiology and Applied Sciences* 9 (9): 118–131.

Campanone, L. and Zaritzky, N. (2005). Mathematical analysis of microwave heating process. *Journal of Food Engineering* 69 (3): 359–368.

Chandrasekaran, S., Ramanathan, S., and Basak, T. (2013). Microwave food processing – a review. *Food Research International* 52 (1): 243–261.

Chayjan, R. and Kaveh, M. (2016). Drying characteristics of eggplant (*Solanum melongena* L.) slices under microwave-convective drying. *Research in Agricultural Engineering* 62 (4): 170–178.

Chee, K.K., Wong, M.K., and Lee, H.K. (1996). Optimization of microwave-assisted solvent extraction of polycyclic aromatic hydrocarbons in marine sediments using a microwave extraction system with high-performance liquid chromatography-fluorescence detection and gas chromatography-mass spectrometry. *Journal of Chromatography A* 723 (2): 259–271.

Ciriminna, R., Fidalgo, A., Avellone, G. et al. (2019). Integral extraction of Opuntia ficus-indica peel bioproducts via microwave-assisted hydrodiffusion and hydrodistillation. *ACS Sustainable Chemistry & Engineering* 7 (8): 7884–7891.

Constantin, A. and Csatlos, C. (2010). Research on the influence of microwave treatment on milk composition. *Bulletin of the Transilvania University of Brasov* 3: 52.

Coronel, P., Simunovic, J., and Sandeep, K. (2003). Temperature profiles within milk after heating in a continuous-flow tubular microwave system operating at 915 MHz. *Journal of Food Science* 68 (6): 1976–1981.

Curet, S., Rouaud, O., and Boillereaux, L. (2014). Estimation of dielectric properties of food materials during microwave tempering and heating. *Food and Bioprocess Technology* 7 (2): 371–384.

Darvishi, H., Asl, A.R., Asghari, A. et al. (2014). Study of the drying kinetics of pepper. *Journal of the Saudi Society of Agricultural Sciences* 13 (2): 130–138.

Das, P., Mishra, S.R., and Nayak, B.S. (2017). Process analytical technique (PAT): an integral part of pharmaceutical process automation. *International Journal of Pharmaceutics and Drug Analysis* 5 (2): 31–35.

Datta, A.K. and Nelson, S. (2000). *Fundamental Physical Aspects of Microwave Absorption and Heating in Handbook of Microwave Technology for Food Applications*. CHIPS.

Decareau, R.V. (1985). *Microwaves in the Food Processing Industry*. New York: Academic Press.

Dıaz, G.R., Martınez-Monzo, J., Fito, P., and Chiralt, A. (2003). Modelling of dehydration-rehydration of orange slices in combined microwave/air drying. *Innovative Food Science & Emerging Technologies* 4 (2): 203–209.

Ekezie, F.-G.C., Sun, D.-W., Han, Z., and Cheng, J.-H. (2017). Microwave-assisted food processing technologies for enhancing product quality and process efficiency: a review of recent developments. *Trends in Food Science & Technology* 67: 58–69.

Eliasson, L., Isaksson, S., Lövenklev, M., and Ahrné, L. (2015). A comparative study of infrared and microwave heating for microbial decontamination of paprika powder. *Frontiers in Microbiology* 6: 1071.

Gallawa, J. C. (2005). Who invented microwaves? *Includes portrait. Available at:* http://www.gallawa.com/microtech/history.html.

Hasted, J.B. (1973). *Aqueous Dielectrics*. Chapman and Hall.

Helmar, S. and Marc, R. (2007). *The Microwave Processing of Foods*, 20–312. Woodhead Publishing Limited and CRC Press.

Hojjati, M., Noguera-Artiaga, L., Wojdyło, A., and Carbonell-Barrachina, Á.A. (2015). Effects of microwave roasting on physicochemical properties of pistachios (*Pistacia vera* L.). *Food Science and Biotechnology* 24 (6): 1995–2001.

Içöz, D., Sumnu, G., and Sahin, S. (2004). Color and texture development during microwave and conventional baking of breads. *International Journal of Food Properties* 7 (2): 201–213.

Iuliana, C., Rodica, C., Sorina, R., and Oana, M. (2015). Impact of microwaves on the physicochemical characteristics of cow milk. *Romanian Reports in Physics* 67 (2): 423–430.

Jeevitha, G.C., Sowbhagya, H.B., and Hebbar, H.U. (2016). Application of microwaves for microbial load reduction in black pepper (*Piper nigrum* L.). *Journal of the Science of Food and Agriculture* 96 (12): 4243–4249.

Jerome, R.E., Singh, S.K., and Dwivedi, M. (2019). Process analytical technology for bakery industry: a review. *Journal of Food Process Engineering* 42 (5): e13143.

Kalla, A.M. and Devaraju, R. (2017). Microwave energy and its application in food industry: a reveiw. *Asian Journal of Dairy and Food Research* 36 (1): 37–44.

Kelen, A., Ress, S., Nagy, T. et al. (2006). Mapping of temperature distribution in pharmaceutical microwave vacuum drying. *Powder Technology* 162 (2): 133–137.

Kodba, Z.C. and Marsel, J. (1999). Microwave assisted extraction and sonication of polychlorobiphenils from river sediments and risk assesment by toxic equivalency factors. *Chromatographia* 49 (1–2): 21–27.

Kozempel, M.F., Annous, B.A., Cook, R.D. et al. (1998). Inactivation of microorganisms with microwaves at reduced temperatures. *Journal of Food Protection* 61 (5): 582–585.

Kumar, C. and Karim, M. (2019). Microwave-convective drying of food materials: a critical review. *Critical Reviews in Food Science and Nutrition* 59 (3): 379–394.

Lakshmi, S., Chakkaravarthi, A., Subramanian, R., and Singh, V. (2007). Energy consumption in microwave cooking of rice and its comparison with other domestic appliances. *Journal of Food Engineering* 78 (2): 715–722.

Lanqingzi, D.Y.F.J.L. (2005). Application of microwave on plant fiber blanching. *Journal of Radiation Research and Radiation Processing* 23 (6): 351–354.

Li, B. and Sun, D.-W. (2002). Novel methods for rapid freezing and thawing of foods – a review. *Journal of Food Engineering* 54 (3): 175–182.

Liazid, A., Guerrero, R., Cantos, E. et al. (2011). Microwave assisted extraction of anthocyanins from grape skins. *Food Chemistry* 124 (3): 1238–1243.

Lin, S. and Brewer, M. (2005). Effects of blanching method on the quality characteristics of frozen peas. *Journal of Food Quality* 28 (4): 350–360.

Lucchesi, M.E., Chemat, F., and Smadja, J. (2004). Solvent-free microwave extraction of essential oil from aromatic herbs: comparison with conventional hydro-distillation. *Journal of Chromatography A* 1043 (2): 323–327.

Manickavasagan, A., Jayas, D., and White, N. (2006). Non-uniformity of surface temperatures of grain after microwave treatment in an industrial microwave dryer. *Drying Technology* 24 (12): 1559–1567.

Maskan, M. (2001). Drying, shrinkage and rehydration characteristics of kiwifruits during hot air and microwave drying. *Journal of Food Engineering* 48 (2): 177–182.

Megahey, E., McMinn, W., and Magee, T. (2005). Experimental study of microwave baking of Madeira cake batter. *Food and Bioproducts Processing* 83 (4): 277–287.

Metaxas, A. and Meredith, R. (1983). *Industry Microwave Heating*. London: Peter Peregrinus Ltd.

Mudgett, R. (1985). Dielectric properties of foods. In: *Microwaves in the Food Processing Industry*, 15–37. Academic Press Inc.

Mujumdar, A.S. (2000). *Drying Technology in Agriculture and Food Sciences*. Science Publishers, Inc.

Mujumdar, A.S. (2014). *Handbook of Industrial Drying*. CRC Press.

Nelson, S.O. and Datta, A.K. (2001). Dielectric properties of food materials and electric field interactions. *Food Science and Technology*: 69–114.

Nelson, S.O. and Kraszewski, A. (1990). Grain moisture content determination by microwave measurements. *Transactions of ASAE* 33 (4): 1303–1305.

Nott, K.P. and Hall, L.D. (1999). Advances in temperature validation of foods. *Trends in Food Science & Technology* 10 (11): 366–374.

Ohlsson, T. and Bengtsson, N. (1975). Dielectric food data for microwave sterilization processing. *Journal of Microwave Power* 10 (1): 93–108.

Ohlsson, T. and Thorsell, U. (1984). Problems in microwave reheating of chilled foods. *Foodservice Research International* 3 (1): 9–16.

Omolola, A.O., Jideani, A.I., and Kapila, P.F. (2014). Modeling microwave drying kinetics and moisture diffusivity of Mabonde banana variety. *International Journal of Agricultural and Biological Engineering* 7 (6): 107–113.

Onuska, F. and Terry, K. (1993). Extraction of pesticides from sediments using a microwave technique. *Chromatographia* 36 (1): 191–194.

Orsat, V., Yang, W., Changrue, V., and Raghavan, G. (2007). Microwave-assisted drying of biomaterials. *Food and Bioproducts Processing* 85 (3): 255–263.

Orsat, V., Raghavan, G., and Krishnaswamy, K. (2017). Microwave technology for food processing: an overview of current and future applications. In: *The Microwave Processing of Foods* (eds. M. Regier, K. Knoerzer and H. Schubert), 100–116. Elsevier.

Patil, H., Shah, N., Hajare, S. et al. (2019). Combination of microwave and gamma irradiation for reduction of aflatoxin B1 and microbiological contamination in peanuts (*Arachis hypogaea* L.). *World Mycotoxin Journal* 12 (3): 269–280.

Pina-Pérez, M.C., Benlloch-Tinoco, M., Rodrigo, D., and Martinez, A. (2014). Cronobactersakazakii inactivation by microwave processing. *Food and Bioprocess Technology* 7 (3): 821–828.

Proestos, C. and Komaitis, M. (2008). Application of microwave-assisted extraction to the fast extraction of plant phenolic compounds. *LWT - Food Science and Technology* 41 (4): 652–659.

Puschner, H. (1966). Body tissue in microwave range. *Heating with Microwaves. Translated by Grubba, E*, 226–238.

Ramesh, M., Wolf, W., Tevini, D., and Bognar, A. (2002). Microwave blanching of vegetables. *Journal of Food Science* 67 (1): 390–398.

Ranjan, S., Dasgupta, N., Walia, N. et al. (2017). Microwave blanching: an emerging trend in food engineering and its effects on *Capsicum annuum* L. *Journal of Food Process Engineering* 40 (2): e12411.

Rattanadecho, P. and Makul, N. (2016). Microwave-assisted drying: a review of the state-of-the-art. *Drying Technology* 34 (1): 1–38.

Rifna, E.J. and Dwivedi, M. (2021). Optimization and validation of microwave–vacuum drying process variables for recovery of quality attribute and phytochemical properties in pomegranate peels (Punica granatum L. cv. Kabul). *Journal of Food Measurement and Characterization* 15 (3): 1–19.

Rifna, E. and Dwivedi, M. (2021). The microbiological safety of food powders. In: *Food Powders Properties and Characterization* (ed. E. Ermiş), 169–193. Springer.

Rifna, E., Singh, S.K., Chakraborty, S., and Dwivedi, M. (2019). Effect of thermal and non-thermal techniques for microbial safety in food powder: recent advances. *Food Research International* 126: 108654.

Ruiz-Ojeda, L.M. and Peñas, F.J. (2013). Comparison study of conventional hot-water and microwave blanching on quality of green beans. *Innovative Food Science & Emerging Technologies* 20: 191–197.

Şakıyan, Ö. (2015). Optimization of formulation of soy-cakes baked in infrared-microwave combination oven by response surface methodology. *Journal of Food Science and Technology* 52 (5): 2910–2917.

Salazar-González, C., San Martín-González, M.F., López-Malo, A., and Sosa-Morales, M.E. (2012). Recent studies related to microwave processing of fluid foods. *Food and Bioprocess Technology* 5 (1): 31–46.

Schiffmann, R. (1986). Food product development for microwave processing. *Food Technology* 40 (6): 94–98.

Schiffmann, R. (1990). Microwave foods: basic design considerations. *TAPPI Journal* 73 (3): 209–212.

Schiffmann, R. F., Mirman, A. H., & Grillo, R. J. (1981). *U.S. Patent No. 4,271,203*. Washington, DC: U.S. Patent and Trademark Office.

Seyhun, N., Ramaswamy, H., Sumnu, G. et al. (2009). Comparison and modeling of microwave tempering and infrared assisted microwave tempering of frozen potato puree. *Journal of Food Engineering* 92 (3): 339–344.

Si, X., Chen, Q., Bi, J. et al. (2016). Infrared radiation and microwave vacuum combined drying kinetics and quality of raspberry. *Journal of Food Process Engineering* 39 (4): 377–390.

Tang, J., Hong, Y.-K., Inanoglu, S., and Liu, F. (2018). Microwave pasteurization for ready-to-eat meals. *Current Opinion in Food Science* 23: 133–141.

Teseme, W.B. and Weldeselassie, H.W. (2020). Review on the study of dielectric properties of food materials. *American Journal of Engineering and Technology Management* 5 (5): 76–83.

Therdthai, N., Tanvarakom, T., Ritthiruangdej, P., and Zhou, W. (2016). Effect of microwave assisted baking on quality of rice flour bread. *Journal of Food Quality* 39 (4): 245–254.

To, E., Mudgett, R., Wang, D. et al. (1974). Dielectric properties of food materials. *Journal of Microwave Power* 9 (4): 303–315.

Trilokia, M., Sood, M., Bandral, J.D. et al. (2019). Changes in quality of microwave blanched vegetables: a review. *IJCS* 7 (2): 732–737.

Vadivambal, R. and Jayas, D. (2007). Changes in quality of microwave-treated agricultural products – a review. *Biosystems Engineering* 98 (1): 1–16.

Vadivambal, R. and Jayas, D. (2010). Non-uniform temperature distribution during microwave heating of food materials – a review. *Food and Bioprocess Technology* 3 (2): 161–171.

Xanthakis, E., Gogou, E., Taoukis, P., and Ahrné, L. (2018). Effect of microwave assisted blanching on the ascorbic acid oxidase inactivation and vitamin C degradation in frozen mangoes. *Innovative Food Science & Emerging Technologies* 48: 248–257.

Yolacaner, E.T., Sumnu, G., and Sahin, S. (2017). Microwave-assisted baking. In: *The Microwave Processing of Foods* (eds. M. Regier, K. Knoerzer and H. Schubert), 117–141. Elsevier.

Zhang, M., Tang, J., Mujumdar, A., and Wang, S. (2006). Trends in microwave-related drying of fruits and vegetables. *Trends in Food Science & Technology* 17 (10): 524–534.

Zhu, Z. and Guo, W. (2017). Frequency, moisture content, and temperature dependent dielectric properties of potato starch related to drying with radio-frequency/microwave energy. *Scientific Reports* 7 (1): 1–11.

11

Radio Frequency

Shunshan Jiao[1], Eva Salazar[2], and Shaojin Wang[3]

[1] *Department of Food Science and Technology, School of Agriculture and Biology, Shanghai Jiao Tong University, Shanghai, China*
[2] *Departamento de Ciencia y Tecnologıa de Alimentos, UCAM Universidad Católica San Antonio de Murcia, Murcia, Spain*
[3] *College of Mechanical and Electronic Engineering, Northwest A&F University, Yangling, Shaanxi, China*

1 Introduction

Radio frequency (RF) energy is one kind of electromagnetic (EM) waves with a frequency range of 3 kHz–300 MHz. Among them, 13.56, 27.12, and 40.68 MHz are the selected ISM (Industrial, scientific, and medical) frequencies and used worldwide. RF energy can heat foods through molecular friction caused by ionic oscillation and dipole rotation. RF heating, also called dielectric heating, can heat foodstuffs rapidly and volumetrically with deeper heat penetration. Besides, it has relatively uniform EM field distribution, simple system configuration, high-energy efficiency, and it can be designed as a continuous process and scaled up easily. As a novel thermal processing method, RF heating has obtained lots of attentions in food processing recently, and it has shown great potential in some of those applications, such as thawing, disinfestation, pasteurization, and stabilization. Currently, there are some industrial applications of RF heating around the world, including post-baking, thawing, and disinfestation. However, RF heating generally requires sample having regular shapes and flat surfaces to get better heating uniformity. Nonuniform heating and reliable large-scale industrial RF units are big challenges for commercially applying RF heating in various food processes.

This chapter briefly introduces the principle of RF dielectric heating and then focuses on the applications of RF heating in various food processes, including thawing, drying, disinfestation, microbial and enzyme inactivation.

Electromagnetic Technologies in Food Science, First Edition. Edited by Vicente M. Gómez-López and Rajeev Bhat.

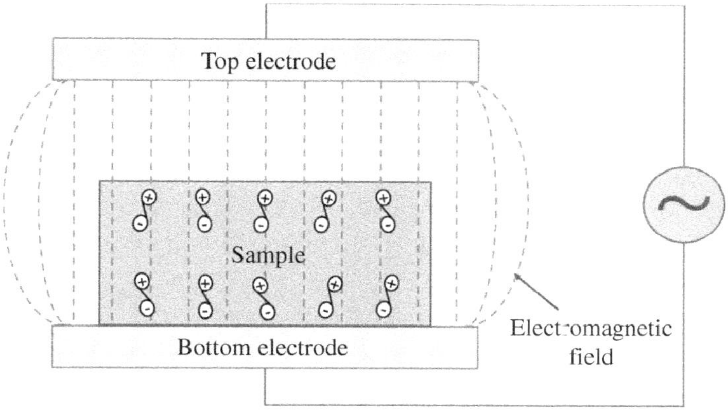

Figure 1 The simplified schematic diagram of and RF heating system. (*See insert for color representation of this figure*). *Source:* Orfeuil (1987).

2 Principle of RF Heating

RF treatment is a dielectric heating, which can only heat dielectric materials. RF heating systems are commonly used with two parallel electrodes, and foods are placed between electrodes (Figure 1). All foods contain water and some foods contain ions, water is a dipole molecule. Under high alternating EM fields, water and other polar molecules would rotate and ions would oscillate up and down at very high speeds. The movements of molecules and ions would cause frictions and then generate heat inside the foods. Therefore, ionic conduction and dipole rotation are the dominant mechanisms for RF heating (Ryynänen 1995).

2.1 Dielectric Properties

Dielectric properties (DPs) of material are the key factor for RF heating, and include two parts: permeability and permittivity. The permeability of foods is the same as air, so it has no influence on dielectric heating. Permittivity is a complex quantity and used to describe the DPs that influence reflection of EM waves at interfaces and the attenuation of the wave energy within materials:

$$\varepsilon = \varepsilon_0 \varepsilon_r = \varepsilon_0 \left(\varepsilon_r' - j\varepsilon_r'' \right) = \varepsilon_r' - j\varepsilon_r'' \tag{1}$$

where ε is the permittivity and ε_r is the relative complex permittivity, which describes permittivity related to free space, and ε_0 is a constant value of $8.854 \times 10^{-12}\,\mathrm{F\,m^{-1}}$, ε_r' and ε_r'' are relative dielectric constant and relative dielectric loss factor. ε' and ε'' are dielectric constant and dielectric loss factor, and $j = \sqrt{-1}$. ε_r' or ε' is related to the ability of material to store electric energy, which would affect the EM field distribution. ε_r'' or ε'' describes the ability of materials to dissipate energy due to various polarization mechanisms. It would influence energy absorption and attenuation, which would finally create heat generation in foods. Relative dielectric constant and relative dielectric loss factor are commonly used in practical RF heating applications.

2.2 Governing Equation

Figure 1 shows the simplified schematic of RF heating system for most commonly used flat capacitor, and the capacitance can be expressed as (Orfeuil 1987):

$$C = \frac{\varepsilon_0 \varepsilon_r' A}{d} \tag{2}$$

where C is the capacitance (F), A is the plate surface area (m^2), and d is the electrode gap (m).

For a perfect capacitor, there is no power absorption between two electrodes, and the current has a phase angle of $\pi/2$ with respect to the voltage. But if there is a dielectric material placed between two electrodes, the material would become a resistance, and the current flowing through the resistance is in phase with the applied voltage (Orfeuil 1987). The current through the capacitor (I_C) can be calculated as:

$$I_C = \omega VC = 2\pi f VC \tag{3}$$

where ω is the angular frequency (Hz), V is the applied voltage (V), and f is the frequency (Hz).

The current through the resistance (I_R) is given by:

$$I_R = I \sin\delta = I_C \tan\delta \tag{4}$$

where the tangent of dielectric loss angle ($\tan\delta$) is equal to the ratio of dielectric loss factor to dielectric constant, and total power dissipated (Q) in the load can be expressed as:

$$Q = VI_R = VI_C \tan\delta = 2\pi f V^2 C \tan\delta \tag{5}$$

For voltage, it has the following relationship with electric field strength:

$$E = V/d \tag{6}$$

where E is electric field strength ($V\,m^{-1}$). Then substitute Eq. (6) into Eq. (5) becomes:

$$Q = 2\pi\varepsilon_0\varepsilon_r' f E^2 dA \tan\delta = 2\pi\varepsilon_0\varepsilon_r'' f E^2 dA \tag{7}$$

where $d \times A$ is the volume (m^3) of load. For the power dissipated per unit volume, Eq. (7) can be expressed as:

$$P = 2\pi f \varepsilon_0 \varepsilon_r'' E^2 = 5.56 \times 10^{-11} f \varepsilon_r'' E^2 \tag{8}$$

where P is the power dissipation per unit volume in the heated material ($W\,m^{-3}$). From Eq. (8), the dissipated power is proportional to the applied frequency, the relative dielectric loss factor of material, and the square of the applied electric field strength.

The temperature increase (ΔT) derived from RF dielectric heating can be calculated from Eq. (9) (Nelson 1996):

$$\rho C_p \frac{\Delta T}{\Delta t} = 2\pi f \varepsilon_0 \varepsilon_r'' E^2 = 5.56 \times 10^{-11} f \varepsilon_r'' E^2 \tag{9}$$

where ρ (kg m^{-3}) is the density, C_p (J kg°C^{-1}) is the specific heat, ΔT (°C) is the temperature rise, and Δt (s) is the time increment.

It is obviously that RF heating rate in foodstuffs not only depends on the power dissipation but also depends on the properties of material itself, such as density and specific heat. These properties together with DPs are very important for developing or modeling the RF heating process.

2.3 Penetration Depth

Power penetration depth (d_p) is defined as the distance through which EM power is reduced to $1/e$ (e = 2.718) or 36.9% of its original magnitude (Metaxas and Meredith 1993). Penetration depth for RF heating can be expressed as

$$d_p = \frac{c}{2\pi f\sqrt{2\varepsilon_r'\left[\sqrt{1+\left(\varepsilon_r''/\varepsilon_r'\right)^2}-1\right]}} \tag{10}$$

Penetration depth is determined by frequency and DPs of materials. For different kinds of foods, penetration depth varies with different DPs. Since RF wave has lower frequencies compared with microwave, and the penetration depth is inversely proportional to frequency, RF penetration depth is usually larger than that of microwave. RF energy can penetrate thicker products and therefore RF heating is more suitable to treat large bulk samples.

3 Applications of RF Heating in Food Processing

3.1 Thawing

Conventional methods of thawing products in the food industry employ heat transfer by air or water and the use of vacuum. In all cases, thawing takes place slowly, which implies higher costs and possible damage to the product. Additionally, this fact leads to a decrease in both physical and chemical qualities of previously frozen foods (Cai et al. 2019). These modifications are related to the increase in surface temperature, a decrease in the water retention capacity, and the potential increase in the microbial load (Jiao et al. 2018).

RF heating is among the emerging technologies that have been developed to minimize the process time involved in the defrosting stage. The principle of RF thawing is explained by an increase in temperature due to dielectric and electromagnetic heating. Both phenomena facilitate the conversion of electrical and electromagnetic energy into thermal energy, especially the latter in frozen products. The change in phase from ice to water increases the absorption of electromagnetic energy and the conversion into heat, as the DPs of the product are modified (Llave and Erdogdu 2020). This direct transformation of electromagnetic energy into thermal energy achieves uniform defrosting in short times (Tang et al. 2020).

Therefore, innovative technologies, such as RF thawing, can reduce processing times. Consequently, it leads to energy savings and other benefits, such as water savings, compared to the conventional techniques, achieving an acceptable uniformity. Thus, RF heating for food thawing has been widely studied, and the technology has been successfully applied on an industrial scale (Llave and Erdogdu 2020). However, it is still necessary to optimize the processing parameters to improve the heat distribution in the product.

Raw meat and seafood are the foodstuffs groups in which most of the RF application studies for thawing have been carried out (Farag et al. 2008, 2009, 2011; Bedane et al. 2017; Kim et al. 2016; Llave et al. 2014; Palazoğlu and Miran 2017; Yang et al. 2019). Regarding raw meat, Farag et al. (2008) developed a pilot scale RF-thawing protocol for beef mixtures (4 kg blocks). When compared with the conventional method, authors observed that the thawing time was reduced 30 times, while the temperature distribution was similar to the air thawing method. In any case, the process was more efficient from the energy point of view compared to the conventional method. In a subsequent investigation, Farag et al. (2009) studied the effect of defrosting rate on water holding properties in lean beef meat with different levels of shredding (comminuted, minced, and whole). They found that drip and micronutrient losses were reduced in thawing by RF heating compared to the conventional air thawing method. With regard to DPs, higher post-thaw values were observed for the RF heating method. Both studies are according to a later work in which the effect of the product composition on the effectiveness of defrosting beef meat was investigated (Farag et al. 2011). Findings also indicated that defrosting lean meat with RF thawing results in satisfactory end point temperature distribution compared to conventional air defrosting. However, this work also stresses the influence of sample composition on the effectiveness of the operation. In the case of samples containing fat mixture, the temperature distribution was significantly less uniform than that observed in lean, highlighting the influence of the product components on the effectiveness of the process.

A current research on RF thawing of lean beef blocks evaluated the effect of processing conditions on heat uniformity distribution by comparing batch and continuous process systems (Bedane et al. 2017). It was observed that continuous processing system slightly improved heat uniformity in the sample. The distance between the electrodes of the equipment and the product and the speed of the process were also main factors to influence the performance of the thawing operation. Thus, Bedane et al. (2017) reported that the heating times were faster in the upper and lower areas while gradually increasing in the middle or central area because of the rapid melting of the ice crystals on the product surface. The uniformity of heating was improved in batch RF thawing of the lean beef block as the gap between the electrodes increased. The same was observed in conveyor RF thawing. In the latter case, the uniformity index was also higher with increasing process speed.

In pork sirloin, Kim et al. (2016) carried out a study aimed at improving the uniformity of the RF-thawing operation, using a device with curved electrodes in order to obtain an equipotential surface similar to that of foods with a curved surface. The results showed that the curved electrode was effective in improving temperature uniformity, despite the fact that heat transfer to the frozen meat caused a surface overheating.

Concerning marine products, Llave et al. (2014) developed a RF-thawing research of tuna fish frozen blocks (skinned and boned). This work evaluated the effect of the electrode size on the efficiency of the thawing operation. Results showed that an electrode with a size similar to that of the sample produces a uniform end point temperature distribution. Furthermore, they also studied the effect of the composition of the muscle, suggesting that the uniformity of the temperature distribution was greater in the samples of tuna with low fat content compared to the fattier muscles. This work describes RF-thawing operating times three times shorter compared to the conventional thawing method.

In a comparative study on the effectiveness of treating two types of tempering (RF and microwave) in tiger shrimp blocks, Palazoğlu and Miran (2017) reported that the central parts of the frozen block are heated more slowly by RF energy, while the corners and the edges heated more quickly in RF systems. In both treatments, the internal temperature distribution was reasonably homogeneous, and overheating of the product surface was lower in RF heating. Therefore, RF energy allows temperature close to 0 °C to be reached, affecting the quality of the product to a lesser extent.

Yang et al. (2019) also studied the temperature distribution of minced silver carp blocks and surimi (fish cooked gels) in RF thawing and its effect on the quality. In this study, it was found that temperature has a significant effect on DPs. Thus, the DPs of the samples increased rapidly when the temperature reached −5 °C and the depth of penetration into the frozen minced fish decreased significantly with increasing temperature. Among other findings, the results for surimi showed that the texture of the product was not affected in the RF thawing.

As described in the cited studies, RF thawing of food products has potential advantages in achieving shorter thawing time, less energy consumption, and better homogeneity compared to conventional thawing methods. Although it is expected that the quality of foods subjected to RF thawing would be higher, it is necessary to expand the study on the chemical, sensory, and texture properties of the treated foods to determine the optimal conditions for RF thawing (Guo et al. 2019). Studying RF thawing applications at different frequencies could be proposed since most of the studies have been developed using the RF systems at the frequency of 27.12 MHz.

3.2 Drying

RF heating is an alternative to technologies that use heat aimed at preserving food, which is of great interest in the food industry, due to the great potential for rapid and uniform heating, obtaining safe and high qualitied food (Ştefănoiu et al. 2016). RF heating significantly increases the performance of the drying operation (Orsat and Raghavan 2005) compared to other types of heating operations (Marshall and Metaxas 1999).

Accordingly, RF-drying technology significantly reduces food-drying time until reaching the desired moisture level compared to conventional drying techniques. Awuah et al. (2015) report the drying times in conventional dryers up to 150% higher compared to RF drying systems. Zhou et al. (2018) have also observed in in-shell walnuts that the drying time was reduced by 70% with RF energy, as compared to tthat with a hot air technique. In addition to the reduction in drying time, the benefits of drying foodstuffs using RF energy include homogeneity of heating, greater depth of penetration, and more stable control of product temperature (Wang et al. 2014; Zhou et al. 2018). This implies obtaining higher quality foods in terms of the nutritional and the organoleptic point of view.

With regard to sensory quality, RF-drying preserves the color of foods products especially those in which the color is due to the composition in heat-sensitive pigments. Conventional drying can degrade heat-labile substances on the surface while RF drying, which takes place from the inside to the surface of the product, greatly minimizes the effect of heat on pigments (Chou and Chua 2001).

By producing a homogeneous action in the application of the RF-drying treatment, the surface cracking that takes place because of the stresses caused by the loss of cellular turgor, very pronounced in fresh vegetables, is also reduced. The impact of this phenomenon is lower than in conventional drying, since the diffusion of water from the center of the product to the surface takes place in a homogeneous way. In hot air drying, the moisture reduction mechanism for dehydrating causes unequal tensions in the product that leads to differential contractions resulting in a considerable alteration of the appearance (Chou and Chua 2001).

Despite RF-drying benefits described above, it should be noted that many factors have an effect on heating uniformity (Jiao et al. 2018), such as physical, dielectric and thermal properties, electrode gap, chemical properties of the medium, and design of the RF-equipment (Fu 2004; Altemimi et al. 2019). Parameters of the product itself, such as moisture or salt content, and conditions of processing, for instance, the intensity and the operating temperature, affect significantly the effectiveness of the dehydration treatment by modifying the DPs.

The effects of moisture content and temperature have been studied in peanut kernels by Zhang et al. (2016a). These authors measured the DPs of the product between 10 and 4500 MHz, at temperatures between 25 and 85 °C and moisture contents between 10% and 30% on a wet basis. In this work, it has been described that both the dielectric constant and the loss factor of peanut kernels decreased significantly with increasing frequency in the RF range (10–300 MHz). Moreover, the dielectric constant and the loss factor increased with increasing moisture content and temperature, being the increase more marked at higher temperatures and higher moisture contents. Furthermore, a decrease of the penetration depth was also observed with increasing the frequency, the humidity constant, and the temperature. Ling et al. (2018) observed an increase in the heating rate for RF-drying of rice bran by incorporating metal salts into the process.

RF-drying technique has been applied in soybeans packed in plastic containers (Huang et al. 2015b), where a computer simulation model was validated to predict the uniformity of heating. This validated model was applied to simulate the effects of several factors, such as the area of the top plate, moisture content of the sample, vertical position of the sample, and dielectric materials around the samples. The experimental and simulation results were similar, showing higher temperature values in the middle and the lower layers compared to those of the upper layers. Additionally, it was observed that the heating was higher in corners and edges than in the central areas in all layers. Among others, this study highlights the lack of uniformity in RF heating (Huang et al. 2015b). Another limitation of RF-drying is related to the cost of operation. RF systems are too expensive to dry large volumes of agricultural products, such as cereals (Jiao et al. 2018).

To increase the performance of dehydration processes and improve the temperature distributions in dried products, different RF-drying applications have been studied combined with other traditional dehydration methods. RF-conventional combined techniques include tandem and parallel methods (Altemimi et al. 2019). Tandem drying or hybrid drying techniques involves subjecting the product to a drying method followed by one or more drying methods, while in parallel drying the process is carried out by two or more technologies that are applied simultaneously (Zhou and Wang 2019).

Recently, Zhou et al. (2019) studied the effectiveness of RF-vacuum drying followed by hot air drying and osmotic dehydration in kiwifruits slices. No differences were observed between hot air and osmotic techniques in terms of temperature and time, although the nutritional and sensory quality was significantly affected by the drying method.

As an example of parallel drying, it can be cited the hot air-assisted RF (HA-RF) drying technology, which has been the most widely studied technique. Thus, this technique has been studied in different food products, such as broad bean (Ptasznik et al. 1990), in-shell macadamia nut (Wang et al. 2014), stem lettuce (Roknul et al. 2014), duck egg (Lin et al. 2014), and in shell-walnut (Zhang et al. 2016b; Zhou et al. 2018). More recently, HA-RF heating has also been used as an effective and high efficiency final stage drying method for fruits and vegetables (Gong et al. 2020; Zhang H et al. 2019).

In general, all these studies are in agreements with the fact that the degree of drying increases and products with acceptable quality are obtained. However, results have shown a high variability in terms of uniformity of the treatment.

3.3 Disinfestation

Disinfestation is an important procedure to guarantee safety and quality of food products. Methyl bromide fumigation is one of the current prevailing disinfestation methods, but it was banned in many countries because of its damage to environment and human health. Meanwhile, increased insect resistance to chemicals has also forced food industry to explore and apply nonchemical alternative disinfestation methods (Ling et al. 2020). Thermal treatments are gradually used to replace chemical fumigation due to their easy applications and nonchemical characteristics. However, traditional thermal treatments often cause undesirable damages to materials due to its long heating time and high temperature (Wang et al. 2001). To overcome the shortage of traditional thermal processing, RF heating has been explored and developed. RF energy can selectively heat and kill the insects with minimal damage on the quality of products (Jiao et al. 2018). The thermal effect on insect mortality could be attributed to the thermal degradation of carbohydrates, proteins, DNA, RNA, and lipids of the insects (Yu et al. 2017). Applications of RF heating as a postharvest disinfestation method could be back to the 1920s. However, the detailed research results were not reported until Nelson (1996) first investigated the DPs of insects at different frequencies and temperatures proposed for controlling several insects in grains. In the past two decades, many studies have been carried out to promote the application of RF disinfestation treatments in agricultural products, including fresh fruits and vegetables, dried fruits and nuts, and grains (cereals, legumes, oilseeds, etc.). Meanwhile, studies have also been conducted to measure DPs of insects and host food materials and investigate different methods to improve RF heating uniformity for insect control.

3.3.1 For Fresh Fruits

One of the major problems in the storage and marketing of fresh fruits and vegetables is the infestation of insects. Ling et al. (2015) measured DPs of three stone fruits (nectarine, peach, and plum). Differential heating for postharvest insect control might be possible since the loss factor of the targeted insect was about 1.8 times of stone fruits at 27 MHz and 20 °C. RF heating could raise the temperature of the insects more quickly than food materials (Ikediala et al. 2000; Piyasena et al. 2003; Wang et al. 2003). The feasibility of RF disinfestation has been explored in various fresh fruits, including codling moth in cherries (Ikediala et al. 2002) or apples (Birla et al. 2004), Mediterranean fruit flies in oranges (Birla et al. 2005), and Mexican fruit fly in persimmons (Monzon et al. 2007) or mangoes (Sosa-Morales et al. 2009).

To overcome the markedly high temperature differential problem within and among fresh fruits during RF heating treatment, rotation and movement, adjustment of surrounded medium and hot water assisted methods have been designed and applied. Ikediala et al. (2002) developed a saline water immersion technique with RF heating as a postharvest treatment against codling moth in cherries. The results showed that 100% larvae mortality in cherries was obtained after RF treatment in 0.15% NaCl water to 50 °C and holding 5 minutes. Compared with MeBr fumigated samples, the RF treated samples showed similar or better quality attributes. The NaCl solution (0.035%) water immersion method during RF heating had no significantly negative effects on quality of persimmons except for color (Monzon et al. 2007).

In addition, hot water preheating was another method to overcome the nonuniform heating problem. Tiwari et al. (2008) investigated the hot water-assisted RF heating treatment for the control of Mexican fruit fly in "Fuyu" persimmons. They preheated persimmons at 40 °C water and raised the temperature up to 46, 48, and 50 °C with RF energy and held at different temperature for certain times. Based on their results, RF treatments for 46 °C + 25 minutes holding, or 48 °C + 6 minutes holding, or 50 °C + 2 minutes holding had the potential to provide 100% insect mortality with acceptable fruit quality. The similar results could also be found in mangoes (Sosa-Morales et al. 2009). However, this method caused unacceptable damage on cherry (Hansen et al. 2005), particularly to the stems, which are an essential quality for the marketing of fresh cherries. Hansen et al. (2006) tried to improve RF heating uniformity by a pulse mode to control the fifth instars of codling moth in apples. Although the pulse mode of 30s-on/30s-off for 29 and 50 minutes holding in hot water killed all infesting codling moth larvae, the quality of apples was unacceptable. That means the thermal requirements for insect control may exceed the injury threshold of the apples.

Furthermore, Birla et al. (2004) designed a fruit mover, which could provide a means to rotate and move fruit in water, to improve RF heating uniformity of large fresh fruits. With this technique, temperature distributions in oranges and apples were significantly improved. The same methods could also be found in the further research of Wang et al. (2006a) and Birla et al. (2005) in fresh apples and oranges, respectively. According to Birla et al. (2008a), although immersing fruits in water helped to reduce uneven heating within the fruits, the different heating rate at different horizontal positions could not be addressed. Hence, through practical experiments and the computer simulation of RF heating of model fruit immersed in water, they suggested that movement and rotation of the spherical object is the only plausible solution for improving RF heating uniformity. Additionally, Birla et al. (2008b) also found that dissimilarity in peel and pulp DPs greatly influenced the RF heating behavior of the fruits. Core heating pattern also happened in apple, peeled orange, and grapefruit, regardless of the movement and rotation in the water. Hence, for these fruits, it is necessary to combine RF heating with traditional surface heating, such as hot water preheating at certain temperature and for a certain time, to establish an energy efficient, rapid, and uniform RF heating process.

3.3.2 For Grains

Contamination of insects is one of major concern for grains processing industry because its influence on the marketability and the nutritional values (Banga et al. 2018). In the past decade, RF heating technique has been studied for various purposes in food grains and their products, including disinfestation, enzyme inactivation, drying, and roasting (Jiao et al. 2016a; Ling et al. 2020). Among them, disinfestation is the most extensive and in-depth research fields of RF heating technology in food grains, including cereals, legumes, oilseeds, and grain products. RF disinfestation has been proved effective to many insects in food grains and their products, including weevil (Jiao et al. 2017), moth (Yang et al. 2018), and beetle (Yu et al. 2017), etc.

Rice and wheat are two most important cereal grain crops. The feasibility of RF heating on rice weevil (*Sitophilus oryzae* L.) in rough, brown, and milled rice has been studied. Zhou et al. (2015) found RF heating uniformity of milled rice could be improved by 50 °C hot air surface heating, movement of conveyor belt (12.5 $m\,h^{-1}$), mixing and holding at 50 °C hot air for 5 minutes. No negative effect or significant difference was found on all the measured quality parameters (moisture, protein, fat, starch, hardness, and color) of RF-treated milled rice. Later, Zhou and Wang (2016a) validated the developed RF treatment protocols for disinfesting rough, brown, and milled rice. Results showed that 100% mortality of adult rice weevils (*Sitophilus oryzae* L.) was achieved after RF heated to 50 °C and then holding in hot air for 6 minutes. In addition, Zhou and Wang (2016b) simulated the continuous industrial processing with a pilot-scale RF system, a complete mortality of adult rice weevils was obtained and the quality attributes were acceptable. Apart from adult rice weevils, Jiao et al. (2017) also validated the killing effect of RF treatment (50 °C, 5 minutes) on immature rice weevils and found that the treatment had no significant influence on fissure ratio and broken rate of milled rice.

Apart from rice weevils, grain moths (*Sitotroga cerealella*) and lesser grain borers (*Rhyzopertha dominica*) in milled rice could also be controlled by RF treatment (Hou et al. 2019; Lagunas-Solar et al. 2007; Yang et al. 2018). Furthermore, all life stages of rusty grain beetle (*Cryptolestes ferrungineu* S.) in wheat could also be heated selectively using RF energy with minimal effects on physicochemical properties of wheat (Shrestha et al. 2013; Shrestha and Baik 2015). By using a computer simulation model, Jiao S et al. (2015) investigated the influence of wheat sample size and vertical position on RF heating uniformity and suggested larger sample size had better RF heating uniformity, which means RF energy is more suitable for treating bulk materials.

Insect infestation can also be a major problem in harvesting, processing, and marketing of legumes. Cowpea weevil (*Callosobruchus maculatus* F.), a serious internal pest of several legume crops, and Indian meal moth (*Plodia interpunctella*), a common pest of many stored products, have particularly economic importance (Wang et al. 2010). According to Guo et al. (2010) and Wang et al. (2008), RF energy has large penetration depth in chickpeas and low uniformity index in RF-treated lentils. The same results were also obtained in black-eyed peas and mung beans (Jiao et al. 2011). Wang et al. (2010) showed that RF heating to 60 °C and holding for 10 minutes by forced hot air could obtain 100% mortality of cowpea weevils (*Callosobruchus maculatus*) in lentils and maintain the physical properties. Similar experiment was also designed for the insect control of the mung beans (Song et al. 2020) and coffee beans (Pan et al. 2012). Jiao et al. (2012) showed that heating uniformity and quality of lentils in industrial-scale continuous RF treatment were acceptable, and the average heating efficiency and throughput of the RF system were 76.7% and 209 kg h^{-1}, respectively.

Huang et al. (2015a) developed a computer simulation model for drying soybeans when subjected to RF heating using commercial finite element software COMSOL. According to the computer simulation of RF selective heating of insects in soybeans, Huang et al. (2015a) found that insect larvae were differentially heated with 5.9–6.6 °C higher than host soybeans when RF heated from 25 to 50 °C. Huang et al. (2016) also studied the impact of polystyrene containers on RF heating uniformity for dry soybeans by computer simulation and found that RF heating uniformity could be effectively improved by increasing wall thickness of the container.

Oilseeds are another important part of grains and have been part of the diet for a long time. Red flour beetle *Tribolium castaneum* (Herbst) is the most common insect pest in the oilseeds (Ling et al. 2020). Yu et al. (2016) reported that 100% mortality of *Tribolium castaneum* infesting the canola seeds could be achieved without a significant degradation of the seed quality at 60 °C with proper design of RF heating treatments. Yu et al. (2017) also found that the thermal mortalities of the test insects increased significantly after the seed temperature reached 60 °C.

Many studies also investigated the RF disinfestation on grain products, including wheat, rice, legume, and corn flours. Guo et al. (2010) evaluated the penetration depth of four legume flour samples to estimate the optimal thickness of treatment beds in RF systems (Li et al. 2015). Li et al. (2015) reported that 100% mortality of rice weevil (*Sitophilus oryzae* L.) was obtained when placed 4.5 kg *indica* and *japonica* rice (*Oryza sativa* L.) flours in 5 kW, 40.68 MHz RF systems with a 16 cm electrode gap.

3.3.3 For Dried Fruits and Nuts

Dried fruits and nuts are essential foods in the Mediterranean diet and have complementary nutritional profiles and bioactivities (Alasalvar et al. 2020; Hernández-Alonso et al. 2017). One of the main problems in production, storage, marketing, and exporting of dried fruits or nuts is the loss caused by insect infestation. RF disinfestation has been applied in many nuts, such as walnut (Wang et al. 2006), chestnut (Hou et al. 2015), and pistachios (Ling et al. 2016). Wang et al. (2002) found that 100% mortality of the fifth-instar navel orangeworm could be obtained after RF heating in-shell walnuts to 55 °C and holding for 5 minutes by hot air. The RF disinfestation had no effent on rancidity, sensory qualities, and shell characteristics. Mixing the walnuts during RF treatment could improve the heating uniformity (Wang et al. 2006b). Industrial-scale RF treatments for insect control in walnuts have been developed, and results showed that the average energy efficiency of two RF units in series could reach 79.5% when heating walnuts at 1561.7 kg h^{-1} (Wang et al. 2007a,b). According to the DPs of raisins, dates, apricots, figs, and prunes marsured by Alfaifi et al. (2013), differential heating for postharvest insect pest control is possible since the dielectric constant and loss factor of the tested dried fruits is lower than those of pests. Alfaifi et al. (2014) established a computer model to investigate the RF heating uniformity of raisins packed in a rectangular plastic container, corners, and edges were heated more than the centers in each layer of the RF-treated raisins.

Overall, the RF disinfestation method would be mostly applied to grains, nuts, and dried fruits because of the insensitivity of those low-moisture foods to heat compared to fresh fruits (Hou et al. 2016). Besides, low-moisture foods in powdery or granular generally have better RF heating uniformity than that of fresh fruits, which are generally in irregular shape and surface.

3.4 Microbial Inactivation

RF heating is reported to be a green microbial inactivation technology, which can control food-borne pathogens or fungi without using chemical fungicides (Guo et al. 2019). The main mechanism of RF pasteurization/sterilization is generally considered to be thermal effects of electromagnetic energy. The applicability of RF pasteurization has been validated in various types of food materials, including fruits and vegetables, meat, egg, dairy, aquatic products, grains, nuts, and spices.

3.4.1 For Fruits and Vegetables

RF pasteurization has been successfully used in many fruits and vegetables and their products. Sisquella et al. (2013) reported that the RF treatment combined with hot water (40 °C) for 4.5 minutes could control the brown rot caused by *Monilinia* spp. in nectarines. Liu et al. (2015) showed RF treatment (6 kW, 27.12 MHz) combined with hot air drying of 60 °C for 20 minutes reduced the total colony number of vacuum packaged Caixin by 3–4 log without significantly changing physicochemical and sensory properties. Furthermore, Xu et al. (2017) combined ZnO nanoparticles with RF treatment for sterilization of vacuum packaged carrots, which extended shelf-life of prepared carrots up to 60 days. For liquid fruits or vegetables products, Lyu et al. (2018) found that RF pasteurization completely inactivate microbes in kiwi puree, and Zhang et al. (2017) reported that the combination of RF heating and nisin was effective to inactivate *Alicyclobacillus* spores in kiwi fruit juice. Additionally, Zhao et al. (2017) found that the number of survival microorganisms in prepackaged broccoli powder ($a_w = 0.586$) was reduced by 4.2 log CFU g^{-1} without a significant color degradation after RF heating for 5 minutes with rotation every 30 seconds.

3.4.2 For Meat, Poultry Dairy, and Aquatic Products

The first study related to RF pasteurization of meat could date back to 1953 (Pircon et al. 1953), which showed RF heating was a promising technique to control pathogens in ground meat due to its rapid and volumetric heating. Orsat et al. (2004) showed RF heating at 85 °C helped reduce the total number of bacteria and 28 days of shelf-life was obtained for ham wrapped in high barrier plastic films and stored under vacuum and refrigeration condition. Guo et al. (2006) compared the microbial inactivation effects of RF cooking (heating until the center temperature of the samples reached 72 °C) and water bath cooking (started at 65 °C and gradually increased to 90 °C) on ground beef samples and found that the two methods had similar effect on reducing *Escherichia coli* counts and maintaining of shelf-life. However, RF treatment had a shorter cooking time and a higher heating uniformity. Schlisselberg et al. (2013) studied the inactivation efficacy of a modern technique based on a combination of convection and controlled RF energy against pathogens in ground meat. The technique resulted in similar or even better effects on selected foodborne pathogens, including *E. coli*, *Salmonella* Typhimurium, and *Listeria monocytogenes* as well as spores of *Bacillus cereus* and *Bacillus thuringiensis*. Nagaraj et al. (2016) reported RF heating with 2.5% potassium lactate and 1.5% potassium bicarbonate resulted in more than 5 log reduction of *E. coli* in ground beef homogenate. Additionally, RF heating also possessed the ability to inactivate *B. cereus* and *Clostridium perfringens* in pork luncheon meat (Byrne et al. 2006).

The chicken egg is high in nutritional value and can be found at various scales of production. Unfortunately, there exists a risk for eggs to be contaminated with *Salmonella* spp. (Lau et al. 2016). Lau et al. (2017) investigated deionized water-assisted RF heating for enhancing microbial safety of shell eggs. According to their model, a scaled-up water-assisted RF heating could achieve a minimum of 3 log reductions of *Salmonella* in the yolk within 37 minutes. Through computational investigation of the effect of orientation and rotation of shell egg on RF heating uniformity, Palazoglu and Miran (2019) found that more even heating with a lesser chance of coagulation can be achieved when shell egg is oriented horizontally, rather than vertically. In addition, Luechapattanaporn et al. (2005) validated the RF sterilization process to control *Clostridium botulinum* in scrambled eggs.

Awuah et al. (2005) validated the effectiveness of RF heating in inactivating surrogates of both *Listeria* and *E. coli* cells in milk. Furthermore, it was reported that RF treatment at 90 °C for 5 minutes could thermally destruct *Cronobacter sakazakii* and *Salmonella* spp. in nonfat dry milk (Michael et al. 2014). Similar results were also reported in powdered infant formula milk (Zhang Y et al. 2020).

Aquatic products are also perishable during transportation and storage because of the high moisture and soluble protein content. Xu et al. (2018) reported that the RF sterilization better preserved the natural color and desired flavor of *Nostoc sphaeroides* compared to high pressure steam sterilization. Meanwhile, RF-heated *N. sphaeroides* showed better physical properties, such as texture, water distribution, and rheological properties. Uemura et al. (2017) found *Bacillus subtilis* spores in vacuum packed fish decreased by 5 logarithmic orders using RF heating to 130 °C for 19 minutes and by 4 logarithmic orders using conventional heating method. The RF-treated fish meat was brighter and backbone was softer than those of conventional heated ones.

3.4.3 For Grains, Nuts, and Spices

In recent years, studies on RF treatments for pathogen control in bulk low-moisture foods, such as grains, nuts, and spices, increase rapidly owing to the unique advantages of volumetric heating and large penetration depth of RF heating. Most of them focused on *Salmonella* spp., *E. coli*, *Staphylococcus aureus*, and *Enterococcus faecium* (Choi et al. 2018; Wei et al. 2018; Zhang L et al. 2019). Some studies also tried to improve RF heating uniformity in low-moisture foods (Jiao Y et al. 2014, 2015).

Kim et al. (2012) used RF heating to inactivate pathogens, such as *S.* Typhimurium and *E. coli* O157:H7, on black and red pepper spice. Results showed that RF heating for 50 seconds reduced *S.* Typhimurium and *E. coli* O157:H7 in black peppers by 2.80–4.29 log CFU g^{-1} and RF heating of red peppers for 40 seconds reduced pathogens by 3.38 log to more than 5 log CFU g^{-1} (below the detection limit) without affecting the color quality change. Zhang et al. (2020) found that RF heating (70 °C, 3 minutes) could generate sublethal injured cells (SICs) of *S.* Typhimurium in red pepper powder when the initial samples with $a_w \geq 0.53$.

The SICs could also be found in corn flour inoculated by *Salmonella* Enteritidis treated at 80 °C for 10 minutes (Ozturk et al. 2019). Zhang et al. (2019) validated the feasibility of RF pasteurization on *S. aureus* ATCC 25923 in in-shell walnuts.

Additionally, many grains, such as corn, peanut, wheat, soybean and oilseeds, are easy to be contaminated by fungi, especially *Aspergillus flavus* and *A. parasiticus*, which can produce aflatoxin (Nesci et al. 2016). Jiao et al. (2016b) found that 3 and 4 log reduction values of *A. flavus* could be obtained when RF heating wheat or corn seeds with 15.0% of moisture content to 65 °C with holding in hot air for 10 minutes.

The above studies in low-moisture foods indicate that although pathogens' inactivation in low-moisture foods is more difficult than in a high-moisture environment, RF pasteurization can be effective in inactivating pathogens in low-moisture conditions.

3.5 Enzyme Inactivation

3.5.1 Blanching

Blanching is an important procedure to inactivate enzymes in canned, frozen, or dehydrated fruits and vegetables. High activity of enzymes, such as polyphenol oxidase (PPO), peroxidase (POD), lipoxygenase (LOX), catalase (CAT), and pectinase, is mainly responsible for browning reaction and the development of off-odors or off-flavors (Severini et al. 2005). Traditional blanching methods including water boiling and steam heating can significantly reduce enzyme activity and prevent browning level of food samples in following process. However, blanching inhomogeneity, water wastage, and reduction of nutritional quality (e.g. vitamin, polyphenol, protein, polysaccharide) limited the application of water and steam blanching. Some innovative technologies have been investigated as alternative blanching technologies to achieve better blanching effect and maintain the food quality. Some of these includes ultrasonic assisted heat treatment (Gamboa-Santos et al. 2013), ultrahigh temperature steam blanching (Bai et al. 2013), pulsed electric fields (Marsellés-Fontanet and Martín-Belloso 2007), ohmic blanching (Kaur and Singh 2016), microwave blanching (Delfiya et al. 2018), infrared blanching (Guiamba et al. 2015), and RF blanching (Gong et al. 2019). RF blanching is a novel blanching technology, which can reduce the loss of nutrients, having great potential in pretreatment of fruits and vegetables. RF blanching has better heating uniformity compared with infrared blanching and microwave blanching, and RF heating can inactivate enzymes in fruits and vegetables without affecting the quality attributes (Manzocco et al. 2008).

Table 1 summarizes the researches of RF blanching for fruits and vegetables. Lopez and Baganis (1971) reported that RF frequency at 60 MHz could inactivate POD, polyphenolase, pectinesterase, CAT, and α-amylase in apple juice milk, and sweet potato due to heating effects. Manzocco et al. (2008) demonstrated that RF heating could effectively inactivate PPO and LOX under appropriate conditions and preserve more sweetness in apples than water blanching. Similarly, Tian et al. (2018) found that RF heating could inactivate PPO in apple tissue, including membrane-bound polyphenol oxidase (mPPO) and soluble polyphenol oxidase (sPPO). Zhang et al. (2018) found that the relative activity of PPO of potatoes was reduced to less than 5% after heated from 25 to 85 °C by RF energy, but when heating temperature increasing from 65 to 85 °C, the potato parenchyma cell became shrunken and starch partly gelatinized according to scanning electron microscopy

Table 1 Summarization of RF heating blanching in fruits and vegetables processing.

Products	Target enzymes	Frequency (MHz)	Power (kW)	RF heating treatment description	Processing time and temperature	Residual activity	References
Apples	PPO LOX	27.12	3.5	With apples submitted to a RF treatment of 6.2 kV	Heat apple core to 98 °C and surface to 52 °C (3 min), thermally equilibrate at room temperature for 30 min (surface to 70 °C)	Complete inactivation	Manzocco et al. (2008)
Potato (potato cuboids, mashed potato, and enzyme extract)	PPO	27.12	6	With centrifuge tube containing potato sample surrounded by the expanded polyethylene foam and placed at the center between two plates	Heat from 25 °C to 65, 70, 75, 80, and 85 °C	65 °C: 28.89% 70 °C: 8.58% 75 °C: 4.19% 80 °C: 3.69% 85 °C: 0.19%	Zhang et al. (2018)
Apple slices	mPPO sPPO	27.12	3.5	With apple slices with thicknesses of 1 cm placed between two plates	Heat from 25 to 70 °C (2 min), hold for 5, 10, 15, 20 min	10 min: sPPO: 45%, mPPO: 13% 20 min: sPPO: 10%, mPPO: 0	Tian et al. (2018)
Carrot cubes	POD	27.12	6	With carrot cubes filled in the polypropylene containers	Heat from 20 to 80 °C (3 min)	2.99%	Gong et al. (2019)
Cabbage	POD	27.12	6	With cabbage slices filled in a cylindrical container	Heat from 20 to 90 °C (100 s)	≤10%	Xu et al. (2019)
Pitaya	PPO CAT	27.12	6	With pitaya strips laid on the center of the lower single layer electrode plate with a plastic frame	Heat sample surface to 92.93 °C and core to 99.4 °C (10 min)	PPO: 14% CAT: 29.8%	Shen et al. (2020)
Stem lettuce	POD	27.12	6	With stem lettuce cuboid placed in a centrifuge tube surrounded by the expanded polyethylene foam	Heat from 25 to 65, 70, 75, 80, and 85 °C	65 °C: 66.03% 85 °C: 6.46%	Yao et al. (2020)
Apple slice	PPO	27.12	6	With apple sample placed in the expanded polyethylene container	Heat from 25 to 85 °C (120 s)	Complete inactivation	Zhang H et al. (2020)

CAT, catalase; LOX, lipoxygenase; mPPO, membrane-bound polyphenol oxidase; POD, peroxidase; PPO, polyphenol oxidase; sPPO, soluble polyphenol oxidase.

micrographs. Shen et al. (2020) observed that 86.0% of PPO and 70.2% of CAT in pitaya fruits were inactivated when blanched in a 6 kW and 27.12 MHz RF equipment.

POD was one of the most heat-stable enzymes in plants, and Gong et al. (2019) found that POD activity in carrot cubes was reduced to 5–10% after RF blanching for 3.0–7.0 minutes, and carrots cubes treated by RF blanching had better hardness, redness, and V_c content than those treated by water blanching. Xu et al. (2019) investigated the effects of electromagnetic field-assisted blanching on physicochemical properties of cabbage, including both RF (27.12 MHz) and microwave (915, 2450 MHz) heating. They found these blanching methods significantly improved nutrient properties of cabbage compared with traditional hot water blanching, and cabbage treated by RF energy had higher chlorophyll and ascorbic acid contents, better flavor, and richer volatile compounds than those blanched by microwave. Yao et al. (2020) also found that RF heating efficiently inactivated POD in stem lettuce.

3.5.2 Stabilization

The by-products of grain milling, such as rice bran and wheat germ, usually have high contents of lipid and high activity of oxidative and hydrolytic enzymes, which lead to short shelf-life of these by-products and limit their deep processing and utilization in food industry. Therefore, stabilization is necessary to inactive enzymes, increase storability, and prolong shelf-life.

Thermal treatments are commonly used as stabilization methods, including extrusion (Gómez et al. 2012; Rafe and Sadeghian 2017), fluidized bed heating (Gili et al. 2018), hot air heating (Wanyo et al. 2014), ohmic heating (Loypimai et al. 2009), infrared radiation (Gili et al. 2017; Irakli et al. 2018), and microwave heating (Lavanya et al. 2019). However, these methods have several drawbacks, such as low efficiency, great damage on quality, high cost, or poor uniformity. Sreenarayanan and Chattopadhyay (1985) stabilized rice bran using RF heating (13.56 MHz) for the first time and found free fatty acid content of rice bran increased much slower than untreated samples, indicating that RF heating significantly reduce lipase (LA) activity in rice bran. Recently, Ling et al. (2018) investigated the effects of temperature, moisture, and metal salt content on DPs of rice bran associated with RF heating. Based on the result, Ling et al. (2018a) heated rice bran to 100 °C by RF energy followed by holding in hot air oven at 100 °C for 15 minutes and found LA and LOX activities decreased to 19.2% and 5.5% of their original values. Besides, Ling et al. (2018b) studied the DPs and heating uniformity of wheat germ stabilized by RF energy and found that LA activities of wheat germ decreased to approximate 70–90% when heated to 90 °C in RF system under heating rate of 6 °C min^{-1}; However, RF heating to 100 °C with holding for 15 minutes in hot air oven or heating to 110 °C with holding for 5 minutes in hot air oven could effectively inactivate LA activity in wheat germ to 18.2% or 22.5% of their original values (Ling et al. 2019). Liao et al. (2020a) found that lipase activity significantly decreased, RF heating to 100 °C with 15-minutes holding at 100–105 °C or to 110 °C with 6-minutes holding at 110–115 °C could reduce LA and POD activities to approximate 10%, and reduce PPO and LOX activities to less than 5%, but HA-RF inactivation of LA, POD, PPO, and LOX was less effective in wheat germ with less moisture content. Liao et al. (2020b) found that RF heating to 100 °C with 15-minutes holding at 100–105 °C or heating to 110 °C with 6-minutes holding at 110–115 °C for rice bran could reduce LA and POD activities to 20–30%.

These studies indicated that RF heating had a great potential in enzymatic inactivation as a blanching method for vegetables and fruits and as a stabilization method for grain milling by-product with less influence on quality of products than traditional blanching methods.

4 Conclusions and Future Outlook

As a novel thermal food processing method, RF heating is obtaining great attentions in recent years due to its rapid and volumetric heating characteristics. The RF technology has been tentatively applied in various food processes, including thawing, drying, disinfestation, enzyme, and microbial inactivation.

1) *Thawing*: RF-thawing treatment of food products has potential advantages in achieving shorter thawing time and better homogeneity compared to conventional thawing methods, which also has a significant impact on energy consumption. However, more efforts are needed on exploring how to improve RF heating uniformity during thawing process, especially for irregular frozen meats and in phase change process.
2) *Drying*: RF heating is a fast and volumetric heating method and can combine with other methods to be a high efficiency drying method, such as vacuum or hot air. HA-RF drying has been proved as an effective drying method for many agricultural products. However, improvement of the energy efficiency and drying uniformity are the big challenges for HA-RF drying.
3) *Disinfestation*: Many studies demonstrate that applying RF energy in disinfestation of fruits or vegetables is not advisable because of the nonuniform heating and thermal sensitivity of fruits and vegetables. However, RF heating shows great potential in disinfestation of low-moisture agricultural products, such as grains and legumes, and some applications have been conducted in large-scale and hopefully the commercial applications could be realized in the near future.
4) *Microbial inactivation*: Currently, there are lacks of effective pasteurization methods for low-moisture foods due to low thermal conductivity of samples and exponentially increased thermal resistance of pathogens in low-moisture environments. RF heating is more effective in controlling pathogens than conventional thermal processing and has great potential to be a novel low-moisture foods pasteurization method.
5) *Enzyme inactivation*: RF heating can effectively inactivate enzymes with less influence on quality of products, and thus holds great potential as a novel blanching method for vegetables and fruits and a stabilization method for grain milling by-product. Scale-up study and designing large-scale RF systems should be the research emphasis for the future study.

With regard to future perspectives, RF treatments have obtained increasing attentions in food processing fields recently and showed great potential in several food processing applications, such as disinfestation of grains and legumes, thawing of frozen meats in regular shape, pasteurization of low-moisture foods, and stabilization of by-products of grain milling. Therefore, more efforts are needed for conducting more in-depth studies to accelerate the commercial applications of RF heating in those areas, such as scale-up study, large-scale RF unit development, and efficient use of energy. Exploring the mechanism of RF pasteurization, combining RF heating with other technologies, and promoting the small-scale home-use RF unit for cooking and thawing could be focused in future research.

References

Alasalvar, C., Salvadó, J.S., and Ros, E. (2020). Bioactives and health benefits of nuts and dried fruits. *Food Chemistry* 314: 126192.

Alfaifi, B., Wang, S., Tang, J. et al. (2013). Radio frequency disinfestation treatments for dried fruit: dielectric properties. *LWT Food Science and Technology* 50 (2): 746–754.

Alfaifi, B., Tang, J., Jiao, Y. et al. (2014). Radio frequency disinfestation treatments for dried fruit: model development and validation. *Journal of Food Engineering* 120: 268–276.

Altemimi, A., Aziz, S.N., Al-Hilphy, A.R. et al. (2019). Critical review of radio-frequency (RF) heating applications in food processing. *Food Quality and Safety* 3: 81–91.

Awuah, G.B., Ramaswamy, H.S., Economides, A., and Mallikarjunan, K. (2005). Inactivation of *Escherichia coli* K-12 and *Listeria innocua* in milk using radio frequency (RF) heating. *Innovative Food Science and Emerging Technologies* 5 (4): 396–402.

Awuah, G.B., Ramaswamy, H.S., and Tang, J. (2015). *Radio-Frequency Heating in Food Processing Principles and Applications*. Florida: CRC Press.

Bai, J., Sun, D., Xiao, H. et al. (2013). Novel high-humidity hot air impingement blanching (HHAIB) pretreatment enhances drying kinetics and color attributes of seedless grapes. *Innovative Food Science and Emerging Technologies* 20: 230–237.

Banga, K.S., Kotwaliwale, N., Mohapatra, D., and Giri S.K. (2018). Techniques for insect detection in stored food grains: an overview. *Food Control* 94: 167–176.

Bedane, T.F., Chen, L., Marra, F., and Wang, S. (2017). Experimental study of radio frequency (RF) thawing of foods with movement on conveyor belt. *Journal of Food Engineering* 201: 17–25.

Birla, S.L., Wang, S., Tang, J., and Hallman, G. (2004). Improving heating uniformity of fresh fruit in radio frequency treatments for pest control. *Postharvest Biology and Technology* 33 (2): 205–217.

Birla, S.L., Wang, S., Tang, J. et al. (2005). Quality of oranges as influenced by potential radio frequency heat treatments against Mediterranean fruit flies. *Postharvest Biology and Technology* 38 (1): 66–79.

Birla, S.L., Wang, S., and Tang, J. (2008a). Computer simulation of radio frequency heating of model fruit immersed in water. *Journal of Food Engineering* 84 (2): 270–280.

Birla, S.L., Wang, S., Tang, J., and Tiwari, G. (2008b). Characterization of radio frequency heating of fresh fruits influenced by dielectric properties. *Journal of Food Engineering* 89 (4): 390–398.

Byrne, B., Dunne, G., and Bolton, D.J. (2006). Thermal inactivation of *Bacillus cereus* and *Clostridium perfringens* vegetative cells and spores in pork luncheon roll. *Food Microbiology* 23 (8): 803–808.

Cai, L., Cao, M., Regenstein, J., and Cao, A. (2019). Recent advances in food thawing technologies. *Comprehensive Reviews in Food Science and Food Safety* 18: 953–970.

Choi, E.J., Yang, H.S., Park, H.W., and Chun, H.H. (2018). Inactivation of *Escherichia coli* O157:H7 and *Staphylococcus aureus* in red pepper powder using a combination of radio frequency thermal and indirect dielectric barrier discharge plasma non-thermal treatments. *LWT Food Science and Technology* 93: 477–484.

Chou, S.K. and Chua, K.J. (2001). New hybrid drying technologies for heat sensitive foodstuffs. *Trends in Food Science & Technology* 12 (10): 359–369.

Delfiya, A., Mohapatra, D., Kotwaliwale, N., and Mishra, A.K. (2018). Effect of microwave blanching and brine solution pretreatment on the quality of carrots dried in solar-biomass hybrid dryer. *Journal of Food Processing and Preservation* 42 (2): e13510.

Farag, K.W., Lyng, J.G., Morgan, D.J., and Cronin, D.A. (2008). A comparison of conventional and radio frequency tempering of beef meats: effects on product temperature distribution. *Meat Science* 80 (2): 488–495.

Farag, K.W., Duggan, E., Morgan, D.J. et al. (2009). A comparison of conventional and radio frequency defrosting of lean beef meats: effects on water binding characteristics. *Meat Science* 83 (2): 278–284.

Farag, K.W., Lyng, J.G., Morgan, D.J., and Cronin, D.A. (2011). A comparison of conventional and radio frequency thawing of beef meats: effects on product temperature distribution. *Food and Bioprocess Technology* 4: 1128–1136.

Fu, Y.C. (2004). Fundamentals and industrial applications of microwave and radio frequency in food processing. In: *Food Processing: Principles and Applications* (eds. J.S. Smith and Y.H. Hui), 79–100. Iowa: Blackwell.

Gamboa-Santos, J., Soria, A.C., Villamiel, M., and Montilla, A. (2013). Quality parameters in convective dehydrated carrots blanched by ultrasound and conventional treatment. *Food Chemistry* 141 (1): 616–624.

Gili, R.D., Palavecino, P.M., Penci, M.C. et al. (2017). Wheat germ stabilization by infrared radiation. *Journal of Food Science and Technology* 54 (1): 71–81.

Gili, R.D., Penci, M.C., Torrez, I.M.R. et al. (2018). Effect of wheat germ heat treatment by fluidised bed on the kinetics of lipase inactivation. *Food and Bioprocess Technology* 11 (5): 1002–1011.

Gómez, M., González, J., and Oliete, B. (2012). Effect of extruded wheat germ on dough rheology and bread quality. *Food and Bioprocess Technology* 5 (6): 2409–2418.

Gong, C., Zhao, Y., Zhang, H. et al. (2019). Investigation of radio frequency heating as a dry-blanching method for carrot cubes. *Journal of Food Engineering* 245: 53–56.

Gong, C., Liao, M., Zhang, H. et al. (2020). Investigation of hot air-assisted radio frequency as a final stage drying of pre-dried carrot cubes. *Food and Bioprocess Technology* 13: 419–429.

Guiamba, I.R.F., Svanberg, U., and Ahrne, L. (2015). Effect of infrared blanching on enzyme activity and retention of carotene and vitamin C in dried mango. *Journal of Food Science* 80 (6): E1235–E1242.

Guo, Q., Piyasena, P., Mittal, G.S. et al. (2006). Efficacy of radio frequency cooking in the reduction of *Escherichia coli* and shelf stability of ground beef. *Food Microbiology* 23 (2): 112–118.

Guo, W., Wang, S., Tiwari, G. et al. (2010). Temperature and moisture dependent dielectric properties of legume flour associated with dielectric heating. *LWT Food Science and Technology* 43 (2): 193–201.

Guo, C., Mujumdar, A.S., and Zhang, M. (2019). New development in radio frequency heating for fresh food processing: a review. *Food Engineering Reviews* 11 (1): 29–43.

Hansen, J.D., Drake, S.R., Heidt, M.L. et al. (2005). Evaluation of radio frequency-hot water treatments for postharvest control of codling moth in 'Bing' sweet cherries. *HortTechnology* 15 (3): 613–616.

Hansen, J.D., Drake, S.R., Watkins, M.A. et al. (2006). Radio frequency pulse application for heating uniformity in postharvest codling moth (Lepidoptera: Tortricidae) control of fresh apples (*Malus domestica* Borkh.). *Journal of Food Quality* 29 (5): 492–504.

Hernández-Alonso, P., Camacho-Barcia, L., Bulló, M., and Salas-Salvadó, J. (2017). Nuts and dried fruits: an update of their beneficial effects on type 2 diabetes. *Nutrients* 9 (7): 673–706.

Hou, L., Hou, J., Li, Z. et al. (2015). Validation of radio frequency treatments as alternative non-chemical methods for disinfesting chestnuts. *Journal of Stored Products Research* 63: 75–79.

Hou, L., Johnson, J.A., and Wang, S. (2016). Radio frequency heating for postharvest control of pests in agricultural products: a review. *Postharvest Biology and Technology* 113: 106–118.

Hou, L., Liu, Q., and Wang, S. (2019). Efficiency of industrial-scale radio frequency treatments to control *Rhyzopertha dominica* (Fabricius) in rough, brown, and milled rice. *Biosystems Engineering* 186: 246–258.

Huang, Z., Chen, L., and Wang, S. (2015a). Computer simulation of radio frequency selective heating of insects in soybeans. *International Journal of Heat and Mass Transfer* 90: 406–417.

Huang, Z., Zhu, H., Yan, R., and Wang, S. (2015b). Simulation and prediction of radio frequency heating in dry soybeans. *Biosystems Engineering* 129: 34–47.

Huang, Z., Zhang, B., Marra, F., and Wang, S. (2016). Computational modelling of the impact of polystyrene containers on radio frequency heating uniformity improvement for dried soybeans. *Innovative Food Science and Emerging Technologies* 33: 365–380.

Ikediala, J.N., Tang, J., Drake, S.R., and Neven, L.G. (2000). Dielectric properties of apple cultivars and codling moth larvae. *Transactions of ASAE* 43 (5): 1175–1184.

Ikediala, J.N., Hansen, J.D., Tang, J. et al. (2002). Development of a saline water immersion technique with RF energy as a postharvest treatment against codling moth in cherries. *Postharvest Biology and Technology* 24 (1): 25–37.

Irakli, M., Kleisiaris, F., Mygdalia, A., and Katsantonis, D. (2018). Stabilization of rice bran and its effect on bioactive compounds content, antioxidant activity and storage stability during infrared radiation heating. *Journal of Cereal Science* 80: 135–142.

Jiao, S., Johnson, J.A., Tang, J. et al. (2011). Dielectric properties of cowpea weevil, black-eyed peas and mung beans with respect to the development of radio frequency heat treatments. *Biosystems Engineering* 108 (3): 280–291.

Jiao, S., Johnson, J.A., Tang, J., and Wang, S. (2012). Industrial-scale radio frequency treatments for insect control in lentils. *Journal of Stored Products Research* 48: 143–148.

Jiao, Y., Tang, J., and Wang, S. (2014). A new strategy to improve heating uniformity of low moisture foods in radio frequency treatment for pathogen control. *Journal of Food Engineering* 141: 128–138.

Jiao, S., Deng, Y., Zhong, Y. et al. (2015). Investigation of radio frequency heating uniformity of wheat kernels by using the developed computer simulation model. *Food Research International* 71: 41–49.

Jiao, Y., Shi, H., Tang, J. et al. (2015). Improvement of radio frequency (RF) heating uniformity on low moisture foods with Polyetherimide (PEI) blocks. *Food Research International* 74: 106–114.

Jiao, S., Zhu, D., Deng, T., and Zhao, Y. (2016a). Effects of hot air-assisted radio frequency heating on quality and shelf-life of roasted peanuts. *Food and Bioprocess Technology* 9 (2): 308–319.

Jiao, S., Zhong, Y., and Deng, Y. (2016b). Hot air-assisted radio frequency heating effects on wheat and corn seeds: quality change and fungi inhibition. *Journal of Stored Products Research* 69: 265–271.

Jiao, S., Sun, W., Yang, T. et al. (2017). Investigation of the feasibility of radio frequency energy for controlling insects in milled rice. *Food and Bioprocess Technology* 10 (4): 781–788.

Jiao, Y., Tang, J., Wang, Y., and Koral, T.L. (2018). Radio-frequency applications for food processing and safety. *Annual Review of Food Science and Technology* 9: 105–127.

Kaur, N. and Singh, A.K. (2016). Ohmic heating: concept and applications – a review. *Critical Reviews in Food Science and Nutrition* 56 (14): 2338–2351.

Kim, S.Y., Sagong, H.G., Choi, S.H. et al. (2012). Radio-frequency heating to inactivate *Salmonella* Typhimurium and *Escherichia coli* O157:H7 on black and red pepper spice. *International Journal of Food Microbiology* 153 (1–2): 171–175.

Kim, J., Park, J.W., Park, S. et al. (2016). Study of radio frequency thawing for cylindrical pork sirloin. *Journal of Biosystems Engineering* 41 (2): 108–115.

Lagunas-Solar, M.C., Pan, Z., Zeng, N.X. et al. (2007). Application of radiofrequency power for non-chemical disinfestation of rough rice with full retention of quality attributes. *Applied Engineering in Agriculture* 23 (5): 647–654.

Lau, S.K., Thippareddi, H., Jones, D. et al. (2016). Challenges in radiofrequency pasteurization of shell eggs: coagulation rings. *Journal of Food Science* 81 (10): E2492–E2502.

Lau, S.K., Thippareddi, H., and Subbiah, J. (2017). Radiofrequency heating for enhancing microbial safety of shell eggs immersed in deionized water. *Journal of Food Science* 82 (12): 2933–2943.

Lavanya, M.N., Saikiran, K.C.S., and Venkatachalapathy, N. (2019). Stabilization of rice bran milling fractions using microwave heating and its effect on storage. *Journal of Food Science and Technology* 56 (2): 889–895.

Li, Y., Chen, S., and Yao, M. (2015). Effects of radio frequency heating on disinfestation and sterilization of rice flour. *Taiwanese Journal of Agricultural Chemistry and Food Science* 53: 125–134.

Liao, M., Damayanti, W., Zhao, Y. et al. (2020a). Hot air-assisted radio frequency stabilizing treatment effects on physicochemical properties, enzyme activities and nutritional quality of wheat germ. *Food and Bioprocess Technology* 13: 901–910.

Liao, M., Damayanti, W., Xu, Y. et al. (2020b). Hot air-assisted radio frequency heating for stabilization of rice bran: enzyme activity, phenolic content, antioxidant activity and microstructure. *LWT Food Science and Technology* 131: 109754.

Lin, W., Zhang, M., Fang, Z.X., and Liu, Y.P. (2014). Effect of salt and sucrose content on the dielectric properties of salted duck egg white protein relevant to radio frequency drying. *Drying Technology* 32 (15): 1777–1784.

Ling, B., Tiwari, G., and Wang, S. (2015). Pest control by microwave and radio frequency energy: dielectric properties of stone fruit. *Agronomy for Sustainable Development* 35 (1): 233–240.

Ling, B., Hou, L., Li, R., and Wang, S. (2016). Storage stability of pistachios as influenced by radio frequency treatments for postharvest disinfestations. *Innovative Food Science and Emerging Technologies* 33: 357–364.

Ling, B., Liu, X., Zhang, L., and Wang, S. (2018). Effects of temperature, moisture, and metal salt content on dielectric properties of rice bran associated with radio frequency heating. *Scientific Reports* 8 (1): 1–12.

Ling, B., Lyng, J.G., and Wang, S. (2018a). Effects of hot air-assisted radio frequency heating on enzyme inactivation, lipid stability and product quality of rice bran. *LWT Food Science and Technology* 91: 453–459.

Ling, B., Lyng, J.G., and Wang, S. (2018b). Radio-frequency treatment for stabilization of wheat germ: dielectric properties and heating uniformity. *Innovative Food Science and Emerging Technologies* 48: 66–74.

Ling, B., Ouyang, S., and Wang, S. (2019). Radio-frequency treatment for stabilization of wheat germ: storage stability and physicochemical properties. *Innovative Food Science and Emerging Technologies* 52: 158–165.

Ling, B., Cheng, T., and Wang, S. (2020). Recent developments in applications of radio frequency heating for improving safety and quality of food grains and their products: a review. *Critical Reviews in Food Science and Nutrition* 60 (15): 2622–2642.

Liu, Q., Zhang, M., Xu, B. et al. (2015). Effect of radio frequency heating on the sterilization and product quality of vacuum packaged Caixin. *Food and Bioproducts Processing* 95: 47–54.

Llave, Y. and Erdogdu, F. (2020). Radio frequency processing and recent advances on thawing and tempering of frozen food products. *Critical Reviews in Food Science and Nutrition* https://doi.org/10.1080/10408398.2020.1823815.

Llave, Y., Terada, Y., Fukuoka, M., and Sakai, N. (2014). Dielectric properties of frozen tuna and analysis of defrosting using a radio-frequency system at low frequencies. *Journal of Food Engineering* 139: 1–9.

Lopez, A. and Baganis, N.A. (1971). Effect of radio-frequency energy at 60 MHz on food enzyme activity. *Journal of Food Science* 36 (6): 911–914.

Loypimai, P., Moonggarm, A., and Chottanom, P. (2009). Effects of ohmic heating on lipase activity, bioactive compounds and antioxidant activity of rice bran. *Australian Journal of Basic and Applied Sciences* 3 (4): 3642–3652.

Luechapattanaporn, K., Wang, Y.F., Wang, J. et al. (2005). Sterilization of scrambled eggs in military polymeric trays by radio frequency energy. *Journal of Food Science* 70 (4): E288–E294.

Lyu, X., Peng, X., Wang, S. et al. (2018). Quality and consumer acceptance of radio frequency and traditional heat pasteurised kiwi puree during storage. *International Journal of Food Science and Technology* 53 (1): 209–218.

Manzocco, L., Anese, M., and Nicoli, M.C. (2008). Radiofrequency inactivation of oxidative food enzymes in model systems and apple derivatives. *Food Research International* 41 (10): 1044–1049.

Marsellés-Fontanet, Á.R. and Martín-Belloso, O. (2007). Optimization and validation of PEF processing conditions to inactivate oxidative enzymes of grape juice. *Journal of Food Engineering* 83 (3): 452–462.

Marshall, M.G. and Metaxas, A.C. (1999). Radio frequency assisted heat pump drying of crushed brick. *Applied Thermal Engineering* 19 (4): 375–388.

Metaxas, A.C. and Meredith, R.J. (1993). *Industrial Microwave Heating*. London: Peter Peregrinus Ltd.

Michael, M., Phebus, R.K., Thippareddi, H. et al. (2014). Validation of radio-frequency dielectric heating system for destruction of *Cronobacter sakazakii* and *Salmonella* species in nonfat dry milk. *Journal of Dairy Science* 97 (12): 7316–7324.

Monzon, M.E., Biasi, B., Mitcham, E.J. et al. (2007). Effect of radiofrequency heating on the quality of 'Fuyu' persimmon fruit as a treatment for control of the Mexican fruit fly. *HortScience* 42 (1): 125–129.

Nagaraj, G., Purohit, A., Harrison, M. et al. (2016). Radiofrequency pasteurization of inoculated ground beef homogenate. *Food Control* 59: 59–67.

Nelson, S.O. (1996). Review and assessment of radio-frequency and microwave energy for stored-grain insect control. *Transactions of ASAE* 39 (4): 1475–1484.

Nesci, A., Passone, M.A., Barra, P. et al. (2016). Prevention of aflatoxin contamination in stored grains using chemical strategies. *Current Opinion in Food Science* 11: 56–60.

Orfeuil, M. (1987). *Electric Process Heating: Technologies, Equipment, Applications*. Battelle Press (chapter 7).

Orsat, V. and Raghavan, G.S.V. (2005). Radio-frequency processing. In: *Emerging Technologies for Food Processing* (ed. D.-W. Sun), 445–468. Academic Press.

Orsat, V., Bai, L., Raghavan, G.S.V., and Smith, J.P. (2004). Radio-frequency heating of ham to enhance shelf-life in vacuum packaging. *Journal of Food Process Engineering* 27 (4): 267–283.

Ozturk, S., Liu, S., Xu, J. et al. (2019). Inactivation of *Salmonella* Enteritidis and *Enterococcus faecium* NRRL B-2354 in corn flour by radio frequency heating with subsequent freezing. *LWT Food Science and Technology* 111: 782–789.

Palazoğlu, T.K. and Miran, W. (2017). Experimental comparison of microwave and radio frequency tempering of frozen block of shrimp. *Innovative Food Science and Emerging Technologies* 41: 292–300.

Palazoglu, T.K. and Miran, W. (2019). Computational investigation of the effect of orientation and rotation of shell egg on radio frequency heating rate and uniformity. *Innovative Food Science and Emerging Technologies* 58: 102238.

Pan, L., Jiao, S., Gautz, L. et al. (2012). Coffee bean heating uniformity and quality as influenced by radio frequency treatments for postharvest disinfestations. *Transactions of the ASABE* 55: 2293–2300.

Pircon, L.J., Loquercio, P., and Doty, D.M. (1953). High-frequency cooking, high-frequency heating as a unit operation in meat processing. *Journal of Agricultural and Food Chemistry* 1 (13): 844–847.

Piyasena, P., Dussault, C., Koutchma, T. et al. (2003). Radio frequency heating of foods: principles, applications and related properties – a review. *Critical Reviews in Food Science and Nutrition* 43 (6): 587–606.

Ptasznik, W., Zygmunt, S., and Kudra, T. (1990). Simulation of RF-assisted convective drying for seed quality broad bean. *Drying Technology* 8 (5): 977–992.

Rafe, A. and Sadeghian, A. (2017). Stabilization of Tarom and Domesiah cultivars rice bran: physicochemical, functional and nutritional properties. *Journal of Cereal Science* 74: 64–71.

Roknul, A.S.M., Zhang, M., Mujumdar, A.S., and Wang, Y. (2014). A comparative study of four drying methods on drying time and quality characteristics of stem lettuce slices (*Lactuca sativa* L.). *Drying Technology* 32 (6): 657–666.

Ryynänen, S. (1995). The electromagnetic properties of food materials: a review of basic principles. *Journal of Food Engineering* 26: 409–429.

Schlisselberg, D.B., Kier, E., Kalily, E. et al. (2013). Inactivation of foodborne pathogens in ground beef by cooking with highly controlled radio frequency energy. *International Journal of Food Microbiology* 160 (3): 219–226.

Severini, C., Baiano, A., De Pilli, T. et al. (2005). Combined treatments of blanching and dehydration: study on potato cubes. *Journal of Food Engineering* 68 (3): 289–296.

Shen, Y., Zheng, L., Gou, M. et al. (2020). Characteristics of pitaya after radio frequency treating: structure, phenolic compounds, antioxidant, and antiproliferative activity. *Food and Bioprocess Technology* 13 (1): 180–186.

Shrestha, B. and Baik, O.D. (2015). Dielectric behaviour of whole-grain wheat with temperature at 27.12 MHZ: a novel use of a liquid dielectric test fixture for grains. *International Journal of Food Properties* 18 (1): 100–112.

Shrestha, B., Yu, D., and Baik, O.D. (2013). Elimination of *Cryptolestes ferrungineus* S. in wheat by radio frequency dielectric heating at different moisture contents. *Progress in Electromagnetics Research* 139: 517–538.

Sisquella, M., Casals, C., Picouet, P. et al. (2013). Immersion of fruit in water to improve radio frequency treatment to control brown rot in stone fruit. *Postharvest Biology and Technology* 80: 31–36.

Song, X., Ma, B., Kou, X. et al. (2020). Developing radio frequency heating treatments to control insects in mung beans. *Journal of Stored Products Research* 88: 101651.

Sosa-Morales, M.E., Tiwari, G., Wang, S. et al. (2009). Dielectric heating as a potential post-harvest treatment of disinfesting mangoes, part II: development of RF-based protocols and quality evaluation of treated fruits. *Biosystems Engineering* 103 (3): 287–296.

Sreenarayanan, V.V. and Chattopadhyay, P.K. (1985). Rice bran stabilization by dielectric heating. *Journal of Food Processing and Preservation* 10 (2): 89–98.

Ştefănoiu, G.A., Tănase, E.E., Miteluţ, A.C., and Popa, M.E. (2016). Unconventional treatments of food: microwave vs. radiofrequency. *Agriculture and Agricultural Science Procedia* 10: 503–510.

Tang, T., Aldhaeebi, M.A., Olaimat, M. et al. (2020). Radio frequency thawing chamber for rapidly thawing applications. *Microwave and Optical Technology Letters* 63: 175–180.

Tian, Y., Yan, W., Tang, Y. et al. (2018). Inactivation of membrane-bound and soluble polyphenol oxidases in apple (*Malus domestica* Borkh) by radio frequency processing for improved juice quality. *Journal of Food Process Engineering* 41 (8): e12923.

Tiwari, G., Wang, S., Birla, S.L., and Tang, J. (2008). Effect of water-assisted radio frequency heat treatment on the quality of 'Fuyu' persimmons. *Biosystems Engineering* 100 (2): 227–234.

Uemura, K., Kanafusa, S., Takahashi, C., and Kobayashi, I. (2017). Development of a radio frequency heating system for sterilization of vacuum-packed fish in water. *Bioscience Biotechnology and Biochemistry* 81 (4): 762–767.

Wang, S., Tang, J., and Cavalieri, R.P. (2001). Modeling fruit internal heating rates for hot air and hot water treatments. *Postharvest Biology and Technology* 22 (3): 257–270.

Wang, S., Tang, J., Johnson, J.A. et al. (2002). Process protocols based on radio frequency energy to control field and storage pests in in-shell walnuts. *Postharvest Biology and Technology* 26 (3): 265–273.

Wang, S., Tang, J., Cavalieri, R.P., and Davies, D.C. (2003). Differential heating of insects in dried nuts and fruits associated with radio frequency and microwave treatments. *Transactions of ASAE* 46 (4): 1175–1182.

Wang, S., Birla, S.L., Tang, J., and Hansen, J.D. (2006a). Postharvest treatment to control codling moth in fresh apples using water assisted radio frequency heating. *Postharvest Biology and Technology* 40 (1): 89–96.

Wang, S., Tang, J., Sun, T. et al. (2006b). Considerations in design of commercial radio frequency treatments for postharvest pest control in in-shell walnuts. *Journal of Food Engineering* 77 (2): 304–312.

Wang, S., Monzon, A., Johnson, J.A. et al. (2007a). Industrial-scale radio frequency treatments for insect control in walnuts I: heating uniformity and energy efficiency. *Postharvest Biology and Technology* 45 (2): 240–246.

Wang, S., Monzon, M., Johnson, J.A. et al. (2007b). Industrial-scale radio frequency treatments for insect control in walnuts II: insect mortality and product quality. *Postharvest Biology and Technology* 45 (2): 247–253.

Wang, S., Yue, J., Chen, B., and Tang, J. (2008). Treatment design of radio frequency heating based on insect control and product quality. *Postharvest Biology and Technology* 49 (3): 417–423.

Wang, S., Tiwari, G., Jiao, S. et al. (2010). Developing postharvest disinfestation treatments for legumes using radio frequency energy. *Biosystems Engineering* 105 (3): 341–349.

Wang, Y., Zhang, L., Johnson, J. et al. (2014). Developing hot air-assisted radio frequency drying for in-shell macadamia nuts. *Food and Bioprocess Technology* 7 (1): 278–288.

Wanyo, P., Meeso, N., and Siriamornpun, S. (2014). Effects of different treatments on the antioxidant properties and phenolic compounds of rice bran and rice husk. *Food Chemistry* 157: 457–463.

Wei, X., Lau, S.K., Stratton, J. et al. (2018). Radio-frequency processing for inactivation of *Salmonella enterica* and *Enterococcus faecium* NRRL B-2354 in black peppercorn. *Journal of Food Protection* 81 (10): 1685–1695.

Xu, J., Zhang, M., Bhandari, B., and Kachele, R. (2017). ZnO nanoparticles combined radio frequency heating: a novel method to control microorganism and improve product quality of prepared carrots. *Innovative Food Science and Emerging Technologies* 44: 46–53.

Xu, J., Zhang, M., An, Y. et al. (2018). Effects of radio frequency and high pressure steam sterilisation on the colour and flavour of prepared *Nostoc sphaeroides*. *Journal of the Science of Food and Agriculture* 98 (5): 1719–1724.

Xu, J., Wang, B., and Wang, Y. (2019). Electromagnetic fields assisted blanching-effect on the dielectric and physicochemical properties of cabbage. *Journal of Food Process Engineering* 42 (8): e13294.

Yang, C., Zhao, Y., Tang, Y. et al. (2018). Radio frequency heating as a disinfestation method against *Corcyra cephalonica* and its effect on properties of milled rice. *Journal of Stored Products Research* 77: 112–121.

Yang, H., Chen, Q.,.H., Cao, H. et al. (2019). Radiofrequency thawing of frozen minced fish based on the dielectric response mechanism. *Innovative Food Science and Emerging Technologies* 52: 80–88.

Yao, Y., Wei, X., Pang, H. et al. (2020). Effects of radio-frequency energy on peroxidase inactivation and physiochemical properties of stem lettuce and the underlying cell-morphology mechanism. *Food Chemistry* 322: 126753.

Yu, D., Shrestha, B., and Baik, O.D. (2016). Radio frequency (RF) control of red flour beetle (*Tribolium castaneum*) in stored rapeseeds (*Brassica napus* L.). *Biosystems Engineering* 151: 248–260.

Yu, D., Shrestha, B., and Baik, O.D. (2017). Thermal death kinetics of adult red flour beetle *Tribolium castaneum* (Herbst) in canola seeds during radio frequency heating. *International Journal of Food Properties* 20 (12): 3064–3075.

Zhang, S., Zhou, L., Ling, B., and Wang, S. (2016a). Dielectric properties of peanut kernels associated with microwave and radio frequency drying. *Biosystems Engineering* 145: 108–117.

Zhang, B., Zheng, A., Zhou, L. et al. (2016b). Developing hot air-assisted radio frequency drying protocols for in-shell walnuts. *Emirates Journal of Food and Agriculture* 28 (7): 459–467.

Zhang, J., Gao, Z., Liu, X. et al. (2017). The effect of RF treatment combined with nisin against *Alicyclobacillus* spores in kiwi fruit juice. *Food and Bioprocess Technology* 10 (2): 340–348.

Zhang, Z., Wang, J., Zhang, X. et al. (2018). Effects of radio frequency assisted blanching on polyphenol oxidase, weight loss, texture, color and microstructure of potato. *Food Chemistry* 248: 173–182.

Zhang, H., Gong, C., Wang, X. et al. (2019). Application of hot air-assisted radio frequency as second stage drying method for mango slices. *Journal of Food Process Engineering* 42: e12974.

Zhang, L., Lyng, J.G., Xu, R. et al. (2019). Influence of radio frequency treatment on in-shell walnut quality and *Staphylococcus aureus* ATCC 25923 survival. *Food Control* 102: 197–205.

Zhang, H., Zhao, Y., Gong, C., and Jiao, S. (2020). Effect of radio frequency heating stress on sublethal injury of *Salmonella* Typhimurium in red pepper powder. *LWT Food Science and Technology* 117: 108700.

Zhang, Y., Xie, Y., Tang, J. et al. (2020). Thermal inactivation of *Cronobacter sakazakii* ATCC 29544 in powdered infant formula milk using thermostatic radio frequency. *Food Control* 114: 107270.

Zhao, Y., Zhao, W., Yang, R. et al. (2017). Radio frequency heating to inactivate microorganisms in broccoli powder. *Food Quality and Safety* 1 (1): 93–100.

Zhou, L. and Wang, S. (2016a). Verification of radio frequency heating uniformity and *Sitophilus oryzae* control in rough, brown, and milled rice. *Journal of Stored Products Research* 65: 40–47.

Zhou, L. and Wang, S. (2016b). Industrial-scale radio frequency treatments to control *Sitophilus oryzae* in rough, brown, and milled rice. *Journal of Stored Products Research* 68: 9–18.

Zhou, X. and Wang, S. (2019). Recent developments in radio frequency drying of food and agricultural products: a review. *Drying Technology* 37 (3): 271–286.

Zhou, L., Ling, B., Zheng, A. et al. (2015). Developing radio frequency technology for postharvest insect control in milled rice. *Journal of Stored Products Research* 62: 22–31.

Zhou, X., Gao, H., Mitcham, E.J., and Wang, S. (2018). Comparative analyses of three dehydration methods on drying characteristics and oil quality of in-shell walnuts. *Drying Technology* 36 (4): 477–490.

Zhou, X., Ramaswamy, H., Qu, Y. et al. (2019). Combined radio frequency-vacuum and hot air dehydration for kiwifruits: drying uniformity, energy efficiency and product quality. *Innovative Food Science and Emerging Technologies* 56: 102182.

12

Infrared Spectroscopy

Daniel Cozzolino

Centre for Nutrition and Food Sciences, Queensland Alliance for Agriculture and Food Innovation, The University of Queensland, St. Lucia, Brisbane, Queensland, Australia

1 Introduction

The advantages of vibrational spectroscopic techniques (e.g. infrared [IR] spectroscopy) have been long recognized by a wide range of industries (e.g. food, pharmaceutical, beverage, environment, etc.) (Cen and He 2007; Karoui et al. 2010; Weeranantanaphan et al. 2011; Liu et al. 2013; Arendse et al. 2018; Bureau et al. 2019; Beć et al. 2020; Chapman et al. 2021). This has been supported by the wide range of applications, including the prediction of chemical parameters (e.g. protein, dry matter, starch, alcohol, and sugars), the evaluation of product freshness, authenticity of ingredients and products (e.g. geographical origin and production method, and traceability), detection and monitoring of spoilage and monitoring process (online analytics) (Rolinger et al. 2020), among other applications (Cen and He 2007; Walsh and Kawano 2009; Karoui et al. 2010; Weeranantanaphan et al. 2011; Liu et al. 2013; Wang et al. 2016; Arendse et al. 2018; Bureau et al. 2019; Cortés et al. 2019; Beć et al. 2020; Chapman et al. 2021; Eifert et al. 2020).

It is in this context that infrared spectroscopy (IR) has been one of the most attractive and commonly used methods of analysis due to its intrinsic characteristics and properties such as simultaneous, rapid, and nondestructive analysis of a wide range of properties in many type of samples (e.g. powders, liquids, solids, and gas) (Cen and He 2007; Walsh and Kawano 2009; Karoui et al. 2010; Weeranantanaphan et al. 2011; Liu et al. 2013; Wang et al. 2016; Arendse et al. 2018; Bureau et al. 2019; Cortés et al. 2019; Beć et al. 2020; Chapman et al. 2021; Eifert et al. 2020; Rolinger et al. 2020).

The tangible technological revolution through the last decades has determined an increase in computer power, the development of new algorithms and mathematical processing methods or techniques, the miniaturization and portability of spectrophotometers, that has boosted the utilization of infrared spectroscopy in a wide range of industries and fields.

Electromagnetic Technologies in Food Science, First Edition. Edited by Vicente M. Gómez-López and Rajeev Bhat.

However, adoption and applications of these technologies have been extensively found in middle- to high-income countries due to easy to access to both instrumentation and training.

This chapter presents and introduces with the main concepts of infrared spectroscopy (near-infrared [NIR] and mid-infrared [MIR]), the origins of the electromagnetic radiation, handheld and portable instrumentation, and both hyperspectral imaging (HSI)/multispectral imaging (MSI) systems.

2 The Electromagnetic Radiation

MIR and NIR spectroscopy are classified as molecular/vibrational spectroscopy techniques and they are used to evaluate and study the interactions of electromagnetic waves with the sample (Walsh and Kawano 2009; Pasquini 2018 Beć et al. 2020). The electromagnetic spectrum comprises of different types of radiation, ranging from radio waves (lower end of the spectrum) through to γ rays (upper end) (Cen and He 2007; Walsh and Kawano 2009; Karoui et al. 2010; Weeranantanaphan et al. 2011; Liu et al. 2013; Wang et al. 2016; Arendse et al. 2018; Bureau et al. 2019; Cortés et al. 2019; Beć et al. 2020; Chapman et al. 2021; Eifert et al. 2020; Rolinger et al. 2020). Molecules absorb IR owing to the vibrational movements of their chemical bonds (e.g. bending, stretching, rocking, wagging, or scissoring) (Cen and He 2007; Walsh and Kawano 2009; Karoui et al. 2010; Weeranantanaphan et al. 2011; Liu et al. 2013; Wang et al. 2016; Arendse et al. 2018; Bureau et al. 2019; Cortés et al. 2019; Beć et al. 2020; Chapman et al. 2021; Eifert et al. 2020; Rolinger et al. 2020). These movements occur at specific energy levels where chemical bonds absorb IR at specific wavenumbers (MIR) or wavelengths (NIR) correlating with their different energy levels. Thus, infrared (IR) spectroscopy is the measure of this absorption, resulting in a spectrum with peaks representing the chemical bonds present in a given sample (Cen and He 2007; Walsh and Kawano 2009; Karoui et al. 2010; Weeranantanaphan et al. 2011; Liu et al. 2013; Wang et al. 2016; Arendse et al. 2018; Bureau et al. 2019; Cortés et al. 2019; Beć et al. 2020; Chapman et al. 2021; Eifert et al. 2020; Rolinger et al. 2020).

NIR has been established as an efficient analytical technique in several industries when combined with multivariate data methods and techniques (e.g. data mining and data processing), as the weaker bands in the NIR region are less intense than those in the MIR (between 10 and 100 times) (Cen and He 2007; Walsh and Kawano 2009; Karoui et al. 2010; Weeranantanaphan et al. 2011; Liu et al. 2013; Wang et al. 2016; Arendse et al. 2018; Bureau et al. 2019; Cortés et al. 2019; Beć et al. 2020; Chapman et al. 2021; Eifert et al. 2020; Rolinger et al. 2020). The NIR region of the spectrum is to be found between 700 and 2500 nm (Cen and He 2007; Walsh and Kawano 2009; Karoui et al. 2010; Weeranantanaphan et al. 2011; Liu et al. 2013; Wang et al. 2016; Arendse et al. 2018; Bureau et al. 2019; Cortés et al. 2019; Beć et al. 2020; Chapman et al. 2021; Eifert et al. 2020; Rolinger et al. 2020). The intrinsic characteristic of the NIR energy facilitates that samples might be analyzed directly without the need of preparation or processing (e.g. drying, homogenization, and grinding) (Cen and He 2007; Walsh and Kawano 2009; Karoui et al. 2010; Weeranantanaphan et al. 2011; Liu et al. 2013; Wang et al. 2016; Arendse et al. 2018; Bureau et al. 2019; Cortés et al. 2019; Beć et al. 2020; Chapman et al. 2021; Eifert et al. 2020; Rolinger et al. 2020).

The NIR spectra results from overtone and combination bands of fundamental vibrations of C—H, O—H, and N—H bonds originated in the infrared fundamentals region (Cen and He 2007; Walsh and Kawano 2009; Karoui et al. 2010; Weeranantanaphan et al. 2011; Liu et al. 2013; Wang et al. 2016; Arendse et al. 2018; Bureau et al. 2019; Cortés et al. 2019; Beć et al. 2020; Chapman et al. 2021; Eifert et al. 2020; Rolinger et al. 2020). However, the interpretation of NIR spectra can be hindered as a consequence of many overlapping bands and by the fact that an individual chemical or property might absorb at several wavelengths (Cen and He 2007; Walsh and Kawano 2009; Karoui et al. 2010; Weeranantanaphan et al. 2011; Liu et al. 2013; Wang et al. 2016; Arendse et al. 2018; Bureau et al. 2019; Cortés et al. 2019; Beć et al. 2020; Chapman et al. 2021; Eifert et al. 2020; Rolinger et al. 2020). Figure 1 shows the NIR spectra of fruit juice samples analyzed using a 1-mm path length cuvette. The NIR spectra of water, glucose, and fructose are also reported. The figure illustrates the complexity of the NIR spectra of a given sample compared with the spectra of its constituents.

The MIR region (4000–400 cm^{-1}) has been very important in the development of several applications in different fields (e.g. biology, environment, food, medicine, and pharmaceutical) (Cen and He 2007; Walsh and Kawano 2009; Karoui et al. 2010; Weeranantanaphan et al. 2011; Liu et al. 2013; Wang et al. 2016; Arendse et al. 2018; Bureau et al. 2019; Cortés et al. 2019; Beć et al. 2020; Chapman et al. 2021; Eifert et al. 2020; Rolinger et al. 2020) The MIR region is very important in routine analysis as it contains the so-called biochemical or chemical fingerprint region (1800–900 cm^{-1}) (Cen and He 2007; Walsh and Kawano 2009; Karoui et al. 2010; Weeranantanaphan et al. 2011; Liu et al. 2013; Wang et al. 2016; Arendse et al. 2018; Bureau et al. 2019; Cortés et al. 2019; Beć et al. 2020; Chapman et al. 2021; Eifert et al. 2020; Rolinger et al. 2020). This region is known to be the range of the electromagnetic spectrum where energy is absorbed by a number of common biochemical and

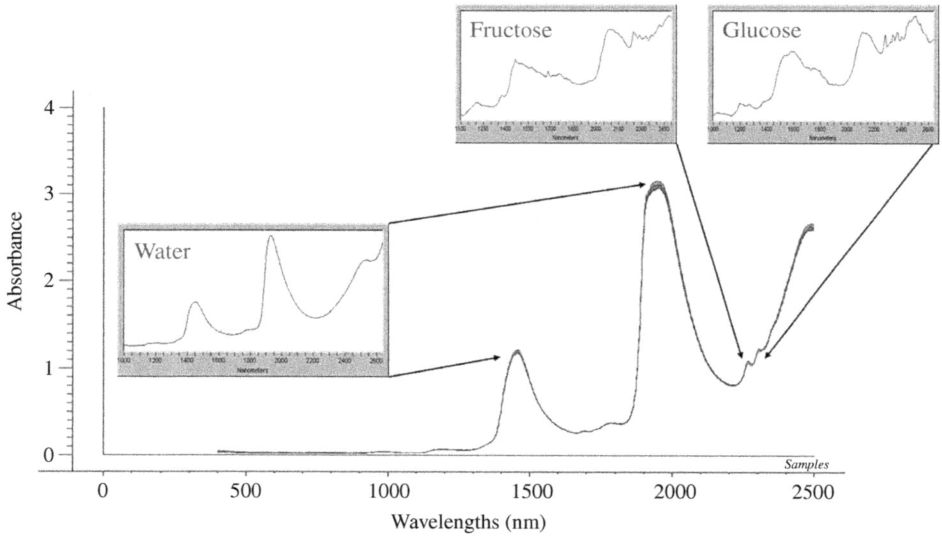

Figure 1 Complexity of the near-infrared spectra. The near-infrared transmission spectra of a fruit juice sample and its key components (water, glucose, and fructose).

chemical compounds present in complex sample matrices. Some examples include the Amide I, II, and III groups around 1650, 1550, and 1260 cm^{-1}, respectively; carbohydrates and polysaccharides around 1155 cm^{-1}, glycogen around 1030 cm^{-1}, both lipids and ester groups around 1750 cm^{-1}, asymmetric phosphate stretching vibrations (PO_2^-) around 1225 cm^{-1}, and symmetric phosphate stretching vibrations (PO_2^-; ~1080 cm^{-1}) and protein phosphorylation around 970 cm^{-1} (Cen and He 2007; Walsh and Kawano 2009; Karoui et al. 2010; Weeranantanaphan et al. 2011; Liu et al. 2013; Wang et al. 2016; Arendse et al. 2018; Bureau et al. 2019; Cortés et al. 2019; Beć et al. 2020; Chapman et al. 2021; Eifert et al. 2020; Rolinger et al. 2020). The range between 1500 and 400 cm^{-1} is defined as the fingerprint region (Cen and He 2007; Walsh and Kawano 2009; Karoui et al. 2010; Weeranantanaphan et al. 2011; Liu et al. 2013; Wang et al. 2016; Arendse et al. 2018; Bureau et al. 2019; Cortés et al. 2019; Beć et al. 2020; Chapman et al. 2021; Eifert et al. 2020; Rolinger et al. 2020). The absorption values or peaks in the fingerprint region are largely originated from the fundamental vibrations, which are predominantly sensitive to large wavenumber shifts, minimizing against unambiguous identification of specific functional groups (Cen and He 2007; Walsh and Kawano 2009 Karoui et al. 2010; Weeranantanaphan et al. 2011; Liu et al. 2013; Wang et al. 2016; Arendse et al. 2018; Bureau et al. 2019; Cortés et al. 2019; Beć et al. 2020; Chapman et al. 2021; Eifert et al. 2020; Rolinger et al. 2020).

3 Sample Presentation

In recent years, modern MIR instruments have incorporated sampling accessories such as attenuated total reflectance (ATR) (Cen and He 2007; Walsh and Kawano 2009; Karoui et al. 2010; Weeranantanaphan et al. 2011; Liu et al. 2013; Wang et al. 2016; Arendse et al. 2018; Bureau et al. 2019; Cortés et al. 2019; Beć et al. 2020; Chapman et al. 2021; Eifert et al. 2020; Rolinger et al. 2020). The incorporation of ATR as an accessory has enhanced the use of MIR spectroscopy in routine analysis, allowing for the development of new applications as it removes any requirement for sample preparation (Cen and He 2007; Walsh and Kawano 2009; Karoui et al. 2010; Weeranantanaphan et al. 2011; Liu et al. 2013; Cozzolino 2014; Wang et al. 2016; Arendse et al. 2018; Bureau et al. 2019; Cortés et al. 2019; Beć et al. 2020; Chapman et al. 2021; Eifert et al. 2020; Rolinger et al. 2020). The ATR cells allowed for the analysis of liquid and solid samples without the need to build KBr discs. Another commonly sampling method is the so-called diffuse reflectance infrared Fourier transform spectroscopy (DRIFTS). This method is usually applied to analysis powders and rough substances without any preparation (Cen and He 2007; Walsh and Kawano 2009; Karoui et al. 2010; Weeranantanaphan et al. 2011; Liu et al. 2013; Cozzolino 2014; Wang et al. 2016; Arendse et al. 2018; Bureau et al. 2019; Cortés et al. 2019; Beć et al. 2020; Chapman et al. 2021; Eifert et al. 2020; Rolinger et al. 2020).

Two main sample presentation modes are utilized to collect the NIR spectra of a given sample, namely reflectance and transmittance (Cen and He 2007; Walsh and Kawano 2009; Karoui et al. 2010; Weeranantanaphan et al. 2011; Liu et al. 2013; Cozzolino 2014; Wang et al. 2016; Arendse et al. 2018; Bureau et al. 2019; Cortés et al. 2019; Beć et al. 2020; Chapman et al. 2021; Eifert et al. 2020; Rolinger et al. 2020) (see Figure 2). In reflectance (e.g. generally used to analyze solids or powder samples), both the light source and detector

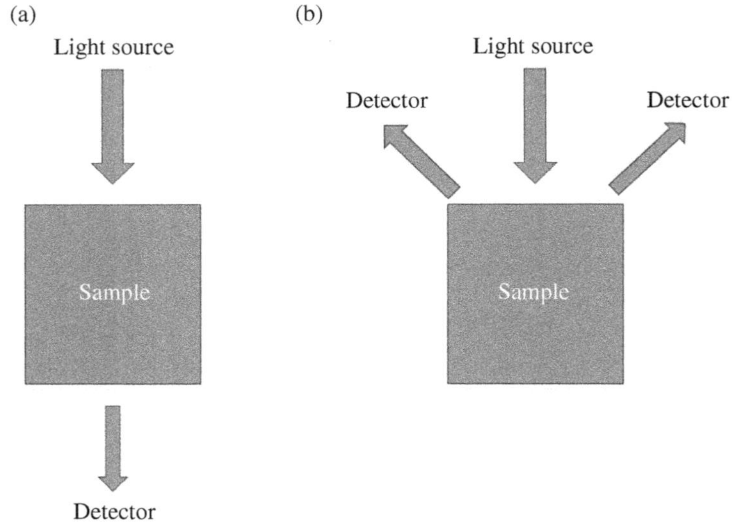

Figure 2 Spectra collection modes using in near-infrared spectroscopy analysis. (a) Transmission (b) reflectance. *Source:* Adapted from Cen and He (2007); Walsh and Kawano (2009); Karoui et al. (2010); Weeranantanaphan et al. (2011); Liu et al. (2013); Cozzolino (2014); Wang et al. (2016); Arendse et al. (2018); Bureau et al. (2019); Cortés et al. (2019); Beć et al. (2020); Chapman et al. (2021); Eifert et al. (2020); Rolinger et al. (2020).

are mounted under a specific angle (e.g. 45°, to avoid specular reflection). In transmittance (e.g. generally used to analyze liquid samples) mode, the light source is positioned opposite to the detector, while in interactance mode the light source and detector are positioned parallel to each other in such a way that light due to specular reflection cannot directly enter the detector (Karoui et al. 2010; Weeranantanaphan et al. 2011; Liu et al. 2013; Chapman et al. 2021). In modern instrumentation, both reflectance and transmittance modes integrated spheres may also be used to collect light and increase the signal-to-noise ratio. Transmission measurements, on the other hand, need very high light intensities which can easily burn the sample surface and alter its spectral properties (Karoui et al. 2010; Weeranantanaphan et al. 2011; Liu et al. 2013; Chapman et al. 2021; Cozzolino and Murray 2012; Porep et al. 2015).

4 Mid-Infrared Spectroscopy – Instrumentation

The use of the Fourier transform in MIR spectroscopy came into existence to overcome the drawbacks of the classic dispersive instruments (Karoui et al. 2010; Chapman et al. 2021). With the use of Fourier transform infrared (FTIR), travel through or be absorbed by the sample, depending upon the frequency and molecular structure in the sample matrix (Karoui et al. 2010; Chapman et al. 2021). To overcome the issues that possess the utilization of dispersive instruments, the Michelson interferometers have been incorporated in modern MIR instruments Karoui et al. 2010; Chapman et al. 2021). In the Michelson interferometer, the first step of processing is associated with splitting the light beam with a partial reflecting mirror surface (beam splitter) to facilitate the smooth travel in two different pathways before its recombined Karoui et al. 2010; Chapman et al. 2021). Once the light beam merges, the wave property of light causes constructive or destructive interference. If the passing length of the two beams is similar, the returning waves are in phase, and the interference is constructive. Thus, the amplitudes of the peaks and troughs in each sine wave are aligned, so that peak intensities are double when they superimposed additively in the new and recombinant wave (Karoui et al. 2010, Chapman et al. 2021). The path length in one arm of the interferometer is progressively shortened or lengthened using a moving mirror, one split beam must travel further than the other so that when they recombine, the pattern is destructive (Karoui et al. 2010; Chapman et al. 2021). Thereafter, the recombined beam passes through the sample before reaching the detector and the intensity of light at specific frequencies is reduced according to the nature and concentration of the absorbing species present (Karoui et al. 2010).

5 Near-Infrared Spectroscopy – Instrumentation

NIR spectrophotometers are conveniently classified according to the type of monochromators (Nicolai et al. 2007; Cozzolino 2015; Cozzolino and Roberts 2016; Pallone et al. 2018; Beć and Huck 2019; Beć et al. 2020). In a filter instrument, the monochromator is a wheel holding a number of absorption or interference filters, while its spectral resolution is limited (Nicolai et al. 2007; Cozzolino 2015; Cozzolino and Roberts 2016; Pallone et al. 2018; Beć and Huck 2019; Beć et al. 2020). On the other hand, in a scanning monochromator instrument, a grating or a prism is utilized to separate the individual frequencies of the radiation either entering or leaving the sample. The wavelength divider rotates allowing radiation of the individual wavelengths to reach the detector. In Fourier transform spectrophotometers, interferometers are used to generate modulated light, while the time domain signal of the light reflected or transmitted by the sample onto the sample can be converted into a spectrum via a fast Fourier transform (Nicolai et al. 2007; Cozzolino 2015; Cozzolino and Roberts 2016; Pallone et al. 2018; Beć and Huck 2019; Beć et al. 2020). Laser-based systems do not contain a monochromator but have different laser light sources or a tunable laser.

Acousto-optic tunable filter (AOTF) instruments use a diffraction-based optical band-pass filter that can be rapidly tuned to pass various wavelengths of light by varying the frequency of an acoustic wave propagating through an anisotropic crystal medium. In recent years, liquid crystal tunable filter (LCTF) instruments have been introduced (Nicolai et al. 2007; Cozzolino 2015; Cozzolino and Roberts 2016; Pallone et al. 2018; Beć and Huck 2019; Beć et al. 2020). They use a birefringent filter to create constructive and destructive interference based on the retardation, in phase between the ordinary and extraordinary light rays passing through a liquid crystal. In this way, they act as an interference filter to pass a single wavelength of light. By combining several electronically tunable stages in series, high spectral resolution can be achieved (Nicolai et al. 2007; Cozzolino 2015; Cozzolino and Roberts 2016; Pallone et al. 2018; Beć and Huck 2019; Beć et al. 2020).

6 Portability (Handheld Instruments)

Several studies highlight the ability of NIR spectroscopy to analyze different sample matrices using portable instrumentation (Sorak et al. 2012; O'Brien et al. 2012; Oliveira et al. 2014; Teixeira Dos Santos et al. 2013; Kaur et al. 2017; Crocombe 2018). In recent years, more attention has been given to the application of miniature and portable instrumentation, and this has led to the increased availability of small and precise handheld NIR spectrometers to the market (Sorak et al. 2012; O'Brien et al. 2012; Teixeira Dos Santos et al. 2013; Kaur et al. 2017; Wang et al. 2015; Crocombe 2018).

The main advantages (e.g. portability, flexibility, and easy to use) of this type of instrumentation provide the ability to develop novel applications targeting specific steps in a wide range of industries. In addition, these advantages make this instrumentation ideal for in field use to monitor and measure fruit composition (Cen and He 2007; Nicolai et al. 2007; Magwaza et al. 2012; Oliveira et al. 2014; Santos Neto et al. 2017; Arendse et al. 2018; Crocombe 2018).

Photodiode array (PDA) spectrophotometers have been introduced. They consist of a fixed grating that focuses the dispersed radiation onto an array of silicon (Si, 350–1100 nm) or Indium Gallium Arsenide (InGaAs, 1100–2500 nm) photodiode detectors (Karoui et al. 2010; Weeranantanaphan et al. 2011; Crocombe 2018; Chapman et al. 2021). A shift toward PDA systems due to their high acquisition speed (the integration time is typically 50 ms but can be as low as a few milliseconds) and the absence of moving parts.

Miniaturized, portable, and low cost versions (approximately less than $10 000 US) are available from companies, such as Ocean Optics (Dunedin, FL), Zeiss (Jena, Germany), Oriel (Stratford, CT), and Axsun Technologies (Billerica, MA) (Crocombe 2018; Chapman et al. 2021). Micro-electro-mechanical systems (MEMS) combine mechanical parts, sensors, actuators, and electronics on a common substrate through the use of microfabrication technology (Crocombe 2018). This technology represents a paradigm shift for industrial spectroscopy and enables a variety of new industrial applications. New spectrophotometers have been made using MEMS and became available in recent years (e.g. Axsun Technologies). Most applications of NIR spectroscopy which are described in the literature essentially rely on defined measurements (Crocombe 2018; Chapman et al. 2021).

7 Hyperspectral and Multispectral Image

Infrared spectroscopy applications have been primarily rely on spot measurements using either the NIR or visible (vis) regions of the electromagnetic spectrum (Gowen et al. 2007; Manley 2014; Cozzolino and Roberts 2016; Baiano 2017; Liu et al. 2017; Sendin et al. 2017; Hussain et al. 2018; Su and Sun 2018; Amodio et al. 2019; Lu et al. 2020). More recently, a diverse range of hyperspectral devices, including cameras and spectral imaging devices, are now readily available providing exciting new possibilities in food analysis and processing (Gowen et al. 2007; Manley 2014; Cozzolino and Roberts 2016; Baiano 2017; Liu et al. 2017; Sendin et al. 2017; Hussain et al. 2018; Su and Sun 2018; Amodio et al. 2019; Lu et al. 2020). This technology can acquire either single or multiple images at discrete wavelengths, presenting the potential for the detection of specific attributes relating to quality in an extensive range of raw materials and products using in the manufacture of foods (Gowen et al. 2007; Manley 2014; Cozzolino and Roberts 2016; Baiano 2017; Liu et al. 2017; Sendin et al. 2017; Hussain et al. 2018; Su and Sun 2018; Amodio et al. 2019; Lu et al. 2020). Overall, the availability of HSI and MSI systems allowed to obtain spatial, spectral, and multi-constituent information about the sample being analyzed (Gowen et al. 2007; Manley 2014; Cozzolino and Roberts 2016; Baiano 2017; Liu et al. 2017; Sendin et al. 2017; Hussain et al. 2018; Su and Sun 2018; Amodio et al. 2019; Lu et al. 2020).

Spectral imaging can be classified as either HSI or MSI (Gowen et al. 2007; Manley 2014; Cozzolino and Roberts 2016; Baiano 2017; Liu et al. 2017; Sendin et al. 2017; Hussain et al. 2018; Su and Sun 2018; Amodio et al. 2019; Lu et al. 2020). MSI involves the acquisition of spectral images at few discrete and narrow wavebands (bandwidths of between 5 and 50 nm) and it is considered an improvement of HSI as this technology is cost-effective. The result is the ability of MSI to simultaneously predict multiple components, providing a key advantage and rendering a promising future outlook with a single, automated image acquisition (Gowen et al. 2007; Manley 2014; Cozzolino and Roberts 2016; Baiano 2017; Liu et al. 2017; Sendin et al. 2017; Hussain et al. 2018; Su and Sun 2018; Amodio et al. 2019; Lu et al. 2020). HSI relies on one of two sensing modes as in-line scanning (push broom) mode or as filter-based imaging mode (Gowen et al. 2007; Manley 2014; Cozzolino and Roberts 2016; Baiano 2017; Liu et al. 2017; Sendin et al. 2017; Hussain et al. 2018; Su and Sun 2018; Amodio et al. 2019; Lu et al. 2020). In-line scanning mode involves the imaging system scanning moving product items. This results in three-dimensional (3D) hyperspectral images, also called hypercubes (Gowen et al. 2007; Manley 2014; Cozzolino and Roberts 2016; Baiano 2017; Liu et al. 2017; Sendin et al. 2017; Hussain et al. 2018; Su and Sun 2018; Amodio et al. 2019; Lu et al. 2020). In filter-based imaging mode, spectral images are obtained from stationary targets for a waveband sequence using either a LCTF or an AOTF (Gowen et al. 2007; Manley 2014; Cozzolino and Roberts 2016; Baiano 2017; Liu et al. 2017; Sendin et al. 2017; Hussain et al. 2018; Su and Sun 2018; Amodio et al. 2019; Lu et al. 2020). In-line scanning is the most common as it is relatively easy to implement, particularly for real-time and online applications. In comparison, filter-based HSI systems are not suitable for online applications and require more complicated calibrations (Gowen et al. 2007; Manley 2014; Cozzolino and Roberts 2016; Baiano 2017; Liu et al. 2017; Sendin et al. 2017; Hussain et al. 2018; Su and Sun 2018; Amodio et al. 2019; Lu et al. 2020).

HSI methods require a high-performance digital camera that has the ability to cover the spectral region of interest, in a wide dynamic range with high signal-to-noise ratio, and good quantum efficiency (Gowen et al. 2007; Manley 2014; Cozzolino and Roberts 2016; Baiano 2017; Liu et al. 2017; Sendin et al. 2017; Hussain et al. 2018; Su and Sun 2018; Amodio et al. 2019; Lu et al. 2020). An imaging spectrograph, with the ability to disperse the line images into different wavelengths, is essential (Gowen et al. 2007; Manley 2014; Cozzolino and Roberts 2016; Baiano 2017; Liu et al. 2017; Sendin et al. 2017; Hussain et al. 2018; Su and Sun 2018; Amodio et al. 2019; Lu et al. 2020). Optical resolution and spectral response efficiency with minimal aberrations are also key features of a HSI system. Finally, a highly stable, DC-regulated light source with a smooth spectral response is critical (Gowen et al. 2007; Manley 2014; Cozzolino and Roberts 2016; Baiano 2017; Liu et al. 2017; Sendin et al. 2017; Hussain et al. 2018; Su and Sun 2018; Amodio et al. 2019; Lu et al. 2020).

HSI is particularly attractive for food applications because of the integration of imaging ability with spectroscopy, which enables simultaneous acquisition of both spectral and spatial information from the target (Amigo et al. 2013; Cheng et al. 2013, 2017). The gains achieved by this amalgamation can allow the monitoring of highly spatially variable raw and food products (Gowen et al. 2007; Manley 2014; Cozzolino and Roberts 2016; Baiano 2017; Liu et al. 2017; Sendin et al. 2017; Hussain et al. 2018; Su and Sun 2018; Amodio et al. 2019; Lu et al. 2020). However, this technology is not widely used by the industry, with key considerations to user uptake depending on several reasons that include cost and instrument availability, whether the application is online or in-field and training of staff (Gowen et al. 2007; Manley 2014; Cozzolino and Roberts 2016; Baiano 2017; Liu et al. 2017; Sendin et al. 2017; Hussain et al. 2018; Su and Sun 2018; Amodio et al. 2019; Lu et al. 2020).

8 Conclusions and Outlook

As a fast and easy-to-operate technique, IR spectroscopy has already gained wide acceptance for routine analysis in several industries such as pharmaceutical, food, and beverage, and petrochemical. Considering the continuing improvements in hardware and software design, and the analytical requirements of real-time or multi-variable analysis by modern industry, it is expected that IR spectroscopy will progressively become a routine method for process monitoring and process control in a wide range of raw materials, ingredients, and products.

Several critical aspects and limitations, for example instrument availability and price, type of application and understanding of the technology, might act as barriers for the wide adoption of IR technologies by the industry. Many IR instrument companies in the market provide with a variety of spectrophotometers including bench top, online, and portable instruments. Selection of a suitable IR spectrophotometer instrument for purchase is an important investment decision. In the process of considering the purchase of a spectrophotometer, it is necessary to define some fundamental specifications, for example wavelength (wavenumber) scanning range, wavelength data point interval (number of absorption data points), noise, stability, and measurement time and the type of sample to be analyzed or the application. Spectrophotometers for industrial applications should be a simple device that will be able to collect, store and retrieve data, and provide with easy to use instrument control software. Other important advantages of these techniques over the traditional chemical and chromatographic methods are the rapidity and ease of use in routine operations. Moreover, IR is a nondestructive technique which requires minimal or zero sample preparation. Therefore, molecular spectroscopy can be suggested as the first line of tools to be used in natural products research and industry applications. Instrumental cost and complexity of IR spectroscopy methods are reduced compared with classical chromatographic methods.

One of the most important critical aspects of the development of IR spectroscopy is the need for appropriate training. Routine analyses can be performed by an analyst with a high school education level. However, knowledge of the chemistry, physics, and mathematics will be needed to further application development. This has become the bottleneck for the further development of these technologies.

Regardless of these issues, both MIR and NIR spectroscopy have become the workhorse technologies to measure composition, faults, and overall quality in a wide range of products and samples as well as tools in the so-called online analytics (process monitoring) in many industrial applications.

Acknowledgments

The support of the University of Queensland and the Department of Agriculture and Food of the Queensland Government are acknowledged.

Conflict of Interest

The author declares no conflict of interest.

References

Amigo, J.M., Martí, I., and Gowen, A. (2013). Hyperspectral imaging and chemometrics: a perfect combination for the analysis of food structure, composition and quality. In: *Data Handling in Science and Technology*, 1e, vol. 28 (ed. F. Marini), 343–370. Elsevier.

Amodio, M.L., Chaudhry, M.M.A., and Colelli, G. (2019). Spectral and hyperspectral technologies as an additional tool to increase information on quality and origin of horticultural crops. *Agronomy* 10: 7. https://doi.org/10.3390/agronomy10010007.

Arendse, E., Fawole, O.A., Magwaza, L.S., and Opara, U.L. (2018). Non-destructive prediction of internal and external quality attributes of fruit with thick rind: a review. *Journal of Food Engineering* 217: 11–23.

Baiano, A. (2017). Applications of hyperspectral imaging for quality assessment of liquid based and semi-liquid food products: a review. *Journal of Food Engineering* 214: 10–15.

Beć, K.B. and Huck, C.W. (2019). Breakthrough potential in near infrared spectroscopy; spectra simulation. A review of recent developments. *Frontiers in Chemistry* 7: 48.

Beć, K.B., Grabska, J.E., and Huck, C.W. (2020). Near-infrared spectroscopy in bio-applications. *Molecules* 25: 2948. https://doi.org/10.3390/molecules25122948.

Bureau, S., Cozzolino, D., and Clark, C.J. (2019). Contributions of Fourier-transform mid infrared (FT-MIR) spectroscopy to the study of fruit and vegetables: a review. *Postharvest Biology and Technology* 148: 1–14.

Cen, H. and He, Y. (2007). Theory and application of near infrared reflectance spectroscopy in determination of food quality. *Trends in Food Science & Technology* 18: 72–83.

Chapman, J., Power, A., Netzel, M.E. et al. (2021). Challenges and opportunities of the fourth revolution: a brief insight into the future of food. *Critical Reviews in Food Science and Nutrition* https://doi.org/10.1080/10408398.2020.1863328.

Cheng, J.-H., Dai, Q., Sun, D.-W. et al. (2013). Applications of non-destructive spectroscopic techniques for fish quality and safety evaluation and inspection. *Trends in Food Science & Technology* 34 (1): 18–31.

Cheng, J.-H., Nicolai, B., and Sun, D.-W. (2017). Hyperspectral imaging with multivariate analysis for technological parameters prediction and classification of muscle foods: a review. *Meat Science* 123: 182–191.

Cortés, V., Blasco, J., Aleixos, N. et al. (2019). Monitoring strategies for quality control of agricultural products using visible and near-infrared spectroscopy: a review. *Trends in Food Science & Technology* 85: 138–148.

Cozzolino, D. (2014). Sample presentation, sources of error and future perspectives on the application of vibrational spectroscopy in the wine industry. *Journal of the Science of Food and Agriculture* 95: 861–868.

Cozzolino, D. (2015). Foodomics and infrared spectroscopy: from compounds to functionality. *Current Opinion in Food Science* 4: 39–43.

Cozzolino, D. and Murray, I. (2012). A review on the application of infrared technologies to determine and monitor composition and other quality characteristics in raw fish, fish products, and seafood. *Applied Spectroscopy Reviews* 47 (3): 207–218.

Cozzolino, D. and Roberts, J. (2016). Applications and developments on the use of vibrational spectroscopy imaging for the analysis, monitoring and characterisation of crops and plants. *Molecules* 21: 755. https://doi.org/10.3390/molecules21060755.

Crocombe, R.A. (2018). Portable spectroscopy. *Applied Spectroscopy* 72: 1701–1751.

dos Santos Neto, J.P., de Assis, M.W.D., Casagrande, I.P. et al. (2017). Determination of 'Palmer' mango maturity indices using portable near infrared (VIS-NIR) spectrometer. *Postharvest Biology and Technology* 130: 75–80.

Eifert, T., Eisen, K., Maiwald, M., and Herwig, C. (2020). Current and future requirements to industrial analytical infrastructure – part 2: smart sensors. *Analytical and Bioanalytical Chemistry* 412: 2037–2045.

Gowen, A.A., O'Donnell, C.P., Cullen, P.J. et al. (2007). Hyperspectral imaging – an emerging process analytical tool for food quality and safety control. *Trends in Food Science & Technology* 18: 590–598.

Hussain, A., Pu, H., and Sun, D.-W. (2018). Innovative non-destructive imaging techniques for ripening and maturity of fruits – a review of recent applications. *Trends in Food Science & Technology* 72: 144–152.

Karoui, R., Downey, G., and Blecker, C. (2010). Mid-infrared spectroscopy coupled with chemometrics: a tool for the analysis of intact food systems and the exploration of their molecular structure–quality relationships – a review. *Chemical Reviews* 10 (10): 6144–6168.

Kaur, H., Kunnemeyer, R., and McGlone, A. (2017). Comparison of hand-held near infrared spectrophotometers for fruit dry matter assessment. *Journal of Near Infrared Spectroscopy* 25 (4): 267–277.

Liu, D., Zeng, X.-A., and Sun, D.-W. (2013). NIR spectroscopy and imaging techniques for evaluation of fish quality – a review. *Applied Spectroscopy Reviews* 48 (8): 609–628.

Liu, Y., Pu, H., and Sun, D.-W. (2017). Hyperspectral imaging technique for evaluating food quality and safety during various processes: a review of recent applications. *Trends in Food Science & Technology* 69: 25–35.

Lu, Y., Saeys, W., Kim, M. et al. (2020). Hyperspectral imaging technology for quality and safety evaluation of horticultural products: a review and celebration of the past. *Postharvest Biology and Technology* 170: 1–19.

Magwaza, L.S., Opara, U.L., Nieuwoudt, H. et al. (2012). NIR spectroscopy applications for internal and external quality analysis of citrus fruit—A review. *Food Bioprocess Technology* 5: 425–444. https://doi.org/10.1007/s11947-011-0697-1.

Manley, M. (2014). Near-infrared spectroscopy and hyperspectral imaging: non-destructive analysis of biological materials. *Chemical Society Reviews* 43: 8600–8620.

Nicolai, B.M., Beullens, K., Bobelyn, E. et al. (2007). Non-destructive measurement of fruit and vegetable quality by means of NIR spectroscopy: a review. *Postharvest Biology and Technology* 46: 99–118.

O'Brien, N.A., Hulse, C.A., Friedrich, D.M. et al. (2012). Miniature near-infrared (NIR) spectrometer engine for handheld applications. In: *Next-Generation Spectroscopy Technology V*, Proceedings of SPIE, vol. 8374 (eds. M.A. Druy and R.A. Crocombe), 3–9.

de Oliveira, G.A., Bureau, S., Renard, C.M.-G.C. et al. (2014). Comparison of NIRS approach for prediction of internal quality traits in three fruit species. *Food Chemistry* 143: 223–230.

Pallone, J.A.L., Caramês, E.T.S., and Alamar, P.D. (2018). Green analytical chemistry applied in food analysis: alternative techniques. *Current Opinion in Food Science* 22: 115–121.

Pasquini, C. (2018). Near infrared spectroscopy: a mature analytical technique with new perspectives – a review. *Analytica Chimica Acta* 1026: 8–36.

Porep, J.U., Kammerer, D.R., and Carle, R. (2015). On-line application of near infrared (NIR) spectroscopy in food production. *Trends in Food Science & Technology* 46 (2): 211–230.

Rolinger, L., Rudt, N., and Hubbuch, J. (2020). A critical review of recent trends, and a future perspective of optical spectroscopy as PAT in biopharmaceutical downstream processing. *Analytical and Bioanalytical Chemistry* 412: 2047–2064.

Sendin, K., Williams, P.J., and Manley, M. (2017). Near infrared hyperspectral imaging in quality and safety evaluation of cereals. *Critical Reviews in Food Science and Nutrition* 58 (4): 575–590.

Sorak, D., Herberholz, L., Iwascek, S. et al. (2012). New developments and applications of handheld Raman, mid-infrared, and near infrared spectrometers. *Applied Spectroscopy Reviews* 47: 83–115.

Su, W.-H. and Sun, D.-W. (2018). Fourier transform infrared and Raman and hyperspectral imaging techniques for quality determinations of powdery foods: a review. *Comprehensive Reviews in Food Science and Food Safety* 17 (1): 104–122.

Teixeira Dos Santos, C.A., Lopo, M., Páscoa, R.N.M.J., and Lopes, J.A. (2013). A review on the applications of portable near-infrared spectrometers in the agro-food industry. *Applied Spectroscopy* 67: 1215–1233.

Walsh, K.B. and Kawano, S. (2009). Near infrared spectroscopy. In: *Optical Monitoring of Fresh and Processed Agricultural Crops* (ed. M. Zude), 192–239. Boca Raton, FL: CRC Press.

Wang, H., Peng, J., Xie, C. et al. (2015). Fruit quality evaluation using spectroscopy technology: a review. *Sensors* 15: 11889–11927.

Wang, L., Sun, D.-W., Pu, H., and Cheng, J.-H. (2016). Quality analysis, classification, and authentication of liquid foods by near-infrared spectroscopy: a review of recent research developments. *Critical Reviews in Food Science and Nutrition* 57: 1524–1538.

Weeranantanaphan, J., Downey, G., Allen, P., and Sun, D.-W. (2011). A review of near infrared spectroscopy in muscle food analysis: 2005–2010. *Journal of Near Infrared Spectroscopy* 19 (2): 61–104.

13

Raman Spectroscopy

Dana Alina Magdas and Camelia Berghian-Grosan

National Institute for Research and Development of Isotopic and Molecular Technologies, Cluj-Napoca, Romania

1 Introduction

Raman spectroscopy is based on the inelastic scattering of monochromatic light (laser beam) by the molecules of the investigated samples. Most of the incident radiation is elastically scattered (Rayleigh scattering), with the same frequency and just a small fraction of it is inelastically scattered (Raman scattering). Raman scattering can appear in two ways as function of the initial energy state of the molecules. Thus, when the incident light interacts with molecules being in the ground state, the frequency of the scattered light is lower in comparison to the incident one (Stokes Raman scattering). If the molecules are already in an excited state, the frequency of the scattered radiation will be higher as compared to the initial one (anti-Stokes Raman scattering). In most cases, the anti-Stokes-Raman scattering is generally smaller than Stokes-Raman scattering because for most of the samples their molecules are initially in their ground state.

In the last years, an increased interest for the development of rapid and reliable approaches for food and beverages control was recorded from research and control laboratories. Also, a high number of samples that are analyzed for control purposes impose the implementation of affordable analytical methods from economical point of view. This tendency went hand in hand with the development of portable devices that allow the on-site verification and also require minimal experimental skills of the users. Apart of this, the requirement for the application of so called "green" approaches which need from a minimal use of solvents to none became of high priority. All these simultaneous requirements are perfectly achieved by the approach, which include the association between Raman spectroscopy and either statistical data processing or application of machine learning (ML) algorithms and artificial intelligence tools. In order to extract the highest information content from Raman spectra, the use of advanced data processing tools became mandatory especially due to the big data sets, which are generated but also because of the subtle changes which appear sometimes among the sample categories which are compared.

Electromagnetic Technologies in Food Science, First Edition. Edited by Vicente M. Gómez-López and Rajeev Bhat.

2 Raman Applications in Food and Beverages Studies

Despite the high potential of Raman spectroscopy for food and beverages investigations, this analytical technique represents an emerging one for this type of studies. Nevertheless, its proved potential conducted to an increased attention among scientific community and we are convinced that this tendency will be kept in the future. Among the food and beverage matrices which were investigated through Raman spectroscopy a special attention was given to: honey, edible oils, wines, and fruit spirits. This technique was mainly applied for the main component's quantification but also for differentiation either with respect to geographical and botanical origin or to vintage (for wines). The main studies which were performed for these matrices are hereafter discussed.

2.1 Honey

Honey represents one of the most expensive form of sugar, being easy to be adulterated and as a consequence is one of the most falsified food commodities in the world. Especially at the EU level, the increased demand of honey on the market is not satisfied by the internal production; therefore, important quantities of honey are imported from external markets, especially from China.

The rareness of honey having some specific geographical or floral origins can be reflected in a significant price increase; therefore, the temptation of some producers or sellers to bring in the market mislabeled honey was recorded (Magdas et al. 2021). In order to discover or prevent these unfair practices, a very strict control, based on reliable analytical approaches, should be enforced. For this reason, in the last years, a special attention was given by researchers for the development of new analytical approaches that could distinguish unequivocally among distinct honey classes. From these analytical tools, it can be found: chromatographic methods (Chan et al. 2013), stable isotope analysis (Rogers et al. 2010), isotope ratios and metal concentration measurements (Magdas et al. 2021), and proton nuclear magnetic resonance (NMR) profiling (Guyon et al. 2020).

The application of Raman spectroscopy in honey studies was successfully used during the last years for different aims like the identification and quantification of some specific compounds and also to estimate the chemical composition of different honey types, Table 1. For these purposes, different experimental set-ups were developed or used. Among these, we can mention here: different sample preparation step or the use of distinct laser lines. For instance, the 1064 nm laser line was used in some studies with the purpose of adulteration detection, physicochemical characterization, and also to assess honey origin (Pierna et al. 2011; Paradkar and Irudayaraj 2001; Anjos et al. 2018). Other authors used the 532 nm laser excitation line, either with the aim of origin discrimination of distinct honey types (Corvucci et al. 2015) or as a tool to estimate the sugar composition in honey and similar sugar mixtures (Sugar and Bour 2016). The 785 nm laser line was used not only to quantify the sugar components in honey samples (Bogdanov 2011) but also to detect the honey adulteration (Isengard et al. 2001).

For the successful application of Raman spectroscopy in honey studies, it should be taking into account the limitation of this method and to find the most suitable countermeasures to overcome them. In the honey case, the main issues are related to the different

Table 1 Brief review of literature data on honey investigation by Raman spectroscopy.

Study	Method	References
Discrimination and classification of beet and cane inverts in honey	FT-Raman 1064 nm Partial least squares and principal component regression analysis were used for quantitative analysis Linear discriminant analysis and canonical variate analysis (CVA) was used for discriminant analysis	Paradkar and Irudayaraj (2001)
Simultaneous determination of fructose and glucose in honey	FT-Raman 1064 nm is combined with a PLS (partial least squares) method (TQ Analyst software) The FT-Raman procedure is evaluated with a standard HPLC-based method	Batsoulis et al. (2005)
Discrimination of Corsican honey from other honey samples	FT-Raman 1064 nm combined with chemometric methods The developed models include the use of the Fisher criterion for wavenumber selection and methods as partial least squares-discriminant analysis (PLS-DA) or support vector machines (SVM)	Pierna et al. (2011)
Detection of honey adulteration by high fructose corn syrup and maltose syrup	Raman 785 nm combined with chemometric methods Partial least squares-linear discriminant analysis (PLS-LDA) was used to develop a binary classification model	Li et al. (2012)
Rapid analysis of sugar in honey. Quantitatively determination of glucose, fructose, sucrose and maltose contents in honey samples without any treatment	Raman 785 nm combined with chemometric methods and artificial neural network (ANN) Principal component analysis (PCA) and partial least squares (PLS) were the chemometric tools	Özbalci et al. (2013)
Discrimination of honey origin using melissopalynology and Raman spectroscopy techniques coupled with multivariate analysis	FT-microRaman 532 nm combined with chemometric methods FT-microRaman in combination with proper PCA models could be successfully adopted to identify the botanical origins Melissopalynologic chemometric model was used for comparison	Corvucci et al. (2015)
Quantitative analysis of sugar composition in honey	Raman 532-nm and Raman optical activity spectra The samples were pre-purified by active carbon	Sugar and Bour (2016)
Rapid prediction of phenolic compounds and antioxidant activity of Sudanese honey	Fourier-transform infrared spectroscopy–attenuated total reflection (FTIR–ATR) and Raman spectroscopy were combined with partial least square regression (PLSR) Raman laser 532 nm	Tahir et al. (2017)
Physicochemical characterization of *Lavandula* spp. honey	FT-Raman 1064 nm Partial least squares (PLS) regression models were performed for the quantitative estimation; correlation with reference methods	Anjos et al. (2018)

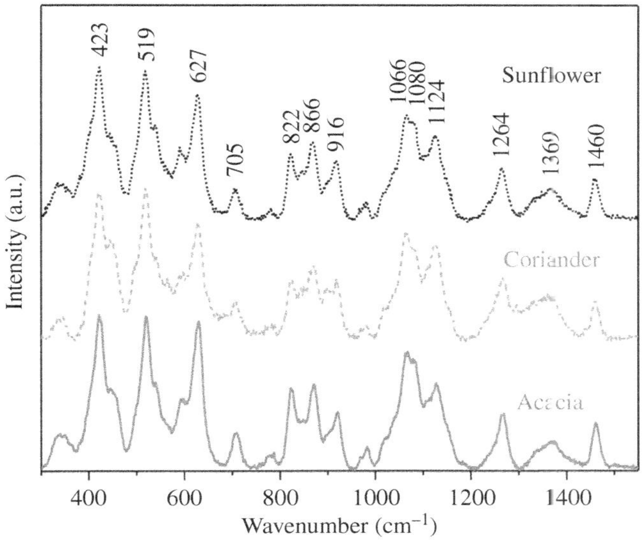

Figure 1 Raman spectra of some honey varieties (acacia, coriander, sunflower) obtained with the 785 nm wavelength laser excitation.

crystallization degrees which can not only affect the shape of Raman spectra and also the fluorescence which appear for some dark color samples.

In order to simultaneously overcome the issues related to crystallization and fluorescence, a preparation method of honey before the Raman analysis was developed and validated by Muller-Molnar et al. (2020). Both issues were overcome through a simple preparation method consisting in the dilution (ratio 1 : 1 w/v) honey versus distilled water. Moreover, a stability tested was performed for the 1 : 1 w/v honey–water solution. It was proved that the obtained solution was stable in time for three days, and during this period no modification of Raman spectra were recorded.

Using this method, the Raman spectra of several honey types were recorded (Figure 1), and their band assignments are presented in Table 2.

Raman spectroscopy in conjunction with soft independent modeling class analogy (SIMCA) and ML algorithms was successfully applied (Figure 2) for the differentiation of 90 mono-floral honey samples originated from two countries, Romania and France. The investigated samples belonged to distinct varieties (acacia, linden, colza, sunflower, coriander, lavender, chestnut, gallium verum, thyme, and raspberry) (Magdas et al. 2020). For this study, the honey samples were prepared according to the previously described method (Muller-Molnar et al. 2020) and then Raman spectra were recorded. The experimental preprocessed data were further used for the development of recognition models using SIMCA method and machine learning algorithms (Figure 2).

The approach allowed a good separation among investigated groups especially for variety differentiation. Thus, obtained separation among investigated honey varieties (acacia, chestnut, colza, lavender, linden, and sunflower) was realized in a percentage of 100% for most of honey varieties, except linden honey, where a separation of 83% was achieved.

Table 2 Main Raman peak positions of honey and their proposed assignments according to literature available data.

Peak position (cm^{-1})	Vibrational assignments
423	δ(C–C–O) of glucose which is covered with the band at $353\,cm^{-1}$, which is associated to the δ(C–C–C) ring vibration from carbohydrates
519	δ(C–C–O) and δ(C–C–C) of carbohydrates
627	Ring deformation
705	ν(C–O), ν(C–C–O), δ(O–C–O)
822	δ(C–H) and δ(CH_2), δ(C–O–H)
866	
917	δ(C–H) and δ(C–OH)
1066	υ(C–O)
1080	δ(C–H) and δ(C–O–H) in carbohydrates and δ(C–N) in proteins and amino acids
1124	υ(C–O) and υ(C–O–C) in carbohydrates and υ(C–N) from proteins and amino acids
1264	υ(C–O–H), υ (amide III)
1369	δ(O–H), δ(C–H)
1460	δ(CH_2) and δ(CCO^- of flavanols and organic acids)

Source: Adapted from Corvucci et al. (2015), Anjos et al. (2018); de Oliveira et al. (2002), Pierna et al. (2011), Özbalci et al. (2013), Tahir et al. (2017), and Mathlouthi et al. (1980).

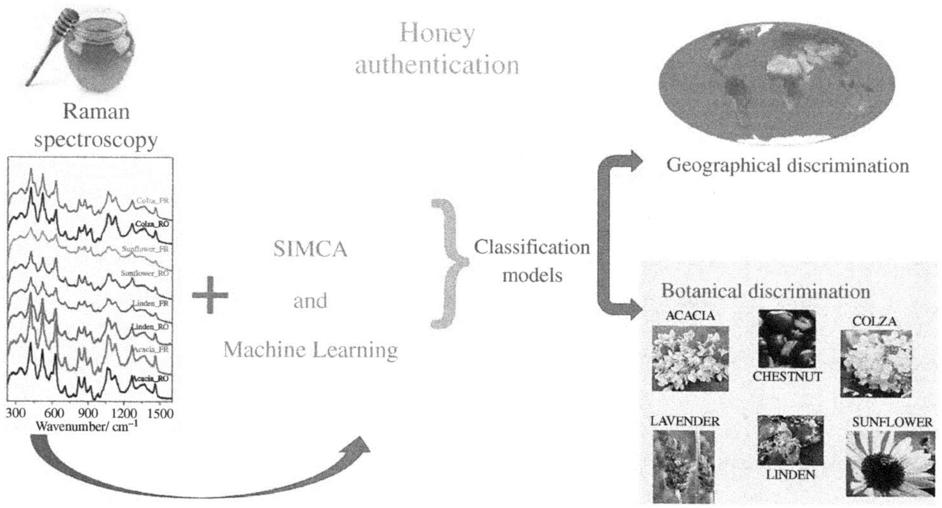

Figure 2 Schematic representation of the used approach for honey classification. (*See insert for color representation of this figure*).

Although very good separation percentages were obtained, the interclass distance was quite low, being situated between 0.4 and 3.1, and it did not allow the use of this approach for honey control purpose. This because for the most of compared classes the achieved interclass distance was lower than required one, which is three. For the differentiation between the two honey samples groups originated from Romania and France a good separation was also obtained (100% and 89%, respectively) but in this case the obtained interclass distance was even lower (0.3).

It was observed a good correlation between the predictions models obtained based on SIMCA and those built on ML algorithms, proved by the fact that most of the honey varieties which were misclassified with ML where those for which in SIMCA model a low interclass distance was obtained. For geographical differentiation of Romanian samples from the French ones, both models provided similar results. A perfect classification was achieved for Romanian samples, while for French samples, some of them were wrong attributed. A possible explanation in this regard could be related by the fact that in geographical model two honey varieties specific only for French samples were employed (lavender and chestnut). Based on the obtained interclass distances, it was proved that Raman spectroscopy is mainly useful for variety differentiation, a varietal effect can overlap the geographical recognition, explaining the wrong classification of the French samples (Magdas et al. 2020).

Due to the fluorescence and crystallization phenomena, this sweet matrix represents a real challenge for researchers if Raman spectroscopy is envisaged for investigation. The development of simple methods for sample preparation can achieved a rapid evaluation of honey. Based on the reported results from literature, it was clearly proved that Raman spectroscopy is a useful tool for the detection and quantification of main honey constituents.

Moreover, the combination of this technique with chemometric and machine learning methods allows the differentiation of honey with respect to its varietal and geographical origin. In this regard, its proved potential for varietal recognition is superior to the geographical one. Therefore, the developed methods give hope for a rapid application of Raman in honey-control analysis.

2.2 Edible Oils

Together with the carbohydrates and proteins, the edible oils are very important for the human diet, being a major source of energy (9 kcal g^{-1}) and linoleic (18,2)/linolenic (18,3) acids (Endo 2018). The edible oils are vegetable oils obtained from plants, especially from seeds and beans while, some of them, as that of olive and palm, are extracted from fruits (Chemat 2017).

The main constituents of the oils are the triglycerides or triacylglycerols (consisting of three molecules of fatty acids (FAs) bonded to a "backbone" – glycerol); they are also accompanied by lower amounts of diglycerides (diacylglycerols), monoglycerides (monoacylglycerols), or free fatty acids (Chemat 2017). The particularities of the different type of the vegetable oils are mainly linked to the FAs composition which is complex (Porta 2006).

To these constituents, several minor compounds as sterols, tocopherols, chlorophylly pigments, and phospholipids can be found, depending of the botanical origin of the seeds/fruits in the edible oils (Endo 2018). These minor compounds can also affect the nutritional values and the oxidative stability of the oils.

Considering the oils composition, some of the chemical characteristics of the edible oils can be used as an indication of their nutritional and quality values. These characteristics are acid and saponification values, fatty acid composition, the trans isomers of the unsaturated FA, triacylglycerol composition, the minor components composition, the presence of synthetic antioxidants added during the technological process, or the presence of undesirable toxic metals (Endo 2018). A series of analytical methods (Codex 2019) as titration (Kardash and Turyan 2005), potentiometric (Kardash-Strochkova et al. 2001), colorimetric (Bartheta et al. 2008), gas chromatography (GC) (AOCS Recommended Practice Ca 5d-01 2001), high-performance liquid chromatography (HPLC) (Endo et al. 2011), or liquid chromatography coupled with mass spectrometry (LC–MS) (Masukawa et al. 2010) are used for the determination of the chemical characteristics and oils quality. To these, nondestructive analytical methods such as NMR (Parker et al. 2014) or other spectroscopic methods like vibrational (infrared [IR], Raman), near infrared (NIR) (Bewig et al. 1994), terahertz (Tz) (Jiang et al. 2012) are often used for characterization and different oils properties determination. Among them, over the years, Raman spectroscopy has proved his efficiency for various investigations of the edible oils such as analysis of the cis/trans isomer composition (Bailey and Horvat 1972; Johnson et al. 2002), total unsaturation (Sadeghi-Jorabchi et al. 1990; Barthus and Poppi 2001), or free fatty acids content (Muik et al. 2003) determination, direct monitoring of lipid oxidation (Muik et al. 2005). Thus, classification of oils and fat (Baeten et al. 1998), authentication (Marigheto et al. 1998), or adulteration detection of different oils varieties has been realized for this important food matrix. More details about the applications of vibrational spectroscopy for the edible oils' analysis can be found in Baranska et al. (2010) and Jimenez-Sanchidrian and Ruiz (2016). Being a complex matrix, the Raman spectroscopy offers large spectral data, to obtain good qualitative and quantitative results, chemometric methods were also associated to Raman spectra for an accurate investigation of oils (Baeten and Aparicio 2000; Zhang et al. 2011).

Below are detailed some of the main applications of Raman spectroscopy in the oils' investigations:

i) *The determination of the degree of unsaturation* is one of the applications for which the Raman spectroscopy can be used in the food industry. The level of unsaturation is of significant importance from nutritional point of view. The unsaturation degree gives an indication about the double bonds number containing in the fatty acids chain; the Raman spectroscopic results showed good correlation with those obtained by gas chromatography or chemical methods, proving that Raman spectroscopy is a valuable alternative technique for this type of analysis (Sadeghi-Jorabchi et al. 1990, 1991; Barthus and Poppi 2001; Farhad et al. 2009; Carmona et al. 2014; Kim et al. 2014). Two regions of the Raman spectra can be used to obtain good results for the degree of unsaturation determination: the 800–1800 cm^{-1} and 2800–3050 cm^{-1} (Baeten et al. 1996; Carmona et al. 2014). Some of the main peaks of the common vegetable oils (i.e. sunflower, olive, walnut, etc.) are located in these regions.

Due to the fact that, for the most of the vegetable oils, the composition in the fatty acids is comparable, the bands assignments have been realized by considering mainly the investigations on the olive oils and the lipid profiles (Baeten et al. 1996, 1998; Sadeghi-Jorabchi et al. 1991). In the first region, the peaks from about 1263, 1301 1440,

1656, and 1748 cm^{-1} can be associated with the in-plane = C—H deformation vibration of the unconjugated cis double bond (—CH=CH—), in-phase CH_2 twisting vibrations, CH_2 scissoring deformation – $\delta(CH_2)$, C=C stretching – ν(C=C) from the fatty acids chain, and the stretching vibration of the C=O bond from the glyceryl esters, respectively (Baeten et al. 1998; Sadeghi-Jorabchi et al. 1991). Some authors used the ratio between the peak area of the 1656 and 1748 cm^{-1} and found some correlation with the degree of saturation (inversely proportional) or the iodine value (directly proportional); the same methodology can be considered for the ratio between the peak area of 1656 and 1440 cm^{-1} or 1301 and 1263 cm^{-1} bands (Sadeghi-Jorabchi et al. 1990, 1991; Baeten et al. 1998; Baranska et al. 2010 and herein citations). In the second region, several bands located at about 2860, 2900, 2940, and 2970 cm^{-1} can be found for the C—H symmetric and asymmetric stretching vibrations – ν(C—H) of methylene and methyl groups; in addition to these peaks, there is also a signal at 2874 cm^{-1} corresponding to the stretching vibrations of the CH_2 groups linked of the unsaturated —CH=CH— function (Baeten et al. 1998; Carmona et al. 2014 and herein references) and one around 3010 cm^{-1} (i.e. 3006 or 3013 cm^{-1}, depending of the fatty acid chain) for the C—H stretching vibrations of the C=C—H groups containing in the fatty acids chains (Sadeghi-Jorabchi et al. 1991). This latest peak together with the band from 1660 cm^{-1} was presented as indicators of different degrees of unsaturation (Li-Chan 1994); the C=C—H vibration was also correlated with the iodine value by Lerner et al. (1992) or was successfully used for the detection and quantification of adulteration of the olive oils (Baeten et al. 1996).

ii) Cis/trans isomers content of the unsaturated esters from oils is also important for the food industry, both in terms of nutritional value and structural stability. Bailey and Horvat (1972) showed that the C=C stretching vibration of a trans isomer appeared at a higher wavenumber (about 14 cm^{-1}) than the appropriate cis isomer, while Sadeghi-Jorabchi et al. (1991) proved that Raman spectra of oils and fats can be easily interpreted in terms of quantitative analysis of total unsaturation and cis/trans isomer content.

iii) *Evaluation of the oxidative stability of oils and minor components identification:* The oxidation of oils and fats leads to a degradation of their quality and can have a negative effect on the human health. Raman spectroscopy has been used, with good results for the investigation of oxidative stability of the vegetative oils, especially olive oils (Muik et al. 2005, 2007; Guzman et al. 2011; Carmona et al. 2014). The minor components (i.e. sterols, terpene, and polyphenols) determination generally requires a pretreatment for their extraction, but their analysis with Raman spectroscopy and chemometrics methods allowed the discrimination between olive and seed oils (Baeten et al. 2001). Free fatty acids determination has also been studied by Raman spectroscopy in combination with partial least squares (PLS) regression; Muik et al. (2003) showed that Raman methodology can be successfully used for olive oils investigation given that it yields to good results and is a fast technique. More details can be found in the review of Jimenez-Sanchidrian and Ruiz (2016).

iv) Due to the high nutritional properties and high price of the olive oils, they have been extensively in the attention of both the counterfeiters and authorities for a long period of time. Raman spectroscopy has been successfully applied for the *authentication and adulteration detection* of the edible oils; as expected, most of the studies have been realized on the olive oils. Baeten et al. (1996) used, for the first time, this technique to detect the virgin olive oil adulteration, but a rapid authentication or the quantitative detection of adulterated olive oil using Raman spectroscopy has also been realized by various research groups (Zou et al. 2009; El-Abassy et al. 2009; Zhang et al. 2011).

However, in the last years, some vegetable cold-pressed oils (i.e. hemp, walnut, linseed, sesame, or sea buckthorn) also gained increased attention of the consumers due to their health benefits or high nutritional properties. To maintain their nutritional and therapeutic properties, the technological process needs cold pressing treatments with a low rate of efficiency comparing with the solvent or expeller extraction. To this, the oils properties are strongly influenced by the quality of the seeds or fruits, and considering that some raw materials (i.e. the sea buckthorn fruits) are available for short periods of time, there is difficulty in obtaining good-quality raw materials. As a result, the oils' price is higher, and the temptation of the producers to dilute or even to replace them with the common edible oils is present. Recently, by combining Raman spectroscopy and machine learning algorithms, it was proved that a rapid classification of some less common oils (sesame, hemp, walnut, linseed, pumpkin, and sea buckthorn) is possible by a simple analysis of their Raman spectra (Berghian-Grosan and Magdas 2020a). Using this approach, the authors demonstrated that in a case of sea buckthorn oils (labeled by the producer as 80% sea buckthorn oil and 20% sunflower oil), the proportion of the cheaper ingredient is higher than on the label; moreover, the producer tentative to substitute the sea buckthorn oil with that obtained from the pumpkin seeds was revealed. Figure 3 shows the Raman spectra of sunflower, pumpkin, and sea buckthorn oils obtained using the 785 nm wavelength laser excitation;

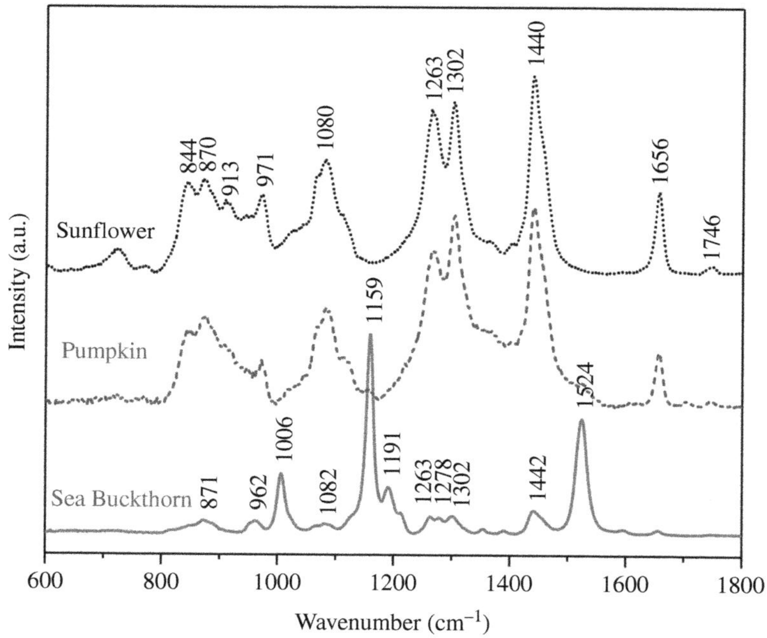

Figure 3 Raman spectra of the sunflower, pumpkin, and sea buckthorn oils obtained with the 785 nm wavelength laser excitation.

Table 3 Raman peaks assignments for sunflower, pumpkin, and sea buckthorn oils.

Raman spectra (cm^{-1}) recorded for			
Sunflower	**Pumpkin**	**Sea buckthorn**	**Vibrational assignments**
844	—	—	C—C/C—O stretching
870	874	871	C—C/C—O stretching
913	—	—	C—C stretching
971	973	962	CH wagging from cis —CH=CH—
—	—	1006	C—CH_3 rocking vibrations
1080	1087	1082	Skeletal C—C stretching
—	—	1159	C—C stretching vibration (C—H in-plane bending mode)
—	—	1191	C—C stretching, C—H in plane bending
1263	1267	1263	CH deformation vibration from cis —CH=CH—
—	—	1278	C—H bending coupled with C=C stretching
1302	1303	1302	CH deformation of CH_2
1440	1439	1442	CH deformation (scissor vibration of CH_2 group)
—	—	1524	C=C stretching vibration
1656	1657	1655	C=C stretching from cis – CH=CH—
1746	1743	—	C=O stretching from RC=OOR

Source: Adapted from Sadeghi-Jorabchi et al. (1990), Baeten et al. (1998), Merlin (1985), Koyama et al. (1984), and Socrates (2001).

their assignments are presented in Table 3. All spectra have been background corrected (to eliminate the fluorescence influence), and they have also normalized in the range of [0, 1].

The discrimination efficiency of Raman spectroscopy for edible oils (sunflower, sesame, hemp, walnut, linseed, and sea buckthorn) differentiation was compared with an acknowledge technique for fatty acids analysis, represented by gas-chromatography equipped with flame ionization detector (GC-FID) (Covaciu et al. 2020). For this purpose, the experimental data were statistically processed using linear discriminant analysis (LDA). The correlation between the classifications obtained by using the two approaches was a very good one, proving thus the suitability of Raman spectroscopy for edible oils differentiation. Moreover, the adulteration of sea buckthorn oil with sunflower oil was proved through the experimental data processing based on artificial neural network (ANN) (Covaciu et al. 2020).

Finally, a schematic representation of the Raman approaches generally used for oils investigation is presented in Figure 4.

The edible oil matrix represents one of the most investigated food commodities by Raman spectroscopy. This technique covers a large domain of evaluation, from the determination of unsaturation degree and minor components (very important from nutritional point of view) to the detection of adulterants and oxidative decomposition which presents valuable probes for the quality control, and even the classification and authentication of some types of oils when it is combined with different statistical methods.

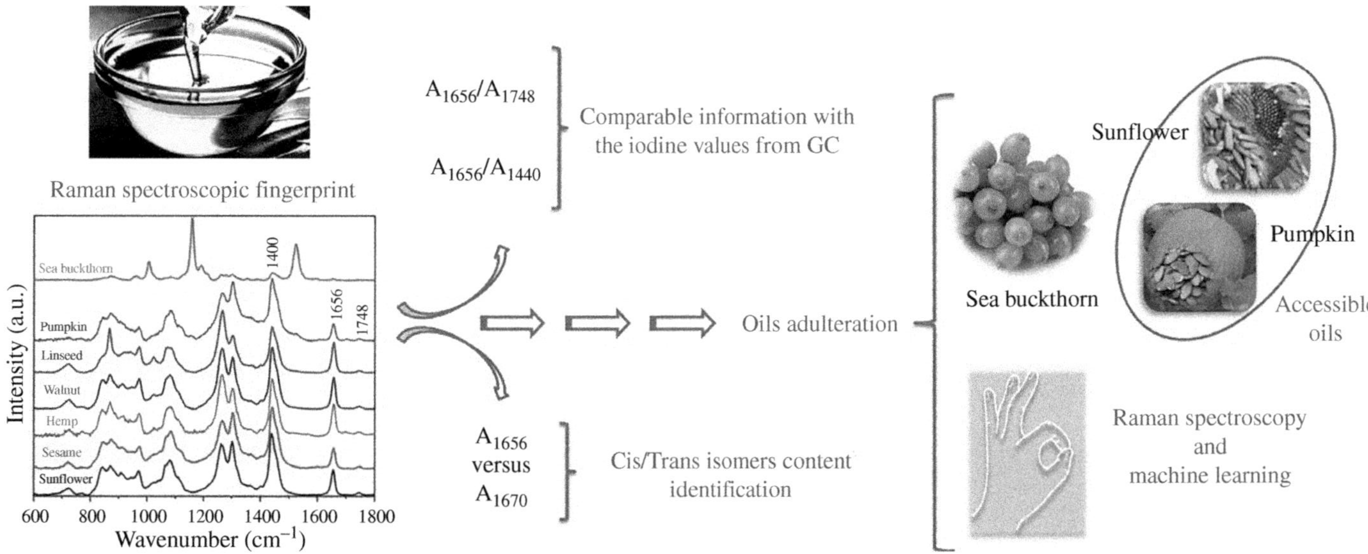

Figure 4 Schematic representation of some Raman approaches employed for the oils investigations. (*See insert for color representation of this figure*).

The comparison between Raman spectroscopy and acknowledged methods for edible oils investigations, like gas chromatography, demonstrated the efficiency of Raman technique for edible oils analysis and classifications. Moreover, the adulteration detection by mixture of cheap oils with most expensive ones can be achieved based on this type of spectroscopy. All these works demonstrate the capacity of Raman spectroscopy to be applied as a rapid control procedure in the oil industry.

2.3 Wines

Wine chemical composition consists in a high number of small molecules which originate from three main sources: the grape variety, the yeast strain which was used for fermentation, and the vessels involved during the wine making process as well as from storage. The effect of these influences will be responsible for all the specificity of a certain wine like quality, cultivar, geographical origin, and vintage (Cuadros-Inostroza et al. 2010). The composition of grapes and the subsequent wine is formed by primary metabolites like sugars, organic, and amino acids as well as secondary metabolites like anthocyannins, flavonoids, and other pigments (Roullier-Gall et al. 2014).

Taking into account that the price of wine is directly reflected in its specificity in terms of cultivar, geographical origin, and vintage, the development of reliable analytical approaches able to differentiate the wine from the market with respect to abovementioned criteria become mandatory. Apart of this, the fact that wine is a such popular commodity that is sell in high quantities, requires fast analytical methods for its quality check and here, spectroscopic methods proved to be very efficient.

As compared with other types of spectroscopies, Raman spectroscopy presents a fundamental advantage in the study of high water content food matrices because of its relatively weak water bending mode in the fingerprint region. Moreover, with the new development in the field, related to the producing of the high resolution portable instruments, this technique becomes more and more attractive for research and control laboratories which are activating in the food and beverages areas, Table 4. This is mainly because this analytical method presents some other advantages such as accessibility for in situ measurements, affordable costs, minimal or no sample preparation step, high sensitivity and specificity, and robustness. These are the main reasons for which Raman spectroscopy was employed in different wine studies aiming: (i) wine compounds i.e. *anthocyanidins, total phenolic content, phenolic acids* determinations (Martin et al. 2015, 2017; Wu et al. 2016a; Zaffino et al. 2015); (ii) the control of fermentation process (Wang et al. 2015; Wu et al. 2015); (iii) metabolomics studies (Mandrile et al. 2016; Magdas et al. 2018a, 2019).

The efficiency of Fourier transform (FT)-Raman spectroscopy for the wine discrimination was discussed in a pilot study, performed on 34 white wine samples (Magdas et al. 2018a), originated from two countries: Romania – 30 samples and France – 4 samples. The French wine sample set was used in this study as control set. The Romanian samples belonged to two varieties Feteasca Regala and Sauvignon Blanc being produced among three regions (Transylvania, Banat, and Moldova) during five consecutive vintages (2011–2015).

In Figure 5, the FT-Raman spectra of a wine data set were plotted. As can be seen, the spectra were recorded between −1000 and 3600 cm^{-1}, comprising both Stokes and

Table 4 Brief review of literature data on wines' investigation by Raman spectroscopy.

Study	Method	References
Quantitative analysis of alcohol–water binary solutions	Raman 488 nm for in situ measurements of alcohols	Numata et al. (2011)
	External standard (acetonitrile) used to eliminate laser power or instrumental effects	
	Proportionality relation between the mass fraction of alcohol and the ratio between alcohol Raman bands and acetonitrile	
Raman spectroscopy of white wines	Three different wavelengths of laser radiation: near-UV (325 nm), visible (532 nm), and near infrared (785 nm)	Martin et al. (2015)
Measurement of fermentation parameters of Chinese rice wine	Raman 785 nm combined with combined with linear and non-linear regression methods	Wu et al. (2015)
Quantitative analysis of multiple components in wine fermentation	Fourier Transform (FT)-Raman spectroscopy (1064 nm) and chemometric techniques	Wang et al. (2015)
Surface-enhanced Raman scattering (SERS) study of anthocyanidins	Micro-Raman (532 nm) and FT Raman (1064 nm) instruments	Zaffino et al. (2015)
Comparison between ATR-IR, Raman, concatenated ATR-IR and Raman spectroscopy for the determination of total antioxidant capacity and total phenolic content of Chinese rice wine	FT-IR spectrometer equipped with an interferometer, a KBr beam splitter and a deuterated triglycine sulfate (DTGS) detector	Wu et al. (2016a)
	Raman equipped with a 785 nm laser source and a cooled charge-coupled device (CCD) detector	
Controlling protected designation of origin of wine	Raman 1064 nm combined with chemometric approach	Mandrile et al. (2016)
	Allow discrimination of different wines produced in the Piedmont area (North West Italy)	
Spectroscopic and theoretical investigations of phenolic acids in white wines	UV–vis, laser induced fluorescence, and Raman (325 nm) spectroscopic techniques coupled with density functional theory calculations of phenolic acids	Martin et al. (2017)
A review on the application of vibrational spectroscopy in the wine industry	Review of the vibrational spectroscopy techniques (near-infrared, mid-infrared, and Raman) employed in the wine industry	Teixeira dos Santos et al. (2017)

anti-Stokes regions of the Raman spectra. The dominated signals in the wine spectra are those given by the ethanol located around: 877, 1047, 1085, 1276, 1454, 2729, 2882, 2933, 2980 cm^{-1} and also the water bands (1636 cm^{-1} bending mode and 3200 cm^{-1} stretching mode).

The differences observed from one spectrum to another are very subtle; therefore, the use of advance data processing methods became mandatory for sample differentiation, according to some predefined criteria. In this study, the experimental spectra were statistically processed using stepwise linear discriminant analysis (SLDA) to differentiate the investigated wines as

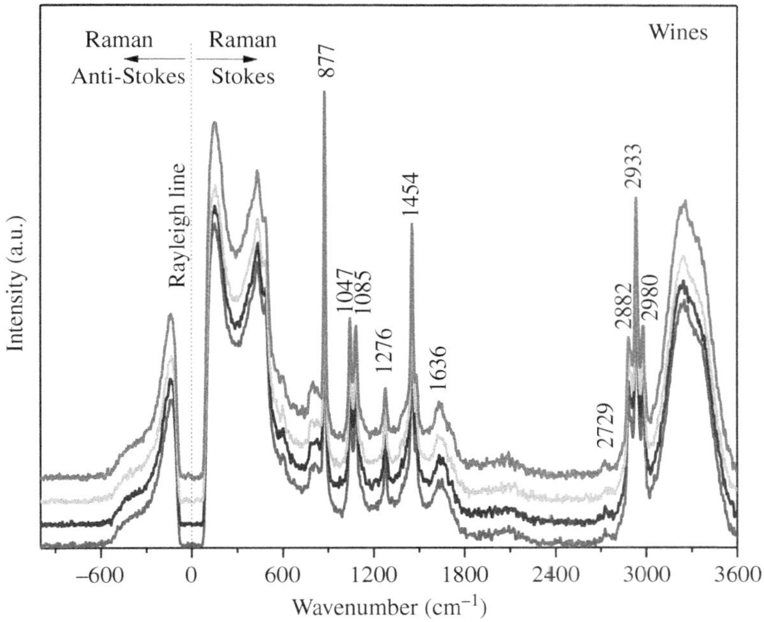

Figure 5 Raman spectra of wines displayed on both Stokes and anti-Stokes regions.

function of their: geographical origin, vintage, and wine cultivar. By using this approach, a perfect classification (100% in both initial and cross-validation procedure) was obtained for the differentiation of geographical origin, vintage, and cultivar for all Romanian wines. The predictors, on which this classification was made, were located in both Stokes and anti-Stokes regions, demonstrated in this case, the efficiency of using the signals from the anti-Stokes region for this purpose. When the control set, formed by four French wines (2 Sauvignon and 2 Chardonnay) was introduced in the classification, the geographical differentiation was achieved in a percentage of 100%, both initial and cross validation; meanwhile, for vintage discrimination, the percentages were 100% in initial and 94.1% cross-validation procedure.

The same sample set was also investigated by mean of SERS (surface-enhanced Raman scattering) technique in order to classify the wines according to: geographical origin, cultivar, and vintage (Magdas et al. 2018b). SERS is a generally recognized technique for the trace detection of a large variety of compounds like the minor constituents of wines which are responsible for wine differentiation. This promising technique is characterized by high sensitivity, rapid, minimal sample preparation needs, and by offering meaningful information. SERS is chemically specific and highly sensitive to the changes in molecular composition (Cinta Pinzaru and Magdas 2018). It was previously assumed that the selective enhancement of the signals given by the minor compounds from wines, like phenolic compounds or acids could have a discrimination potential for wines (Cinta Pinzaru and Magdas 2018; Aguilar-Hernandez et al. 2017). Using the association between SERS technique and chemometry – SLDA, the wine variety differentiation (Feteasca Regala versus Sauvignon Blanc) was achieved in a percentage of 90.3%. For the geographical origin differentiation, among investigated Romanian areas, a percentage of 83.3% was obtained,

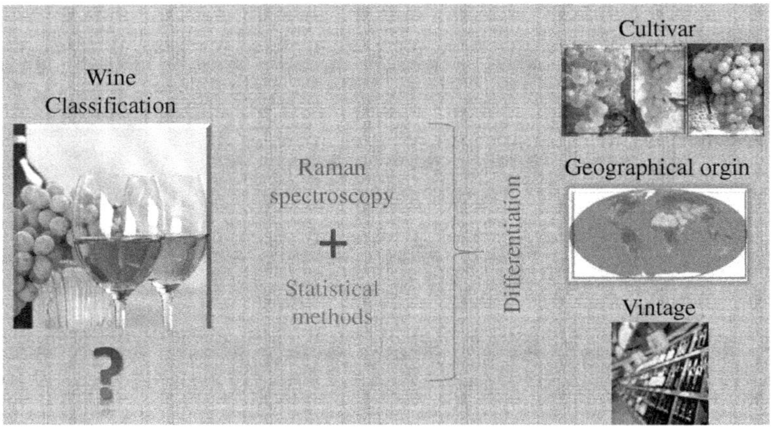

Figure 6 Schematic representation of the performed wine differentiation. (*See insert for color representation of this figure*).

while the separation of French samples from Romanian ones was realized in a percentage of 100% in both, initial and cross-validation procedure. For vintage discrimination, the method appeared to be less sensitive, only the differentiation of 2012 vintage from the other investigated vintages being discriminated. In this case, the obtained percentages were 90% for initial classification and 83.3% in cross validation.

The validation of the results obtained in the pilot study, in which a perfect samples classification was obtained through the association between FT-Raman spectroscopy and SLDA, was made through the employment of an extended wine sample set (Magdas et al. 2019). For this purpose, 126 wines originated from Romania (80 samples) and France (46 samples) were used (Figure 6). The investigated wines belonged to four cultivars, namely, Chardonnay, Pinot Gris, Riesling, and Sauvignon Blanc. It was demonstrated that by using FT-Raman spectroscopy (both anti-Stokes and Stokes spectral ranges) in conjunction with chemometrics (SLDA), the simultaneous discrimination of cultivars Chardonnay, Pinot Gris, Riesling, and Sauvignon can be successfully performed. The sensitivity, in terms of cultivar differentiation, of this approach was proved, even for underling subtle differences which appear between the Chardonnay wines coming from the two countries (Romania and France). In this regard, Chardonnay cultivar appears to be more sensitive to the terroir effects in comparison with Sauvignon.

The wine differentiation, with respect to their geographical origin (Romania versus France), was achieved in a percentage of 100% in both initial and cross-validation procedure. Moreover, the geographical discrimination among the three Romanian regions, namely, Transylvania, Oltenia, and Moldova, was realized in the same percentage. The vintage differentiation, among three distinct production years (2012, 2013, and 2017), was perfectly performed (100% initial and cross validation). For Romanian samples belonging to four consecutive vintages 2012–2015, the differentiation was made in a percentage of 90.7% in initial classification and 77.8% in cross validation, respectively.

Wine authentication based on FT-Raman spectroscopy was also reported for wines produced in North West of Italy, in the Piedmont area (Mandrile et al. 2016). For this aim, more than 300 wine samples obtained from Nebbiolo, Barbera, and Dolcetto grapes were employed. The association between FT-Raman spectroscopy and chemometric analysis proved to be very efficient in the successful classification of wine cultivar; in this case, the validation test provided reliability of 93%. The geographical origin was correctly recognized with a percentage higher than 90%, while for the aging differentiation, a little lower percentage was realized, around 80%.

As was previously demonstrated, based on the reported results from the literature, Raman spectroscopy in association with advanced statistical methods represents a reliable tool for wine classification with respect to geographical origin, cultivar, and vintage. The fact that this technique is a rapid and affordable one and also because of the development of portable devices, it give us the confidence that this method will became extensively used for control purposes in the future.

2.4 Fruit Spirits

Fruit spirits are well-known alcoholic beverages due to their particular composition and flavors. They are composed of high percentages of ethanol and water, but their specific taste is due mainly to minor components (known as congeners) found in these beverages. Although in low concentrations, these compounds allow the distinction between the spirits obtained from different fruits as plum, apple, pear, apricot, cherry, etc.

The alcoholic beverages industry is very important from economical point of view, and the alcoholic drinks counterfeiting is a real problem throughout the world. The quantification of the ethanol and methanol from the alcoholic beverages has a significant importance for the quality control. If the ethanol concentration quantification is more important to assure the consumer confidence about the quality of the products and also for the taxation proposes, the methanol quantification is significant because of its toxicity.

Many methods, as NMR spectroscopy (Fotakis and Zervou 2016), gas or liquid chromatography (Aylott et al. 1994; Wisniewska et al. 2015; Canas et al. 2003), GC–mass spectrometry (Coldea et al. 2014), electrospray ionization mass spectrometry (Moller et al. 2005), paper spray mass spectrometry (Teodoro et al. 2017), inductively coupled plasma atomic emission spectroscopy (Terol et al. 2011; Ceballos-Magana et al. 2012), and capillary electrophoresis (Heller et al. 2011) are used for the ethanol or methanol quantification, minor volatile compounds (congeners) or wood components determination, metals identification, and stable isotope ratio analysis.

Besides these methods, vibrational spectroscopy has also received a special attention due to its potential for a rapid and solvent-free investigation. Thus, infrared analysis in combination with chemometric methods have been intensively used either for ethanol content determination, quality control, classification, or even geographical origin identification (Lachenmeier 2007; Coldea et al. 2013; Anjos et al. 2016; Cirino de Carvalho et al. 2016; Jakubikova et al. 2016). Comparing to infrared, in Raman spectra, a relatively weak bending vibration appeared for water in the fingerprint region, so Raman spectroscopy can be an important tool for the fruit spirits' investigations. Over the years, Raman approaches have been employed especially for determination of the alcohol (ethanol, methanol) content from these beverages (Sanford et al. 2001; Nordon et al. 2005; Cleveland et al. 2007; Vaskova 2014; Pappas et al. 2016; Sramek et al. 2019), but also for the detection of

Table 5 Main Raman bands of the fruit spirits and their assignments according to the available literature data.

Main Raman bands (cm^{-1})	Assignments
437	C—C—O in-plane bending
883	C—C stretching
1050	C—O stretching
1091	CH_3 rocking
1280	CH_2 torsion and rotational vibrations
1456	CH_3 bending
1482	CH_3 bending

Source: Adapted from Mammone et al. (1980), Picard et al. (2007), and Burikov et al. (2010).

carcinogenic substances (i.e. ethyl carbamate) when SERS was involved (Wu et al. 2016b). Moreover, in the last years, in order to identify the fake spirits and to quantify the methanol composition, a procedure based on through-bottle detection of chemical markers with Raman spectroscopy has been tested (Ellis et al. 2019).

As we stated before, the quantification of ethanol and methanol in the spirits has received a special attention. By Raman spectroscopy, these determinations can be made using both external aqueous calibration and standard addition method (Cleveland et al. 2007); the peaks from about 881 cm^{-1} (for ethanol) and 1035 cm^{-1} (for methanol) are generally employed to generate the calibration curves for the alcohols' quantification (Nordon et al. 2005; Cleveland et al. 2007; Vaskova 2014; Sramek et al. 2019). For multiple band normalization method, Cleveland et al. (2007) utilized the peaks from 888, 1062, and 1462 cm^{-1} when the 532 nm laser line has been engaged for the samples' investigations. However, Sanford et al. (2001) have succeeded to create calibration curve for the ethanol content by analyzing the Raman shift of the band from 2941 cm^{-1} (attributed to the C—H stretching vibration of ethanol). Table 5 contains the main peaks of the fruits spirits and their assignments based on the literature data.

These bands represent the vibrations of ethanol molecules, and they are significant intensities in the Raman spectra of the fruit spirits; due to this situation, Raman investigations are generally combined with chemometric methods for good authentication results (Kiefer and Cromwell 2017; Ellis et al. 2019). Moreover, the efficiency of this combination of methods becomes more valuable considering that, on the market, there is not a database for all chemical compounds, the fruit distillates are complex mixtures of organic and inorganic substances, and fluorescence phenomenon embarrassed some Raman investigations.

In a recent study (Berghian-Grosan and Magdas 2020b), the association between Raman spectroscopy and machine learning algorithms for the fruit' spirits differentiation with respect to their trademark, botanical, and geographical origin was employed for the first time. For this propose, different varieties of fruit spirits (plum, apple, pear, apricot, cherry, sour-cherry, grape, and quince) from five producers have been analyzed by Raman spectroscopy (785 nm laser excitation), Figure 7; the obtained Raman data have been further analyzed using five predictive modeling approaches: the decision trees (Quinlan 1986), the

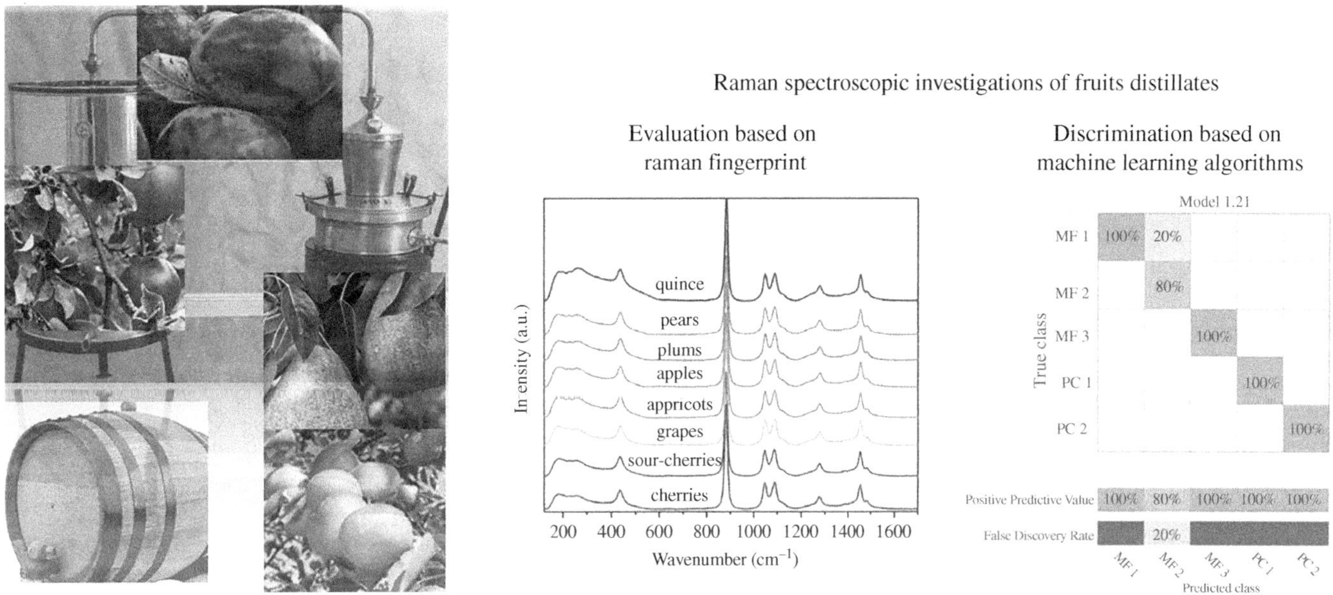

Figure 7 Schematic representation of fruit spirits Raman spectroscopic investigation. (*See insert for color representation of this figure*). *Source:* Adapted from Quinlan (1986), Usoro (2015), Weinberger and Saul (2009), Dietterich (2000).

discriminant analysis (Usoro 2015), the support vector machines (SVMs) (Cortes and Vapnik 1995), the nearest neighbor classifiers – KNN (Weinberger and Saul 2009), and ensemble classifiers (Dietterich 2000).

This approach appeared to be very effective for the trademark fingerprint differentiation (model accuracy – 95.5%); for the geographical discrimination a model with a comparable accuracy (90.9%) has also been obtained in conditions that this investigation was made on some fruit spirits from the neighboring regions of Transylvania. Moreover, a successful fruit variety differentiation was possible inside each producer, after the trademark fingerprint was minimized.

Thus, this study demonstrates the high potential offered by the association between Raman spectroscopy and machine learning algorithms for a rapid investigation of the fruit spirits with regard to trademark and geographical origin. The detection of fraudulent attempts, as substitution of a brand, false declaration of product provenience, or even the denaturation of different varieties of fruit spirits by initial mixture of the raw materials, can be more easily obtained.

Due to the high content of ethanol and large varieties of minor components, the fruit spirits can be seen as a complex and complicated food matrix for Raman spectroscopic investigations.

The application of Raman spectroscopy in the quantification of the main fruit spirits constituents like ethanol is given by the fact that these major signals can be easily followed and quantified as compared with those given by the minor compounds specific for geographical or varietal differentiation. As compared with other matrices that are less processed, the differentiation among fruit distillates varieties is not straightforward. This is mainly because the influences of the technological obtained process are very pronounced, imposing a trademark fingerprint. Nevertheless, this can be an advantage for studies related to brand protection.

Thus, by proving that Raman spectroscopy is efficient not only for the ethanol or methanol quantification, a new step forward was performed and it opens a new way for Raman spectroscopy application in this vast domain of alcoholic beverages.

3 Conclusions and Future

The literature data presented above for some food and beverages matrices (honey, edible oils, wines, and fruit spirits) highlight the growing interests of the researchers' community for the continuous development of rapid and appropriate Raman technique for food industry. Real interest of consumers for good quality products, the necessity to prevent fraud together with the rapid development of portable Raman devices, easy to use, and allowing investigation even through the containers make the Raman spectroscopy an interesting procedure for the food industry.

Despite that Raman spectroscopy is complementary to IR technique, its application in food and beverages industry and control is used in a much lesser extent. The main reason for this was, until the development of Raman portable devices, the price differences existed between the two types of equipment. During the last years, more and more applications for food and beverages studies were developed based on Raman spectroscopy.

The main advantage of Raman spectroscopy is given by the fact that this technique is very suitable for high water content matrices like wines, where its suitability was proved for classifications with respect to geographical origin, cultivar, and vintage. Being a technique sensitive to the investigated matrix and due to the complex structures of these matrices, the phenomena as crystallization and fluorescence need to be taken into account before the beginning of the investigation. These issues can be overcomed either by the appropriate development of sample preparation methods or through the suitable choose of excitation laser line.

However, by combining Raman spectroscopy with chemometric or artificial intelligence methods, a further development of easy procedures for a rapid control of the food and beverages can be established by considering each specific feature of these matrices.

Contribution Statement

Both authors, Dana Alina Magdas and Camelia Berghian-Grosan had an equal contribution in the chapter conception and writing.

Acknowledgments

The authors Dana Alina Magdas and Camelia Berghian-Grosan are grateful to the financial support offered by Executive Agency for Higher Education, Research, Development and Innovation Funding Romania, UEFISCDI, through grant number 260PED/2020, PNCDI III program.

Conflict of Interest

Dana Alina Magdas declares that she has no conflict of interest. Camelia Berghian-Grosan declares that she has no conflict of interest.

References

Aguilar-Hernandez, I., Afseth, N.K., Lopez-Luke, T. et al. (2017). Surface enhanced Raman spectroscopy of phenolic antioxidants: a systematic evaluation of ferulic acid, p-coumaric acid, caffeic acid and sinapic acid. *Vibrational Spectroscopy* 89: 113–122.

Anjos, O., Santos, A.J.A., Estevinho, L.M., and Caldeira, I. (2016). FTIR–ATR spectroscopy applied to quality control of grape-derived spirits. *Food Chemistry* 205: 28–35. https://doi.org/10.1016/j.foodchem.2016.02.128.

Anjos, O., Santos, A.J.A., Paixao, V., and Estevinho, L.M. (2018). Physicochemical characterization of *Lavandula* spp. honey with FT-Raman spectroscopy. *Talanta* 178: 43–48. https://doi.org/10.1016/j.talanta.2017.08.099.

AOCS Recommended Practice Ca 5d-01 (revised 2001). Free fatty acids in crude vegetable oils by capillary gas chromatography. https://www.aocs.org/attain-lab-services/methods/methods/search-results?method=111484&SSO=True.

Aylott, R.I., Clyne, A.H., Fox, A.P., and Walker, D.A. (1994). Analytical strategies to confirm scotch whisky authenticity. *Analyst* 119: 1741–1746. https://doi.org/10.1039/AN9941901741.

Baeten, V. and Aparicio, R. (2000). Edible oils and fats authentication by Fourier transform Raman spectrometry. *Biotechnology, Agronomy, Society and Environment* 4 (4): 196–203.

Baeten, V., Meurens, M., Morales, M.T., and Aparicio, R. (1996). Detection of virgin olive oil adulteration by Fourier transform Raman spectroscopy. *Journal of Agricultural and Food Chemistry* 44: 2225–2230. https://doi.org/10.1021/jf9600115.

Baeten, V., Hourant, P., Morales, M.T., and Aparicio, R. (1998). Oil and fat classification by FT-Raman spectroscopy. *Journal of Agricultural and Food Chemistry* 46: 2638–2646. https://doi.org/10.1021/jf9707851.

Baeten, V., Dardenne, P., and Aparicio, R. (2001). Interpretation of Fourier transform Raman spectra of the unsaponifiable matter in a selection of edible oils. *Journal of Agricultural and Food Chemistry* 49: 5098–5107. https://doi.org/10.1021/jf010146x.

Bailey, G.F. and Horvat, R.J. (1972). Raman spectroscopic analysis of the cis/trans isomer composition of edible vegetable oils. *Journal of the American Oil Chemists' Society* 49: 494–498. https://doi.org/10.1007/BF02582487.

Baranska, M., Schulz, H., Strehle, M., and Popp, J. (2010). Applications of vibrational spectroscopy to oilseeds analysis. In: *Handbook of Vibrational Spectroscopy* (eds. P. Griffiths and J.M. Chalmers), 1–24. Wiley https://doi.org/10.1002/0470027320.s8953.

Bartheta, V.J., Gordon, V., and Daun, J.K. (2008). Evaluation of a colorimetric method for measuring the content of FFA in marineand vegetable oils. *Food Chemistry* 111: 1064–1068.

Barthus, R.C. and Poppi, R.J. (2001). Determination of the total unsaturation in vegetable oils by Fourier transform Raman spectroscopy and multivariate calibration. *Vibrational Spectroscopy* 26: 99–105. https://doi.org/10.1016/S0924-2031(01)00107-2.

Batsoulis, A.N., Siatis, N.G., Kimbaris, A.C. et al. (2005). FT-Raman spectroscopic simultaneous determination of fructose and glucose in honey. *Journal of Agricultural and Food Chemistry* 53: 207–210. https://doi.org/10.1021/jf048793m.

Berghian-Grosan, C. and Magdas, D.A. (2020a). Raman spectroscopy and machine-learning for edible oils evaluation. *Talanta* 218: 121176. https://doi.org/10.1016/j.talanta.2020.121176.

Berghian-Grosan, C. and Magdas, D.A. (2020b). Application of Raman spectroscopy and Machine Learning algorithms for fruit distillates discrimination. *Scientific Reports* 10: 21152. https://doi.org/10.1038/s41598-020-78159-8.

Bewig, K.M., Clarke, A.D., Roberts, C., and Unklesbay, N. (1994). Discriminant analysis of vegetable oils by near-infrared reflectance spectroscopy. *Journal of the American Oil Chemists' Society* 71: 195–200.

Bogdanov, S. (2011). Honey composition. In: *The Honey Book*, 1–5. Bee Product Science. www.bee-hexagon.net.

Burikov, S., Dolenko, T., Patsaeva, S. et al. (2010). Raman and IR spectroscopy research on hydrogen bonding in water–ethanol systems. *Molecular Physics* 108 (18): 2427–2436. https://doi.org/10.1080/00268976.2010.516277.

Canas, S., Belchior, A.P., Spranger, M.I., and Bruno-de-Sousa, R. (2003). High-performance liquid chromatography methodfor analysis of phenolic acids, phenolic aldehydes, and furanic derivatives in brandies. Development and validation. *Journal of Separation Science* 26: 496–502.

Carmona, M.A., Lafont, F., Jimenez-Sanchidrian, C., and Ruiz, J.R. (2014). Raman spectroscopy study of edible oils and determination of the oxidative stability at frying temperatures. *European Journal of Lipid Science and Technology* 116 (11): 1451–1456. https://doi.org/10.1002/ejlt.201400127.

Ceballos-Magana, S.G., Jurado, J.M., Muniz-Valencia, R. et al. (2012). Geographical authentication of tequila according to its mineral content by means of support vector machines. *Food Analytical Methods* 5: 260–265. https://doi.org/10.1007/s12161-011-9233-1.

Chan, C.W., Deadman, B.J., Manley-Harris, M. et al. (2013). Analysis of the flavonoid component of bioactive New Zealand manuka (*Leptospermum scoparium*) honey and the isolation, characterization and synthesis of an unusual pyrrole. *Food Chemistry* 141: 1772–1781.

Chemat, S. (2017). Edible oils: extraction, processing, and applications. In: *Contemporary Food Engineering* (ed. D.-W. Sun), 1–261. Boca Raton, FL: CRC Press.

Cinta Pinzaru, S. and Magdas, D.A. (2018). Ag nanoparticles meet wines: SERS for wine analysis. *Food Analytical Methods* 11 (3): 892–900. https://doi.org/10.1007/s12161-017-1056-2.

Cirino de Carvalho, L., Morais, C.L.M., de Lima, K.M.G. et al. (2016). Determination of the geographical origin and ethanol content of Brazilian sugarcane spirit using near-infrared spectroscopy coupled with discriminant analysis. *Analytical Methods* 28: 5658–5666. https://doi.org/10.1039/C6AY01325B.

Cleveland, D., Carlson, M., Hudspeth, E.D. et al. (2007). Raman spectroscopy for the undergraduate teaching laboratory: quantification of ethanol concentration in consumer alcoholic beverages and qualitative identification of marine diesels using a miniature Raman spectrometer. *Spectroscopy Letters* 40: 903–924.

CODEX STAN 234-1999 (A.1.1. Fats and Oils and Related Products) (2019). *Recommended Methods of Analysis and Sampling*. CODEX Committee on Fats and Oils, 26th Session, Kuala Lumpur, Malaysia.

Coldea, T.E., Socaciu, C., Fetea, F. et al. (2013). Rapid quantitative analysis of ethanol and prediction of methanol content in traditional fruit brandies from Romania, using FTIR spectroscopy and chemometrics. *Notulae Botanicae Horti Agrobotanici* 41: 143–149. https://doi.org/10.15835/nbha4119000.

Coldea, T.E., Socaciu, C., Moldovan, Z., and Mudura, E. (2014). Minor volatile compounds in traditional homemade fruit brandies from Transylvania-Romania, as determined by GC-MS analysis. *Notulae Botanicae Horti Agrobotanici* 42: 530–537. https://doi.org/10.15835/nbha4229607.

Cortes, C. and Vapnik, V. (1995). Support-vector networks. *Machine Learning* 20: 273–297. https://doi.org/10.1007/BF00994018.

Corvucci, F., Nobili, L., Melucci, D., and Grillenzoni, F.V. (2015). The discrimination of honey origin using melissopalynology and Raman spectroscopy techniques coupled with multivariate analysis. *Food Chemistry* 169: 297–304. https://doi.org/10.1016/j.foodchem.2014.07.122.

Covaciu, F.-D., Berghian-Grosan, C., Feher, I., and Magdas, D.A. (2020). Edible oils differentiation based on the determination of fatty acids profile and Raman spectroscopy – a case study. *Applied Sciences* 10 (23): 8347. https://doi.org/10.3390/app10238347.

Cuadros-Inostroza, A., Giavalisco, P., Hummel, J. et al. (2010). Discrimination of wine attributes by metabolome analysis. *Analytical Chemistry* 82: 3573–3580.

Dietterich, T.G. (2000). Ensemble methods in machine learning. In: *Multiple Classifier Systems. MCS 2000*, Lecture Notes in Computer Science, vol. 1857 (eds. J. Kittler and F. Roli), 1–15. Berlin, Heidelberg: Springer https://doi.org/10.1007/3-540-45014-9_1.

El-Abassy, R.M., Donfack, P., and Materny, A. (2009). Visible Raman spectroscopy for the discrimination of olive oils from different vegetable oils and the detection of adulteration. *Journal of Raman Spectroscopy* 40: 1284–1289. https://doi.org/10.1002/jrs.2279.

Ellis, D.I., Muhamadali, H., Xu, Y. et al. (2019). Rapid through-container detection of fake spirits and methanol quantification with handheld Raman spectroscopy. *Analyst* 144: 324–330. https://doi.org/10.1039/C8AN01702F.

Endo, Y. (2018). Analytical methods to evaluate the quality of edible fats and oils: the JOCS standard methods for analysis of fats, oils and related materials (2013) and advanced methods. *Journal of Oleo Science* 67: 1–10. https://doi.org/10.5650/jos.ess17130.

Endo, Y., Ohta, A., Kido, H. et al. (2011). Determination of triacylglycerol composition in vegetable oils using high-performance liquid chromatography: a collaborative study. *Journal of Oleo Science* 60: 451–456.

Farhad, S.F.U., Abedin, K.M., Islam, M.R. et al. (2009). Determination of ratio of unsaturated to total fatty acids in edible oils by laser Raman spectroscopy. *Journal of Applied Sciences* 9 (8): 1538–1543. https://doi.org/10.3923/jas.2009.1538.1543.

Fotakis, C. and Zervou, M. (2016). NMR metabolic fingerprinting and chemometrics driven authentication of Greek grape marc spirits. *Food Chemistry* 196: 760–768. https://doi.org/10.1016/j.foodchem.2015.10.002.

Guyon, F., da Costa, E.C., Maurin, A. et al. (2020). Metabolomics applied to proton nuclear magnetic resonance profile for the identification of seven floral origin of French honeys. *Journal of Food and Nutrition Research* 59 (2): 137–146.

Guzman, E., Baeten, V., Pierna, J.A.F., and Garcıa-Mesa, J.A. (2011). Application of low resolution Raman spectroscopy for the analysis of oxidazed olive oil. *Food Control* 22: 2036–2040. https://doi.org/10.1016/j.foodcont.2011.05.025.

Heller, M., Vitali, L., Oliveira, M.A.L. et al. (2011). Rapid sample screening method for authenticity control of whiskey using capillary electrophoresis with online preconcentration. *Journal of Agricultural and Food Chemistry* 59: 6882–6888.

Isengard, H.D., Schultheiß, D., Radovic, B., and Anklam, E. (2001). Alternatives to official analytical methods used for the water determination in honey. *Food Control* 12: 459–466. https://doi.org/10.1016/S0956-7135(01)00044-5.

Jakubikova, M., Sadecka, J., Kleinova, A., and Majek, P. (2016). Near-infrared spectroscopy for rapid classification of fruit spirits. *Journal of Food Science and Technology* 53: 2797–2803. https://doi.org/10.1007/s13197-016-2254-4.

Jiang, F.L., Ikeda, I., Ogawa, Y., and Endo, Y. (2012). Rapid determination of saponification value and polymer content of vegetable and fish oils by terahertz spectroscopy. *Journal of Oleo Science* 61: 531–535.

Jimenez-Sanchidrian, C. and Ruiz, J.R. (2016). Use of Raman spectroscopy for analyzing edible vegetable oils. *Applied Spectroscopy Reviews* 51 (5): 417–430. https://doi.org/10.1080/05704928.2016.1141292.

Johnson, G.L., Machado, R.M., Freidl, K.G. et al. (2002). Evaluation of Raman spectroscopy for determining cis and trans isomers in partially hydrogenated soybean oil. *Organic Process Research & Development* 6: 637–644. https://doi.org/10.1021/op0202080.

Kardash, E. and Turyan, Y.I. (2005). Acid value determination in vegetable oils by indirect titration in aqueous-alcohol media. *Croatica Chemica Acta* 78 (1): 99–103.

Kardash-Strochkova, E., Turyan, Y.I., and Kuselman, I. (2001). Redox-potentiometric determination of peroxide value inedible oils without titration. *Talanta* 54: 411–416.

Kiefer, J. and Cromwell, A.L. (2017). Analysis of single malt Scotch whisky using Raman spectroscopy. *Analytical Methods* 9: e511–e518. https://doi.org/10.1039/C6AY02907H.

Kim, J., Lee, J.H., and Ko, D.-K. (2014). Determination of degree of unsaturation in edible oils using coherent anti-stokes Raman scattering spectroscopy. *Journal of Raman Spectroscopy* 45: 591–595. https://doi.org/10.1002/jrs.4494.

Koyama, Y., Mukal, Y., Umemura, J. et al. (1984). Raman and infrared spectra of the 7-*cis* and di-*cis* isomers of retinal. *Journal of Raman Spectroscopy* 15: 300–307. https://doi.org/ .1002/jrs.1250150503.

Lachenmeier, D.W. (2007). Rapid quality control of spirit drinks and beer using multivariatedata analysis of Fourier transform infrared spectra. *Food Chemistry* 101: 825–832. https://doi.org/10.1016/j.foodchem.2005.12.032.

Lerner, B., Garry, M., and Walder, F. (1992). Characterization of polyunsaturation in cooking oils by the 910 FT-Raman. In: *Proceedings of the 2nd International on FT-IR* (ed. E.F. Vansant), 75–82. Antwerp, Belgium: University of Antwerp.

Li, S., Shan, Y., Zhu, X. et al. (2012). Detection of honey adulteration by high fructose corn syrup and maltose syrup using Raman spectroscopy. *Journal of Food Composition and Analysis* 28: 69–74. https://doi.org/10.1016/j.jfca.2012.07.006.

Li-Chan, E. (1994). Developments in the detection of adulteration of olive oil. *Trends in Food Science & Technology* 5 (1): 3–11. https://doi.org/10.1016/0924-2244(94)90042-6.

Magdas, D.A., Guyon, F., Feher, I., and Cinta-Pinzaru, S. (2018a). Wine discrimination based on chemometric analysis of untargeted markers using FT-Raman spectroscopy. *Food Control* 85 (473): 385–391.

Magdas, D.A., Cinta Pinzaru, S., Guyon, F. et al. (2018b). Application of SERS technique in white wines discrimination. *Food Control* 92: 30–36.

Magdas, D.A., Cozar, B.I., Feher, I. et al. (2019). Testing the limits of FT-Raman spectroscopy for wine authentication: cultivar, geographical origin, vintage and terroir effect influence. *Scientific Reports* 9: 19954.

Magdas, D.A., Guyon, F., Berghian-Grosan, C., and Muller-Molnar, C. (2020). Challenges and a step forward in honey classification based on Raman spectroscopy. *Food Control*: 107769. https://doi.org/10.1016/j.foodcont.2020.107769.

Magdas, D.A., Guyon, F., Puscas, R. et al. (2021). Applications of emerging stable isotopes and elemental markers for geographical and varietal recognition of Romanian and French honeys. *Food Chemistry* 334: 127599. https://doi.org/10.1016/j.foodchem.2020.127599.

Mammone, J.F., Sharma, S.K., and Nicol, M. (1980). Raman spectra of methanol and ethanol at pressures up to 100 kbar. *Journal of Physical Chemistry* 84: 3130–3134. https://doi.org/10.1021/j100460a032.

Mandrile, L., Zeppa, G., Giovannozzi, A.M., and Rossi, A.M. (2016). Controlling protected designation of origin of wine by Raman. *Food Chemistry* 211: 260–270.

Marigheto, N.A., Kemsley, E.K., Defernez, M., and Wilson, R.H. (1998). A comparison of mid-infrared and raman spectroscopies for the authentication of edible oils. *Journal of the American Oil Chemists' Society* 75: 987–992. https://doi.org/10.1007/s11746-998-0276-4.

Martin, C., Bruneel, J.L., Guyon, F. et al. (2015). Raman spectroscopy of white wines. *Food Chemisty* 181: 235–e240.

Martin, C., Bruneel, J.-L., Castet, F. et al. (2017). Spectroscopic and theoretical investigations of phenolic acids in white wines. *Food Chemistry* 221: 568–575.

Masukawa, Y., Shiro, H., Nakamura, S. et al. (2010). A new analytical method for the quantification of glycidol fatty acid esters in edible oils. *Journal of Oleo Science* 59: 81–88.

Mathlouthi, M., Luu, C., Meffroy-Biget, A.M., and Vinh Luu, D. (1980). Laser-raman study of solute-solvent interactions in aqueous solutions of D-fructose, D-glucose, and sucrose. *Carbohydrate Research* 81: 213–223. https://doi.org/10.1016/S0008-6215(00)85653-0.

Merlin, J.C. (1985). Resonance Raman spectroscopy of carotenoids and carotenoid-containing systems. *Pure and Applied Chemistry* 57: 785–792. https://doi.org/10.1351/pac198557050785.

Moller, J.K.S., Catharino, R.R., and Eberlin, M.N. (2005). Electrospray ionization mass spectrometry fingerprinting of whisky: immediate proof of origin and authenticity. *Analyst* 2005: 890–897. https://doi.org/10.1039/B415422C.

Muik, B., Lendl, B., Molina-Diaz, A., and Ayora-Canada, M.J. (2003). Direct, reagent-free determination of free fatty acid content in olive oil and olives by Fourier transform Raman spectrometry. *Analytica Chimica Acta* 487: 211–220. https://doi.org/10.1016/S0003-2670(03)00560-9.

Muik, B., Lendl, B., Molina-Diaz, A., and Ayora-Canada, M.J. (2005). Direct monitoring of lipid oxidation in edible oils by Fourier transform Raman spectroscopy. *Chemistry and Physics of Lipids* 134: 173–182. https://doi.org/10.1016/j.chemphyslip.2005.01.003.

Muik, B., Lendl, B., Molina-Dıaz, A. et al. (2007). Two-dimensional correlation spectroscopy and multivariate curve resolution for the study of lipid oxidation in edible oils monitored by FTIR and FT-Raman spectroscopy. *Analytica Chimica Acta* 593: 54–67. https://doi.org/10.1016/j.aca.2007.04.050.

Muller-Molnar, C., Berghian-Grosan, C., and Magdas, D.A. (2020). An optimized green preparation method for the successful application of Raman spectroscopy in honey studies. *Talanta* 208: 120432. https://doi.org/10.1016/j.talanta.2019.120432.

Nordon, A., Mills, A., Burn, R.T. et al. (2005). Comparison of non-invasive NIR and Raman spectrometries for determination of alcohol content of spirits. *Analytica Chimica Acta* 548: 148–158. https://doi.org/10.1016/j.aca.2005.05.067.

Numata, Y., Iida, Y., and Tanaka, H. (2011). Quantitative analysis of alcohol-water binary solutions using Raman spectroscopy. *Journal of Quantitative Spectroscopy and Radiative Transfer* 112: 1043–1049. https://doi.org/10.1016/j.jqsrt.2011.01.005.

de Oliveira, L.F.C., Colombara, R., and Edwards, H.G.M. (2002). Fourier transform Raman spectroscopy of honey. *Applied Spectroscopy* 56: 306–311. https://doi.org/10.1366/0003702021954881.

Özbalci, B., Boyaci, İ.H., Topcu, A. et al. (2013). Rapid analysis of sugars in honey by processing Raman spectrum using chemometric methods and artificial neural networks. *Food Chemistry* 136: 1444–1452. https://doi.org/10.1016/j.foodchem.2012.09.064.

Pappas, C., Marianthi, B., Konstantinou, E. et al. (2016). Evaluation of a Raman spectroscopic method for the determination of alcohol content in Greek Spirit Tsipouro. *Current Research in Nutrition and Food Science* 4: 1–9. (Special Issue Confernce October 2016). https://doi.org/10.12944/CRNFSJ.4.Special-Issue-October.01.

Paradkar, M.M. and Irudayaraj, J. (2001). Discrimination and classification of beet and cane inverts in honey by FT-Raman spectroscopy. *Food Chemistry* 76: 231–239.

Parker, T., Limer, E., Watson, A.D. et al. (2014). 60 MHz ^{1}H NMR spectroscopy for the analysis of edible oils. *Trends in Analytical Chemistry* 57: 147–158.

Picard, A., Daniel, I., Montagnac, G., and Oger, P. (2007). In situ monitoring by quantitative Raman spectroscopy of alcoholic fermentation by *Saccharomyces cerevisiae* under high pressure. *Extremophiles* 11: 445–452. https://doi.org/10.1007/s00792-006-0054-x.

Pierna, J.A.F., Abbas, O., Dardenne, P., and Baeten, V. (2011). Discrimination of Corsican honey by FT-Raman spectroscopy and chemometrics. *Biotechnology, Agronomy, Society and Environment* 15: 75–84. http://www.pressesagro.be/base/text/v15n1/75.pdf.

Porta, R.A.D. (2006). *Edible Oils Manual*. Arlington, VA: AOCS Press.

Quinlan, J.R. (1986). Induction of decision trees. *Machine Learning* 1: 81–106. https://doi.org/10.1007/BF00116251.

Rogers, K.M., Somerton, K., Rogers, P., and Cox, J. (2010). Eliminating false positive C4 sugar tests on New Zealand manuka honey. *Rapid Communications in Mass Spectrometry* 24: 2370–2374. https://doi.org/10.1002/rcm.4642.

Roullier-Gall, C., Lucio, M., Noret, L. et al. (2014). How subtle is the "Terroir" effect? Chemistry-related signatures of two "Climats de Bourgogne". *PLoS One* 9 (5): e97615.

Sadeghi-Jorabchi, H., Hendra, P.J., Wilson, R.H., and Belton, P.S. (1990). Determination of the total unsaturation in oils and margarines by Fourier transform Raman spectroscopy. *Journal of the American Oil Chemists' Society* 67: 483–486. https://doi.org/10.1007/BF02540752.

Sadeghi-Jorabchi, H., Wilson, R.H., Belton, P.S. et al. (1991). Quantitative analysis of oils and fats by Fourier transform Raman spectroscopy. *Spectrochimica Acta Part A: Molecular Spectroscopy* 47: 1449–1458. https://doi.org/10.1016/0584-8539(91)80236-C.

Sanford, C.L., Mantooth, B., and Jones, B.T. (2001). Determination of ethanol in alcohol samples using a modular Raman spectrometer. *Journal of Chemical Education* 78: 1221–1225. https://doi.org/10.1021/ed078p1221.

Socrates, G. (2001). *Infrared and Raman Characteristic Group Frequencies: Tables and Charts*, 3e. Chichester, UK: Wiley.

Sramek, J., Svancara, I., and Sys, M. (2019). Determination of ethanol in alcoholic drinks using Raman spectrometry. *Scientific Papers of the University of Pardubice, Series A* 25: 5–14.

Sugar, J. and Bour, P. (2016). Quantitative analysis of sugar composition in honey using 532-nm excitation Raman and Raman optical activity spectra. *Journal of Raman Spectroscopy* 47: 1298–1303. https://doi.org/10.1002/jrs.4960.

Tahir, H.E., Xiaobo, Z., Zhihua, L. et al. (2017). Rapid prediction of phenolic compounds and antioxidant activity of Sudanese honey using Raman and Fourier transform infrared (FT-IR) spectroscopy. *Food Chemistry* 226: 202–211. https://doi.org/10.1016/j.foodchem.2017.01.024.

Teixeira dos Santos, C.A., Pascoa, R.N.M.J., and Lopes, J.A. (2017). A review on the application of vibrational spectroscopy in the wine industry: from soil to bottle. *TrAC Trends in Analytical Chemistry* 88: 100–118. https://doi.org/10.1016/j.trac.2016.12.012.

Teodoro, J.A.R., Pereira, H., Sena, V.M.M. et al. (2017). Paper spray mass spectrometry and chemometric tools for a fast and reliable identification of counterfeit blended Scottish whiskies. *Food Chemistry* 237: 1058–1064. https://doi.org/10.1016/j.foodchem.2017.06.062.

Terol, A., Paredes, E., Maestre, S.E. et al. (2011). Alcohol and metal determination in alcoholic beverages through high-temperature liquid-chromatography coupled to an inductively coupled plasma atomic emission spectrometer. *Journal of Chromatography A* 1218 (22): 3439–3446.

Usoro, A.E. (2015). Multivariable discriminant analysis; application of a three dimensional case on students measurements. *American Journal of Mathematics and Statistics* 5: 123–127. https://doi.org/10.5923/j.ajms.20150503.03.

Vaskova, H. (2014). Spectroscopic determination of methanol content in alcoholic drinks. *International Journal of Biology and Biomedical Engineering* 8: 27–34.

Wang, Q., Li, Z., Ma, Z., and Si, G. (2015). Quantitative analysis of multiple components in wine fermentation using Raman spectroscopy. *Advance Journal of Food Science and Technology* 9: e13–e18.

Weinberger, K.Q. and Saul, L.K. (2009). Distance metric learning for large margin nearest neighbor classification. *Journal of Machine Learning Research* 10: 207–244.

Wisniewska, P., Sliwinska, M., Dymerski, T. et al. (2015). Application of gas chromatography to analysis of spirit-based alcoholic beverages. *Critical Reviews in Analytical Chemistry* 45 (3): 201–225. https://doi.org/10.1080/10408347.2014.904732.

Wu, Z., Xu, E., Long, J. et al. (2015). Measurement of fermentation parameters of Chinese rice wine using Raman spectroscopy combined with linear and non-linear regression methods. *Food Control* 56: e95–e102. https://doi.org/10.1016/j.foodcont.2015.03.015.

Wu, Z., Xu, E., Long, J. et al. (2016a). Comparison between ATR-IR, Raman concatenated ATR-IR and Raman spectroscopy for the determination of total antioxidant capacity and total phenolic content of Chinese rice wine. *Food Chemistry* 194: e671–e679.

Wu, Z., Xu, E., Li, J. et al. (2016b). Highly sensitive determination of ethyl carbamate in alcoholic beverages by surface-enhanced Raman spectroscopy combined with a molecular imprinting polymer. *RSC Advances* 6: 109442. https://doi.org/10.1039/C6RA23165A.

Zaffino, C., Russo, B., and Bruni, S. (2015). Surface-enhanced Raman scattering (SERS) study of anthocyanidins. *Spectrochimica Acta Part A: Molecular and Biomolecular Spectroscopy* 149: e41–e47.

Zhang, X.-F., Zou, M.-Q., Qi, X.-H. et al. (2011). Quantitative detection of adulterated olive oil by Raman spectroscopy and chemometrics. *Journal of Raman Spectroscopy* 42: 1784–1788. https://doi.org/10.1002/jrs.2933.

Zou, M.-Q., Zhang, X.-F., Qi, X.-H. et al. (2009). Rapid authentication of olive oil adulteration by Raman spectrometry. *Journal of Agricultural and Food Chemistry* 57 (14): 6001–6006. https://doi.org/10.1021/jf900217s.

14

Visible Light Imaging

Maimunah Mohd Ali[1] *and Norhashila Hashim*[1,2]

[1] *Department of Biological and Agricultural Engineering, Faculty of Engineering, Universiti Putra Malaysia, Serdang, Selangor, Malaysia*
[2] *SMART Farming Technology Research Centre, Faculty of Engineering, Universiti Putra Malaysia, Serdang, Selangor, Malaysia*

1 Introduction

Visible light is detectable by the human eye and constitutes the major fraction of the electromagnetic spectrum of wavelength region ranging from 400 to 700 nm (Li et al. 2014). The visible region is found between the infrared and ultraviolet regions. Generally, when visible light passes through an object, it is either absorbed, reflected, or transmitted according to the light intensity and nature of the material (Manickavasagan and Jayasuriya 2014). Visible light is a combination of several colors including red, blue, green, yellow, orange, and violet waves (Silva et al. 2019). Each color space is denoted by a specific wavelength, starting from violet at the beginning of the visible spectrum while green and blue are represented in the middle of the visible region, whereas the red wave is in the region of 700 nm (Raponi et al. 2017). Human eyes are most sensitive to yellow and green light within the region of 550 nm, especially under strong illumination settings (Cayuela and Camino 2010).

In an imaging system, light signifies a key role in terms of providing detailed information regarding an object. In a sense, visible light carries information within the range of human vision which is useful to facilitate quality assessment. Visible light imaging, also known as red-green-blue (RGB) imaging is represented as the color intensity based on the three base color spaces (RGB) (Vollmer and Möllmann 2018). Visible light imaging delivers an optical interpretation of a sample by visualizing the presence and exact image which is perceived by human vision. The evolution in imaging technology and computerized systems has led to the application of imaging devices for rapid and nondestructive analysis of food quality detection with minimum human interference. Many studies have been conducted to develop visible light imaging techniques in order to evaluate food quality such as size grading, defect detection, and color evaluation (Du et al. 2020; Ireri et al. 2019; Hussain et al. 2018).

Electromagnetic Technologies in Food Science, First Edition. Edited by Vicente M. Gómez-López and Rajeev Bhat.

Although visible light imaging can perform quality evaluation, the detection of chemical constituents or contaminants would be quite challenging (Ravikanth et al. 2017). To effectively implement this technique, it is crucial to understand the optical properties of the food samples being examined. The determination of the optical properties offers plentiful information relevant to the chemical and physical traits of food samples in order to assist in nondestructive food quality assessment (Zhang et al. 2019). Nevertheless, visible light imaging has proven to be advantageous over conventional methods and routine analyses that are prone to destruction of the sample during testing. The application of visible light imaging in combination with chemometric methods makes it possible for the determination of the analytical composition of food samples. Hence, visible light imaging can be successfully applied for the quantitative and qualitative evaluation of food products considering the potential applications and trends of using visible light imaging for real-time monitoring situations.

2 Principle of Visible Light Imaging

Visible light imaging has been widely used to generate a versatile system for the measurement of various food qualities. The rapid development of visible light imaging in food quality detection is inseparable from the instrumentation system, hardware, and image processing analysis.

2.1 Development and Instrumentation

Visible light imaging systems are commonly used in various food applications under visible illumination. A basic visible light imaging system consists of a camera, a lighting unit, and a computer equipped with image processing and analysis software. Typically, the captured image depends heavily on the quality of the light used for illumination. A well-developed illumination system could aid in the processing time during image acquisition. In real-time applications where time is of the main concern, appropriate illumination could reduce digital image processing tasks and reduce the processing time. Further, it can also improve the system accuracy, contributing to effective image processing and analysis. The most commonly used light illumination sources are fluorescent, halogen lamp, light-emitting diodes, and incandescent bulbs (Zhu et al. 2016). It is necessary to establish a good lighting system as part of visible light imaging in order to eliminate noise and enhance the image quality.

A visible light imaging system utilizes cameras ranging from monochrome for specialized simple recognition tasks to aperture multispectral cameras for the detection of diseases and defects in food. A typical digital camera receives light from the surface of the object under examination and converts the light into electrical signals. The electrical signals are proportional to the light intensity in which the digitized data such as 8 or 16-bits is stored in the computer (Chen et al. 2016). Typically, the bit depth is used to determine the color of a single pixel in an image. Two main types of digital cameras implemented in visible light imaging system are charge-coupled device (CCD) and complementary metal oxide semiconductor (CMOS) cameras. Both CCD and CMOS cameras are silicon-based instruments and are solid-state, which are accessible in a linear or multiple array configurations (Shi et al. 2015). The digital cameras capture images that are created from millions of pixels. Two major aspects of image acquisition using visible light imaging are file compression and image resolution. File compression allows more images to be stored by reducing the size of the image. In terms of image resolution, it is associated with the number of pixels to determine the quality of the image.

In visible light imaging, the camera is utilized along with a frame grabber to support the function of data synchronization, data formatting, digitization, and data transfer during the image acquisition. The frame grabber can obtain either analog or digital images according to the type of camera. Normally, the aspects of a frame grabber needed for visible light imaging applications comprises of camera control, image acquisition, and image preprocessing (Zhang et al. 2014). However, most digital cameras nowadays do not require a frame grabber due to the existing image processing and analysis software supplied with the computer. When the configuration instrumentation of visible light imaging is developed, image processing and analysis are performed with a computer in order to generate information based on the captured images. With a well-designed visible light imaging system, many tedious image processing procedures can be reduced to meet the real-time needs encountered in modern food applications.

2.2 Hardware-Orientated Color System

Visible light imaging is a promising technique with the capability of delivering a detailed characteristic of the color space system at pixel-based levels (Wu and Sun 2013). The purpose of a color system is to aid in the standard color specification which is used for image acquisition and hardware processing and analysis. Various color models can be detected by visible light imaging including RGB color model, L*a*b* color model, HSI (hue, saturation, intensity) color model, and CMY (cyan, magenta, yellow) color model. The RGB color model is commonly used to represent the color intensity where the camera is able to detect natural scenes and visualize the image. In a typical visible light imaging system, the image is denoted in terms of pixel intensity of the three basic color spaces (RGB) (Sharma 2019).

Generally, visible light imaging develops the algorithmic and theoretical concept in order to extract useful information based on the object. In visible light imaging applications, color is a piece of fundamental information stored in terms of pixel intensity in a digital image (Wasnik et al. 2019). For this reason, image processing and analysis have been conducted to extract quantitative color information from the digital images. With this, a noncontact and rapid color measurement was achieved by analyzing each pixel on the entire food surface and evaluating the surface characteristics. A major difference between visible light imaging and conventional color measurement is the amount of spatial information provided (Pathare et al. 2013). High-spatial resolution allows visible light imaging to determine pixel intensity based on the entire surface area, to separate regions of interest flexibly, as well as calculate the average of color intensities (Thomas et al. 2018). In addition, a color distribution map can be generated in order to specify the visual appearance of the food sample.

Theoretically, an image is acquired by the incident light in the visible region passing through a partially reflective area with the scattered photons collected in the camera lens. Color space transformation is the most widely used pixel preprocessing approach for food quality detection in a visible light imaging system. Before the image is processed for further image processing and analysis, the image is captured in the three-dimensional (3D) RGB color space (Magwaza and Opara 2014). Nonetheless, the RGB color space is not quite effectively uniform since it represents the colors perceived visually by human vision. In this sense, the hue, saturation, value (HSV) and CIELAB (Commission Internationale de l'Eclairage L*a*b*) color spaces are recommended in food applications due to the nature of the perceptually uniform color spaces (Pathare et al. 2013). For the image acquisition of visible light imaging, it involves the scrupulous design of image capturing as well as attentive operation to attain high-quality images. Thus, color measurement using visible light imaging has proved highly efficient in reducing human inconsistency and subjectivity.

2.3 Image Processing and Analysis

The core theory of visible light imaging encompasses image processing and analysis that is able to quantify the food samples. Image processing procedures have been employed as an integral part in visible light imaging as it enables rapid signal processing for large sample datasets. The initial step in visible light imaging is image acquisition in which the quality of the image is highly emphasized. It is an important aspect in order to create high-quality images to proceed for the image processing steps. High-quality images could decrease the time taken for image processing as well as reduce the complexity of subsequent image processing procedures (Vithu and Moses 2016). Further, consistent light illumination, noise reduction, and appropriate image acquisition settings could also assist in the image processing steps (El-Mesery et al. 2019).

Image processing in food applications may comprise of three steps, namely, image enhancement, feature extraction, and feature classification. Image enhancement is a process of adjusting images for identification of key features in displaying the desired image (Renugambal and Senthilraja 2015). Image enhancement steps include such as morphological operations, filtering, and histogram equalization and are applied to adjust the discrepancies in the images obtained (Manickavasagan et al. 2014). In this case, the quality of the image is improved before performing image analysis procedures. Feature extraction may be performed to transform raw data into numerical features such as mean, variance, and standard deviation (Mohd Ali et al. 2020). Prior to this step, image segmentation is done to separate the desired image from the background. The image segmentation aims to obtain the region of interest (ROI) which is important for the image feature extraction (Singh et al. 2020). Once the image features have been extracted, feature classification is applied using various classifiers including a neural network, principal component (PC), fuzzy logic, etc. Several image segmentation operations including thresholding, region-based, and gradient-based methods are vital to identify the ROI of an image (Adebayo et al. 2016).

Interpretation and recognition are the final steps in the image processing operation involving visible light imaging. Most pattern recognition is performed using computer learning algorithms such as statistical learning, genetic algorithm, linear regression, and vector machine (Fu and Chen 2019). The goal of the learning algorithm is to simulate the decision-making process of human capability using automated approaches. Regardless of the purpose of qualitative and quantitative description, a training data set is created as a

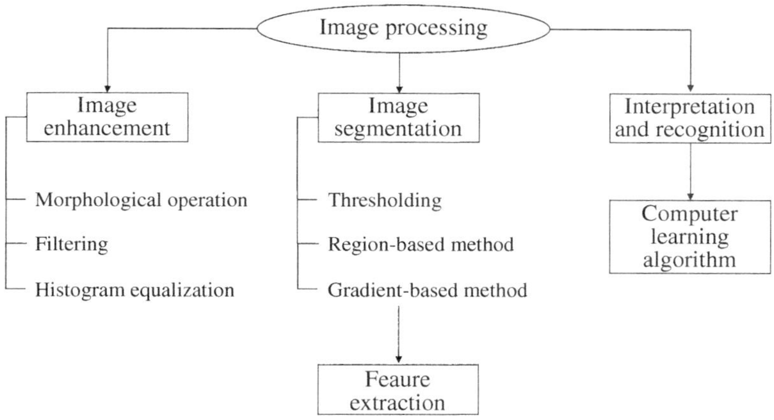

Figure 1 The mechanism of image processing steps involving visible light imaging.

reference base for a specific recognition task. The developed algorithm achieves the recognition accuracy based on the entire data set which is helpful and effective in subsequent image analysis (Dai et al. 2014). A mandatory check in the implementation of image processing operations is required to choose the appropriate image processing algorithms in order to obtain detailed information regarding the food sample. Hence, it is crucial to determine different image preprocessing approaches to validate a suitable data analysis approach. Figure 1 shows the mechanism of image processing steps involving visible light imaging.

3 Applications of Visible Light Imaging in Food

The applications of visible light imaging have been tested and intensively developed over a broad range of food products. This includes fruits, vegetables, meat, fish, poultry, nuts, grains, dairy products, oils, as well as processed foods. Table 1 shows a summary of food quality detection using visible light imaging.

3.1 Fruits and Vegetables

Visible light imaging has been revealed to be a practical technique for inspection of the quality of fruits and vegetables. Wang et al. (2016) assessed the sugar content of Japanese citrus fruit using visible light imaging. Color image parameters were extracted based on the threshold values during the image segmentation process. The segmentation errors were in the range of 0.15–6.03% based on the correlation analysis between the sugar content and the color parameters of the images. The findings demonstrated that the prediction of sugar content in citrus fruit is feasible to be performed in the visible spectrum without the need for an extra light source. Sendin et al. (2018) determined the discrimination of white maize using multispectral imaging in the visible region. Partial least squares-discriminant analysis (PLS-DA) models were established to classify between the healthy and infected maize

Table 1 The applications of visible light imaging for food quality evaluation.

Sample	Quality parameters	Data analysis	References
Citrus fruit	Sugar content	Correlation	Wang et al. (2016)
Maize	Disease classification	PLS-DA	Sendin et al. (2018)
Strawberry	Microbial content	PCA	Liu et al. (2019)
Blueberry	Bruise detection	Monte Carlo simulation	Zhang et al. (2019)
Peppers	Chilling injury	LDA	Delwiche et al. (2019)
Bananito fruit	Maturity detection	PLS	Pu et al. (2019)
Carrot	Cultivar detection	LDA, PLS	Česonienė et al. (2019)
Mandarin	Defect detection	PCA	Zhang et al. (2020)
Jujube	Soluble solid content	LS-SVM	Zhao et al. (2020)
Peach	Fungal contamination	PLS	Liu et al. (2020)
Grass carp fillet	Total volatile basic nitrogen	LS-SVM	Cheng et al. (2016)
Minced beef	Adulteration	PLS	Kamruzzaman et al. (2016)
Chicken breast	pH content	PLS	Jia et al. (2017)
Fish fillet	Freshness	PLS	Khoshnoudi-Nia et al. (2018)
Beef	Adulteration	PLS, SVM, LS-SVM	Zhao et al. (2019)
Pork meat	Protein content	BPNN	Ma et al. (2019)
Beef	Adulteration	PCR	Jiang et al. (2019)
Fish fillet	Freshness	SVM, decision trees, Naïve Bayes	Qin et al. (2020)
Peanut	Oil content	PLS	Jin et al. (2016)
Corn kernel	Aflatoxins content	LS-SVM	Zhu et al. (2016)
Milk powder	Functional properties	PLS	Munir et al. (2018)
Mung beans	Variety classification	LDA	Xie and He (2018)
Wheat	Deoxynivalenol content	PCA	Shi et al. (2020)
Rice	Variety classification	SVM	Wang et al. (2020)
Cheese	Starch content	PLS	Barreto et al. (2018)
Swiss cheese	Texture	PLS, ANN	Vásquez et al. (2018)
Cow ghee	Adulteration	Discriminant analysis	Wasnik et al. (2019)
Olive oil	Adulteration	ANN	Aguilera Puerto et al. (2019)
Olive oil	Oxidative stability	ANN, SVM	Sanaeifar and Jafari (2019)
Extra virgin olive oil	Adulteration	PLS	Song et al. (2020)
Butter cookies	Rancidity	PLS	Xia et al. (2016)

Table 1 (Continued)

Sample	Quality parameters	Data analysis	References
Roasted coffee beans	Bean classification	LDA	Nansen et al. (2016)
Tomato paste	Sucrose adulteration	LS-SVM, PLS, BPNN	Liu et al. (2017)
Honey	Origin identification	SVM, random forest	Minaei et al. (2017)
Potato chip	Acrylamide formation	SVM	Yadav et al. (2018)
Nutmeg	Adulteration	PCA, ANN, PLS-DA	Kiani et al. (2019)
Turmeric	Adulteration	PCA	Chaminda Bandara et al. (2020)

ANN, artificial neural network; BPNN, back-propagation neural network; LDA, linear discriminant analysis; LS-SVM, least squares-support vector machine; PCA, principal component analysis; PCR, principal component regression; PLS, partial least squares; PLS-DA, partial least squares-discriminant analysis; SVM, support vector machines.

with classification accuracies of up to 83%. The presence of lycopene and anthocyanin was detected at the visible wavelength peaks at 505, 525, 570, and 590 nm.

In another approach, Liu et al. (2019) reported the application of hyperspectral imaging to identify the microbial content of strawberries during decay. Principal component analysis (PCA) models were developed to predict the microbial content of strawberries. The results showed a good coefficient of correlation (R^2) of 0.93 with 10 essential principal components (PCs) extracted from the hyperspectral datasets in order to improve the prediction models. Zhang et al. (2019) evaluated the bruising of blueberries using Monte Carlo multilayered simulation based on a light propagation model in the visible region. The Monte Carlo multilayered simulation demonstrated significant differences between the bruised and nonbruised blueberries proving the backward light scattering method and good absorption of the skin of the blueberries. Delwiche et al. (2019) used hyperspectral fluorescence imaging ranging from 464 to 700 nm to monitor the physical changes of fresh-cut peppers stored at 4 °C for 21 days. Linear discriminant analysis (LDA) classifiers successfully distinguished fresh-cut peppers according to the storage days associated with the physiological effects of fluorescence absorption and chilling injury response. Pu et al. (2019) investigated the quality attributes of bananito fruits at three different maturity stages (stages 2, 4, and 6) using hyperspectral imaging in the visible region. Two wavelength selection methods were developed to optimize the PLS models, namely regression vector and variable importance in projection (VIP) scores. Based on the wavelength selection methods, three feature wavelengths were identified from the regression vector (650, 705, and 740 nm) and VIP scores (665, 705, and 740 nm), respectively.

Česonienė et al. (2019) evaluated the potential of visible multispectral imaging to determine the chemical and electro-chemical properties of different carrot cultivars under organic farming conditions. Based on the results, the overall classification accuracies of carrot cultivars achieved 94% using LDA models. PLS models were used to predict electrical conductivity and reduction potential with R^2 values of 0.88 and 0.81, respectively. Similarly, Zhang et al. (2020b) applied visible hyperspectral imaging to identify different type of defects (scarring, decay, thrips scarring, and anthracnose) on mandarins. Two wavelengths at 680 and 715 nm were selected for defect detection with a classification accuracy of 96% using the PCA model (Figure 2). Zhao et al. (2020b) explored the feasibility of hyperspectral imaging ranging from 400 to 700 nm to evaluate the soluble solid content (SSC) of winter jujube fruits. The least squares-support vector machine (LS-SVM) model was used to predict the SSC of the fruit with an R^2 of 0.89 and residual predictive deviation of 3.07, respectively.

Gao et al. (2020) classified aflatoxin content in maize using hyperspectral imaging ranging from 390 to 710 nm. Several classification methods including PCA, random forest, k-nearest neighbor (kNN), and the Relieff algorithm were used for maize classification. Based on the findings, the Relieff classifier achieved the highest classification accuracy of 99%, followed by random forest, kNN, and PCA. Liu et al. (2020) detected fungal contamination in peach using hyperspectral imaging in the visible region. A PLS model was established to predict the fungal colony counts with R^2 of 0.84. The classification of infected and healthy peach samples was visualized using a PCA score plot in order to reduce the dimensionality of hyperspectral data sets. A recent study conducted by Sun et al. (2021), evaluated the application of visible light imaging ranging from 650 to 700 nm to study the light propagation of citrus. The maximal diffuse reflectance distance was placed within 0.56 cm in the juice vesicles according to the combination of light source distance and power intensity.

3.2 Meat, Fish, and Poultry

The visible light imaging technique can be very useful in the quality evaluation of meat, fish, and poultry. Cheng et al. (2016) determined total volatile basic nitrogen in grass carp fillet during spoilage using multispectral imaging at 632 nm. The LS-SVM models were developed to predict total volatile basic nitrogen with an R^2 of 0.90 generated from six different genetic algorithms. Kamruzzaman et al. (2016) studied the potential of visible hyperspectral imaging ranging from 400 to 700 nm to identify adulteration in fresh minced beef with chicken. PLS models were applied to predict the adulteration levels of minced beef samples with an R^2 of 0.97 and transferred in terms of pixel image to obtain a prediction map. Likewise, Paluchowski et al. (2016) reported the application of visible hyperspectral imaging to classify Atlantic cod roe, milt, and liver samples. Classification maps were generated to discriminate between cod roe, milt, and liver using a one-band model and a two-band model with classification specificities of 94% and 98%, respectively. Jia et al. (2017) applied visible hyperspectral imaging to determine the pH of fresh chicken breast fillets. Eight pretreatment algorithms were performed in order to build a new PLS model with an R^2 of 0.94 based on the competitive adaptive reweighted sampling method.

Crichton et al. (2017) classified between fresh and frozen–thawed beef using hyperspectral imaging at 500–700 nm. They successfully distinguished between fresh and frozen–thawed beef with a correct classification rate of 1.00 using the entire visible range. Cheng et al. (2017) evaluated visible images to predict total volatile basic nitrogen of grass carp fish fillet. A PLS model extracted by the gray-level gradient co-occurrence matrix (GLCM) algorithm demonstrated a high R^2 value of 0.98 which was suitable to identify the degree of freshness during fish cold storage. Al-Sarayreh et al. (2018) evaluated adulteration in

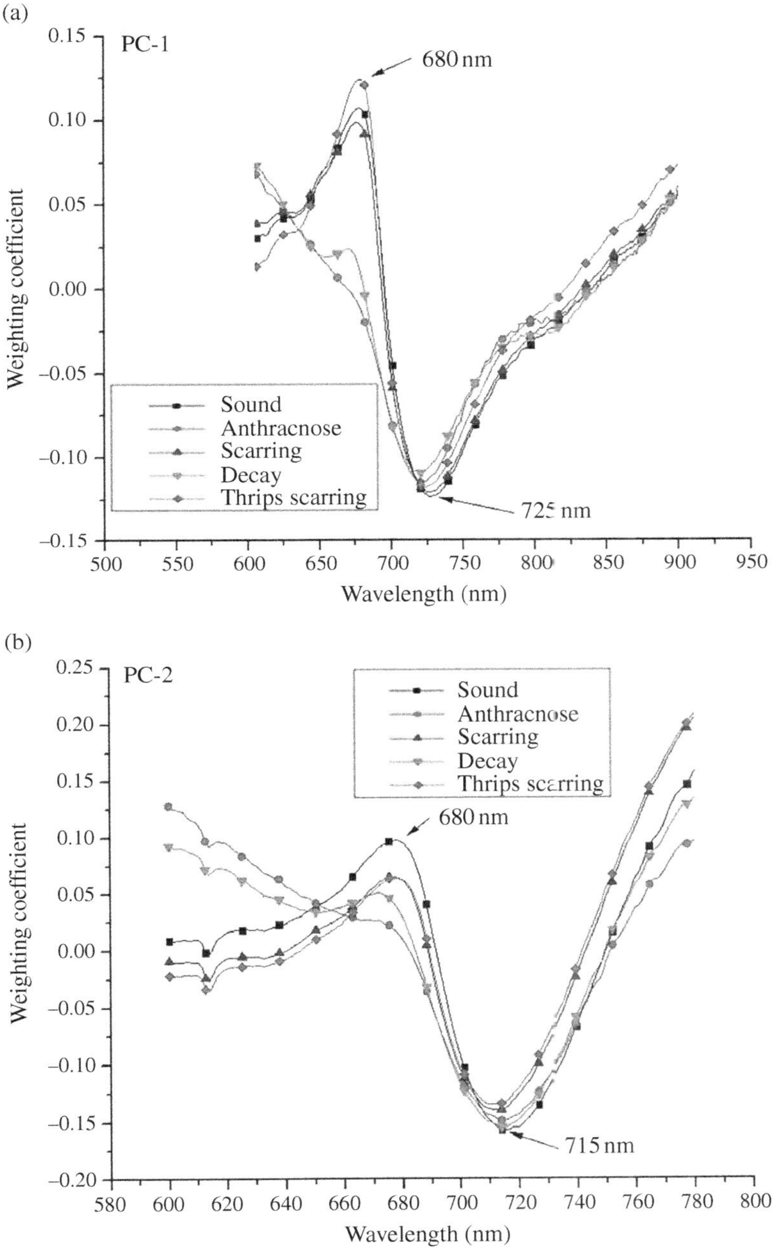

Figure 2 Weighting coefficient curves using visible images from (a) full spectral region (600–900 nm) and (b) visible spectral region (600–780 nm). *Source:* Zhang et al. (2020).

red-meat based on the texture properties acquired from visible images. It was observed that the addition of texture features to the normalized spectral features demonstrated the highest overall classification accuracy of 92%, providing a rapid technique for the detection of adulteration in red-meat products. Khoshnoudi-Nia et al. (2018) investigated the freshness indicators of rainbow trout fish fillets based on visible images. The prediction of total volatile basic nitrogen was determined using a PLS model with the highest R^2 value of 0.89 which showed a promising method for the type of colony counting determination.

Jiang et al. (2018a) aimed to determine the tenderness of intact fresh chicken breast fillets using hyperspectral imaging in the visible region. In that study, the Warner-Bratzler shear force value was measured as a tenderness indicator in order to classify the fillets according to the appropriate tenderness categories. A radial basis function-SVM was applied to predict the fillet tenderness with a correct classification rate of 0.92. Jiang et al. (2018b) used visible hyperspectral imaging ranging from 400 to 700 nm to evaluate color changes and the pH of broiler breast fillets. Distribution maps were generated by pixel prediction in the visible images in which the color changes and pH of fillets were observed. The LDA models successfully predicted the color changes (L*, a*, and b*) with R^2 values higher than 0.87. However, a low R^2 value of 0.52 was observed for the pH of broiler breast fillets. In a similar sense, Shi et al. (2019) studied the freshness of tilapia fillets stored at nonisothermal conditions using visible imaging ranging from 400 to 700 nm. The findings revealed that the nine optimal wavelengths within the visible region were selected using successive projection algorithms to identify the freshness of tilapia fillets.

Zhao et al. (2019) evaluated adulteration in beef using visible hyperspectral imaging coupled with chemometrics. Four chemometric methods were compared including PLS, SVM, LS-SVM, and extreme learning machine. The LS-SVM model showed the best performance among the four multivariate methods for identification of spoiled adulterant in beef samples with an R^2 of 0.97. Figure 3 shows the visualization of adulteration in fresh minced beef samples in which the color progression from dark blue (0%) to dark red (100%), signified the spoiled beef adulteration percentages from low (fresh) to very high (fully spoiled). Ma et al. (2019) assessed the application of visible light imaging ranging from 465 to 630 nm to determine the protein content of pork meat. A back propagation-neural network (BPNN) model was developed combined with absorbance spectra to predict the protein content of pork meat with an R^2 value of cross-validation set of 0.83 and root mean square error of cross validation (RMSECV) of 8.38. Khoshnoudi-Nia and Moosavi-Nasab (2019) reported the spoilage of fish fillets stored at 4 °C for 12 days using multispectral imaging 430–700 nm. Nine optimal wavelengths were selected using a genetic algorithm to simplify the calibration models. Based on the findings, the LS-SVM model showed the best prediction for total volatile basic nitrogen and psychrotrophic plate count with R^2 values higher than 0.85.

Jiang et al. (2019) identified beef adulteration with duck meat using visible hyperspectral images ranging from 400 to 700 nm. Optimal wavelengths were selected by PC loadings to develop a prediction model based on the principal component regression (PCR) method. The visualization validity of the adulteration levels demonstrated good prediction with an R^2 of 0.96 and a limit of detection of 7.6%. Qin et al. (2020) determined the freshness and mislabeling of six species of fish fillets including tilapia, white bass, red snapper, Malabar snapper, summer flounder, and vermilion snapper using visible hyperspectral imaging. Six classifiers were used, namely SVM, kNN, discriminant analysis, decision trees, Naïve

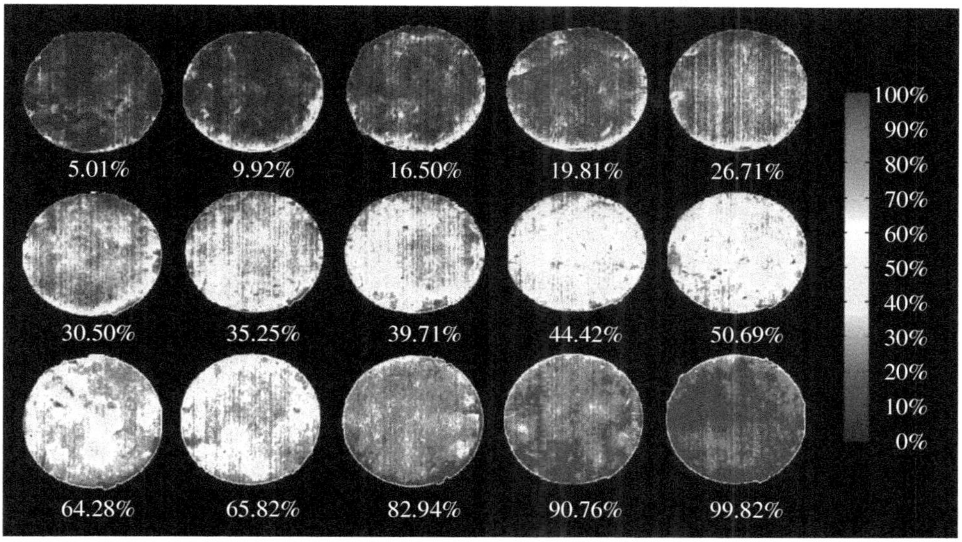

Figure 3 Visualization of adulteration in fresh minced beef samples. (*See insert for color representation of this figure*). *Source:* Zhao et al. (2019).

Bayes, and an ensemble classifier to discriminate the freshness level of fish fillet with the highest accuracies of 100% using full visible reflectance spectra. Rady and Adedeji (2020) evaluated the adulteration of minced beef and pork by using visible hyperspectral imaging ranging from 400 to 700 nm. Based on the results, the classification rates achieved 75–100% for the discrimination of pure and adulterated samples using the optimal wavelengths of the visible region. A recent study performed by Jiang et al. (2020) revealed that visible hyperspectral imaging could be used to investigate the adulteration in minced pork jowl meat in the range of 0–100% (w/w) at 10% increments. PLS models indicated that visible spectra pretreated using standard normal variate showed the best prediction of detection of adulteration with an R^2 value of 0.95.

3.3 Nuts, Grains, and Dairy Products

The application of visible light imaging has also been widely used for quality determination of nuts, grains, and dairy products. Jin et al. (2016) evaluated visible hyperspectral imaging to predict oil content in different peanut varieties. Six optimal wavelength bands within the visible region were selected to develop PLS models which presented a high R^2 value of 0.93. Zhu et al. (2016) evaluated visible hyperspectral images to identify aflatoxins in corn kernels. Based on the results, the contaminated corn kernels displayed distinct and visible contaminant compared to the healthy corn kernels. An LS-SVM model showed the best prediction accuracy of 95% in the integrated analysis between the germ side and the endosperm side. Bogomolov et al. (2017) developed visible reflectance imaging ranging from 400 to 700 nm to identify the fat and protein content in milk. The modeling accuracy generated from the genetic algorithm was enhanced using joint variable selection with root mean square errors lower than 0.10% and 0.08% for the fat and protein content, respectively.

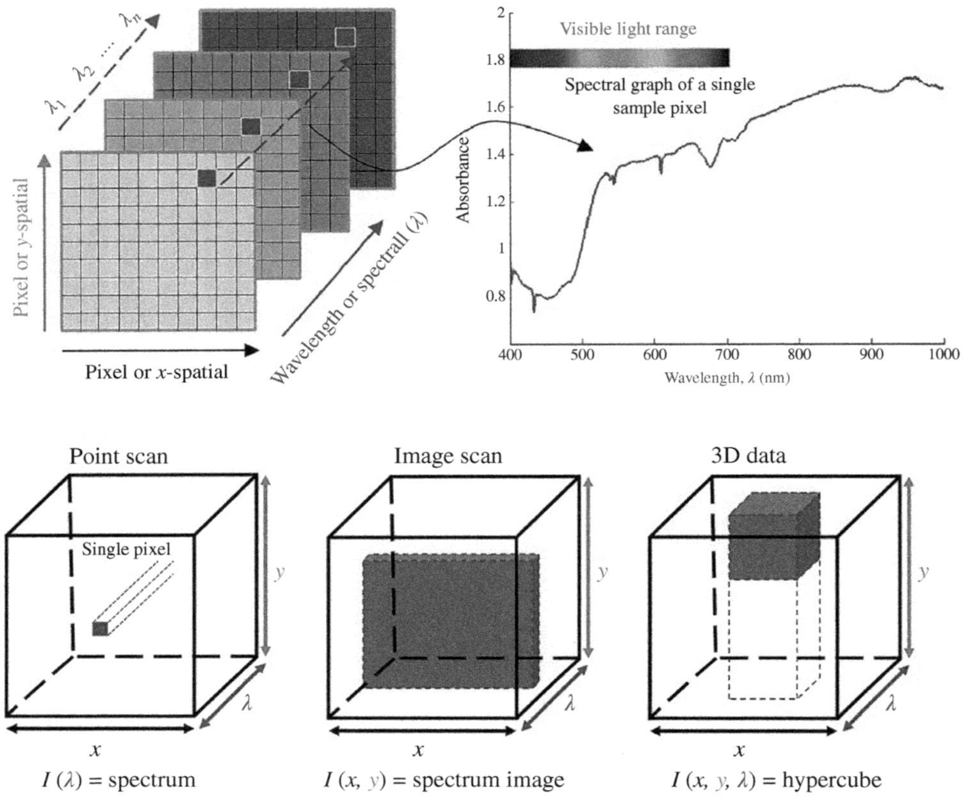

Figure 4 Visible images as a function of wavelength and data structures. (*See insert for color representation of this figure*). *Source:* Munir et al. (2018).

In other research, Munir et al. (2018) used visible hyperspectral imaging to classify milk powders based on different functional qualities. The PLS results described that data information from the visible images was applied to predict the accuracy of the functional property of milk samples. A spatial map based on the visible light range was generated as a function of wavelength. The spectrum variation created a 3D hypercube through the superimposition of spatial images as shown in Figure 4. Each spatial image of the hypercube was composed of pixels describing the reflection of a single visible wavelength. Xie and He (2018) performed the classification of four different varieties of mung beans based on visible images ranging from 400 to 700 nm. Both LDA and extreme learning machine models were established and achieved high correct classification rates of up to 99%. Four wavelength bands (432, 455, 468, and 560 nm) were selected as an effective wavelength for mung bean classification using the modified Gram-Schmidt method.

In addition, Acorsi et al. (2019) assessed the feasibility of visible light imaging combined with unmanned aerial vehicles (UAVs) to determine the fresh and dry biomass of a black oat crop. Based on the prediction models for fresh and dry biomass, the application of visible UAV-based imaging successfully determined the fresh and dry biomass with R^2 values of 0.68–0.92. Alisaac et al. (2019) studied the infection of wheat kernels and flour using visible hyperspectral imaging. It was revealed that the reflectance of wheat kernels was positively correlated to the deoxynivalenol (DON) content. For flour detection, the coefficient correlation (*r*) obtained 0.80 in the visible range. Zhang et al. (2020a) identified the infection of *Fusarium* head blight in wheat kernels using visible light imaging. Several image classifiers including PCA, random forest, SVM, Naïve Bayes, and successive projection algorithms were compared to determine the best classification models. It was revealed that the combination of successive projection algorithms and random forest methods provided the highest classification accuracy of 96% among all the classifiers. In another study, Wang et al. (2020b) compared the visible images of eight millet varieties according to the color and texture features. An attention-convolutional recurrent neural network (CRNN) and SVM models were built for the classification of millet varieties. The overall identification accuracy for millet varieties achieved 88% and 78% for the CRNN and SVM models, respectively.

Zhao et al. (2020a) investigated the potential of visible hyperspectral imaging to identify the severity of wheat infected with powdery mildew. It was observed that the SVM model built from the PCA dimensionality obtained the highest classification accuracy of 93%. Similarly, França-Silva et al. (2020) detected seed-borne fungi on black oat using visible imaging ranging from 400 to 700 nm. A LDA model was developed using the color and texture features of the black oat images. The findings demonstrated good performance of visible imaging in the detection of seed-borne fungi with a classification accuracy of 0.86. Shi et al. (2020) used multispectral imaging in the visible region to determine the DON content contamination in wheat. A PCA model coupled with a genetic algorithm presented the best prediction of DON content contamination with an R^2 value of 0.99. Wang et al. (2020a) reported the application of visible imaging 380–700 nm to evaluate herbicide toxicity in two rice varieties (Xiushui 134 and Zhejing 88). The visible images of rice were compared to identify the differences between the samples. The results revealed that support vector classification models achieved classification accuracy higher than 80%. Zhejing 88 demonstrated better classification results than Xiushui 134, indicating the variation in herbicide tolerance among the rice varieties.

3.4 Fats and Oils

Visible light imaging has been discovered to be a practical technique for quality assessment of fats and oils. Karagiorgos et al. (2017) investigated the adulteration of olive oil with soybean oil using visible imaging which was integrated into an application for smartphones. The visible images of olive oil samples were obtained by classifying the captured images into two regions: oil samples only and olive oils with adulterants. Based on the results, the color difference between the two regions was used to develop a linear regression model, creating an average absolute error of 3%. Romaniello and Baiano (2018) examined the classification of three different flavored olive oils (basil, chili pepper, and garlic with chili pepper) using visible hyperspectral imaging. It was revealed that the wavelength ranging from 400 to 570 nm and 695 nm was appropriate for the classification of flavored olive oils. Barreto et al. (2018) aimed to evaluate the starch content in fresh cheese using visible hyperspectral imaging. The visible images were preprocessed to identify the ROI. The prediction of starch content was obtained using PLS based on calibration and validation models with an R^2 of 0.99 based on five latent variables. Figure 5 shows the main procedures for the prediction of starch content in fresh cheese using visible hyperspectral imaging.

Vásquez et al. (2018) evaluated the texture of Swiss-type cheese during ripening using visible hyperspectral imaging ranging from 400 to 700 nm. The relationship between the texture features of cheese and spectral profiles was modeled using PLS and artificial neural network (ANN) models. Based on the R^2 values of the proposed models, the ANN model ($R^2 = 0.96$) demonstrated better performance compared to the PLS model ($R^2 = 0.94$). Wasnik et al. (2019) developed prediction models for adulteration levels of cow ghee adulterated with 5%, 10%, 15%, and 20% vegetable fat. Visible images were acquired and evaluated in terms of color, pixel intensity, textural, and morphological properties of the cow ghee samples. The adulteration levels were classified using discriminant analysis with an overall classification accuracy of 92%. Minz and Saini (2019) assessed a visible imaging system to determine the effect on color evaluation of mozzarella cheese. The study suggested a fundamental basis for selecting color shades for the RGB cube calibration method using four different types of calibration chart: Macbeth matt, Macbeth gloss, color cube matt, and color cube gloss. From the findings, it was noted that the color cube gloss chart gave the best calibration performance among all the calibration charts, signifying good color evaluation accuracy for mozzarella cheese. In addition, Aguilera Puerto et al. (2019) reported the application of a visible imaging system to classify olive oil according to the quality level. Several texture features based on the visible images were selected including image entropy, homogeneity, contrast, and energy to develop an ANN model with high classification accuracy up to 98%.

Likewise, Beyaz et al. (2019) determined the detection of olive fly stings in virgin olive oil production using a visible imaging system. The fly sting detection was evaluated based on the defective areas and bruises on the fruit surface. To enhance the virgin olive oil production, classification algorithms were built using a LDA model with a success rate of 93%. Sanaeifar and Jafari (2019) demonstrated the use of visible imaging to measure the oxidative stability index of olive oil. Three multivariate algorithms were employed including multiple linear regression (MLR), ANN, and SVM to predict the oxidative stability. Among the three algorithms, the SVM model showed the best results for oxidative stability of olive oil with an R^2 of 0.97 indicating a potential alternative for rapid evaluation of olive oil oxidation. A recent study investigated by Gila et al. (2020) discussed the moisture content in virgin olive oils using a visible imaging system. High correlation was achieved by artificial vision with an R^2 of 0.99 signifying a useful method for assessment of insoluble impurities in virgin olive oils. Song et al. (2020) employed the use of visible imaging coupled with chemometrics to determine the adulterant levels in extra virgin olive oils. PLS models were used based on the visible images for the quantification of vegetable oil in extra virgin olive oils ranging from 5% to 50% (v/v). The results yielded a high R^2 of 0.98 indicating high potential for detection of adulteration in edible oils.

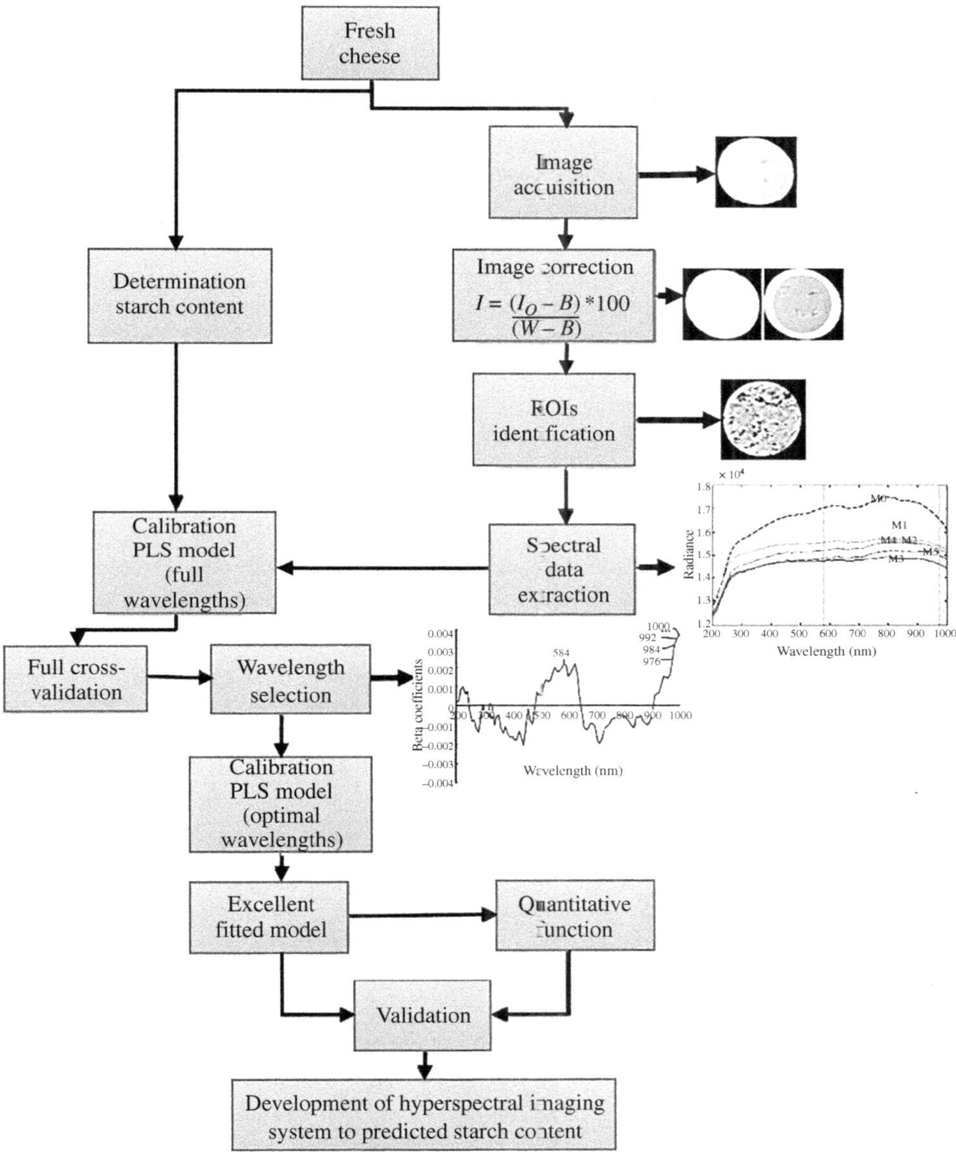

Figure 5 Main procedures for the starch content prediction in fresh cheese. *Source:* Barreto et al. (2018).

3.5 Processed Foods

Visible light imaging has advantageous applicability for processed foods in order to produce high accuracy for various food quality detection processes. Xia et al. (2016) reported the application of visible multispectral imaging ranging from 405 to 700 nm to determine rancidity in butter cookies. PLS models successfully obtained good prediction of the acid value (R^2 = 0.94), peroxide value (R^2 = 0.97), and moisture content (R^2 = 0.91) of butter cookies, respectively. Nansen et al. (2016) classified different brands of roasted coffee beans into four classes including light, medium, medium-dark, and dark roast using visible hyperspectral imaging ranging from 408 to 700 nm. High classification accuracies of up to 90% were obtained using LDA models in the discrimination of coffee bean samples. Liu et al. (2017) examined the sucrose adulteration in tomato paste using visible multispectral imaging ranging from 405 to 700 nm. LS-SVM, PLS, and BPNN models were established to predict the performance of sucrose adulteration. It was signified that LS-SVM achieved the highest prediction value (R^2 = 0.96) compared to the PLS and BPNN models.

Minaei et al. (2017) explored the application of visible hyperspectral imaging for the identification of the origin of floral honey using three different machine learning algorithms (random forest, SVM, and radial basis function network). Based on the results, the radial basis function network demonstrated the best classifier with an accuracy of 94% for the detection of the origin of floral honey. In a similar manner, Yadav et al. (2018) detected acrylamide formation in fried potato chips using the visible imaging technique. The ROI of the potato chip was segmented from the image background followed by the extraction of selected features based on the wavelet transform domain. Figure 6 shows the visible images of potato chips with different conditions including normal, mild acrylamide formation,

Figure 6 Visible images of fried potato chips: (a) normal potato chip, (b) potato chip with mild acrylamide formation, (c) potato chip with moderate acrylamide formation, and (d) potato chip with severe acrylamide formation. (*See insert for color representation of this figure*). *Source:* Yadav et al. (2018).

moderate acrylamide formation, and severe acrylamide formation. A SVM model was developed to evaluate the identification of acrylamide in the potato chips, resulting in an accuracy of up to 98%. Kiani et al. (2019) studied the adulteration in nutmeg mixed with spent powder (5–60%) using visible hyperspectral imaging. Three algorithms were applied including PCA, ANN, and PLS-DA to develop the model for the identification of authentic nutmeg samples from adulterant materials. The ANN model showed the best prediction ability in order to detect adulteration levels in nutmeg as low as 5%, indicating good potential for spice powder authentication.

Apart from that, Udomkun et al. (2019) discovered the feasibility of a visible imaging system to determine oil content and color changes of durian chips during frying. A high linear correlation was obtained with R^2 values higher than 0.92 in order to assess the surface oil content and browning index of fried durian chips. Chaminda Bandara et al. (2020) identified adulteration in turmeric samples using visible multispectral imaging ranging from 405 to 700 nm. A machine learning algorithm based on PCA and Bhattacharyya Distance was established and modeled using a second-order polynomial, resulting in an R^2 of 0.99. In another study, Li et al. (2020) evaluated the foreign matter in dried and pickled mustard using visible multispectral imaging. The pixel intensity was extracted from the images for the development of SVM and BPNN models for the detection of foreign matter in dried and pickled mustard. The BPNN model was selected as the optimal classification model since the identification accuracy recorded the highest classification (98%) for the detection of foreign matter in dried and pickled mustard. Chen et al. (2020) investigated the application of visible hyperspectral imaging to obtain coffee bean images for the detection of defects. It was revealed that the SVM model showed the highest classification accuracy of 95% for the defect detection in coffee beans, which could be useful for commercial use in the future.

4 Advantages and Limitations

Visible light imaging is an emerging and innovative technique, which is reliable for quality inspection of food products. This technique defines the behavior and physical properties of a sample when it is exposed to light. Visible data information depends on several parameters such as the wavelength region of the electromagnetic spectrum and the imaging mode. Data information acquired using visible light imaging comprises of variations because of the external factors and system components. Apart from that, the heterogeneity and condition of the food products could result in differences when retrieving the acquired data. These external factors such as noise and light scattering must be eliminated prior to retrieving data information using the appropriate chemometrics. Qualitative information involving chemical attributes, defects, and adulteration can be obtained using discriminant analysis tools. In contrast, quantitative information such as contamination level and maturity classification can be determined using regression analysis. In combination with image preprocessing and chemometrics, visible light imaging stands out as one of the prominent quality and safety inspection methods for food products.

The evaluation of food quality using visible light imaging yields high speed and accurate results. Physical attributes including shape, size, color, and texture can be determined easily by visible light imaging, whereas internal attributes require multivariate tools and predictive models to facilitate the nondestructive assessment of food products. Visible light imaging has revolutionized the evaluation of the quality of food products by delivering both quantitative and qualitative information simultaneously since it does not need any separate sample preparation. In addition, this technique is relatively safe for the environment and is economical compared to the conventional methods which involve supplementary costs for chemical reagents and labor skills. For this reason, both quantitative and qualitative studies as well as multiple constituents can be performed using similar images. Even though the food samples may be similar in terms of color and morphological properties, the variation of chemical content is sufficient enough to categorize the food classes accurately. In this case, once the predictive models are calibrated and validated, the analysis of quality attributes of the food samples refines the data to perform critical analysis.

Despite the advantages of this technique, visible light imaging also has some limitations or constraints. From the analyst point of view, structured calibration and precise model establishment are required as visible light imaging is an indirect analysis. For massive data sets, it requires high-speed computers and very large capacity storage for data processing and analysis. The identification and detection of chemical compounds also require efficient and powerful analytical devices in order to understand the minor chemical constituents. Nonetheless, visible light imaging is still an effective method due to the significant aspects including the data computing capabilities and integration of chemometrics for data refining. It is suggested that the camera settings and components of visible light imaging system need to be stabilized before conducting data acquisition. Hence, it is critical to comprehend the implications of external factors on visible light imaging for a successful operation and application of quality and safety evaluation of food products.

5 Future Trends

Visible light imaging has proven to be a promising tool for quality and safety inspection of various food products. This technique can be realized using different imaging modes depending on the type of application and equipment availability. Image preprocessing analysis coupled with chemometrics is typically used to evaluate the food quality. Future work should emphasize the application of different chemometric methods in order to enhance the determination of quality and safety inspection of food products. In addition, the development of data mining is suitable to extract the data information by providing additional details of the food samples. In view of the existing applications of visible light imaging, it is expected that the implementation of this technique can be integrated with other imaging systems in the food production lines. Having this possible integration in the near future, the development of in-line visible light imaging as a machine vision device can be achieved, especially in terms of real-time quality monitoring techniques.

Additionally, visible light imaging needs to be associated with controllable process variables to achieve real-time quality control. In some cases, it is worthwhile to select a specific wavelength within the visible range rather than using the full range of the spectrum, so that the data can be processed faster due to the smaller spectral range. In terms of hardware and software for a visible light imaging system, the image processing and analysis are relatively slow, especially with a large number of data sets. Taking the food application as an example, most of the related studies are involved in the evaluation of physical and chemical properties of food products within similar groups. In spite of the foregoing works for the nondestructive evaluation of food quality, various image processing methods can be implemented for reducing random errors such as the GLCM, Otsu method, scale-invariant feature transform (SIFT), wavelet-based decomposition (Patrício and Rieder 2018). Rapid algorithm tools need to be enhanced subjected to real-time quality monitoring using visible light imaging to shorten the time taken for data processing and the image acquisition process.

In the context of detection of food quality, the development of sensing design and analytical features is an important aspect as it may be influenced by environmental factors or phenotypic traits. In the long run, there are gaps to be occupied with the recent advances in the artificial intelligence field. Expanding the use of artificial intelligence as well as integration with visible light imaging are also promising alternatives for the need for agile and effective utilization of the system. Diverse applications in the food industry could be explored with advancements in artificial intelligence which can pave the way to new solutions for the evaluation of the quality of food products using visible light imaging. Hence, it is critical for researchers from various fields to engage in technical encounters related to food quality detection, allowing for the possibility of integrated and applicable tools.

6 Conclusions and Outlook

Visible light imaging has demonstrated the ability to be very efficient and reliable for reducing industrial dependence on human skills. The accuracy and cost-effectiveness of visible light imaging allow the evaluation of various attributes of food quality including shape, size, color, defects, etc. The technology assists in the food quality measurement due to the simplicity and flexibility of the system. Visible light imaging serves as a new alternative and is well suited for the inspection of food products, especially during processing and production lines. Driven by the demand of nondestructive applications in the food industry, visible light imaging has been developing rapidly over the decades in order to determine physical, chemical, and biological properties of a wide range of food products. As the technology evolves, visible light imaging has become commercially available to deliver information based on food products and further develop practical food applications. The advances in instrumentation and hardware components coupled with multivariate methods may drive the future expansion of the visible light imaging technique. Recent applications in visible light imaging have proven to be an alternative for conventional techniques and have become a promising device for real-time food quality and safety inspection. Hence, attempts at new developments should be exploited for practical in-line applications which provide pathways for future studies.

Visible light imaging for the detection of food quality offers a wide variety of food-related determinations. An increasing number of visible light imaging applications have arisen in order to meet the demands of modern technologies in the food industries. With this, huge contribution from this technique is necessary to move forward the developed applications from the laboratory scale to the fieldwork. The success of a visible light imaging technique relies on the commercial values of the equipment to enhance the quality standards of the food products. Particular attention has been paid to the development of future applications of the nondestructive methods reinforced by the availability of the sensing device. To further improve the novelty of the applications, it is essential to increase the accuracy and reliability of the imaging devices, specifically for online and real-time evaluation. Besides, the combination or hybrid of different imaging methods could overcome the shortcomings of a single approach as an alternative to solve related tasks. It is still worth making a venture into the visible light imaging technique since there is a high potential for practical and large-scale development for the detection of food quality.

For this purpose, the need for portable and on-field sensing devices is desired to monitor the quality and processing of food products. Looking into potential future trends in food quality inspection, the integration of hybrid technologies consists of several imaging applications that could be implemented for automated harvesting and grading operation. Apart from that, the integration with the Internet of Things technology could be utilized for monitoring food quality and enhanced the degree in the automated system in yield traceability. In order to achieve real-time operation, the accuracy of visible light imaging applications could be monitored based on the processing speed of algorithms, implementation of the low-cost device, as well as the enhancement in data analysis. In addition, the simplification of the hardware to provide a more robust and user-friendly platform is needed to improve the predictive ability of the systems. The design of a simplified version of hardware was developed in order to allow the expansion of a compact device by reducing the dependency of trained personnel. The recent advances in visible light imaging methods lead to miniaturization and paved the way for developing new approaches of diverse food products. Thus, future efforts should be directed to focus on the behavior of food products in order to assist in postharvest processes, especially in food quality detection.

Acknowledgment

The authors would like to acknowledge and thank the Department of Biological and Agricultural Engineering, Faculty of Engineering, Universiti Putra Malaysia for providing the support and technical facilities. This work was supported by the Putra Grant, GP-IPB (Vot. No.: 9687800).

Conflict of Interest

The authors have declared no conflict of interest.

References

Acorsi, M.G., das Dores Abati Miranda, F., Martello, M. et al. (2019). Estimating biomass of black oat using UAV-based RGB imaging. *Agronomy* 9 (7): 1–14.

Adebayo, S.E., Hashim, N., Abdan, K., and Hanafi, M. (2016). Application and potential of backscattering imaging techniques in agricultural and food processing – a review. *Journal of Food Engineering* 169: 155–164.

Aguilera Puerto, D., Cáceres Moreno, Ó., Martínez Gila, D.M. et al. (2019). Online system for the identification and classification of olive fruits for the olive oil production process. *Journal of Food Measurement and Characterization* 13 (1): 716–727.

Alisaac, E., Behmann, J., Rathgeb, A. et al. (2019). Assessment of fusarium infection and mycotoxin contamination of wheat kernels and flour using hyperspectral imaging. *Toxins* 11 (10): 1–18.

Al-Sarayreh, M., Reis, M.M., Yan, W.Q., and Klette, R. (2018). Detection of adulteration in red meat species using hyperspectral imaging. In: *Image and Video Technology. Lecture Notes in Computer Science (including subseries Lecture Notes in Artificial Intelligence and Lecture Notes in Bioinformatics)* (eds. M. Paul, C. Hitoshi and Q. Huang), 182–196. Cham: Springer.

Barreto, A., Cruz-Tirado, J.P., Siche, R., and Quevedo, R. (2018). Determination of starch content in adulterated fresh cheese using hyperspectral imaging. *Food Bioscience* 21: 14–19.

Beyaz, A., Martínez Gila, D.M., Gómez Ortega, J., and Gámez García, J. (2019). Olive fly sting detection based on computer vision. *Postharvest Biology and Technology* 150: 129–136.

Bogomolov, A., Belikova, V., Galyanin, V. et al. (2017). Reference-free spectroscopic determination of fat and protein in milk in the visible and near infrared region below 1000 nm using spatially resolved diffuse reflectance fiber probe. *Talanta* 167: 563–572.

Cayuela, J.A. and Camino, M.D.C.P. (2010). Prediction of quality of intact olives by near infrared spectroscopy. *European Journal of Lipid Science and Technology* 112: 1209–1217.

Česonienė, L., Masaitis, G., Mozgeris, G. et al. (2019). Visible and near-infrared hyperspectral imaging to describe properties of conventionally and organically grown carrots. *Journal of Elementology* 24 (2): 421–435.

Chaminda Bandara, W.G., Kasun Prabhath, G.W., Sahan Chinthana Bandara Dissanayake, D.W. et al. (2020). Validation of multispectral imaging for the detection of selected adulterants in turmeric samples. *Journal of Food Engineering* 266: 1–12.

Chen, W.T., Yeh, Y.H.F., Liu, T.Y., and Lin, T.T. (2016). An automated and continuous plant weight measurement system for plant factory. *Frontiers in Plant Science* 7: 1–9.

Chen, S.-Y., Chang, C.-Y., Ou, C.-S., and Lien, C.-T. (2020). Detection of insect damage in green coffee beans using VIS-NIR hyperspectral imaging. *Remote Sensing* 12 (15): 2348.

Cheng, J.H., Sun, D.W., Qu, J.H. et al. (2016). Developing a multispectral imaging for simultaneous prediction of freshness indicators during chemical spoilage of grass carp fish fillet. *Journal of Food Engineering* 182: 9–17.

Cheng, J.H., Sun, D.W., and Wei, Q. (2017). Enhancing visible and near-infrared hyperspectral imaging prediction of TVB-N level for fish fillet freshness evaluation by filtering optimal variables. *Food Analytical Methods* 10 (6): 1888–1898.

Crichton, S.O.J., Kirchner, S.M., Porley, V. et al. (2017). Classification of organic beef freshness using VNIR hyperspectral imaging. *Meat Science* 129: 20–27.

Dai, Q., Sun, D.W., Xiong, Z. et al. (2014). Recent advances in data mining techniques and their applications in hyperspectral image processing for the food industry. *Comprehensive Reviews in Food Science and Food Safety* 13 (5): 891–905.

Delwiche, S.R., Stommel, J.R., Kim, M.S. et al. (2019). Hyperspectral fluorescence imaging for shelf life evaluation of fresh-cut Bell and Jalapeno Pepper. *Scientia Horticulturae* 246: 749–758.

Du, Z., Zeng, X., Li, X. et al. (2020). Recent advances in imaging techniques for bruise detection in fruits and vegetables. *Trends in Food Science and Technology* 99: 133–141.

El-Mesery, H.S., Mao, H., and Abomohra, A.E.F. (2019). Applications of non-destructive technologies for agricultural and food products quality inspection. *Sensors (Switzerland)* 19 (4): 1–23.

França-Silva, F., Rego, C.H.Q., Gomes-Junior, F.G. et al. (2020). Detection of *Drechslera avenae* (Eidam) Sharif [*Helminthosporium avenae* (Eidam)] in black oat seeds (*Avena strigosa* Schreb) using multispectral imaging. *Sensors* 20: 1–10.

Fu, X. and Chen, J. (2019). A review of hyperspectral imaging for chicken meat safety and quality evaluation: application, hardware, and software. *Comprehensive Reviews in Food Science and Food Safety* 18 (2): 535–547.

Gao, J., Ni, J., Wang, D. et al. (2020). Pixel-level aflatoxin detecting in maize based on feature selection and hyperspectral imaging. *Spectrochimica Acta Part A: Molecular and Biomolecular Spectroscopy* 234: 1–8.

Gila, A., Bejaoui, M.A., Beltrán, G., and Jiménez, A. (2020). Rapid method based on computer vision to determine the moisture and insoluble impurities content in virgin olive oils. *Food Control* 113: 1–5.

Hussain, A., Pu, H., and Sun, D.W. (2018). Innovative nondestructive imaging techniques for ripening and maturity of fruits – a review of recent applications. *Trends in Food Science and Technology* 72: 144–152.

Ireri, D., Belal, E., Okinda, C. et al. (2019). A computer vision system for defect discrimination and grading in tomatoes using machine learning and image processing. *Artificial Intelligence in Agriculture* 2: 28–37.

Jia, B., Yoon, S.C., Zhuang, H. et al. (2017). Prediction of pH of fresh chicken breast fillets by VNIR hyperspectral imaging. *Journal of Food Engineering* 208: 57–65.

Jiang, H., Yoon, S.C., Zhuang, H. et al. (2018a). Tenderness classification of fresh broiler breast fillets using visible and near-infrared hyperspectral imaging. *Meat Science* 139: 82–90.

Jiang, H., Yoon, S.C., Zhuang, H. et al. (2018b). Non-destructive assessment of final color and pH attributes of broiler breast fillets using visible and near-infrared hyperspectral imaging: a preliminary study. *Infrared Physics and Technology* 92: 309–317.

Jiang, H., Wang, W., Zhuang, H. et al. (2019). Hyperspectral imaging for a rapid detection and visualization of duck meat adulteration in beef. *Food Analytical Methods* 12 (10): 2205–2215.

Jiang, H., Cheng, F., and Shi, M. (2020). Rapid identification and visualization of jowl meat adulteration in pork using hyperspectral imaging. *Foods* 9 (154): 1–16.

Jin, H., Ma, Y., Li, L., and Cheng, J.H. (2016). Rapid and non-destructive determination of oil content of peanut (*Arachis hypogaea* L.) using hyperspectral imaging analysis. *Food Analytical Methods* 9 (7): 1–8.

Kamruzzaman, M., Makino, Y., and Oshita, S. (2016). Rapid and non-destructive detection of chicken adulteration in minced beef using visible near-infrared hyperspectral imaging and machine learning. *Journal of Food Engineering* 170: 8–15.

Karagiorgos, N., Nenadis, N., Trypidis, D., Siozios, K., Siskos, S., Nikolaidis, S., and Tsimidou, M.Z. 2017. An approach for estimating adulteration of virgin olive oil with soybean oil using image analysis. *2017 6th International Conference on Modern Circuits and Systems Technologies, MOCAST 2017*. Institute of Electrical and Electronics Engineers Inc. 1–4.

Khoshnoudi-Nia, S. and Moosavi-Nasab, M. (2019). Prediction of various freshness indicators in fish fillets by one multispectral imaging system. *Scientific Reports* 9 (1): 1–11.

Khoshnoudi-Nia, S., Moosavi-Nasab, M., Nassiri, S.M., and Azimifar, Z. (2018). Determination of total viable count in rainbow-trout fish fillets based on hyperspectral imaging system and different variable selection and extraction of reference data methods. *Food Analytical Methods* 11 (12): 3481–3494.

Kiani, S., van Ruth, S.M., van Raamsdonk, L.W.D., and Minaei, S. (2019). Hyperspectral imaging as a novel system for the authentication of spices: a nutmeg case study. *LWT - Food Science and Technology* 104: 61–69.

Li, L., Zhang, Q., and Huang, D. (2014). A review of imaging techniques for plant phenotyping. *Sensors* 14 (11): 20078–20111.

Li, M., Huang, M., Zhu, Q. et al. (2020). Pickled and dried mustard foreign matter detection using multispectral imaging system based on single shot method. *Journal of Food Engineering* 285: 1–12.

Liu, C., Hao, G., Su, M. et al. (2017). Potential of multispectral imaging combined with chemometric methods for rapid detection of sucrose adulteration in tomato paste. *Journal of Food Engineering* 215: 78–83.

Liu, Q., Sun, K., Zhao, N. et al. (2019). Information fusion of hyperspectral imaging and electronic nose for evaluation of fungal contamination in strawberries during decay. *Postharvest Biology and Technology* 153: 152–160.

Liu, Q., Zhou, D., Tu, S. et al. (2020). Quantitative visualization of fungal contamination in peach fruit using hyperspectral imaging. *Food Analytical Methods* 13 (6): 1262–1270.

Ma, J., Sun, D.W., Pu, H. et al. (2019). Protein content evaluation of processed pork meats based on a novel single shot (snapshot) hyperspectral imaging sensor. *Journal of Food Engineering* 240: 207–213.

Magwaza, L.S. and Opara, U.L. (2014). Investigating non-destructive quantification and characterization of pomegranate fruit internal structure using X-ray computed tomography. *Postharvest Biology and Technology* 95: 1–6.

Manickavasagan, A. and Jayasuriya, H. (2014). Imaging with electromagnetic spectrum: applications in food and agriculture. In: *Imaging with Electromagnetic Spectrum: Applications in Food and Agriculture* (eds. A. Manickavasagan and H. Jayasuriya), 1–204. Springer.

Manickavasagan, A., Al-Mezeini, N.K., and Al-Shekaili, H.N. (2014). RGB color imaging technique for grading of dates. *Scientia Horticulturae* 175: 87–94.

Minaei, S., Shafiee, S., Polder, G. et al. (2017). VIS/NIR imaging application for honey floral origin determination. *Infrared Physics and Technology* 86: 218–225.

Minz, P.S. and Saini, C.S. (2019). Evaluation of RGB cube calibration framework and effect of calibration charts on color measurement of mozzarella cheese. *Journal of Food Measurement and Characterization* 13 (2): 1537–1546.

Mohd Ali, M., Hashim, N., Bejo, S.K., and Shamsudin, R. (2020). Comparison of laser backscattering imaging and computer vision system for grading of seedless watermelons. *Journal of Food Measurement and Characterization* 14: 69–77.

Munir, M.T., Wilson, D.I., Yu, W., and Young, B.R. (2018). An evaluation of hyperspectral imaging for characterising milk powders. *Journal of Food Engineering* 221: 1–10.

Nansen, C., Singh, K., Mian, A. et al. (2016). Using hyperspectral imaging to characterize consistency of coffee brands and their respective roasting classes. *Journal of Food Engineering* 190: 34–39.

Paluchowski, L.A., Misimi, E., Grimsmo, L., and Randeberg, L.L. (2016). Towards automated sorting of Atlantic cod (*Gadus morhua*) roe, milt, and liver – spectral characterization and classification using visible and near-infrared hyperspectral imaging. *Food Control* 62: 337–345.

Pathare, P.B., Opara, U.L., and Al-Said, F.A.J. (2013). Colour measurement and analysis in fresh and processed foods: a review. *Food and Bioprocess Technology* 6 (1): 36–60.

Patrício, D.I. and Rieder, R. (2018). Computer vision and artificial intelligence in precision agriculture for grain crops: a systematic review. *Computers and Electronics in Agriculture* 153: 69–81.

Pu, Y.Y., Sun, D.W., Buccheri, M. et al. (2019). Ripeness classification of bananito fruit (*Musa acuminata*, AA): a comparison study of visible spectroscopy and hyperspectral imaging. *Food Analytical Methods* 12 (8): 1693–1704.

Qin, J., Vasefi, F., Hellberg, R.S. et al. (2020). Detection of fish fillet substitution and mislabeling using multimode hyperspectral imaging techniques. *Food Control* 114: 1–14.

Rady, A. and Adedeji, A.A. (2020). Application of hyperspectral imaging and machine learning methods to detect and quantify adulterants in minced meats. *Food Analytical Methods* 13 (4): 970–981.

Raponi, F., Moscetti, R., Monarca, D. et al. (2017). Monitoring and optimization of the process of drying fruits and vegetables using computer vision: a review. *Sustainability (Switzerland)* 9 (11): 1–27.

Ravikanth, L., Jayas, D.S., White, N.D.G. et al. (2017). Extraction of spectral information from hyperspectral data and application of hyperspectral imaging for food and agricultural products. *Food and Bioprocess Technology* 10 (1): 1–33.

Renugambal, K. and Senthilraja, B. (2015). Application of image processing techniques in plant disease recognition. *International Journal of Engineering Research & Technology* 4 (3): 919–923.

Romaniello, R. and Baiano, A. (2018). Discrimination of flavoured olive oil based on hyperspectral imaging. *Journal of Food Science and Technology* 55 (7): 2429–2435.

Sanaeifar, A. and Jafari, A. (2019). Determination of the oxidative stability of olive oil using an integrated system based on dielectric spectroscopy and computer vision. *Information Processing in Agriculture* 6 (1): 20–25.

Sendin, K., Manley, M., and Williams, P.J. (2018). Classification of white maize defects with multispectral imaging. *Food Chemistry* 243: 311–318.

Sharma, A. (2019). Understanding RGB color spaces for monitors, projectors, and televisions. *Information Display* 35 (2): 17–21.

Shi, L., Seiler, J., and Ussmueller, T. (2015). Concept for a CMOS image sensor suited for analog image pre-processing. In: *Workshop on Heterogeneous Architectures and Design Methods for Embedded Image Systems (HIS 2015)*, 16–21. Springer.

Shi, C., Qian, J., Zhu, W. et al. (2019). Nondestructive determination of freshness indicators for tilapia fillets stored at various temperatures by hyperspectral imaging coupled with RBF neural networks. *Food Chemistry* 275: 497–503.

Shi, Y., Liu, W., Zhao, P. et al. (2020). Rapid and nondestructive determination of deoxynivalenol (DON) content in wheat using multispectral imaging (MSI) technology with chemometric methods. *Analytical Methods* 12: 3390–3396.

Silva, J.R., Rodrigues, W.P., Ruas, K.F. et al. (2019). Light, photosynthetic capacity and growth of papaya (*Carica papaya* L.): a short review. *Australian Journal of Crop Science* 13 (3): 480–485.

Singh, S.K., Vidyarthi, S.K., and Tiwari, R. (2020). Machine learnt image processing to predict weight and size of rice kernels. *Journal of Food Engineering* 274: 1–10.

Song, W., Song, Z., Vincent, J. et al. (2020). Quantification of extra virgin olive oil adulteration using smartphone videos. *Talanta* 216: 1–7.

Sun, C., Aernouts, B., Van Beers, R., and Saeys, W. (2021). Simulation of light propagation in citrus fruit using Monte Carlo multi-layered (MCML) method. *Journal of Food Engineering* 291: 1–12.

Thomas, S., Behmann, J., Steier, A. et al. (2018). Quantitative assessment of disease severity and rating of barley cultivars based on hyperspectral imaging in a non-invasive, automated phenotyping platform. *Plant Methods* 14 (1): 1–12.

Udomkun, P., Innawong, B., and Jeepetch, K. (2019). Computer vision system (CVS) for color and surface oil measurements of durian chips during post-frying. *Journal of Food Measurement and Characterization* 13 (3): 2075–2081.

Vásquez, N., Magán, C., Oblitas, J. et al. (2018). Comparison between artificial neural network and partial least squares regression models for hardness modeling during the ripening process of Swiss-type cheese using spectral profiles. *Journal of Food Engineering* 219: 8–15.

Vithu, P. and Moses, J.A. (2016). Machine vision system for food grain quality evaluation: a review. *Trends in Food Science & Technology* 56: 13–20.

Vollmer, M. and Möllmann, K.-P. (2018). *Infrared Thermal Imaging: Fundamentals, Research and Applications*. WILEY-VCH VerlagGmbH & Co. KGaA.

Wang, X., Wu, C., and Hirafuji, M. (2016). Visible light image-based method for sugar content classification of citrus. *PLoS One* 11 (1): 1–16.

Wang, J., Zhang, C., Shi, Y. et al. (2020a). Evaluation of quinclorac toxicity and alleviation by salicylic acid in rice seedlings using ground-based visible/near-infrared hyperspectral imaging. *Plant Methods* 16: 1–16.

Wang, X., Li, Z., Zheng, D., and Wang, W. (2020b). Nondestructive identification of millet varieties using hyperspectral imaging technology. *Journal of Applied Spectroscopy* 87 (1): 54–61.

Wasnik, P.G., Menon, R.R., Sivaram, M. et al. (2019). Development of mathematical model for prediction of adulteration levels of cow ghee with vegetable fat using image analysis. *Journal of Food Science and Technology* 56 (4): 2320–2325.

Wu, D. and Sun, D.W. (2013). Colour measurements by computer vision for food quality control – a review. *Trends in Food Science and Technology* 29 (1): 5–20.

Xia, Q., Liu, C., Liu, J. et al. (2016). Rapid and non-destructive determination of rancidity levels in butter cookies by multi-spectral imaging. *Journal of the Science of Food and Agriculture* 96 (5): 1821–1827.

Xie, C. and He, Y. (2018). Modeling for mung bean variety classification using visible and near-infrared hyperspectral imaging. *International Journal of Agricultural and Biological Engineering* 11 (1): 187–191.

Yadav, A., Sengar, N., Issac, A., and Dutta, M.K. (2018). Image processing based acrylamide detection from fried potato chip images using continuous wavelet transform. *Computers and Electronics in Agriculture* 145: 349–362.

Zhang, B., Huang, W., Li, J. et al. (2014). Principles, developments and applications of computer vision for external quality inspection of fruits and vegetables: a review. *Food Research International* 62: 326–343.

Zhang, M., Li, C., and Yang, F. (2019). Optical properties of blueberry flesh and skin and Monte Carlo multi-layered simulation of light interaction with fruit tissues. *Postharvest Biology and Technology* 150: 28–41.

Zhang, D., Chen, G., Zhang, H. et al. (2020a). Integration of spectroscopy and image for identifying fusarium damage in wheat kernels. *Spectrochimica Acta Part A: Molecular and Biomolecular Spectroscopy* 236: 1–8.

Zhang, H., Zhang, S., Dong, W. et al. (2020b). Detection of common defects on mandarins by using visible and near infrared hyperspectral imaging. *Infrared Physics and Technology* 108: 1–10.

Zhao, H.T., Feng, Y.Z., Chen, W., and Jia, G.F. (2019). Application of invasive weed optimization and least square support vector machine for prediction of beef adulteration with spoiled beef based on visible near-infrared (Vis-NIR) hyperspectral imaging. *Meat Science* 151: 75–81.

Zhao, J., Fang, Y., Chu, G. et al. (2020a). Identification of leaf-scale wheat powdery mildew (*Blumeria graminis* f. sp. Tritici) combining hyperspectral imaging and an SVM classifier. *Plants* 9 (936): 1–13.

Zhao, Y., Zhang, C., Zhu, S. et al. (2020b). Shape induced reflectance correction for non-destructive determination and visualization of soluble solids content in winter jujubes using hyperspectral imaging in two different spectral ranges. *Postharvest Biology and Technology* 161: 1–11.

Zhu, F., Yao, H., Hruska, Z. et al. (2016). Integration of fluorescence and reflectance visible near-infrared (VNIR) hyperspectral images for detection of aflatoxins in corn kernels. *Transactions of the ASABE* 59 (3): 785–794.

15

Hyperspectral Imaging

Antoni Femenias and Sonia Marín

Applied Mycology Unit, Food Technology Department, University of Lleida, Agrotecnio Center, Lleida, Spain

1 Introduction

In current times, food industry receives a demand of innocuous products associated with an increasing quality exigency from the consumers. In acquiring products, consumers take in consideration visual aspect, price, nutritional value, and taste and they take for granted the safety of those products. To guarantee both safety and quality parameters, traditional methods have been used by food industry. Nevertheless, the tests for visual quality inspection are imprecise, and those used in the laboratory for quality and safety assurance are destructive, time-consuming, costly, and need specialized staff (Ren et al. 2020). Agro-food products have a strict safety legislation and also a high association of the external features of the products with their quality. Those attributes include size, shape, color, maturity, shriveling, defects, or foreign materials (El-Mesery et al. 2019). The visual inspection according to all these features introduces high errors in their classification due to the high variability; it demands time to categorize all the product characteristics and is subjective due to the operator interpretation. As manual inspection has a limited reproducibility, an automatic method based on objective interpretations is needed to achieve a precise examination of the food products. On the other hand, the rapid analysis of some internal characteristics as indicators of food quality, such as starch, moisture, sugar, protein, among other nutritional properties, is interesting (Li et al. 2018).

Alternatives to human eye check are used for plant products sorting. Visual devices based on red green and blue (RGB) wavelengths are an option for color differences in products (Saito et al. 2009). However, they present limitations in internal detection of products' composition and in the detection of damages with similar color as the product. Thus, the acquisition of various monochromatic images introduces more information of the product, usually used in the visible or the near-infrared (NIR) region of the spectra (Delwiche and Gaines 2005). Nevertheless, more recent and advanced methods involving the maximum information are required to control rapidly and accurately all those parameters in

Electromagnetic Technologies in Food Science, First Edition. Edited by Vicente M. Gómez-López and Rajeev Bhat.

industrial stages. Hyperspectral imaging (HSI) has been proposed as a promising method to assess quality and safety of food products. The main advantages of HSI include a minimum sample preparation, avoiding waste of chemicals, and simultaneous characterization of different compounds in a sample (Li et al. 2018). In addition, the HSI introduces the spectral recognition of the sample that makes it suitable for heterogeneous food analysis, as it is able to recognize separately features from different parts of the scanning region (Femenias et al. 2020). High accuracies of HSI in majority compound detection, such as proteins, moisture, soluble solids, etc., have been demonstrated in food products (Caporaso et al. 2018a). This fact makes it more suitable for quality assessment than for safety reasons, as the legal limits of the harmful compounds for human and animal health are low. Nevertheless, investigations in regard to the application of HSI on harmful substances detection are increasing; thus, its application in food industry is closer (Du et al. 2020).

HSI application in food industry would imply a diminished loss in production yield associated with a positive economic impact and a substitution of the expensive laboratory equipment and qualified personnel, being a more sustainable working system. The principal economic impacts would be reflected in the diversion of only those plant products which do not accomplish the quality and safety parameters, thus avoiding the rejection of complete product batches. On the other hand, industries would reduce significantly their expenditure on chemical products in routine analysis which could prevent from negative environmental impact. Thus, the application of HSI technology in postharvest activities and in the food processing industry would offer an alternative to the wet chemistry methods and to the artificial computer vision. The present chapter focuses on the quality and safety evaluation of plant products. The basis of the HSI and the image acquisition is explained and illustrated along with different applications on plant products being discussed.

2 Fundamentals of the Hyperspectral Imaging

HSI technology is based on the light exposition of the sample through the spatial positions of the sample. In HSI, the information is distributed in a 3D conformation, in which the spatial information (image) corresponds to a 2 dimension ($X \times Y$) and the spectral information of each pixel (Z) to the reflectances, absorbances, or transmittances at a specific spectral range (Qin 2010). These 3D data are well-known as hypercube, and it is represented in Figure 1.

Depending on the spectral range calibration, a HSI device can register information from different spectral ranges. Although different spectroscopic types are available, the analytical processing focuses on ultraviolet (UV), visible (vis), or infrared spectra (IR) for their low price and well-proven performance in online scanning (Boldrini et al. 2012).

Fluorescence occurs when some compounds emit light of lower energy than the incident, usually in the vis region, when excited with UV light. The fluorescence intensities are usually collected as an emission spectrum, which is usually proportional to the concentration of fluorophore in the sample (Delwiche et al. 2017). Visible spectrum region is also used to detect compounds from food products in the visible spectrum region, and it is known as color imaging. It is widely used for defects, diseases on aspect of the product assessment.

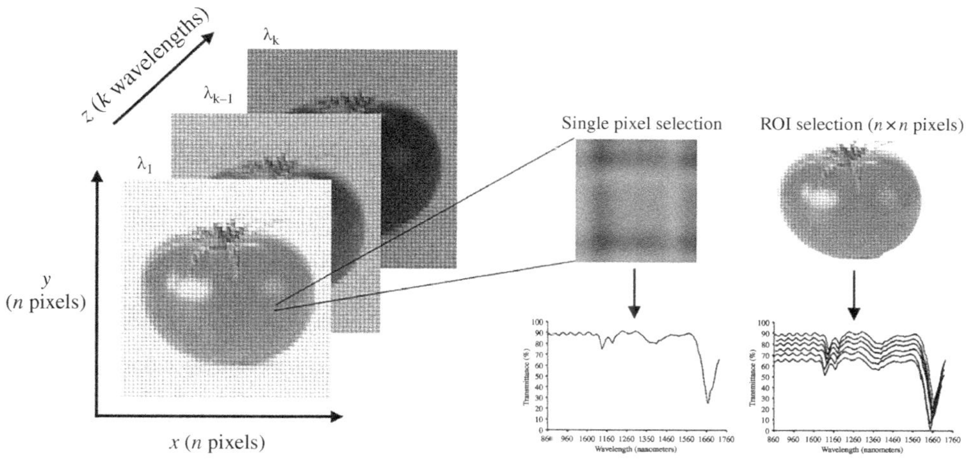

Figure 1 Hyperspectral image illustration for a tomato sample and the relationship between spatial (x, y) and spectral axis (λ).

Despite of these applications, most of the studies combine vis and NIR spectra to introduce more spectroscopic information of the target compounds (Schaare and Fraser 2000). For NIR, the information of the vibration of hydrogen molecular bonds (C—H, C—H, N—H, and S—H) is collected. When the light beam interacts with the measured object, part of it is absorbed by those compounds and the other part is reflected or transmitted. Once the light beam interacts with the object, the reflected part is dispersed and measured by a detector, which registers the spectral information at different wavelengths.

HSI uses light properties to measure the interaction between its incidence and the measured object. Reflectance, transmittance, and absorbance modes are the most commonly used. Differences between the incident illumination and the released radiation are interpreted at a specific spectral range, and they are analyzed by chemometric tools for correlation with sample features. In solid plant products, as in diffuse reflectance the light beam penetrates partially into the sample, it is used to extract their physical and chemical features. Meanwhile, in specular reflectance, the beam does not penetrate into the plant product, so only the information of its surface is measured and the chemical and physical information are overlapped. In transmittance mode, the detector is located in the opposite side of the illumination unit, so worse behaviors are registered as the solid samples are impermeable to the light. However, this mode is useful for liquid and gas sample analyses. On the other hand, the absorbance mode can be calculated by the transformation of the reflectance to absorbance following Eq. (1).

$$A = \log \frac{1}{R} \tag{1}$$

These measurement modes can be attached to three different methods of image production: line-scanning, point-scanning, and focal-plane imaging. The first one is based on the scanning of the X axis line of pixels by the movement of the sample in the Y axis direction. A line-scanning HSI device is represented in Figure 2, which includes all the imaging,

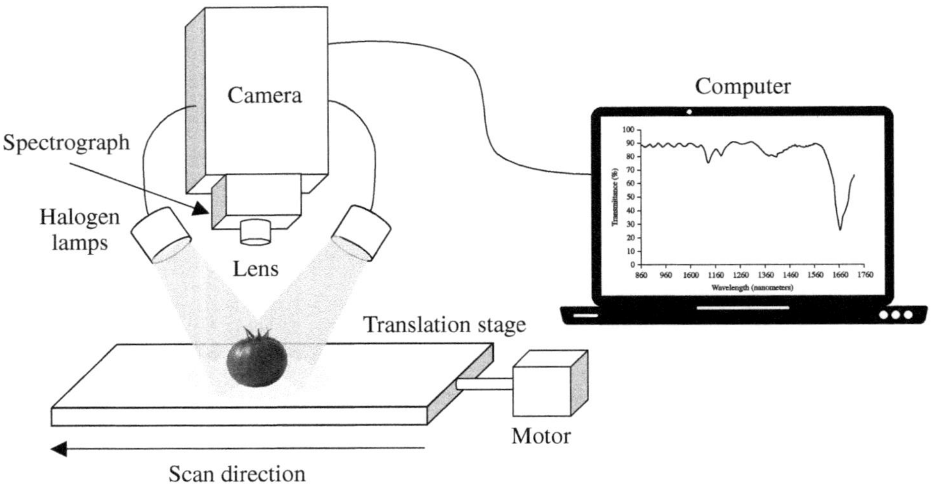

Figure 2 Line-scanning HSI equipment. (*See insert for color representation of this figure*).

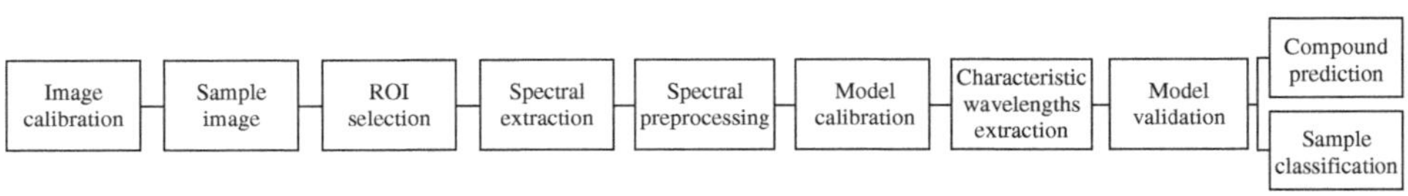

Figure 3 Diagram of the hyperspectral imaging calibration process.

spectrograph, and motor components of the system. In the point-scanning method the spectra are collected pixel by pixel. The scan takes long times, and the spectral resolution is reduced. Finally, the focal-plane imaging captures a single wavelength signal of a plane, for which the direction of scan does not change as the scanning changes only the band number for the configured spectral range. The HSI calibration involves several steps, deeply explained in the subsequent Sections (3–7), for compound predictions and classification of samples. The calibration process is represented in Figure 3.

3 Image Calibration

Even though the images are captured under the same conditions for different times, the data acquisition may present differences due to the camera configuration and the incorrect radiance (Li et al. 2018). Thus, image calibrations are needed to ensure the correct performance of the HSI system. To calibrate the system, dark and white images are captured as references. The raw radiance intensities are corrected with those images, following the subsequent mathematical operation:

$$I = \frac{I_0 - I_b}{I_w - I_b} \tag{2}$$

where the I_0, the I_b, and the I_w are the raw, black, and white intensities, respectively. Those values are used to obtain I, which is the corrected reflectance intensity. The I_b is obtained by the capture of a 0% reflectance image, usually by plugging the lens with an opaque material. Alternatively, the I_w acquisition, for which the highest reflectances are tested (99% reflectance), is usually performed by a Teflon surface image acquisition.

HSI can individually select pixels in the image, so specific areas of the sample can be particularly selected to analyze a specific region of the image, called region of interest (ROI). Most of the agricultural products present high heterogeneity. Those heterogeneities can be handled with HSI, as the different regions of the image can be analyzed individually. For example, in grain industry, the features or compounds found in a kernel can be very different from a kernel from the same batch. HSI is able to overcome those heterogeneities to evaluate the quality and safety parameters in a single kernel, and thus they can be sorted or even removed (Liang et al. 2018; Ropelewska and Zapotoczny 2018).

4 Spectral Pre-processing

The most common pre-treatments applied to plant products spectral data are scatter correction pretreatments, which are applied to minimize the non-linear effect of light scatter due to particle size differences, and first and second derivatives, which are used to avoid noise, multiplicative and additive effects. For scatter correction, the most applied techniques are multiplicative scatter correction (MSC) and standard normal variate (SNV) which are applied to overcome the deviations in the light path lengths caused principally by the effect of the vibrational energy of the sample closer molecular bounds, usually in solids or emulsions (Martens et al. 2003; Yao and Lewis 2010). First derivative computes only for two subsequent spectral points, for which the baseline is removed, while second derivative calculation removes both derivative trends and baseline. However, derivatives are sensitive to noise and can introduce problems to the analysis (Tsai and Philpot 1998). For that reason, smoothing techniques, as Savitzky–Golay (SG) is more commonly used because it can simplify and smooth data by a partial least square (PLS) process. The application of SG filter offers sound reduction, smoothing of spectra, and a single-step derivative calculation (Savitzky and Golay 1964). Finally, averaging is frequently used to reduce troubles in measurements, noise, or excessive variables, and normalization is based on the application of a scale factor in order to obtain variables with the same size (Rinnan et al. 2009).

5 Model Calibration

HSI produces a high load of information as a spectrum is obtained for each pixel in the image. Thus, in order to handle with the data, multivariate statistical methods are used. Principal component analysis (PCA) is able to overcome the high amount of information by producing a compressed new data set with the most relevant information from the orthogonal variables (directions of maximum variance). The PCA results in the projection of the samples according to those directions, which are called principal components (PCs), that can be interpreted by group formation according to the sample features. It is an unsupervised technique as it does not use information of reference before its analysis. On the other hand, as a supervised method, discriminant analysis (DA) is commonly used for NIR calibration in agricultural product applications. In this case, some recognition patters in data are used to build different classes, which are subsequently used to associate new information with those groups. The most used DA method is the linear discriminant analysis (LDA), which is based on a linear combination of the characteristics by increasing the ratio between the variances of the samples and the group variance.

As a considerable amount of information is collected from each sample, a simple linear regression is not suitable to calibrate the technique for future measurements. In order to correlate the high number of variables to the reference method results, PLSs regression is applied, for which a decomposition of variances into X-variables (predictors or dependent variables) and Y-variables (responses or independent variables) is performed. Those variables are explained by separating them into different principal components (PC1, PC2, . . .) so that each of them can describe, as far as possible, the variations of X that explain the variations in Y. Other multivariate regression methods, as multiple linear regression (MLR) and principal component regression (PCR), are alternative techniques to PLS. However, MLR presents problems when noise, interferences, and collinearity between X variables appear. In the case of PCR, it is focused on the maximum variance of X, while the HSI calibration purpose is focused on the analyte information (maximum variance of Y). Hence, PLS is the most widely used procedure to obtain a well-fitted model with good performances for HSI equipment calibrations.

Support vector machines (SVMs) have been used for the pattern recognition and classification and, in combination with least squares (LSs), also for the prediction of chemical compounds. It consists in division of two classes by the hyperplane, which is a vector with the direction between those groups. The introduction of a new observation will assign a position on one of the sides of this hyperplane, which will determine the group which it belongs to (Melgani and Bruzzone 2004). Moreover, LS-SVM is a combined nonlinear method which enables both pattern recognition and regression with some advantages, as a high predictive capability, simplification, and a low over-fitting (Li et al. 2018). Other classification methods, based on pattern recognition, have been proposed for HSI calibration. Nearest neighbors (NN) is a commonly used algorithm in HSI studies for its classification power. It consists in the calculation of the distance between the spectral feature and the previous calibration sample. Thus, a probability index is built for each pixel which establishes the distance from the reference values (Bo et al. 2017). The closer neighbors to the reference positions give the best classifications. An extension to NN is the K-nearest neighbors (KNN) which is a nonparametric technique able to classify elements with irregular limit values. Depending on the number of neighbors selected (k), the NN probability index for classification is built according to the distance from the k elements previously selected. Nevertheless, noise and irrelevant predictors can affect the results (Guo et al. 2018; Ravikanth et al. 2016). Finally, artificial neural networks (ANN) are based on the connection multiple layer perceptron (a unit of neurons) with one or more layers. They are able to solve problems which are not lineal to predict one or more response variables by a flexible function of the input variables. However, it presents problems to describe the relationship between the inputs and the response and tends to over-fit data when reduced data sets are computed (Ozyildirim and Avci 2013).

6 Characteristic Wavelengths Extraction

Depending on the spectral resolution of the HSI device, the acquired spectral information may include hundreds or even thousands of wavelengths for each sample. Moreover, the management of such data set is laborious, and the mathematical computation of the model is time-consuming. Moreover, the application of those methods to the industry needs simplified and rapid procedures. Consequently, wavelength selection is a key step to eliminate those irrelevant bands and noise to perform a fast and reliable analysis. Wavelength selection can be executed by a manual localization of the wavelengths with higher regression coefficients, which are the bands with major influence over the model. Alternatively, the application of complex algorithms to select the optimal bands is also used. The most common ones are competitive adaptive reweighted sampling (CARS), random frog (RF), successive projections algorithm (SPA), genetic algorithm (GA), and uninformative variables elimination (UVE) (Li et al. 2018; Tang 2012; Xiaobo et al. 2010).

7 Model Validation

Once the calibration model is built, the performance of the prediction has to be evaluated. As a regression can be validated in different ways, its performance depends on the validation tool used. The most frequently used validation methods are leverage correction, cross-validation, and independent set validation. The first tends to be too optimistic; for that reason, it is used as a first approximation to know how the model may work. When a model has a limited number of objects to be separated into two sets to calibrate and validate it, a cross-validation is used. Full cross-validation is the ideal validation when data set is small because the segments used are at most the 10% of the objects. Finally, an independent test set validation is the best choice when a large number of samples are available. As different samples from the calibration are used, the model obtained is more accurate for future predictions. Ideally, the set should contain at least the 25% of the total of objects of the model (Esbensen et al. 2002).

The validated model should be tested to assess its performance. Some mathematical parameters determine the fitness of the calibration and validation, although the validation performance is interpreted as the predictive ability. The parameters are the coefficient of validation (R^2_v), standard error of prediction (SEP), the root mean square error of prediction (RMSEP), and the ratio of performance to deviation (RPD) (Chavez et al. 2013). Table 1 shows the mathematical calculations of the performance parameters. The R^2_v is determined by the variance between reference and predicted values divided by the difference between the reference and total variance, and it should be as close as possible to one. The SEP determines the accuracy of the model; however, it needs a bias rectification to correct the difference between the SEP and its real value. The RMSEP is the error we expect to have using the calibrated model in future predictions. High RMSEPs indicate that the model is weak, or it has a non-representative validation set thus, the model should obtain RMSEP closer to 0. As the reference values of two models can be different, the performance of validation can be affected by those variances. The models calibrated with high standard deviations in the reference values usually obtain higher RMSEPs. The RPD is a parameter used

Table 1 Performance statistic parameters of the validation set.

Validation parameters		
Coefficient of determination of prediction	$R^2{}_p$	$R_p^2 = \frac{\sum(\hat{y}_i - \bar{y})^2}{\sum(y_i - \bar{y})^2}$
Bias	Bias	$\text{Bias} = \frac{\sum(\hat{y}_i - y_i)}{n}$
Standard error of prediction (corrected by the bias)	SEPc	$\text{SEPc} = \sqrt{\frac{\sum(\hat{y}_i - y_i - \text{bias})^2}{n-1}}$
Root-mean-square error of prediction	RMSEP	$\text{RMSEP} = \sqrt{\frac{\sum(\hat{y}_i - y_i)^2}{n}}$
Ratio of performance to deviation	RPD	$\text{RPD} = \frac{Sdev_{\text{ref}}}{\text{SEP}}$

n = number of samples; *S*devref = standard deviation of reference; $\hat{y}_i$ = *i*th validation sample predicted value; y_i = *i*th validation measured values; $\bar{y}$ = mean of the n values (measured or predicted).

to compare different model performance as it calculates the ratio between the standard deviation of the reference values and the RMSEP. Thus, it normalizes the efficiency of the model which can be compared with other ones with different features. If the model works appropriately, the RPD values are high (Sileoni et al. 2011; Westad and Marini 2015).

8 Application of HSI for Plant Products Quality Assessment

High-quality fruits and vegetables are demanded by consumers, for which they are willing to pay more in order to obtain a product with better appearance (size, shape, color, gloss, and consistency), texture, flavor, including the safety parameters. HSI has demonstrated its suitability for quality management in fruits and vegetables. Color imaging has been applied frequently in analysis of external appearance attributes. On the other hand, spectroscopic techniques such as NIR, Raman, and nuclear magnetic resonance (NMR) have been used for internal quality assessment and composition of the products. HSI-NIR spectra are correlated with the amount of the organic compounds in the sample. The performance of this technology depends on the amount of samples in the calibration set and the reference procedure (Pathmanaban et al. 2019). On the other hand, Raman HSI is a novel technique based on light scattering in which the spatial recognition has been incorporated. Despite this technique is promising in vegetable and fruit quality assessment, few studies have been published using this technology, probably due to the weak results presented, as it requires a high excitation light and high-performance detector to obtain an accurate signal

Table 2 Applications of hyperspectral imaging for the classification of plant products according to their quality attributes.

Quality attribute	Food product	Spectral range (nm)	Characteristic wavelengths	Accuracy (%)	References
Black heart	White radish	400–1000	4/10	95.6–100	Song et al. (2016)
Black spots	Potato	400–2500	11	94	López-Maestresalas et al. (2016)
Browning	Mushroom	380–1000	8	80.3	Mollazade (2017)
Bruising	Blueberries	970–1400	—	94.5	Zhang et al. (2017)
		950–1650	—	92	(Jiang et al. 2016a)
	Apple	900–2500	11	98	Keresztes et al. (2016)
Firmness	Nectarines	630–900	7	95.7	Munera et al. (2019)
Ingredients	Bread	470–900	15	93.1	Blanch-Perez-del-Notario et al. (2020)
Mold infection	Dried figs	390–1005	27	99.3	Ortaç et al. (2016)
Ripeness	Sweet cherries	900–1700	—	98.0	Wang et al. (2018)
	Pear	425–1000	10	77.8	Khodabakhshian and Emadi (2018)
Roasting defects	Coffee beans	500–1650 1000–1700	—	86.2	Cho et al. (2017)
Pits	Cherries	450–1000	19	94.6	Siedliska et al. (2017)
Food fraud	Black tea	350–1100	3	92.7	Ren et al. (2020)

(Lu et al. 2020). Finally, NMR has been employed to describe the quality attributes of food products as dry matter quantification in potatoes (Hansen et al. 2010). Despite the different techniques available for quality assessment of plant products, vis and NIR spectroscopic techniques are the most widely used.

Most extensively quality attributes analyzed by HSI during the last years are black heart, browning, bruising, mould infection, maturity, defects, soluble solid content, phenols content, pH, acidity, moisture, color, firmness, texture, sugars, proteins, and so on. The most recent studies about hyperspectral screening of those attributes are revised in Tables 2 and 3.

8.1 Discrimination According to Quality Parameters

Quality parameters are a concern to plant products producers. An effective sorting to remove the products which do not achieve the minimum quality standard has been investigated by many authors during the last decades. The emerging of HSI has been proven for rapid online sorting of products in the industry arrival. Thus, the studies aim to obtain

Table 3 Applications of hyperspectral imaging for quantification of quality attributes of plant products.

Food product	Quality attribute	Spectral range (nm)	Characteristic wavelengths	R^2	SEP/RMSEP	References
Soluble solid content	Sweet cherry	400–1650	23	0.88	0.43%	Pullanagari and Li (2020)
		900–1700	—	0.83	1.53° Brix	Wang et al. (2018)
	Kiwifruit	450–1000	27	0.96	0.43° Brix	Zhu et al. (2017)
	Orange fruit	400–1000	13	0.76	32%	Aredo et al. (2019)
	Honey	400–1000	—	0.99	0.45° Brix	Sulistya and Saputro (2018)
	Mushrooms	886–1645	67	0.87	1.60° Brix	Xiao et al. (2020)
Protein content	Almonds	400–1000	—	0.57	0.27%	Gama et al. (2018)
Polyphenols	Cocoa	1000–2500	—	0.70	34.1 mg ferulic acid g^{-1}	Caporaso et al. (2018b)
	Raspberry	950–1650	—	0.75	101 mg malvidin-3-G kg^{-1}	Rodríguez-Pulido et al. (2017)
	Barley malt	1000–2500 220–400	—	0.98	0.32 mg kg^{-1} total phenols	Tschannerl et al. (2019)
pH	Kiwifruit	951–1670	12	0.91	0.015	Zhu et al. (2017)
	Orange fruit	400–1000	15	0.55	0.1	Aredo et al. (2019)
Acidity	Orange fruit	400–1000	15	0.58	37%	Aredo et al. (2019)
Moisture	Dried banana slices	400–1000	—	0.97	0.05 kg H_2O/kg DM	Nguyen-Do-Trong et al. (2018)
	Rice	900–1750	15	0.93	0.026%	Lu et al. (2018)
	Wheat	400–1100	—	0.98	4.34%	Zheng et al. (2016)

Color	Orange fruit	400–1000	$L^* = 3$ $a^* = 8$ $b^* = 5$ CI = 8	$L^* = 0.90$ $a^* = 0.95$ $b^* = 0.92$ CI = 0.97	$L^* = 1.4$ $a^* = 1.2$ $b^* = 1.6$ CI = 0.3	Aredo et al. (2019)
	Tomato	325–985	—	$L^* = 0.86$ $a^* = 0.93$ $b^* = 0.42$ Hue = 0.95 Chroma = 0.51	$L^* = 10.35$ $a^* = 13.3$ $b^* = 10.75$ Hue = 8.90 Chroma = 16.71	van Roy et al. (2017)
Firmness	Dried banana slices	400–1000	—	0.53	1.32 N	Nguyen-Do-Trong et al. (2018)
	Kiwifruit	450–1000	7	0.98	35.8 $N\,cm^{-2}$	Zhu et al. (2017)
	Pear	425–1000	4	0.84	0.78 N	Khodabakhshian and Emadi (2018)
Fermentation index	Cocoa beans	1000–2500	—	0.50	0.27	Caporaso et al. (2018b)
Adulterants	Wheat flour	Raman	102–2865 cm^{-1}	Detection	0.5%	Qin et al. (2016)

DM, dry matter; RMSEP, root mean square error of prediction; SEP, standard error of prediction.

high-accuracy classifications of food products according to different quality parameters (Mo et al. 2017). Table 2 includes the attribute classification studies for different plant products in which the accuracies obtained can be compared. The overall accuracies achieved in those studies are above 77% of correct classification of products according to some quality parameters. In some of these studies, their best results were close to 100% of correct classification; thus, the ability of HSI for quality assessment is confirmed. The studies worked in the NIR spectral region or the combination of vis/NIR spectra which includes the physical and chemical information of the product.

Internal and external defects of plant products are common due to changes in their properties that affect their structure and composition. Internal and surface darkening of the products in the form of black heart, black spots, or browning of the complete product is commonly found in fruits and vegetables. Those defects were sorted with an accuracy larger than 94% for heart and black spots in white radish and potatoes, respectively (López-Maestresalas et al. 2016; Song et al. 2016). López-Maestresalas et al. (2016) reported the classification of mushrooms according to browning, obtaining a correctness of 80.3% with only eight latent variables. Fruit bruising studies used the NIR spectral data to classify apples and blueberries with an accuracy between 92% and 98% (Jiang et al. 2016a; Keresztes et al. 2016; Zhang et al. 2017). Firmness, ripeness, and maturity are parameters that describe the stage of development of the plant product. Munera et al. (2019) classified nectarines according to firmness threshold (35N) with an exactitude of 95.7%. The whole NIR spectral range was used by Wang et al. (2018) who classified correctly the 98% of the sweet cherries according to maturity level. Lower accuracy was obtained by Khodabakhshian and Emadi (2018) (77.8%) in the vis/NIR range. HSI has also been used to detect defects in round-shaped plant products for which the surface of the whole product should be scanned. Internal defects are more difficult to detect than surface defects as they are hidden by the own product structure. A combination of internal and external quality attributes was used by Cho et al. (2017) to detect roasting defects of coffee beans, such as moisture content, browning, chlorogenic acid, trigonelline, and caffeine content, for which the overall classification accuracy was 86.2%. Finally, Siedliska et al. (2017) used the transmittance mode in the vis/NIR region to detect internal intrusions in cherries, such as pits or pit fragments. The study obtained a positive prediction accuracy for those intrusion detection with a 94.6% correct classifications. Finally, food fraud is responsible for significant economic losses in food industry and generates suspicion in consumers. HSI has also been calibrated for a potential discrimination of black tea according to the authenticity of its quality categories. The study of Ren et al. (2020) achieved a 92.7% of correct discrimination among seven quality grades by using only three characteristic wavelengths. All these advances demonstrate the potential of HSI for the discrimination of plant products according to numerous quality attributes to ensure a minimum quality standards in food industry.

8.2 Quantification of Quality Parameters

HSI offers accurate quantification of some constituents in food samples (Jiang et al. 2011). The quantification of some organoleptic attributes (color, texture, acidity, sweetness, flavor, etc.) and composition (energy sources, functional nutrients, and bioactive compounds) and identification of food fraud (geographical origin and variety) are interesting from the perspective of guaranteeing a high-quality product (Table 3).

HSI has been tested in numerous food products for soluble solids quantification, generally for sugar content determination. The fitting obtained for the calibration achieved R^2 between 0.87 and 0.99, except for Aredo et al. (2019) who got slightly lower performances (0.76). The models presented errors between 0.43 and 1.60°Brix. In addition to sugars, other macronutrients, such as proteins, can also be quantified by HSI. Almonds are products that contain a high amount of vegetal protein. Gama et al. (2018) predicted indirectly the amount of crude protein of this product by the calibration of HSI for nitrogen content. The validation results were poor, with a R^2 of 0.57 and a root-mean-square error of cross-validation (RMSECV) of 0.27%. Nevertheless, a robust calibration performance demonstrated the potential ability of HSI to predict the crude protein of almond seeds.

Polyphenols are antioxidant substances found in plant-based products and interesting for their potential health benefits. As well as for these benefits, the determination of the concentration of polyphenols is an important quality parameter because it affects the astringency and bitterness of plant products. Different phenols were determined in cocoa, raspberry, and barley malt. The results demonstrated that the quantification of total phenols (R^2 = 0.98, RMSEP = 32 mg kg^{-1}) achieved better accuracies than for specific polyphenols (R^2 = 0.70, RMSEP = 34.1 mg ferulic acid g^{-1}; R^2 = 0.75, RMSEP = 101 mg malvidin-3-G kg^{-1}) (Rodríguez-Pulido et al. 2017).

Fruit acidity is one of the internal quality parameters that can be measured by HSI avoiding the sample destruction. In fruit industry, pH and titratable acidity are very important parameters to control the product flavor in terms of sourness. Aredo et al. (2019) and Zhu et al. (2017) determined the pH of citric fruits (orange and kiwifruit). Zhu et al. (2017) obtained better results in the NIR region (R^2 = 0.91, RMSEP = 0.015) than Aredo et al. (2019), who obtained a weaker pH prediction for the vis/NIR region (R^2 = 0.55, RMSEP = 0.1). In the last study, titratable acidity was also a target parameter which was attempted to be measured, for which they obtained similar fitting than for pH (R^2 = 0.58, RMSEP = 37%).

Food industries require the monitoring of moisture content, especially in dried products and cereals. This parameter has been evaluated during the last years through HSI in dried banana slices, rice, and wheat. HSI would be able to replace common used methods for moisture determination as electric oven, vacuum drying, far-infrared heating drying, microwave drying, resistance, and capacitance methods (Zheng et al. 2016). The studies obtained good results for moisture content prediction with HSI, with high-accuracy adjustments (R^2 = 0.93–0.98) and low errors of prediction (0.05 g H_2O/100 g DM and 0.026–4.34%) (Lu et al. 2018; Nguyen-Do-Trong et al. 2018; Zheng et al. 2016).

HSI technology is commonly used to determine the external attributes of food products, usually in the vis/NIR region. Color is an important quality parameter to assess the fruit appearance and evaluate the postharvest ripeness. Color assessment has been performed with the calibration of the HSI for the different color parameters in an unambiguous way from the RGB measurements. Orange fruit and tomatoes color have been recently predicted with high performances (Aredo et al. 2019; van Roy et al. 2017). Another external attribute that can be used as a maturity indicator is firmness of the product. Typical firmness measurements involve the quantification of the resistance of a plant product at a pound-force. This technique assumes the destruction of the sample. HSI calibration could avoid those loses by firmness predictive models. Different authors studied the firmness assessment in fruits such as banana, pear, and kiwifruit (Khodabakhshian and Emadi 2018; Nguyen-Do-Trong et al. 2018; Zhu et al. 2017). The studies worked in the vis/NIR spectral range. The best results were obtained for kiwifruit, with a coefficient of determination of 0.98. Nevertheless, the results on pears were also positive with a R^2 of 0.84, while for dried banana slices, the calibration was weaker (R^2 0.53).

Fermentation of cocoa beans depends directly on their composition. Fermentation index (FI) is a quality parameter used to estimate the grade of fermentation by measuring the phenolic compounds or the changes in cocoa bean color. The estimation of the FI is commonly measured by the spectroscopic reading of the dissolvent from the extraction of cocoa beans. HSI would avoid the extraction step and, consequently, the use of pollutant substances and sample destruction. The use of HSI for single cocoa bean FI estimation was studied by Caporaso et al. (2018b) who obtained a R^2 of 0.50 and a RMSEP of 0.27. Their weak prediction could be due to variability between beans due to environmental conditions or post-harvest practices. Nevertheless, the study is considered a first approach of HSI methodology to be potentially applied for single cocoa beans screening.

Finally, adulteration of flours and powdered products is increasing, and it poses a risk in consumer's health. The detection of adulterant substances at low concentrations is a challenge. HSI and Raman spectroscopy tandem were used by Qin et al. (2016) in the identification of benzoyl peroxide in wheat flour. Individual Raman peaks for a concentration of 0.5% of adulterant were obtained, so it is a prospective technique for the identification of adulterant particles mixed with flours at low concentrations. A recent study on adulteration detection with HSI in honey is also considered as an example of the potential of this technique for non-desirable substances detection (Sulistya and Saputro 2018). Based on all the previous studies, HSI is considered a promising technique for the quantification of organoleptic substances, composition, and food adulteration of plant products, and its future implementation in industry would contribute to a fast, non-destructive, and eco-friendly analysis of those attributes.

9 Application of HSI for Safety Assessment in Fruits and Vegetables

Several studies have been published on the use of vis-HSI to assess microbial pathogens and fecal contamination in fruits and vegetables, as well as infestation by insects. Fruit and vegetables are exposed to different sources of microbial contamination, including organic residues from soil and water used for irrigation, insects, and cross-contamination.

HSI studies at laboratory level in vegetables include inoculation of *E. coli* into spinach leaves and analysis by vis/NIR-HSI, obtaining a coefficient of determination R^2 of 0.97, between plate count method and the spectral reflectance (Siripatrawan et al. 2011). On the other hand, Yang et al. (2010) employed a vis-HSI fluorescence system for the detection of fecal contamination on lettuce and spinach leaves, and a two-band ratio, 666/680 nm, to differentiate the contaminated spots from uncontaminated leaf area was used. Compared to fruits like apple, leafy vegetables generally have more strong fluorescence emissions due to high chlorophyll concentrations and the emissions are in proximity to those due to fecal matters in the red and far-red regions; thus, higher spectral-resolution imaging for effective detection of fecal contamination is required (Cho et al. 2018). Regarding fruits, studies in apples include detection of fecal contamination by using vis-HSI in fluorescence mode which achieved a classification accuracy of 100%, while for reflectance mode, the classification accuracy was 99.5%. Selected wavelengths for multispectral detection were 680, 684, 720, and 780 nm, for the detection of fecal contamination spots on apple surfaces (Kim et al. 2007; Yang et al. 2012).

Stricter methodologies of analysis are required for bacterial pathogens determination in fruits and vegetables, that is the reason why most of the studies in these food commodities using HSI are devoted to quality assessment.

10 Application of HSI for Microbiological Quality and Safety Assessment in Cereals, Nuts, and Dried Fruits

Cereals and nuts are low moisture food commodities for which moulds, rather than bacteria and yeast, are the main spoilage agents. The main interest of HSI in this case is related to fungal damage and mycotoxin contamination detection.

10.1 Assessment of Fungal Damage

Fungi colonize cereals both pre- and postharvest leading to dry matter losses and quality degradation and consequently to economic losses. During cereal storage, if moisture content is not low enough, fungal colonization may affect grain appearance, chemical composition, or germination capacity.

The basis of HSI lies in the reflection changes in both the vis/NIR and short wavelength infrared (SWIR) ranges. The commonly investigated NIR range provides chemical information about cereal composition because most absorption bands observed in this region arise from overtones of C–H, O–H, and N–H stretching vibrations (Moller and Munch 2009). The trough at 1200 nm is related to the C–H second overtone, the main one at 1450 nm is linked to the O–H and N–H first overtone and to the C–H combination band in the aromatic ring. Spectral differences of cereal kernels in the vis region, such as 480–600 nm range, are associated with bands attributed to color changes in fungal-infected grains. The most significant bands related to fungal infection, however, are around 870–1250 nm, corresponding to first overtone of the N–H bound in most amino acids and aromatic rings, radical structures in cell wall components, such as furanic or phenolic compounds, that might be interpreted as deteriorative alterations of kernels (Fernández-Ibáñez et al. 2009). The damage to the kernels can also be associated with other specific absorbance regions in the NIR such as 1400–1510 nm related to the first overtone of O–H and N–H bonds (Berardo et al. 2005) (Figure 4). On the other hand, it can be linked to the 1750–1800 nm to C–H stretching mode corresponding at CH_2 group, and 1960–2190 nm could be attributed to the first overtone of the O–H stretching modes of glucose, N–H in amino acids, and C–H combination bands in cis instauration (Osborne and Fearn 1986; Berardo et al. 2005).

Many studies have focused in fungal contamination in wheat grain. Shahin and Symons (2011) used vis/NIR-HSI (400–1000 nm) to detect *Fusarium*-damaged grains of wheat. LDA

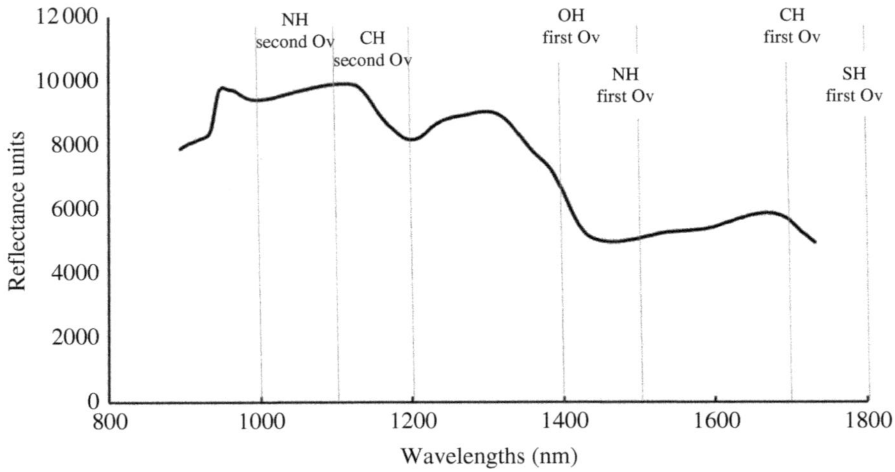

Figure 4 Schematic representation of relevant overtone absorptions in the SWIR region. *Source:* Adapted from Nicola Berardo (2005).

was applied, and an overall classification accuracy of ≥92% was obtained. Later, the same authors applied PLS-DA and a reduced number of wavelengths (494, 578, 639, and 678 nm) for detecting *Fusarium* sp.; however, a lower overall accuracy was achieved (90%) (Shahin and Symons 2012). Mildew damage was also detected by vis/NIR-HSI, in wheat grains using a PLS model analysis based on four wavelengths (450, 561, 861, and 917 nm) where $R^2 = 0.89$ and RMSE (root mean square error) = 0.68 (calibration model), and $R^2 = 0.87$ and RMSE = 0.74 (validation model) (Shahin et al. 2010). Classification accuracy between 97.3% and 100.0% was achieved for healthy and fungal-infected wheat grains using the 700–1100 nm range for the detection of infection by *Penicillium* spp., *Aspergillus glaucus*, and *A. niger* (Singh et al. 2012). The researchers concluded that wavelength 870 nm was the most significant. Finally, Serranti et al. (2013) used SWIR-HSI (1000–1700 nm) to classify bulk wheat kernels into different types, such as vitreous, yellow berry, and *Fusarium*-damaged kernels. The results demonstrated that the classification was good (96%), both using wavelengths from 1013 to 1650 nm, and selecting "only" 3 windows of 4 wavelengths each (1209–1230, 1489–1510, and 1601–1622 nm).

In the vis range, *Fusarium*-damaged kernels have higher reflectance because of the white or pinkish fungal tissue on the kernel surface. In the SWIR, changes are observed in bands related to moisture, carbohydrate, and protein content, as *Fusarium*-damaged kernels contain less water, carbohydrates, and proteins (Peiris et al. 2009).

In maize, Del Fiore et al. (2010) evaluated the presence of *Aspergillus flavus*, *A. niger*, *A. parasiticus*, and *Fusarium verticillioides* by vis/NIR-HSI. Results revealed that between 500 and 700 nm, the absorbance increased, while fungal contamination increased, however, between 850 and 950 nm, the absorbance values decreased. Pearson and Wicklow (2006) also tested the capability of multispectral imaging in transmittance mode in identifying the fungus infected corn kernels with "extensive discoloration" symptoms from asymptomatic kernels. They used 11 single wavelengths between 780 and 1020 nm, and using DA, classification accuracies of 93% and 90% were achieved for asymptomatic and "extensive discoloration" groups, respectively. Williams et al. (2012) worked in maize grains inoculated with spores of *F. verticillioides* and analyzed using SWIR-his. They observed changes in the composition of the grain components, especially with prominent peaks at 1405 and 2136 nm, which are related to the absorbance for starch (O–H stretch) and protein (N–H stretch and a C=O stretch), respectively, substances used by the fungus for growth. It has been observed that fungal presence is related to increase in absorbance; consequently, spectral reflectance is inversely proportional to the infection level. This would be related to the formation of cellular structures of the fungus (as spores and mycelium), and synthesis of metabolites and changes in the chemical composition and color of the grains (Fernández-Ibañez et al. 2009; Da Fiore et al. 2010).

More recently, Siripatrawan and Makino (2015) used vis/NIR-HSI to evaluate rice grains inoculated with a non-toxigenic strain of *Aspergillus oryzae*. Partial least squares regression (PLSR) was used to predict fungal growth from the HSI reflectance spectra, obtaining a high R^2 (0.97) between formed colonies and spectral data. Regarding nuts, Jiang et al. (2016b) applied two-fold PCA on SWIR-NIR spectra to identify moldy peanuts and got satisfactory results of 99%. Moreover, healthy, damaged, and moldy peanuts were classified using vis/NIR-HSI with a kernel accuracy of 92.1% (Liu et al. 2020).

In general, vis/short-wavelength near infrared (SWNIR)-HSI was enough for fungal detection. Although some of the studies were carried out in inoculated grains, those in naturally damaged wheat or peanuts confirmed the usefulness of the technique to discriminate between healthy and damaged or moldy kernels. Some of the authors suggested that simplified models, those based in multispectral images would fulfill the requirements for grain selection, and more suitable for online applications, due to their relatively low instrument cost and high analytical speed. Overall, it highlighted the advantage of the HSI detection on microbial contamination when symptoms are not yet visible by direct observation.

10.2 Assessment of Mycotoxin Contamination

Most published researches dealt with aflatoxins (AFs) and deoxynivalenol (DON) detection, the two most relevant mycotoxins regarding human exposure. Although mycotoxin content, as low as ppb or ppm, cannot be detected given the low sensibility of these techniques, it can be indirectly estimated based on the changes observed in the spectra as a result of a modified composition of the food. For example, wheat kernel epidermises infected with DON toxin are white- or red-headed, and their surfaces become wrinkled after infection. Simultaneously, the contents or properties of organic substances such as proteins, carbohydrates, and lipids in wheat infected by *Fusarium* exhibit changes, and such changes can be reflected in the spectral characteristics (Liang et al. 2018). Thus, the usefulness of spectral techniques depends on the correlation between toxin presence and surface changes of grains, and compositional variation.

10.2.1 Aflatoxins

The aim of the existing studies of AFs in cereals has been, in general, to detect contaminated kernels among the others, as a starting point for sorting operations. Earlier studies in AF detection in cereals and nuts were conducted using fluorescence-HSI. Yao et al. (2013)

estimated AF concentration in maize grains from ears inoculated with *A. flavus* spores in the field, using fluorescence HSI (400-1000nm) obtaining a R^2 of 0.72 using a MLR model for concentrations between <1 and >100 ng g^{-1}, and the classification accuracy ranged from 0.84 to 0.91 when a threshold of either 20 or 100 ng g^{-1} was used. They concluded that the range between 450 and 500 nm was figured out as the best spectral region to detect AF contamination. Results from the DA classification indicated overall higher classification accuracy for a 100 μg kg^{-1} threshold (94.4%).

AFs have also been investigated in nuts and dried fruits. Kalkan et al. (2011) detected the presence of AFs by using fluorescence multispectral imaging (MSI) (400–600 nm) in laboratory artificially contaminated hazelnut samples with *A. parasiticus*. A classification accuracy of 92.3% was achieved for AF-contaminated and uncontaminated hazelnuts. After the removal of contaminated nuts, the AF concentration was decreased from 608 to 0.84 μg kg^{-1}, confirming the usefulness of this technique. The algorithm was also used to classify fungal contaminated and uncontaminated bulk samples, and an accuracy of 95.6% was achieved. The same authors proved the potential of this technique for classification of dry pepper and figs according to their AF contamination level. Xing et al. (2017) investigated the application of fluorescence HSI to assess the spectral difference between peanuts inoculated with toxigenic and atoxigenic *A. flavus* and healthy kernels. For the kernels without skin, the inoculated kernels had fluorescence peaks shifted to longer wavelengths with lower intensity than healthy kernels. In addition, the fluorescence intensity of peanuts without skin was higher than that of peanuts with skin (10 times). DA showed that the inoculation group could be separated from uninoculated controls with 100% accuracy. However, the groups inoculated with toxigenic and atoxigenic strains could be separated with only ~80% accuracy.

To date, studies also include vis/NIR-HSI or SWIR-HSI equipments, as described below artificially spiked maize kernels were scanned using Vis/NIR-HSI. PCA score plots revealed a distinct separation of low contaminated kernels (25, 40, and 70 μg kg^{-1}) from highly contaminated ones (200, 300, and 500 μg kg^{-1}) without any overlapping of data. Classification accuracies ranged from 94.7 to 98.3 depending on the preprocessing and classification technique. The developed PLS model depicted good prediction accuracy (R^2 = 0.82, SEP = 79.4 μg kg^{-1}, RPD = 2.4) during cross-validation. The wavelength 508 nm was selected based on PCA loadings as the most influent to distinguish between unspiked and spiked maize kernels (Chakraborty et al. 2020). In another study by vis/NIR-HSI in which maize kernels were prepared by inoculating the toxigenic *A. flavus* suspension artificially at an early dough/late milk stage of kernel development, in order to simulate natural contamination, Zhu et al. (2016) obtained 90% overall accuracies from the germ side when taking 20 and 100 mg kg^{-1} as thresholds, separately. It was suggested that the range 800–700 nm had a major impact in kernels classification. Overall, the germ side performed better than the endosperm side. When using fluorescence HSI data paired to the reflectance data, the results showed that individual fluorescence and reflectance image data achieved generally similar classification accuracies. Using the images collected from the germ sides, the integrated form of fluorescence and reflectance was able to produce better results than using only one type of spectra. The best overall prediction accuracy of 95.33% was obtained using the integrated information from the germ side of corn kernels and the threshold of 100 mg kg^{-1}. Moreover, the authors calculated the mean AF concentration of the prediction samples and found it to be reduced from 2662.01 mg kg^{-1} to 64.04, 87.33, and 7.59 mg kg^{-1} after removing contaminated kernels identified by fluorescence, reflectance, and integrating both, respectively, from the germ side.

Pearson et al. (2001) evaluated the presence of AF in maize grains inoculated in the field and analyzed using vis/NIR-HSI and SWIR-HSI (550–1700 nm), and more than 95% of the grains were correctly classified as containing either high (>100 μg kg^{-1}) or low (<10 μg kg^{-1}) levels of AF. Classification accuracy for kernels between 10 and 100 μg kg^{-1} was only about 25%. Besides that, they determined that the wavelength of 735 nm was related to the presence of this toxin. Kandpal et al. (2015) employed the SWIR-HSI system combined with PLS-DA model to check the aflatoxin B_1 (AFB1) in maize kernels previously spiked at 10, 100, 500, and 1000 μg kg^{-1}. The established model successfully categorized the infected and healthy grain. The overall classification accuracies were 90.7%, 92.3%, and 96.9% for yellow, white, and purple corns, respectively. The major differences were noticed between control and 1000 μg kg^{-1} group around O–H region (1400 nm) and C–H regions (1150, 1250, and 1700 nm).

Wang et al. (2014) reported a study discriminating the AFB1 contamination levels on corn surface, in which the contaminated samples were artificially spiked at 10, 20, 100, and 500 mg kg^{-1} contamination levels. Working with SWIR-HSI, an overall classification accuracy of 88% was obtained in the prediction set. In subsequent studies, *A. flavus* spore suspension was inoculated an early dough stage in the field to produce AFB1-contaminated kernels, and SWIR-HSI was applied; overall classification accuracies over 92.3% were achieved using the threshold of 20 mg kg^{-1}. It could be observed that superior model performance could be generally obtained in identifying the AFB1 < 10 mg kg^{-1} or AFB1 > 100 mg kg^{-1} corn kernels compared to those with 10 mg kg^{-1} > AFB1 < 100 mg kg^{-1}. The 1729 and 2344 nm wavelengths were identified to characterize AFB1. Also, no significant differences of model performance were observed between the germs-up and germs down placement of corn kernels (Wang et al. 2015a). In the third study of the same authors, the classification accuracy for kernels contains lower levels of AFB1, that is, either high (>100 μg kg^{-1}) or low (<10 μg kg^{-1}) levels of AFB1 in three varieties of maize kernels which reached 96.15%, 80%, and 82.61%, respectively. In addition, for those kernels with germ side down, the classification accuracy was slightly better than that of kernels with germ up. Then, the repeatability of the method was also verified using an individual variety of thirty maize kernels for which the classification accuracies were 83.33% and 88.88% for kernels with germ up and down, respectively. However, for kernels with low AF levels (between 10 and 100 μg kg^{-1}), the classification accuracy ranged from 50% to 85.71%, respectively (Wang et al. 2015b).

More recently, Chu et al. (2017) reported a study using SWIR-HSI to detect AFB1 contamination in corn kernels which were artificially infected by inoculating *A. flavus* suspension in the field and obtained an overall classification accuracy of 82.50% when discriminating the corn kernels into three groups of <20 mg kg^{-1}, 20–100 mg kg^{-1}, and >100 mg kg^{-1}.

Although results from all these studies are promising, it must be noted that most of them were carried out on artificially contaminated foodstuffs and at high contamination levels. Thus, the application in the real situation is not fully proven. Latest studies in NIR-HSI in peanuts showed the potential of the technique, not only for kernel classification but also at pixel level (Han and Gao 2019).

10.2.2 *Fusarium* Toxins

Dealing with *Fusarium* toxins, in particular with DON, has become a harder task in more recent years. In this case, researchers have aimed to both analyze bulk samples for DON quantification and also for sorting kernels according to their contamination levels. As DON, the most common *Fusarium* toxin in food products does not show fluorescence properties, reflectance HSI has mainly been used.

Barbedo et al. (2015) built an algorithm in the vis/NIR, which aimed the estimation of DON contamination of bulk wheat samples. Full-image DON estimation was performed (25–50 kernels per image), so individual kernel analysis was not feasible. They studied the correlation between FI (index factor based on a reduced number of wavelengths) and DON concentrations, and they obtained a strong correlation of 0.84. However, at low DON levels, the correlations with FI seemed to be weaker because visual discrimination is more complex. A later study of Barbedo et al. (2017) focused on DON screening of wheat samples and developed a new algorithm. Two wavelengths (627 and 1411 nm) were selected because they seemed to converge with DON presence and an index issued based on these wavelengths, which was able to classify the samples into the three classes proposed (<500 $\mu g\,kg^{-1}$, 500–1250 $\mu g\,kg^{-1}$, and >1250 $\mu g\,kg^{-1}$) with an accuracy of 72%. Furthermore, the accuracy increased to 81% when the classes were reduced into two, separated by the UE legal limit (1.25 $mg\,kg^{-1}$). Liang et al. (2018) also used algorithms for DON detection in bulk wheat samples containing 250–5000 $\mu g\,kg^{-1}$ in the vis/NIR. First, samples were analyzed by LC-MS, and they were divided into three groups. Images were acquired for each level of contamination (70 wheat kernels each). Optimal wavelengths were inspected to reduce dimensionality. The best selected algorithm is obtaining an accuracy of 100% and 97.72% in the training and the testing set, respectively.

Furthermore, data for both naturally contaminated wheat kernels and wheat flour obtained using vis/NIR-HSI and SWIR-NIR were investigated in order to predict DON content. The results suggested that it is possible to detect contaminated wheat kernel samples in the vis-NIR bands (100% for both the training and test sets) and to detect contaminated wheat flour samples from SWIR data (100% for the training set and 96% for the test set) with high accuracy (Liang et al. 2020). SWIR-NIR applied to naturally contaminated wheat samples led PLS prediction which achieved a RMSEP of 405 $\mu g\,kg^{-1}$ and 1174 $\mu g\,kg^{-1}$ for a cross-validated model and an independent set validated model, respectively. Moreover, the classification accuracy obtained by LDA was 62.7% for two categories depending on the EU maximum level (1250 $\mu g\,kg^{-1}$) (Femenias et al. 2020) using as reference the HPLC-analyzed DON value in a sample from the same batch. Later, the same authors reported prediction models which presented a RMSEP of 501 $\mu g\,kg^{-1}$ for DON. Classification achieved an encouraging accuracy of 85.4% for an independent validation set of samples, using as reference the HPLC analyzed DON in the same sample analyzed by SWIR-NIR. The results confirm that HSI-NIR may be a suitable technique for classification of samples according to the EU legal limit for DON (Femenias et al. 2020).

Tekle et al. (2015) aimed to measure DON content from oat single kernels average NIR spectra wavelength range of 1000–2500 nm by a PLS regression model. For 112 images in the validation set, an R_V^2 of 0.81 was obtained. An alternative PLS-LDA using the ratio of damaged pixels in the kernel was also performed. The correlation between predicted and measured DON in this case was 0.79, so it was proved that both options were valid for DON prediction. Infected kernels showed higher intensities at 1925, 2070, and 2140 nm.

Regarding other toxins, Firrao et al. (2012) evaluated the presence of fumonisin in milled maize using HPLC and multispectral images (720–940 nm), with a R^2 of 0.68 for fumonisin concentration (linear regression model fitting). Furthermore, on the basis of the predictions provided by a ANN, the samples were assigned to one of three classes named high (>2.8 $mg\,kg^{-1}$), medium (0.5–2.8 $mg\,kg^{-1}$), and low (≤0.5 $mg\,kg^{-1}$), for their predicted fumonisin content. For its part, the presence of ochratoxin A in stored barley was evaluated using NIR-HSI (reflectance mode at 1000–1600 nm). PCA allowed to select key wavelengths: 1260, 1310, and 1360 nm to evaluate *A. glaucus*, *Penicillium* spp., and non-ochratoxin A producing *Penicillium verrucosum*–infected grains, and 1310, 1360, and 1480 nm to differentiate ochratoxin A-contaminated grains. Fungal-infected grains could be classified as such with more than 80% (initial periods of infection) and 100% classification accuracy (four weeks after infection) (Senthilkumar et al. 2016).

Most efforts have been devoted to DON detection in wheat due to the significance and prevalence of this toxin in the European diet; however, the promising results obtained should trigger further research in other existing combinations cereal/mycotoxin of particular interest.

11 Conclusions and Future Outlook

The major application of HSI is in in-line food surface inspection of food batches for bruises and other damages, as well as assessment of major components for classification according to ripeness, or grading. Further research is needed for in-line sorting of contaminated raw materials, which would largely contribute to increase the sustainability of the food system, as it would prevent from rejection of full batches. HSI can be used for suitable calibration of multispectral devices in the food industry, which would be a more economically feasible alternative.

In off-line sample analysis, a number of advantages must be highlighted, including no sample preparation requirement, being a chemical-free and nondestructive technique, and the amount of information provided, as the spectral print can be correlated with multiple quality and safety parameters. However, further research is required for those contaminants which are present at low concentration as HSI systems do not have good detection limits compared to chemical-based analytical methods. Accurate reference calibration and robust model transfer algorithms are required, as well as increased speed in acquisition and analysis of the hyperspectral data.

In conclusion, this technology should improve quality and safety management in the food industry, as well as help to solve presently witnessed sustainability challenges, as long as future technology advances allow for lower-cost scan technologies.

Acknowledgments

This work was supported by the Spanish Ministry of Economy and Competitiveness (MINECO/AEI/FEDER, UE, project AGL2017-87755-R). Antoni Femenias acknowledges the financial support of the University of Lleida (Pre-doctoral grant).

References

Aredo, V., Velásquez, L., Carranza-Cabrera, J., and Siche, R. (2019). Predicting of the quality attributes of orange fruit using hyperspectral images. *Journal of Food Quality and Hazards Control 6* (3): 82–92. https://doi.org/10.18502/jfqhc.6.3.1381.

Barbedo, J.G.A., Tibola, C.S., and Fernandes, J.M.C. (2015). Detecting *Fusarium* head blight in wheat kernels using hyperspectral imaging. *Biosystems Engineering 131*: 65–76. https://doi.org/10.1016/j.biosystemseng.2015.01.003.

Barbedo, J.G.A., Tibola, C.S., and Lima, M.I.P. (2017). Deoxynivalenol screening in wheat kernels using hyperspectral imaging. *Biosystems Engineering 155*: 24–32. https://doi.org/10.1016/j.biosystemseng.2016.12.004.

Berardo, N., Pisacane, V., Battilani, P. et al. (2005). Rapid detection of kernel rots and mycotoxins in maize by near-infrared reflectance spectroscopy. *J Agric Food Chem* 53: 8128–8134. https://doi.org/10.1021/jf0512297.

Blanch-Perez-del-Notario, C., Saeys, W., and Lambrechts, A. (2020). Fast ingredient quantification in multigrain flour mixes using hyperspectral imaging. *Food Control* 118: 107366. https://doi.org/10.1016/j.foodcont.2020.107366.

Bo, C., Lu, H., and Wang, D. (2017). Spectral-spatial K-nearest neighbor approach for hyperspectral image classification. *Multimedia Tools and Applications*: 1–18. https://doi.org/10.1007/s11042-017-4403-9.

Boldrini, B., Kessler, W., Rebnera, K., and Kessler, R.W. (2012). Hyperspectral imaging: a review of best practice, performance and pitfalls for in-line and on-line applications. *Journal of Near Infrared Spectroscopy 20* (5): 483–508. https://doi.org/10.1255/1003.

Caporaso, N., Whitworth, M.B., and Fisk, I.D. (2018a). Protein content prediction in single wheat kernels using hyperspectral imaging. *Food Chemistry 240*: 32–42. https://doi.org/10.1016/j.foodchem.2017.07.048.

Caporaso, N., Whitworth, M.B., Fowler, M.S., and Fisk, I.D. (2018b). Hyperspectral imaging for non-destructive prediction of fermentation index, polyphenol content and antioxidant activity in single cocoa beans. *Food Chemistry 258* (2017): 343–351. https://doi.org/10.1016/j.foodchem.2018.03.039.

Chakraborty, S.K., Mahanti, N.K., Mansuri, S.M. et al. (2020). Non-destructive classification and prediction of aflatoxin-B1 concentration in maize kernels using Vis–NIR (400–1000 nm) hyperspectral imaging. *J Food Sci Technol.* https://doi.org/10.1007/s13197-020-04552-w.

Chavez, P.-F., De Bleye, C., Sacré, P.-Y. et al. (2013). Validation methodologies of near infrared spectroscopy methods in pharmaceutical applications. *European Pharmaceutical Review 18* (1): 3–6.

Cho, J.S., Bae, H.J., Cho, B.K., and Moon, K.D. (2017). Qualitative properties of roasting defect beans and development of its classification methods by hyperspectral imaging technology. *Food Chemistry 220*: 505–509. https://doi.org/10.1016/j.foodchem.2016.09.189.

Cho, H., Kim, M.S., Kim, S. et al. (2018). Hyperspectral determination of fluorescence wavebands for multispectral imaging detection of multiple animal fecal species contaminations on romaine lettuce. *Food Bioproc. Tech.* 11: 774–784.

Chu, X., Wang, W., Yoon, S.C. et al. (2017). Detection of aflatoxin B1 (AFB1) in individual maize kernels using short wave infrared (SWIR) hyperspectral imaging. *Biosyst Eng.* 157: 13–23.

Del Fiore, A., Reverberi, M., Ricelli, A. et al. (2010). Early detection of toxigenic fungi on maize by hyperspectral imaging analysis. *Int J Food Microbiol* 144: 64–71.

Delwiche, S.R. and Gaines, C.S. (2005). Wavelength selection for monochromatic and bichromatic sorting of *Fursarium*-damaged wheat. *Applied Engineering in Agriculture 21* (4): 681–688. https://doi.org/10.13031/2013.18557.

Delwiche, S.R., Qin, J., Chao, K. et al. (2017). Line-scan hyperspectral imaging techniques for food safety and quality applications. *Applied Sciences 7* (2): 125. https://doi.org/10.3390/app7020125.

Du, Z., Zeng, X., Li, X. et al. (2020). Recent advances in imaging techniques for bruise detection in fruits and vegetables. *Trends in Food Science and Technology 99* (2019): 133–141. https://doi.org/10.1016/j.tifs.2020.02.024.

El-Mesery, H.S., Mao, H., and Abomohra, A.E.F. (2019). Applications of non-destructive technologies for agricultural and food products quality inspection. *Sensors (Switzerland) 19* (4): 1–23. https://doi.org/10.3390/s19040846.

Esbensen, K.H., Guyot, D., Westad, F., and Houmoller, L.P. (2002). *Multivariate data analysis-in practice: an introduction to multivariate data analysis and experimental design*. CAMO. https://books.google.es/books?id=Qsn6yjRXOaMC.

Femenias, A., Gatius, F., Ramos, A.J. et al. (2020). Use of hyperspectral imaging as a tool for *Fusarium* and deoxynivalenol risk management in cereals: a review. *Food Control 108*: 106819. https://doi.org/10.1016/j.foodcont.2019.106819.

Fernández-Ibáñez, V., Soldado, A., Martínez-Fernández, A., and de la Roza-Delgado, B. (2009). Application of near infrared spectroscopy for rapid detection of aflatoxin B1 in maize and barley as analytical quality assessment. *Food Chem* 113: 629–634. https://doi.org/10.1016/j.foodchem.2008.07.049.

Firrao, G., Torelli, E., Gobbi, E. et al. (2012). Prediction of milled maize fumonisin contamination by multispectral image analysis. *J Cereal Sci* 52: 327–330.

Gama, T., Wallace, H.M., Trueman, S.J. et al. (2018). Hyperspectral imaging for non-destructive prediction of total nitrogen concentration in almond kernels. *Acta Horticulturae* 1219: 259–264. https://doi.org/10.17660/ActaHortic.2018.1219.40.

Guo, Y., Cao, H., Han, S. et al. (2018). Spectral-spatial hyperspectralImage classification with K-nearest neighbor and guided filter. *IEEE Access 6*: 18582–18591. https://doi.org/10.1109/ACCESS.2018.2820043.

Han, Z. and Gao, J. (2019). Pixel-level aflatoxin detecting based on deep learning and hyperspectral imaging. *Comput Electr Agr* 164: 104888.

Hansen, C.L., Thybo, A.K., Bertram, H.C. et al. (2010). Determination of dry matter content in potato tubers by low-field nuclear magnetic resonance (LF-NMR). *Journal of Agricultural and Food Chemistry 58* (19): 10300–10304. https://doi.org/10.1021/jf101319q.

Jiang, Y., Zhang, R., Yu, J. et al. (2011). Applications of visible and near-infrared hyperspectral imaging for non-destructive detection of the agricultural products. *Advanced Materials Research 317–319*: 909–914. https://doi.org/10.4028/www.scientific.net/AMR.317-319.909.

Jiang, Y., Li, C., and Takeda, F. (2016a). Nondestructive detection and quantification of blueberry bruising using near-infrared (NIR) hyperspectral reflectance imaging. *Sci. Rep.* 6: 1–14. https://doi.org/10.1038/srep35679.

Jiang, J., Qiao, X., and He, R. (2016b). Use of near-infrared hyperspectral images to identify moldy peanuts. *J Food Eng* 169: 284–290.

Kalkan, K., Beriat, P., Yardimci, Y., and Pearson, T.C. (2011). Detection of contaminated hazelnuts and ground red chili pepper flakes by multispectral imaging. *Comput Electr Agr* 77: 28–34.

Kandpal, L.M., Lee, S., Kim, M.S. et al. (2015). Short wave infrared (SWIR) hyperspectral imaging technique for examination of aflatoxin B1 (AFB1) on corn kernels. *Food Control* 51: 171–176.

Keresztes, J.C., Goodarzi, M., and Saeys, W. (2016). Real-time pixel based early apple bruise detection using short wave infrared hyperspectral imaging in combination with calibration and glare correction techniques. *Food Control 66*: 215–226. https://doi.org/10.1016/j.foodcont.2016.02.007.

Khodabakhshian, R. and Emadi, B. (2018). Application of Vis/SNIR hyperspectral imaging in ripeness classification of pear. *International Journal of Food Properties 20* (3): S3149–S3163. https://doi.org/10.1080/10942912.2017.1354022.

Kim, M.S., Chen, Y.R., Cho, B.K. et al. (2007). Hyperspectral reflectance and fluorescence line-scan imaging for online defects and fecal contamination inspection of apples. *Sens Instrum Food Qual Saf* 1: 151–159.

Li, X., Li, R., Wang, M. et al. (2018). *Hyperspectral Imaging and Their Applications in the Nondestructive Quality Assessment of Fruits and Vegetables.* Food and Environment: *Hyperspectral Imaging in Agriculture* https://doi.org/10.5772/intechopen.72250.

Liang, K., Liu, Q.X., Xu, J.H. et al. (2018). Determination and visualization of different levels of deoxynivalenol in bulk wheat kernels by hyperspectral imaging. *Journal of Applied Spectroscopy 85* (5): 953–961. https://doi.org/10.1007/s10812-018-0745-y.

Liang, K., Huang, J., He, R. et al. (2020). Comparison of Vis-NIR and SWIR hyperspectral imaging for the non-destructive detection of DON levels in Fusarium head blight wheat kernels and wheat flour. *Infrared Phys Technol* 106: 103281.

Liu, Z., Jiang, J., Qiao, X. et al. (2020). Using convolution neural network and hyperspectral image to identify moldy peanut kernels. *LWT* 132: 109815.

López-Maestresalas, A., Keresztes, J.C., Goodarzi, M. et al. (2016). Non-destructive detection of blackspot in potatoes by Vis-NIR and SWIR hyperspectral imaging. *Food Control 70*: 229–241. https://doi.org/10.1016/j.foodcont.2016.06.001.

Lu, B., Sun, J., Yang, N. et al. (2018). Quantitative detection of moisture content in rice seeds based on hyperspectral technique. *Journal of Food Process Engineering 41* (8): 1–7. https://doi.org/10.1111/jfpe.12916.

Lu, Y., Saeys, W., Kim, M. et al. (2020). Hyperspectral imaging technology for quality and safety evaluation of horticultural products: a review and celebration of the past 20-year progress. *Postharvest Biology and Technology 170* (111318) https://doi.org/10.1016/j.postharvbio.2020.111318.

Martens, H., Nielsen, J.P., and Engelsen, S.B. (2003). Light scattering and light absorbance separated by extended multiplicative signal correction. Application to near-infrared transmission analysis of powder mixtures. *Analytical Chemistry 75* (3): 394–404. https://doi.org/10.1021/ac020194w.

Melgani, F. and Bruzzone, L. (2004). Classification of hyperspectral remote sensing with support vector machines. *IEEE Transactions on Geoscience and Remote Sensing 42*: 1778–1790.

Mo, C., Kim, G., and Lim, J. (2017). Online hyperspectral imaging system for evaluating quality of agricultural products. *Optical Measurement Systems for Industrial Inspection X 10329*: 103293G. https://doi.org/10.1117/12.2269895.

Mollazade, K. (2017). Non-destructive Identifying level of browning development in button mushroom (*Agaricus bisporus*) using hyperspectral imaging associated with chemometrics. *Food Analytical Methods 10* (8): 2743–2754. https://doi.org/10.1007/s12161-017-0845-y.

Moller, B. and Munch, L. (2009). *Infrared Spectroscopy for Food Quality Analysis and Control* (ed. D.-W. Sun), 275–319. USA: Academic Press.

Munera, S., Blasco, J., Amigo, J.M. et al. (2019). Use of hyperspectral transmittance imaging to evaluate the internal quality of nectarines. *Biosystems Engineering 182*: 54–64. https://doi.org/10.1016/j.biosystemseng.2019.04.001.

Nguyen-Do-Trong, N., Dusabumuremyi, J.C., and Saeys, W. (2018). Cross-polarized VNIR hyperspectral reflectance imaging for non-destructive quality evaluation of dried banana slices, drying process monitoring and control. *Journal of Food Engineering 238*: 85–94. https://doi.org/10.1016/j.jfoodeng.2018.06.013.

Ortaç, G., Bilgi, A.S., Taşdemir, K., and Kalkan, H. (2016). A hyperspectral imaging based control system for quality assessment of dried figs. *Computers and Electronics in Agriculture 130*: 38–47. https://doi.org/10.1016/j.compag.2016.10.001.

Osborne, B.G. and Fearn, T. (1986). Near infrared spectroscopy in food analysis. *Longman Scientific and Technical.*: 36–39.

Ozyildirim, B.M. and Avci, M. (2013). Generalized classifier neural network. *Neural Networks 39*: 18–26. https://doi.org/10.1016/j.neunet.2012.12.001.

Pathmanaban, P., Gnanavel, B.K., and Anandan, S.S. (2019). Recent application of imaging techniques for fruit quality assessment. *Trends in Food Science and Technology 94*: 32–42. https://doi.org/10.1016/j.tifs.2019.10.004.

Pearson, T.C. and Wicklow, D.T. (2006). Detection of corn kernels infected by fungi. *Transactions of the ASABE* 49: 1235–1245.

Pearson, T.C., Wicklow, D.T., Maghirang, E.B. et al. (2001). Detecting aflatoxin in single corn kernels by transmittance and reflectance spectroscopy. *Trans ASAE* 44: 1247–1254.

Peiris, K.H.S., Pumphrey, M.O., and Dowell, F.E. (2009). NIR absorbance characteristics of deoxynivalenol and of sound and fusarium-damaged wheat kernels. *Journal of Near Infrared Spectroscopy 17*: 213–221.

Pullanagari, R.R. and Li, M. (2020). Uncertainty assessment for firmness and total soluble solids of sweet cherries using hyperspectral imaging and multivariate statistics. *Journal of Food Engineering 289* https://doi.org/10.1016/j.jfoodeng.2020.110177.

Qin, J. (2010). Hyperspectral imaging instruments. In: *Hyperspectral Imaging for Food Quality Analysis and Control*, 1e (ed. D.-W. Sun), 129–172. Elsevier https://doi.org/10.1016/B978-0-12-374753-2.10005-X.

Qin, J., Chao, K., Kim, M.S., and Cho, B.K. (2016). Line-scan macro-scale Raman chemical imaging for authentication of powdered foods and ingredients. *Food and Bioprocess Technology 9* (1): 113–123. https://doi.org/10.1007/s11947-015-1605-x.

Ravikanth, L., Singh, C.B., Jayas, D.S., and White, N.D.G. (2016). Performance evaluation of a model for the classification of contaminants from wheat using near-infrared hyperspectral imaging. *Biosystems Engineering 147*: 248–258. https://doi.org/10.1016/j.biosystemseng.2016.04.001.

Ren, G., Liu, Y., Ning, J., and Zhang, Z. (2020). Hyperspectral imaging for discrimination of Keemun black tea quality categories: multivariate calibration analysis and data fusion. *International Journal of Food Science and Technology*: 1–8. https://doi.org/10.1111/ijfs.14624.

Rinnan, Å., van den Berg, F., and Engelsen, S.B. (2009). Review of the most common pre-processing techniques for near-infrared spectra. *TrAC - Trends in Analytical Chemistry 28* (10): 1201–1222. https://doi.org/10.1016/j.trac.2009.07.007.

Rodríguez-Pulido, F.J., Gil-Vicente, M., Gordillo, B. et al. (2017). Measurement of ripening of raspberries (*Rubus idaeus* L) by near infrared and colorimetric imaging techniques. *Journal of Food Science and Technology 54* (9): 2797–2803. https://doi.org/10.1007/s13197-017-2716-3.

Ropelewska, E. and Zapotoczny, P. (2018). Classification of *Fusarium*-infected and healthy wheat kernels based on features from hyperspectral images and flatbed scanner images: a comparative analysis. *European Food Research and Technology 244* (8): 1453–1462. https://doi.org/10.1007/s00217-018-3059-7.

van Roy, J., Keresztes, J.C., Wouters, N. et al. (2017). Measuring colour of vine tomatoes using hyperspectral imaging. *Postharvest Biology and Technology 129*: 79–89. https://doi.org/10.1016/j.postharvbio.2017.03.006.

Saito, S., Ishibashi, J., Miyamoto, T. et al. (2009). Reduction of wheat DON and NIV concentrations with optical sorters. *Transactions of the ASABE 52* (3): 859–866. https://doi.org/10.13031/2013.27374.

Savitzky, A. and Golay, M.J.E. (1964). Smoothing and differentiation of data by simplified least squares procedures. *Analytical Chemistry 36* (8): 1627–1639. https://doi.org/10.1021/ac60214a047.

Schaare, P.N. and Fraser, D.G. (2000). Comparison of reflectance, interactance and transmission modes of visible-near infrared spectroscopy for measuring internal properties of kiwifruit (*Actinidia chinensis*). *Postharvest Biology and Technology 20* (2): 175–184. https://doi.org/10.1016/S0925-5214(00)00130-7.

Senthilkumar, T., Jayas, D.S., White, N.D.G. et al. (2016). Detection of fungal infection and ochratoxin A contamination in stored barley using near-infrared hyperspectral imaging. *Biosyst Eng* 147: 162–173.

Serranti, S., Cesare, D., and Bonifazi, G. (2013). The development of a hyperspectral imaging method for the detection of Fusarium-damaged, yellow berry and vitreous Italian durum wheat kernels. *Biosyst Eng* 115: 20–30.

Shahin, M.A. and Symons, S.J. (2011). Detection of Fusarium damaged kernels in Canada Western Red Spring wheat using visible/near-infrared hyperspectral imaging and principal component analysis. *Comput Electron Agr* 75: 107–112. https://doi.org/10.1016/j.compag.2010.10.004.

Shahin, M.A. and Symons, S.J. (2012). Detection of Fusarium damage in Canadian wheat using visible/near-infrared hyperspectral imaging. *J Food Meas Charact* 6: 3–11. https://doi.org/10.1007/s11694-012-9126-z.

Shahin, M.A., Hatcher, D.W., and Symons, S.J. (2010). Assessment of mildew levels in wheat samples based on spectral characteristics of bulk grains. *Qual Assur Saf Crop Food* 2: 133–140. https://doi.org/10.1111/j.1757-837X.2010.00070.x.

Siedliska, A., Baranowski, P., Zubik, M., and Mazurek, W. (2017). Detection of pits in fresh and frozen cherries using a hyperspectral system in transmittance mode. *Journal of Food Engineering 215*: 61–71. https://doi.org/10.1016/j.jfoodeng.2017.07.028.

Sileoni, V., Van Den Berg, F., Marconi, O. et al. (2011). Internal and external validation strategies for the evaluation of long-term effects in NIR calibration models. *Journal of Agricultural and Food Chemistry 59* (5): 1541–1547. https://doi.org/10.1021/jf104439x.

Singh, C.B., Jayas, D.S., Paliwal, J., and White, N.D.G. (2012). Fungal damage detection in wheat using short-wave near-infrared hyperspectral and digital colour imaging. *Int J Food Prop* 15: 11–24.

Siripatrawan, U. and Makino, Y. (2015). Monitoring fungal growth on brown rice grains using rapid and non-destructive hyperspectral imaging. *Int J Food Microbiol* 199: 93–100.

Siripatrawan, U., Makino, Y., Kawagoe, Y., and Oshita, S. (2011). Rapid detection of *Escherichia coli* contamination in packaged fresh spinach using hyperspectral imaging. *Talanta* 85: 276–281.

Song, D., Song, L., Sun, Y. et al. (2016). Black heart detection in white radish by hyperspectral transmittance imaging combined with chemometric analysis and a successive projections algorithm. *Applied Sciences (Switzerland) 6* (9) https://doi.org/10.3390/app6090249.

Sulistya, S.O. and Saputro, A.H. (2018). Soluble solid content prediction system of honey based on spectral transmittance profile of hyperspectral imaging. In: *ISSIMM 2018 – 3rd International Seminar on Sensors, Instrumentation, Measurement and Metrology, Proceeding*, 25–29. https://doi.org/10.1109/ISSIMM.2018.8727644.

Tang, H. (2012). Optimal wavelength selection algorithm of non-spherical particle size distribution based on the light extinction data. *Thermal Science 16* (5): 1353–1357. https://doi.org/10.2298/TSCI1205353T.

Tekle, S., Mage, I., Segtnan, V.H., and Bjornstad, A. (2015). Near-infrared hyperspectral imaging of *Fusarium*-damaged oats (*Avena sativa* L.). *Cereal Chemistry 92* (1): 73–80. https://doi.org/10.1094/CCHEM-04-14-0074-R.

Tsai, F. and Philpot, W. (1998). Derivative analysis of hyperspectral data. *Remote Sensing of Environment 66* (1): 41–51. https://doi.org/10.1016/S0034-4257(98)00032-7.

Tschannerl, J., Ren, J., Jack, F. et al. (2019). Potential of UV and SWIR hyperspectral imaging for determination of levels of phenolic flavour compounds in peated barley malt. *Food Chemistry 270*: 105–112. https://doi.org/10.1016/j.foodchem.2018.07.089.

Wang, W., Heitschmidt, G.W., Ni, X. et al. (2014). Identification of aflatoxin B1 on maize kernel surfaces using hyperspectral imaging. *Food Control* 42: 78–86.

Wang, W., Lawrence, K.C., Ni, X. et al. (2015a). Near-Infrared Hyperspectral Imaging for detecting aflatoxin B1 of maize kernels. *Food Control* 51: 347–355.

Wang, W., Ni, X., Lawrence, K.C. et al. (2015b). Feasibility of detecting Aflatoxin B1 in single maize kernels using hyperspectral imaging. *J. Food Eng* 166: 182–192.

Wang, T., Chen, J., Fan, Y. et al. (2018). SeeFruits: design and evaluation of a cloud-based ultra-portable NIRS system for sweet cherry quality detection. *Computers and Electronics in Agriculture 152*: 302–313. https://doi.org/10.1016/j.compag.2018.07.017.

Westad, F. and Marini, F. (2015). Validation of chemometric models – a tutorial. *Analytica Chimica Acta 893*: 14–24. https://doi.org/10.1016/j.aca.2015.06.056.

Williams, P., Geladi, P., Britz, T., and Manley, M. (2012). Investigation of fungal development in maize kernels using NIR hyperspectral imaging and multivariate data analysis. *J Cereal Sci* 55: 272–278.

Xiao, K., Liu, Q., Wang, L. et al. (2020). Prediction of soluble solid content of *Agaricus bisporus* during ultrasound-assisted osmotic dehydration based on hyperspectral imaging. *LWT - Food Science and Technology 122*: 109030. https://doi.org/10.1016/j.lwt.2020.109030.

Xiaobo, Z., Jiewen, Z., Povey, M.J.W. et al. (2010). Variables selection methods in near-infrared spectroscopy. *Analytica Chimica Acta 667* (1–2): 14–32. https://doi.org/10.1016/j.aca.2010.03.048.

Xing, F., Wang, L., Liu, X. et al. (2017). Aflatoxin B1 inhibition in Aspergillus flavus by Aspergillus niger through down-regulating expression of major biosynthetic genes and AFB1 degradation by atoxigenic A. flavus. *Int J Food Microbiol* 256: 1–10.

Yang, C.C., Jun, W., Kim, M.S. et al. (2010). Classification of fecal contamination on leafy greens by hyperspectral imaging. In: *Sensing for Agriculture and Food Quality and Safety II* (eds. L.S. Kim, S.Y. Ma and K. Chao), 76760F. Orlando, Florida: Proceedings of SPIE.

Yang, C.C., Kim, M.S., Kang, S. et al. (2012). Red to far-red multispectral fluorescence image fusion for detection of fecal contamination on apples. *J Food Eng* 108: 312–319.

Yao, H. and Lewis, D. (2010). Spectral preprocessing and calibration techniques. In: *Hyperspectral Imaging for Food Quality Analysis and Control* (ed. D.-W. Sun), 45–78. Academic Press, Elsevier https://doi.org/10.1016/B978-0-12-374753-2.10002-4.

Yao, H., Hruska, Z., Kincaid, R. et al. (2013). Detecting maize inoculated with toxigenic and atoxigenic fungal strains with fluorescence hyperspectral imagery. *Biosyst Eng* 115: 125–135.

Zhang, M., Li, C., Takeda, F., and Yang, F. (2017). Detection of internally bruised blueberries using hyperspectral transmittance imaging. *Transactions of the ASABE 60* (5): 1489–1502. https://doi.org/10.13031/trans.12197.

Zheng, X., Peng, Y., Li, Y. et al. (2016). Non-destructive prediction of moisture of wheat seed kernel by using VIS/NIR hyperspectral technology. In: *2016 American Society of Agricultural and Biological Engineers Annual International Meeting, ASABE 2016*, 10–16. https://doi.org/10.13031/aim.20162461233.

Zhu, F., Yao, H., Hruska, Z. et al. (2016). Integration of fluorescence and reflectance visible near-infrared (VNIR) hyperspectral images for detection of aflatoxins in corn kernels. *Trans ASABE* 59: 785–794.

Zhu, H., Chu, B., Fan, Y. et al. (2017). Hyperspectral imaging for predicting the internal quality of kiwifruits based on variable selection algorithms and chemometric models. *Scientific Reports 7* (1): 1–13. https://doi.org/10.1038/s41598-017-08509-6.

16

Future Challenges of Employing Electromagnetic Spectrum

Bibhuti B. Mishra and Prasad S. Variyar

Homi Bhabha National Institute, Food Technology Division, Bhabha Atomic Research Centre, Mumbai, India

1 Introduction

The electromagnetic spectrum covers a wide range of frequencies starting from γ rays of 10^{21} Hz frequency (wavelength ~10^{-13} m) to radio waves of 10^{6} Hz frequency (wavelength ~1 m). The application of different part of the spectrum in food processing started at a modest rate around 1950s (Mudgett 1988). The mode of interaction of different frequencies is different with food material yielding a variety of benefits. It can be used for sterilization to drying applications and also for food quality inspection (Rosenthal 1992). However, the journey of its use was not easy and has come through technological difficulties and safety concerns. The most challenging was the γ irradiation as it is ionizing with energy in the range of MeV (Figure 1). After elaborate safety studies, its use was recognized by the United Nations with the establishment of Joint Expert Committee on Food Irradiation (Craven et al. 2018). The committee after a series of meetings concluded in 1980 that "irradiation of foods up to the dose of 10 kGy introduces no special nutritional or microbiological problems." In 1999, the World Health Organization (WHO) assured its safety and stated that γ irradiation should be considered similar to cooking on aspects of safety (Farkas and Mohácsi-Farkas 2011). Irrespective of many challenges, its use is increasing being indispensable in terms of high penetration power and many other advantages.

The use of UV-C of EM spectrum with energies in range of approximately 10 eV (wavelength of about 10^{-7} m) for liquid food processing and surface decontamination was well accepted due to technological advantages of avoiding undesirable effects of heat treatments, low installation and operation costs, and less energy usage (Guerrero-Beltrán and Barbosa-Cánovas 2004). However, the equipment design and process parameter (exposure, dose, wavelength, and UV source) selection are the bigger difficulties and need to be specifically developed for different types of liquid food depending on the nature of microbial load (Delorme et al. 2020).

Electromagnetic Technologies in Food Science, First Edition. Edited by Vicente M. Gómez-López and Rajeev Bhat.

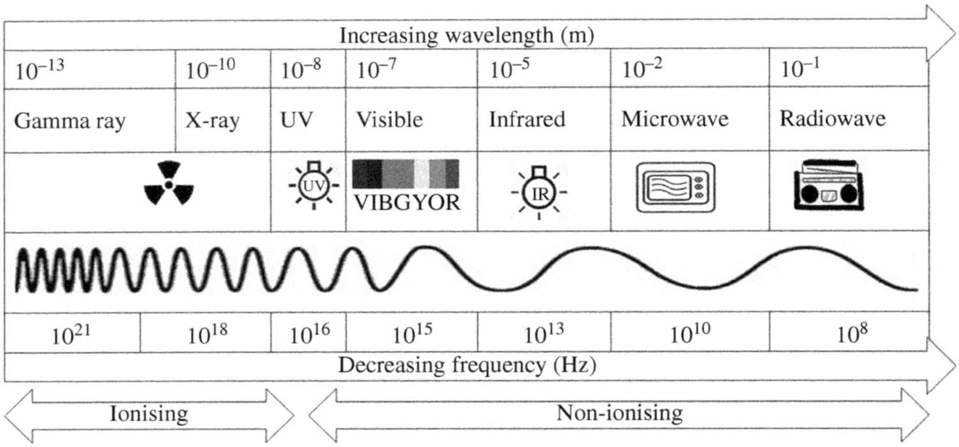

Figure 1 The electromagnetic spectrum composition with wavelength, frequency, and ionizing characteristics. (*See insert for color representation of this figure*).

Application of infrared (IR) spectrum with energy less than 1 eV (equivalent to molecular interatomic bonding) in food industry was reported in the 1950s for drying applications (Rastogi 2012). Despite greater potential, its use has been underestimated in the food industry, compared with other industries. In general, the food substances absorb far-infrared (FIR) energy most efficiently through the mechanism of changes in the molecular vibrational state, which can lead to radiative heating (Rosenthal 1992). During the early 1970s, its use started to pick up as a novel method for industrial baking, cooking, and drying. Ever since, the application areas are growing with development of advanced near-infrared (NIR) spectroscopy for proximate analysis of food samples and online checking of food composition in industry. NIR spectroscopy as well as tetrahertz spectroscopy is having high scope as a quality assurance tool for compositional analysis of food ingredients for in/online measurements in food industry (Pan et al. 2017).

Microwave frequencies (2450 and 915 MHz) used in processing of food are part of special frequency bands reserved for industrial, scientific, and medical (ISM) applications. In 1945, the heating effect of microwave was discovered followed by US patent for microwave cooking. The application of microwave in newer areas includes cooking, drying, blanching, pasteurization, sterilization, as well as in extraction of bioactive compounds (James et al. 2006).

The frequency of electromagnetic spectrum in the range of 1–300 MHz is termed as radiofrequency (RF) and selected frequencies, namely 13.56, 27.12, and 40.68 MHz, are permitted for ISM applications (Afsah-Hejri et al. 2019). The RF heating field is constantly evolving with challenges to ensure uniform heating since 1960, when it was used for thawing of frozen foods at industrial scale using a 25 kW source (Jiao et al. 2018). RF heating, as a thermal processing alternative, has few unique advantages of volumetric heating over conventional methods in industrial drying and frozen food thawing (Guo et al. 2019). More research is underway for food safety with different food matrices because of its complex heating mechanism (Altemimi et al. 2019).

Another important challenge for adoption of new EM technologies lies in commercial acceptance considering the economic burden on the existing industries. In general, there is a reluctance by the industry for adoption of newer technologies as it needs setting up of new machineries while disrupting existing infrastructure. Further, newer EM technologies have to be competitive in terms of capital investment and also be economical in terms of energy usage. The lesser energy usage will also have additional advantage of being environmental friendly and reduce concerns about global warming. The development of portable and movable food processing machines employing EM spectrum will help in processing of agricultural produce near to the farm and thus help in uplifting regional economy in low and middle-income areas. The present chapter details the advantages and disadvantages of the above technologies and the challenges posed in their scale-up and economic viability when applied at industrial level. The latest developments and difficulties related to commercial adoption, equipment design, and new application areas are also discussed.

2 Challenges in γ Irradiation Processing of Food

The patents on use of ionizing radiation for food processing were granted in 1905. Its use at industrial scale started only after 1960 and has ever since been increasing at a modest rate. The International Consultative Group on Food Irradiation (ICGFI) was established in 1984 by joint initiatives from Food and Agriculture Organization (FAO), International Atomic Energy Agency (IAEA), and WHO to focus on international cooperation on food irradiation. About 4 million tons of food was processed using γ radiation worldwide in 2005. Commercial use of γ radiation processing of food has significantly increased particularly in Asia after 2010 with 0.2 million tons of food being processed in China by γ radiation (Carreño 2018). As per recent information available, 22 countries, including the United States, Japan, China, India, and South Korea, have around 515 γ radiation plants that processes various food commodities (Eustice 2018). With growing need for food processing and preservation, the challenges and opportunities for γ radiation processing of food have increased many fold in recent times.

2.1 Sources of Radiation: Cobalt 60 and Cesium 137, Electron Beam, and X-ray

The irradiation sources permitted for food irradiation by joint expert committee of FAO/IAEA/WHO are as follows:

1) γ rays from Co60 (1.17 and 1.33 MeV, $T_{1/2} = 5.2$ years)
2) γ rays from Cs137 (0.662 MeV, $T_{1/2} = 30$ years)
3) X-rays produced from machine source (Up to 5 MeV electron beam accelerators)
4) High-energy electrons produced by accelerators (maximum energy 10 MeV)

The most used radionuclide for the irradiation of food is cobalt-60. The technology for its production, fabrication, and encapsulation is well established. Cobalt-60 is produced by neutron bombardment on metal cobalt-59 in a particular type of nuclear reactor and then doubly encapsulated in stainless steel pencils in specialized facility to prevent any leakage during its use. As the supply of Co-60 is limited, there is a need to increase its production to meet the growing demand for this isotope with increasing number of irradiators being commissioned world over (Eustice 2018).

Cs-137 is a fission product in spent fuel of nuclear power reactor and is extracted from fuel in a reprocessing plant and is extracted as cesium chloride. As it is obtained in the powdered form, additional precaution for handling is warranted. In this direction, recent development in the area of vitrification of 137Cs-sodium borosilicate glass for improving its safe handling in pharmaceutical applications (Dash et al. 2009) is worth mentioning. The developed process, however, needs to be scaled up to higher capacity for industrial applications and for development of potable irradiators. If successfully used in food irradiation, it will be a milestone in recycling of nuclear waste and reducing the load on Co60 manufacturing. Being an alternate radioisotope source of γ radiation with higher half-life and lesser energy, it gives specific advantage for design of small irradiators.

In addition to radioisotope sources, electron beam accelerators and X-ray machines are becoming popular due to many advantages. Electron beam machines use linear accelerators (LINACs) that produce electron beams of desired energy (up to 10 MeV). It is powered by electricity and can be switched on or off and is not related to nuclear radiation (Craven et al. 2018). However, electrons have low penetration power compared with γ radiation and cannot be applied to bulky food material.

To address the penetration limitation of electron, X-ray machines are being developed suitable for food irradiation. X-ray processing is permitted to energies below 5 MeV in most countries (Farkas and Mohácsi-Farkas 2011). In the United States and India, up to 7.5 MeV is already approved. The electron beam (5 MeV) bombarded on a metal target such as tungsten, which in turn produces X-ray. Higher-energy electron beam may induce radioactivity in metal target plate due to which up to 5 MeV energy electron is permitted. Suitable target material needs to be developed, which can tolerate higher-energy electron beam without becoming radioactive (Table 1). Further, it has the disadvantage of low efficiency as electron to X-ray conversion is less than 8%. The method needs to be further developed to improve the efficiency for X-ray irradiation for making it.

2.2 Scope for Future Research in γ Radiation

The γ radiation applications are established for a variety of purposes including sprout inhibition, insect disinfestation, shelf-life extension, microbial decontamination, and sterilization of food (Figure 2). The referred dose ranges suggested should not be assumed to work for all food products. The sensitivity of different food items in terms of the effect of radiation on sensory parameters differs. The irradiation conditions (dose rate and temperature), packaging size (box dimensions), and temperature (ambient or frozen food) need to be optimized depending on food product category (fruit, vegetable, grain, or meat) and state (solid or liquid) to improve its application for the desired purpose (Carreño 2018). Research needs to be conducted on case-by-case basis for ensuring suitability for use of γ radiation. The effect of ionizing radiation with food with different composition and density is different (Dourado et al. 2019). In addition, newly invented packaging materials need to be tested for compatibility with irradiation.

Table 1 Research approaches to deal with future challenges associated with application of electromagnetic spectrum (EMS) for food processing.

EMS applications	Future challenges	Research approaches for the problem
γ radiation processing of food	Personnel and environmental safety	Safety feature improvement through design of irradiator and continuous monitoring of environmental exposure
	Safety and availability of Co60 radiation source	Increase in manufacturing of sealed Co60 and prevention of leakage
	Manufacturing safety and handling of Cs137	Immobilization or vitrification of Cs137 to improve safety
X-ray processing of food	Personnel and environmental safety	Design feature improvement with appropriate shielding for restricting environmental exposure
	Energy efficiency	Use of novel target material for conversion of electron beam to X-ray
UV processing of food	Personnel safety	Use of appropriate shielding for protection
	Scale-up of processing	Design of processing equipment with positioning of UV lamp
	Low penetration power	Surface decontamination of solid food
		Use of thinner tube and allowing mixing of liquid food during UV exposure
	Efficiency for microbial decontamination	Process optimization for determination of treatment parameters
IR processing of food	IR emitter efficiency	Research on improvement of electric power conversion during heating
	IR filament material	Increase in melting point, thermal stability, and electrical resistivity
	IR lamp design	Suitable position for maximum use of heat energy, increase in lamp life, suitable reflector development
	Drying efficiency improvement	Combination with other drying methods and ventilation system for moisture removal
Microwave processing	Industrial microwave processing unit	Design and safety issues related to leakage and personnel safety
	Nonuniform temperature distribution	Novel cavity design including use of movable tray and fans for deflection
	Microwave baking	Appropriate color development
Radiowave processing	Nonuniform heating	Design of drying cavity, use of pulse mode for exposure, configuration of electrodes
	Industrial unit	Design for improvement of performance
	Drying efficiency improvement	Combination with other drying methods

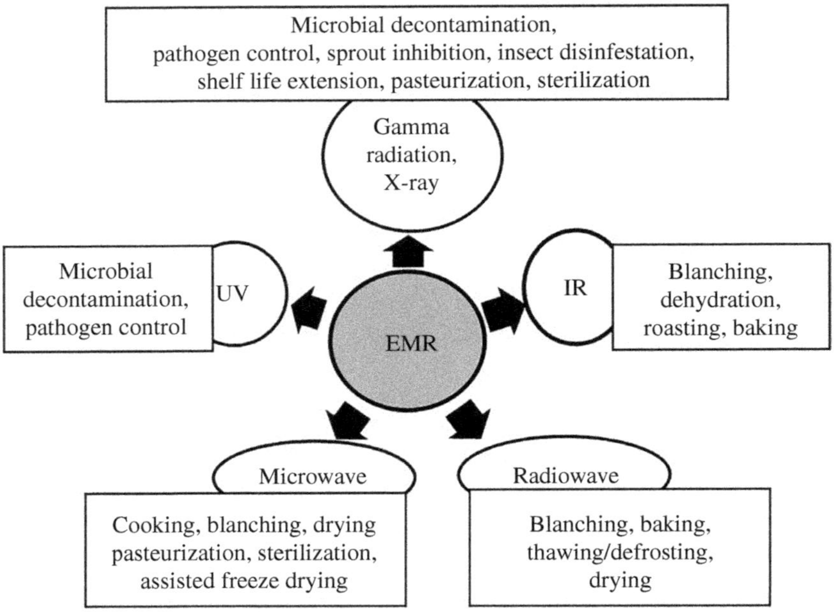

Figure 2 The electromagnetic spectrum applications for food processing and preservation.

2.3 Economic Considerations for Setting Up Facilities

The food irradiation facilities for industrial-scale processing costs about 2 million USD excluding land cost. Half of this cost is the price of the cobalt-60 source. The land of about 0.4 hectare is needed for the plant itself. The cobalt-60 activity reduces due to radioactive decay at the rate of about 12% a year. The radiation source needs to be replenished periodically at interval of 2 years in a regulated manner to maintain the source strength. A typical plant has a source strength of 1000 kilo Curie (Eustice 2018). The capital cost and land requirement make the plant available only at limited places in trading route. The radiation processing adds to the cost of product by about 5–10%, which is marginal and insignificant considering the benefits. However, due to higher capital cost, such facilities can only be made in limited places as per the demand. This makes the cost of logistics an additional factor to be considered for economic calculations. In poor and developing countries, high cost of it makes it uneconomical.

3 Challenges in Using UV Light for Processing of Food

According to the United States Food and Drug Administration (US-FDA), UV-C at 200–280 of electromagnetic spectrum is germicidal and effectively inactivates bacteria and viruses (Figure 1). UV-C is proven to reduce or eliminate *E. coli*, *Salmonella*, *Listeria*, and other foodborne pathogens (Koutchma 2009). The mechanism of action involves mutations in DNA due to formation of pyrimidine dimers followed by impaired replication leading to cellular damage and death of microorganisms. The UV processing has the advantage over thermal processing in terms of retention of original flavor. Since it has lower penetration power, it is highly suitable for surface decontamination in solid foods. Its applicability has been widely studied in liquid foods. UV-C has been used to disinfect drinking water, sanitize wastewater, and sterilize food contact surfaces. In juices and similar foods where thermal treatment affects the quality, UV processing treatment is emerging as a suitable alternative. The effectiveness of UV gradually reduces, as the juice gets turbid and denser. However, in clear juices, the effectiveness for microbial decontamination is very effective (Shah et al. 2016). The juices from different fruits and vegetables can be exposed to UV light at various doses for appropriate level of microbial decontamination in a specialized UV treatment machine. Like any other processing method, the parameters including exposure time and flow rate need to be optimized for desirable level of microbial decontamination (Table 1). In addition, increased shelf life, sensory, and nutritional characteristics need to be evaluated periodically during storage. The major challenge in using UV for industrial-scale processing is the design of processing equipment.

3.1 Design of UV Processing Equipment

Recent developments in engineering of UV irradiation devices are making it a viable option for commercial application in food processing (Koutchma 2008). The design features of processing equipment need to be decided considering the lesser penetration power of UV. For pasteurization of juices and beverages, a thinner flow tube or plate near the UV source will ensure the maximum exposure. The mixing of liquid during flow exposes the contaminating microorganism to UV increasing its efficacy.

In addition, novel methods and models need to be developed to measure the inactivation rates or dose response behavior of food pathogens in highly absorptive and viscous liquid foods such as juices and beverages. The correct choice and/or design of the UV reactor, its flow characteristics, and UV source can reduce the interference of high UV absorptivity and viscosity associated with some liquid food products, thereby improving inactivation efficiency (Shah et al. 2016).

UV treatment has also been used as a post lethality treatment in controlling microbial contamination on meats and shell egg surfaces and as a means for the shelf life extension of fresh produce (Manzocco and Nicoli 2014). UV light processing can improve safety of selected solid and liquid foods without appreciable loss in quality or nutrient content. However, to improve the efficacy of UV light for food application, the following areas of research need to be conducted (Delorme et al. 2020). To predict UV disinfection rates on food surfaces, more kinetic inactivation data need to be obtained for pathogen and spoilage microorganisms (Table 1). This will require taking into account interactions between microorganisms and surface materials, such as shielding effects from incident UV and their dependency on surface structure or topography. The development of validation methods for food processing facilities requires identification of surrogate microorganisms and suitable actinometers. Research in the indicated areas can ensure the effectiveness of UV light for microbial inactivation, stimulate the growing interest in the nonthermal technologies, and assist in the successful commercialization of UV light for food processing applications (Fan et al. 2017; Deng et al. 2020).

3.2 UV for Disinfestation of Contact Surfaces in Food Processing Facilities

Since UV light penetrates differently depending on the type of liquid, as explained above, further studies on UV radiation regarding dosage, color, and appearance, flavor, microbial quality, and nutritional quality of the final product are required (Delorme et al. 2020). At present, the application of UV light for disinfection of food products is no longer used, but it could easily be applied to liquid and solid food products. Each food processing method is indeed different, and the performance of UV lamps for treatment of liquids or solids should be studied to obtain basic information regarding microbial disinfection. In addition, sensory evaluation, nutritional quality, and shelf life should be considered. All new processes must be tested at low scale before being scaled up at the industrial level. However, validation of methods to ensure microbiological effectiveness and the optimization of critical process factors must be studied. Ultimately, UV-treated products should satisfy FDA requirements along with HACCP programs to deliver safe food products to consumers (Delorme et al. 2020).

4 Challenges in Using Infrared (IR) for Processing of Food

IR region of electromagnetic spectrum consists of wavelength region from 0.7 μm to 1 mm and is also termed as thermal radiation, as absorption of IR results in heating effect (Figure 1). The IR applications in food processing are based on its heating effect. IR spectrum is classified into three regions, namely NIR short wave (0.7–1.4 μm), mid-infrared medium wave (MIR, 1.4–3.0 μm), and FIR long wave (3.0–1000 μm) region (Krishnamurthy et al. 2008). Short-wave IR needs prolonged drying time for food materials. The IR absorption range of water is in medium-wave range, therefore for food drying applications, medium-wave range of IR heats up water for evaporative dehydration. Medium-wave emitters are preferred for drying and curing of food products. In general, FIR radiation is advantageous for food processing because most food components absorb radiative energy in the FIR region. IR drying is known to offer many advantages over hot air oven drying. Due to uniform heating, the product temperature is maintained during drying, which results in better quality of finished products. Instruments can be designed as compact equipment as well as scalable to high capacities with significant energy savings (Rastogi 2012). Since it consumes less electric power compared with hot air oven drying, it is considered more environmental friendly.

The IR heating system is being used in many food processing operations including blanching, dehydration, roasting, and baking. IR heating has attracted widespread attention for surface heating applications such as prevention of growth of yeasts and mold on food surfaces in a variety of foods such as cheese, dehydration of fruits and vegetables, disinfestation of cereals and pulses, decontamination of the surface of eggs, as well as decontamination of dry fruits (Figure 2). Though it has many advantages over conventional drying process, it has many limitations in terms of need and expectations of food industry. The current research on applications of IR in food science is focused on further improving its efficiency by using combination with other processes such as vacuum drying and by change of design features of IR drying systems (Aboud et al. 2019).

4.1 Limitations of Infrared Processing

The heating effect of IR radiation is due to its interaction with the exposed thin layer on the food surface. The penetration power of IR radiation is known to be low as it cannot penetrate deep inside the food (Rastogi 2012). Interaction with IR heats up only a few mm thickness below the surface of the food material. As the IR energy is absorbed on the surface, it allows only a shallow layer to be dried. The absorbed energy is then transferred by conduction to other areas within the food material. The layer close to the IR source dries more rapidly compared with the one that was deep inside. As the sample thickness increases, this conduction is limited, and thus the total energy absorbed is limited. The variation in moisture distributions during drying is also observed with increasing thickness. The penetrative radiation energy of IR does not make significant contribution to internal heating. Due to this reason, most of industrial applications of IR drying involve heating of thin layer materials.

The efficient moisture removal process by use of suitable exhaust will improve the drying rate and product quality. In case of flakes or powdered material, drying by external agitation or vibration of the drying bed will improve the drying kinetics. As reported for papaya fruit cubes, the mixing or turning of food during IR drying also helps in each part receiving uniform heat (Mishra et al. 2015). Depending upon the nature of food, the innovations in design of IR drying system will help in meeting versatile usage in different types of food (Aboud et al. 2019). In order to achieve optimum use of energy in IR heating, a combination with microwave or other common conductive and convective modes of heating holds significant promise.

4.2 Selection of Infrared Emitters for Drying Applications

The IR heaters used as source of IR vary based on the usage. The different types of heaters are available, namely ceramic IR heaters, quartz heaters, halogen heaters, metal rod heaters, foil heaters, gas heaters, and heaters with exposed heating coils (Yadav et al. 2020). The IR sources used in drying of food are IR lamps and ceramic heaters powered by electricity or gas. The cost of gas heaters compared with electrical IR heaters is high, but their operating cost is lesser. IR heaters are more widely used due to ease of control and faster heating. The efficiency of electric and gas IR heaters is up to 70% and 50%, respectively. The emitters produce IR radiation in a range of 343–1100 °C for gas and 1100–2200 °C for electric heaters. The increase in efficiency of both the type of heaters is a major objective for the development of new IR heating sources. Incandescent lamps are used as the source of short wave, while quartz tubes and resistance elements are used as medium- and long-wave emitters, respectively.

4.3 Future Scopes for IR Lamp Design Features

The IR lamps used in drying applications usually have an average life of 5000 hours. With prolonged usage, the glass tube through which the IR from heating filament passes becomes partially opaque. This results in less IR being transferred for drying and optimum efficiency becomes compromised (Yadav et al. 2020). The current research in this area is oriented toward lamps of high durability life and high-quality glass material for lamp cover (Table 1). In addition, another important feature is the reflector used in the back side of lamp or heating filament to reflect maximum IR toward drying bed. Most of the dryers have lamps on one side of the drying chamber. The food material for drying is kept on one side. However, IR filament emits radiation on all sides. The reflector at the back of lamp or filament is used to direct all IR toward food surface on drying bed. The reflectors are currently made of reflective aluminum and also with silver or gold-coating materials. Aluminum coating is less expensive than silver and gold and slightly less effective in many regions of the IR. Gold is used because of its oxidation resistance and very high IR reflectivity (~95%). The white ceramic reflector can reflect to about 70% of incident IR. Another type of reflector called ruby (reddish color) is also used to mitigate the IR intensity by coating all over the quartz tube. It is reported that a good-quality reflector can save power usage by 40%. Another problem is the gradual discoloration of reflector with use, which results in IR absorption. Attention needs to be paid for better-quality coating formulation for prevention of discoloration during usage. The better quality of reflector with maximum reflection is an important research area for maximizing energy efficiency and reducing energy wastage. Another important area of engineering is the shape of an optimized reflector to focus the maximum IR to the drying area for maximum energy usage. The elliptical concave reflectors are known to have better effectiveness to pinpoint IR beam to a smaller area, whereas parabolic reflectors are better in terms of reflecting uniform parallel beam of IR on the drying surface.

4.4 Novel IR Filament Material

The IR lamp consists of filament that heats up to emit IR. The most common filament material used for electrical IR heaters is tungsten wire because of its high melting point (3695 K) and electrical resistivity. Other materials used as heating filament are carbon and other alloys (of iron, chromium, and aluminum). Carbon filaments heat up much more quickly (Yadav et al. 2020). The carbon filament is also woven or helically coiled in quartz tubes for better heating efficiency. The current research is focused on development of better heating element with high melting point, thermal stability, and electrical resistivity (Table 1). The type of IR emitted depends on the temperature of filament. The filament temperature is 3000 K in NIR lamp, whereas short-wave and medium-wave IR has filament temperature of 2300 and 900 K, respectively. The materials used in manufacture of filament need to have thermal stability and tensile strength to work at such higher temperature.

4.5 Future of IR Drying

Within the electromagnetic spectrum, the usability of IR for drying application in thinner food material is excellent. It is being used in industry and continues to excel with new technological advancements in novel filaments, heater, and dryer design. However, for thicker food material, IR has limitation of lesser penetration (Rastogi 2012). In order to overcome this challenge and effectively utilize the potential of IR for thicker food, it would be necessary to use combination of IR heating with adequate ventilation system, air circulation, or vacuum drying to accelerate the moisture removal from food. The combination with microwave and other modes of heating may give better dehydration for thicker food items.

Another important challenging area is selective IR heating. The use of IR spectrum, after passing through suitable filters to allow radiation within a specific spectral range for specific application, is referred as selective heating (Yadav et al. 2020). As different food components have different IR absorption range, it can be highly effective for specific applications. For example, microbial inactivation in food can be achieved more in far IR region without much drying effect, as water absorbs IR in medium range. Specific applications of IR for selective heating of foods could be very useful and open up avenues for future research in this area.

Another challenging area is the linking the heating of IR and drying kinetics of different kinds of food. Though it imperative that higher heating will dehydrate faster, but for fruits and vegetables and other food items with moderate moisture content, the parameters need to be defined and documented. For retention of nutritional and flavor compounds, the temperature and drying duration will vary for different fruits and vegetables. The ongoing research on effect of IR on nutritional or sensorial characteristic and physicochemical properties and publications may aid in providing IR as an alternate method of food processing.

4.6 Scopes for Near-infrared (NIR) Spectroscopy in Industrial Food Processing

NIR spectroscopy is a nondestructive technique becoming highly popular in food processing laboratory and industry due to its low running costs and non-requirement of sample preparation (Dixit et al. 2017). It is based on the absorption of electromagnetic radiation of food at wavelengths in the range of 780–2500 nm. The NIR spectra of foods comprise broad bands arising from overlapping absorptions corresponding C—H, O—H, and N—H chemical bonds. The analysis is very simple and very fast (approximately 1 seconds) and can be carried out on/inline within a processing plant. Usually it is employed for simultaneous determination of different qualitative and quantitative sample parameters such as moisture, fat, protein, salt, ash content, and others. It is also used routinely for the compositional, functional, and sensory evaluation of food ingredients, process intermediates, and final products. However, the instrument needs to be calibrated using reference method. For food samples for NIR measurement reflectance mode is normally used and multivariate linear regression model is used as calibration method. The predictions for sensory characteristics such as color and tenderness are also successfully made using NIR spectroscopy. Further advancements may be required to meet the requirement of different industrial processes. This technique has been projected by researchers as a reliable method for continuous industrial quality control and process monitoring (Dixit et al. 2017). However, challenge still remains due to issues including measurements on moving conveyor belts, in tubes with continuously flowing food material, fermentation processes monitoring, and temperature fluctuations. For such purposes, different designs of NIR spectrometers are being developed with provision of hyperspectral imaging, portability, innovative probes (fiber optical, tube integrated, or contact probes), and automated sample cell loading. There is a need for further large-scale online studies in the food industry to verify reliability and accuracy of NIR spectroscopy under process conditions. Recent advancements made in chemometrics have immensely increased the potential of the NIR technology as a reliable on/inline monitoring tool for food industry.

5 Challenges in Microwave Processing of Food

The use of microwave for thermal processing of food is in use for long time. The specific advantages of microwave processing such as lesser thermal processing time, instant turn on and off, and optimal energy usage are some of the features that may lead to its success in food industry. Interaction of microwave radiation with dielectric components of food results in heating (Orsat et al. 2017). The polar molecules in food rotate to align with the electromagnetic field of microwave. As a result, friction takes place, converting electromagnetic energy into heat. Microwave covers the range of the electromagnetic spectrum from 300 to 300 GHz. The microwave frequency range is also used for telecommunications such as mobile phones and radar transmissions. In order to avoid interference problems, special frequency bands are reserved for ISM applications (called ISM frequencies).The only frequencies allowed for thermal processing of food are 2450 and 915 MHz, which are part of ISM frequencies. Domestic microwave ovens operate at 2450 MHz, and industrial systems use either 2450 or 915 MHz. At these permitted frequencies, the food heats up in a nonuniform manner due to standing microwaves in the cavity. The peaks of the wave, instead of moving forward, oscillates up or down. The part of the food coinciding with peak/trough of wave heats up maximum, whereas other parts remain colder. This nonuniform temperature distribution results in hot and cold spots in food. To reduce this effect, microwave ovens have a rotating plate. Fans are also used to deflect the microwave to bring uniform distribution of waves in the cavity. It is challenging to design and build industrial microwave heater of high capacity by keeping these two major factors into consideration. However, because of its advantages, interest in its use in industry has grown significantly in recent years. In recent years increasing demand for this technology has led to the development of novel microwave processing methods for drying, cooking, defrosting, pasteurization, and sterilization (Table 1). Microwave also finds application for extraction of bioactive compounds (Orsat et al. 2017). The challenges in development of industrial microwave processing in food industry are due to design and safety issues related to leakage and considerable cost during scaling up of microwave equipment (Wray and Ramaswamy 2015).

5.1 Microwave Cooking

Microwave cooking is commonly used in household kitchens and restaurants for long time. However, at industrial scale, it could be successfully employed only in the processing of bacon, meat patties, and meat balls (Figure 2). It is a challenging task to expand the utility of microwave usage to other food items at larger scale. Conventional cooking can cause overcooking on the exterior of meat, whereas microwaves penetrate fast to begin internal cooking more quickly, reducing cooking time by as much as 50%, and increasing throughput by as much as 30%. Microwave cooking of bacon can be achieved at lower energy consumption with less product shrinkage and desirable sensory quality (James et al. 2006). Bacon precooking and crisping are one of the common applications of industrial microwave processing in fast food restaurants. A problem with microwave cooking is that the surface of the food does not brown because of absence of Maillard reactions. In such cases microwave oven is often supplemented with hot air oven, which can brown the food surface for desired color development. Combinations of conventional and microwave ovens are a good solution at industrial scale, where the application of microwaves allows for acceleration of the cooking process and the surface water is evaporated through convectional airflow.

5.2 Microwave Blanching

Microwave blanching is gaining considerable interest, mainly due to short processing time and reduced loss of nutrients in fruits and vegetables. It has been demonstrated that microwave blanching can improve the microwave freeze–drying process (Bhatta et al. 2020). Blanching is known to increase the ionic content of foods, resulting in improved heating uniformity in microwave processing. Irrespective of it advantages, it is still a challenge to develop an industrial-scale microwave blanching system like any other equipment for microwave applications. The industrial scaling-up of microwave blanching remains a mostly unexplored area.

5.3 Microwave Pasteurization/Sterilization

Microwave pasteurization/sterilization of packaged meal, another area of wide interest, has significant potential for industrial application (Orsat et al. 2017). A continuous process for the cooking and vacuum packing of food wherein the packed meal is safe and shelf-stable during subsequent storage is desirable (Figure 2). Though there are reports about its working for meals including chicken nuggets and potato, it has expanded to include other food servings. Commercialization of continuous microwave-assisted pasteurization and sterilization of meals and meal components has also seen slow progress since the 1970s. During 1980–2000, over 200 machines were sold based on continuous microwave sterilization process. Such machines were used for pasta meals during 1990. In Europe, an industrial continuous in-pack meal pasteurization system was in use that utilized a one-way steam valve on the food package. Currently the research is also directed toward improving nutrient retention in food products sterilized by microwave.

5.4 Microwave-assisted Drying

Microwave drying is very efficient and is commercially used in applications for snacks, as well as spices and other ingredients. It is also used to finish dry pasta and instant noodles. Industrial Microwave Systems estimate that microwave use as a pre-dryer or post-dryer can increase overall production capacities by 25–33% and can produce a return on capital investment within as little as 12–24 months. The microwave can also be combined with hot air drying in a microwave-assisted hot air drying system, which has the advantages of dual heating for effective dehydration of thick food materials (Orsat et al. 2017). Microwave drying has better advantages over conventional drying, because heat is generated deep within the material to be dried. Viscous and bulk products with poor thermal conductivity have slower drying rate in conventional vacuum drying. In microwave vacuum drying, however, such food materials can be dehydrated in lesser time. Currently the development cost of such a system is very high and is thus used only for high-end drying applications of heat-sensitive products. The reduction in cost of the system by a suitable design is the focus of current research.

5.5 Microwave-assisted Freeze Drying

In conventional freeze drying, moisture is removed by sublimation of water with internal heating of frozen material in vacuum. In case of microwave-assisted freeze drying, microwave is used for internal heating. It has advantage over conventional freeze drying as microwave causes faster drying due to volumetric heating of frozen food (Orsat et al. 2017). In conventional freeze drying system, drying takes place layer by layer starting from the outside, whereas the microwave system generates heat within the product and sublimation takes place in the complete product. It has been tested successfully using fruits and vegetables including strawberry, raspberry, banana, and radish. However, design and development of an industrial-scale microwave freeze drying system have many challenges. The design faults can lead to plasma in the drying chamber due to ionization of residual gases in the vacuum chamber. Plasma can cause burns on the product surface, damaging the final product. To avoid plasma discharges, there is a need for optimization of the process parameters chamber pressure and microwave power. Due to these issues, the system is in continuous development stage for scale-up and commercial usage.

5.6 Future of Applications of Microwave

The microwave processing of foods at industrial scale using the volumetric dielectric heating as a specific requirement will continue to grow with advancement of technological capabilities (Guo et al. 2017). The possibility of using microwave in innovative ways and exploring novel applications will remain an exciting area of research (Table 1). It has immense scope for development and will attain a strong presence in the food processing industry in the years to come. A substantial amount of research is still required to use it in combination with other processing technologies. The challenging tasks of high cost of equipment manufacturing and adherence to safety measures are some of the current issues that sooner or later will possibly be addressed.

6 Future Scopes for Radiofrequency Processing of Food

The frequency of electromagnetic spectrum in range of 1–300 MHz is termed as RF and selected frequencies, namely 13.56, 27.12, and 40.68 MHz, are permitted for ISM applications. In RF heating food is placed between two electrodes wherein the high-frequency alternating electric field causes polarization of the molecules (Huang et al. 2016). As the molecules rotate to align along the electric field, friction takes place resulting in heating of the food material. RF heating is another alternative to conventional thermal treatment in which electromagnetic energy is transferred directly to the product. RF heating represents considerable potential for additional research and novel applications in food processing industry. Compared with microwave, RF has longer wavelength and can penetrate deeper into the food products resulting in more even heating (Altemimi et al. 2019). RF works well with large quantities of food material having high ionic conductivity, and microwave works well with smaller quantities of food material with dipolar characteristics. Moreover, because of the capacity of RF power to penetrate deeper into foods than conventional microwaves, the heat is generated inside the food and distributed evenly. The construction of a large-scale RF heater is comparatively easier and the final product quality is also reported to be better. Another advantage of this eco-friendly technology is its higher energy use efficiency, as there is no heat loss in the process. Current industrial applications of RF heating include blanching, baking, thawing/defrosting, and drying of food materials. However, its industrial use in pasteurization and sterilization is limited and needs to be developed further.

6.1 Improvement of RF-H Uniformity

The industrial applications of RF heating are limited due to nonuniform heating in many food items. The product and packaging material can be affected due to uneven temperature in heating chamber. There are many other factors associated with this issue including dielectric properties of food, distance between the product and electrode, chemical properties of the medium, and equipment design (Table 1). To address the dielectric heating issues, methods need to be explored including use of hot air, hot water, or saline water in the medium. It is also possible to try using pulse mode of RF supply to improve uniformity in heating. The heating uniformity can also be improved by changing the geometry of containers. For example, insect disinfestation of raisins was successfully achieved by rounding the corners of the containers and reducing sharp edges on packets (Alfaifi et al. 2016). The configuration of electrodes can be modified to address heat uniformity issues. A reduction in the length of the electrode in container was also shown to improve heating uniformity. Computer simulations have been used for achieving uniform heating of raisins treated by RF-H.

6.2 Future Research on RF Heating Applications in Food

RF heating is promising alternative for thermal heating as it provides better-quality food products with lesser energy consumption. However, most of reported studies were performed using laboratory-scale equipment. There is a need to develop scaled-up process for commercial operations. The challenges of RF heating applications in fresh food processing are to ensure heating uniformity and reduction in equipment cost. Ensuring successful application of RF for fresh food processing in future would require extensive research and innovations. Computer simulation models need to be developed for better understanding of the electromagnetic field distribution (Jiao et al. 2018). This will help to address RF heating uniformity issue by adjustment of parameters of RF system for different food materials. Effective application of RF for blanching and sterilization will require greater R&D efforts. The biological effects of RF heating on live cells, microorganisms, as well as chemical effects on proteins, enzymes, and carbohydrates are not well understood. In addition, the applicability of different RF frequency to different food applications is a vast area to be explored for specific benefits. For example, 40.68 MHz can be suitable for blanching of fruits and vegetables, whereas 13.56 MHz is suitable for thawing/defrosting of frozen foods. Another important area is RF-assisted heating applications, where RF heating can be used for heating water and steam generated can be used for blanching, cooking, or sterilization (Figure 2).

7 Current Problems and Future Prospects of Tetrahertz (THz) Technology

The electromagnetic spectrum in the range of 0.3–3 THz has unique properties suitable for various applications including inspection of packaged good and contamination detection in food and water. THz spectroscopy is a safe, nondestructive, noninvasive, and convenient method for food testing and analysis (Kawase 2012). It has immense potential for detection of melamine, pesticides, antibiotics in food and dairy, and honey adulteration. However, THz spectroscopy instrument is costly due to high cost of THz source and detector. Due to the high cost, the majority of the THz systems are used in research and academic areas. Further research and development with less costly source and detector can improve the cost factor of THz system. Increase in demand may lower the cost due to higher volume of manufacturing. Another hurdle is the strong absorption of THz radiation by water present in food. Modern THz systems are developed to solve this problem by using higher radiant powers. However, there is still difficulty in identification of compounds with high degree of hydration. Another challenge is the scattering effect due to nonhomogeneous food of irregular shape and size, which affect the monitoring process. The penetration power of THz reduces with thickness of food sample. The liquid foods worsen the problem further due to limited penetration in water. In addition, another difficulty to work on is the labor-intensive optimization of process for inspection of food material. Recently, significant progress has been made in THz technology and THz imaging research with better limit of detection and higher accuracy (Afsah-Hejri et al. 2019). This imaging technology needs to be developed further for commercial use. Further studies are needed for increasing its application base in food monitoring and inspection.

8 Regulations for Use of EM Spectrum

The international bodies including IAEA, FAO, and WTO have jointly laid out policies, harmonized standards, guidelines, and procedures for safe adoption of new technologies worldwide. The Codex Alimentarius Commission (CAC) is the body responsible for implementation of the Joint FAO/WHO Food Standards Program. The Joint FAO of the United Nations and IAEA Division of Nuclear Techniques in Food and Agriculture (the "JointFAO/IAEADivision") supports and implements activities related to the improvement of food safety and control systems. Its activities are therefore closely related to the work of the CAC. The Commission meets in regular session once a year alternating between Geneva and Rome. The ISO 14470:2011 standard specifies requirements for the development, validation, and routine control of the process of irradiation using ionizing radiation for the treatment of food and establishes guidelines for meeting the requirements. This was last reviewed and confirmed in 2018. The ISO/ASTM 51900:2009 (Reapproved 2017) standard guide for dosimetry in radiation research on food and agricultural products was published in January 2018 and is intended to provide direction on dosimetry for experiments in food and agricultural research and on the reporting of dosimetry results.

In the United States, the title 21 of Code of Federal Regulations (CFR) of US-FDA in part 179 includes the general and permanent rules for use of electromagnetic spectrum for food

Table 2 USFDA federal rules pertaining to use of electromagnetic spectrum in food processing.

21 CFR Part 179	Irradiation in the production, processing, and handling of food
Section 179.21	Sources of radiation used for inspection of food, for inspection of packaged food, and for controlling food processing
Section 179.25	General provisions for food irradiation
Section 179.26	Ionizing radiation for the treatment of food
Section 179.30	Radiofrequency radiation for the heating of food, including microwave frequencies
Section 179.39	Ultraviolet radiation for the processing and treatment of food
Section 179.41	Pulsed light for the treatment of food
Section 179.45	Packaging materials for use during the irradiation of prepacked foods

applications (Table 2). The different sections included the rules for use of ionizing radiation, UV, radiofrequencies, and pulsed light for food processing.

9 Conclusion and Outlook

Electromagnetic spectrum applications encompass all aspects of food science and are slowly finding use in niche areas of application. The future challenges are manifold as establishing a newer technology is difficult starting from equipment design and manufacturing and the equipment are generally costlier. It becomes very difficult to bring in new technology replacing an established practice. Thus researchers and engineers need to work harder for coming up with appropriate solutions to make new technologies commercially successful. Many types of equipment successful in lab scale are finding technological challenges during scale-up to industrial level. In years to come, the similar trend will continue and many new applications of electromagnetic spectrum will replace conventional processing methods providing distinct advantages to the growing food industry. The γ radiation, X-ray, and UV processing methods are already popular and accepted as safer nonthermal methods for food preservation for many types of food items. With inputs from R&D sector, these technologies will find new applications during years to come. The microwave and RF heating have the potential to replace the conventional heating in many processes or will be used in combination due to its advantage of volumetric heating. However, to realize the potential of electromagnetic spectrum, it is essential to reduce the cost of new equipment and overcome the barrier of extra economic burden on existing infrastructure. EM technologies also have the bigger goal to be competitive in terms of capital investment compared with existing alternatives. Another important consideration would be the economic burden for low-income countries for high-end novel technologies. For low and middle-income countries, special considerations need to be made taking into account the limited infrastructure support available locally. The emphasis is underway to qualify these equipment to be economical in terms of energy usage and environmental friendliness.

The development of portable and movable devices employing EM spectrum will help in processing of agricultural produce near the farm and help in uplifting regional economy in low and middle-income areas. To meet the sustainability challenges, the emphasis needs to be focused on integration of renewable energy sources for operation and maintenance of food processing machines. R&D in food processing with maximum utilization of natural resources and recycling of raw materials will be very essential for management of resources and avoidance of losses due to wastage. The application of electromagnetic radiation in the area of technology development for food processing although faces several challenges has a promising future for enhancing global food security.

References

Aboud, S.A., Altemimi, A.B., Al-Hilphy, A.R.S. et al. (2019). A comprehensive review on infrared heating applications in food processing. *Molecules* 24: 4125.

Afsah-Hejri, L., Hajeb, P., Ara, P. et al. (2019). A comprehensive review on food applications of terahertz spectroscopy and imaging. *Comprehensive Reviews in Food Science and Food Safety* 18 (5): 1563–1621.

Alfaifi, B., Tang, J., Rasco, B. et al. (2016). Computer simulation analyses to improve radio frequency (RF) heating uniformity in dried fruits for insect control. *Innovative Food Science and Emerging Technologies* 37: 125–137.

Altemimi, A., Aziz, S.N., Al-Hilphy, A.R.S. et al. (2019). Critical review of radio-frequency (RF) heating applications in food processing. *Food Quality and Safety* 3: 81–91.

Bhatta, S., Janezic, T.S., and Ratti, C. (2020). Freeze-drying of plant-based foods. *Foods* 9 (1): 2–22.

Carreño, I. (2018). International standards and regulation on food irradiation. In: *Food Irradiation Technologies, Concepts, Applications and Outcomes* (eds. I.C.F.R. Ferreira, A.L. Antonio, S.C. Verde and T. Page), 5–24. London, UK: Royal Society of Chemistry.

Craven, E., Mittendorfer, J., and Howard, C. (2018). Software for food irradiation simulation and equipment validation. In: *Food Irradiation Technologies, Concepts, Applications and Outcomes* (eds. I.C.F.R. Ferreira, A.L. Antonio, S.C. Verde and T. Page), 105–122. London, UK: Royal Society of Chemistry.

Dash, A., Varma, R.N., Ram, R. et al. (2009). Fabrication of Cesium-137 brachytherapy sources using vitrification technology. *Cancer Biotherapy and Radiopharmaceuticals* 24: 489–502.

Delorme, M.M., Guimarães, J.T., Coutinho, N.M. et al. (2020). Ultraviolet radiation: an interesting technology to preserve quality and safety of milk and dairy foods. *Trends in Food Science and Technology* 102: 146–154.

Deng, L.Z., Tao, Y., Mujumdar, A.S. et al. (2020). Recent advances in non-thermal decontamination technologies for microorganisms and mycotoxins in low-moisture foods. *Trends in Food Science and Technology* 106: 104–112.

Dixit, Y., Casado-Gavalda, M.P., Cama-Moncunill, R. et al. (2017). Developments and challenges in online NIR spectroscopy for meat processing. *Comprehensive Reviews in Food Science and Food Safety* 16 (6): 1172–1187.

Dourado, C., Pinto, C., Barba, F.J. et al. (2019). Innovative non-thermal technologies affecting potato tuber and fried potato quality. *Trends in Food Science and Technology* 88: 274–289.

Eustice, R.F. (2018). Global status and commercial applications of food irradiation. In: *Food Irradiation Technologies, Concepts, Applications and Outcomes* (eds. I.C.F.R. Ferreira, A.L. Antonio and S.C. Verde), 397–493. London, UK: Royal Society of Chemistry.

Fan, X., Huang, R., and Chen, H. (2017). Application of ultraviolet-C technology for surface decontamination of fresh produce. *Trends in Food Science and Technology* 70: 9–19.

Farkas, J. and Mohácsi-Farkas, C. (2011). History and future of food irradiation. *Trends in Food Science and Technology* 22 (2): 121–126.

Guerrero-Beltrán, J.A. and Barbosa-Cánovas, G.V. (2004). Review: advantages and limitations on processing foods by UV light. *Food Science and Technology International* 10 (3): 0137–11.

Guo, Q., Sun, D., Cheng, J. et al. (2017). Microwave processing techniques and their recent applications in the food industry. *Trends in Food Science and Technology* 67: 236–247.

Guo, C., Mujumdar, A.S., and Zhang, M. (2019). New development in radio frequency heating for fresh food processing: a review. *Food Engineering Reviews* 11: 29–43.

Huang, Z., Marra, F., and Wang, S. (2016). A novel strategy for improving radiofrequency heating uniformity of dry food products using computational modeling. *Innovative Food Science and Emerging Technologies* 34: 100–111.

James, C., Barlow, K.E., James, S.J. et al. (2006). The influence of processing and product factors on the quality of microwave pre-cooked bacon. *Journal of Food Engineering* 77 (4): 835–843.

Jiao, Y., Tang, J., Wang, Y. et al. (2018). Radio-frequency applications for food processing and safety. *Annual Review of Food Science and Technology* 9: 105–127.

Kawase, M. (2012). Application of terahertz waves to food science. *Food Science and Technology Research* 18 (5): 601–609.

Koutchma, T. (2008). UV light for processing foods. *Ozone Science and Engineering* 30: 1–6.

Koutchma, T. (2009). Advances in ultraviolet light technology for non-thermal processing of liquid foods. *Food and Bioprocess Technology* 2: 138–155.

Krishnamurthy, K., Khurana, H.K., Jun, S. et al. (2008). Infrared heating in food processing: an overview. *Comprehensive Reviews in Food Science and Food Safety* 7: 2–13.

Manzocco, L. and Nicoli, M.C. (2014). Surface processing: existing and potential applications of ultraviolet light. *Critical Reviews in Food Science and Nutrition* 55 (4): 469–484.

Mishra, B.B., Gautam, S., Chander, R. et al. (2015) Characterization of nutritional, organoleptic and functional properties of intermediate moisture shelf stable ready-to-eat *Carica papaya* cubes. *Food Bioscience* 10: 69–79.

Mudgett, R. (1988). Electromagnetic energy and food processing. *Journal of Microwave Power and Electromagnetic Energy* 23 (4): 225–230.

Orsat, V., Raghavan, G.S.V., and Krishnaswamy, K. (2017). Microwave technology for food processing: an overview of current and future applications. In: *The Microwave Processing of Foods* (eds. M. Regier, K. Knoerzer and H. Schubert), 100–116. Cambridge, US: Woodhead.

Pan, Y., Sun, D.-W., and Han, Z. (2017). Applications of electromagnetic fields for nonthermal inactivation of microorganisms in foods: an overview. *Trends in Food Science and Technology* 64: 13–22.

Rastogi, N.K. (2012). Recent trends and developments in infrared heating in food processing. *Critical Reviews in Food Science and Nutrition* 52 (9): 737–760.

Rosenthal, I. (1992). *Electromagnetic Radiations in Food Science, Advanced Series in Agricultural Sciences, 19*. Berlin, Germany: Springer-Verlag.

Shah, N.N.A.K., Shamsudin, R., Rahman, R.A. et al. (2016). Fruit juice production using ultraviolet pasteurization: a review. *Beverages* 2: 22.

Wray, D. and Ramaswamy, H.S. (2015). Novel concepts in microwave drying of foods. *Drying Technology* 33 (7): 769–783.

Yadav, G., Gupta, N., Sood, M. et al. (2020). Infrared heating and its application in food processing. *The Pharma Innovation Journal* 9 (2): 142–151.

Index

Note: page numbers in *italics* refer to figures, those in **bold** refer to tables.

Electromagnetic Technologies in Food Science, First Edition. Edited by Vicente M. Gómez-López and Rajeev Bhat.

e

f

t

y

z

Chapter 1

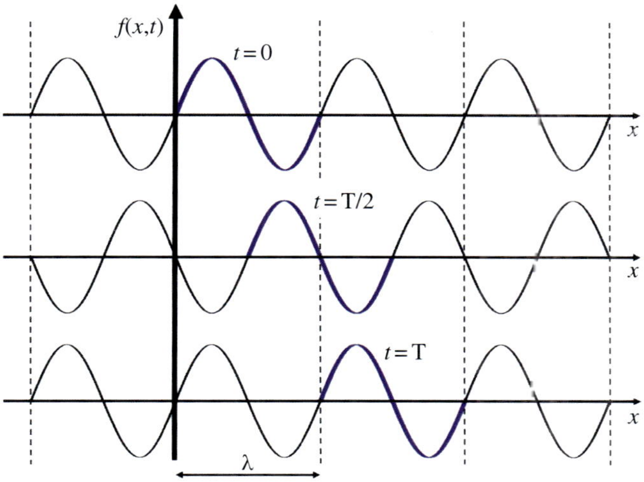

Figure 1 Example of elongation $f(x,t)$ of a harmonic wave along a spatial coordinate x at three different times.

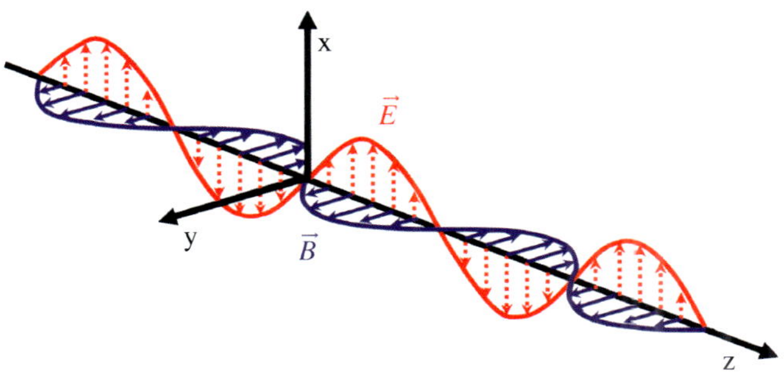

Figure 3 Snapshot of a plane harmonic three-dimensional EM wave according to Eq. (4).

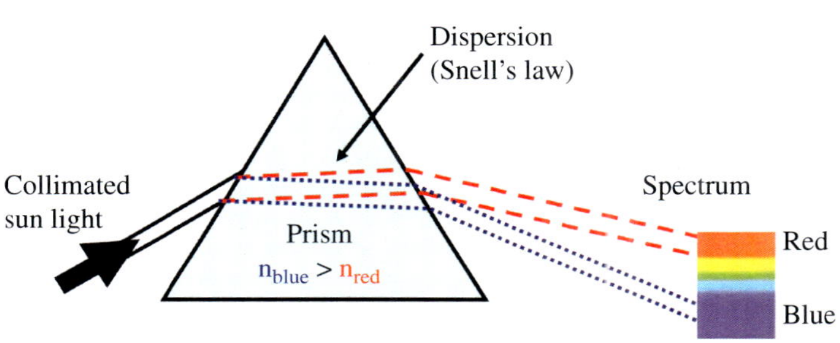

Figure 4 Light passing through a prism is deflected into a spectrum due to the wavelength (frequency)-dependent index of refraction.

Electromagnetic Technologies in Food Science, First Edition. Edited by Vicente M. Gómez-López and Rajeev Bhat.

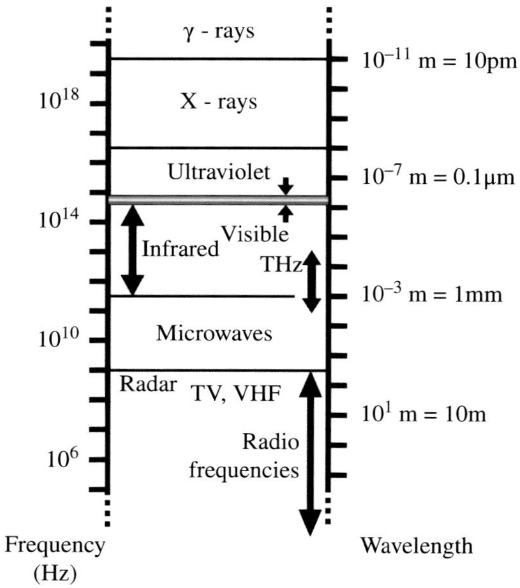

Figure 6 Various regions of the electromagnetic spectrum spanning more than 15 orders of magnitude in frequency or wavelength which are utilized in food science. The visible part of the spectrum is only a tiny part. The regions were mostly defined based on available sources and/or detectors.

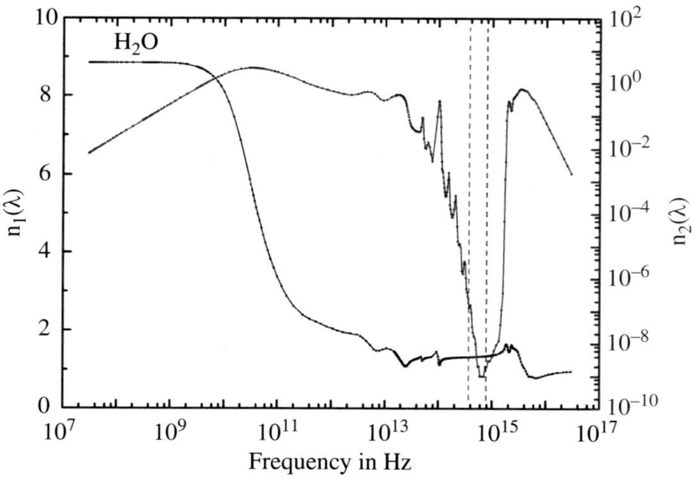

Figure 12 Index of refraction of pure water as a function of frequency. *Source:* Palik (1991).

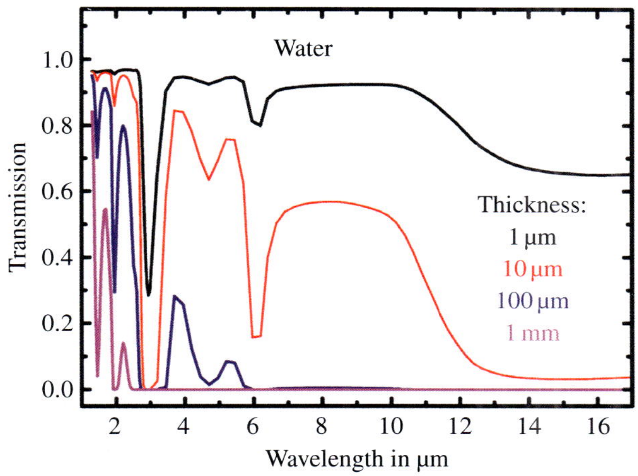

Figure 13 Transmission spectra of pure water in the near-infrared spectra range. Water very effectively absorbs IR radiation already within distances of less than 1 mm.

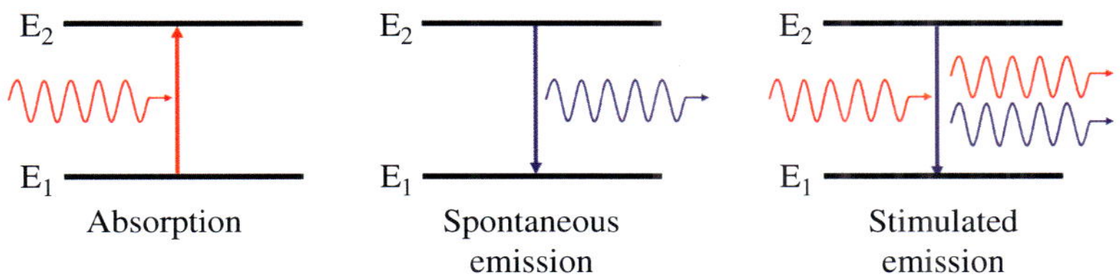

Figure 14 The three elementary microscopic interaction processes of EM radiation with atoms are absorption, spontaneous emission, and stimulated emission.

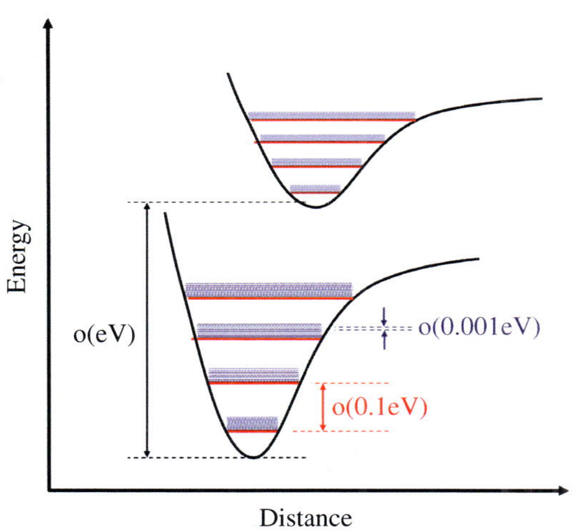

Figure 16 Schematic overview of energy levels of diatomic molecules: for each of the two drawn electronic levels with minima there are many discrete vibrational levels (with typical distances of 0.1 eV) and on top of each vibrational level, there are many discrete rotational energy levels (with typical distances of 0.001 eV).

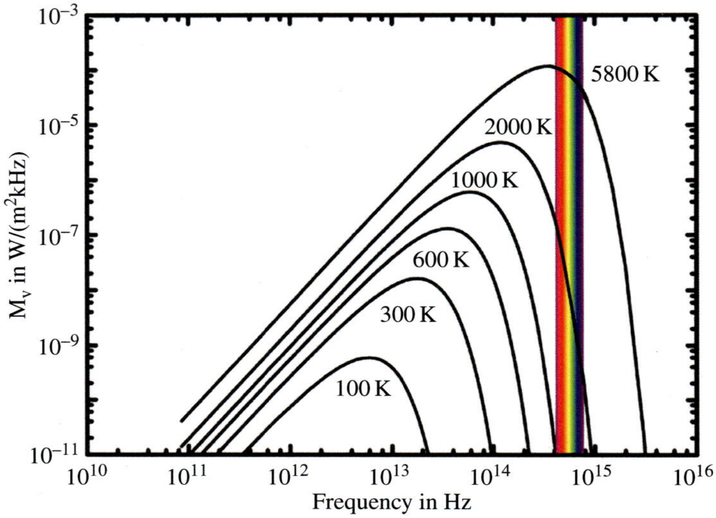

Figure 18 Blackbody radiation excitance spectra for object temperatures varying between 100 and 5800 K.

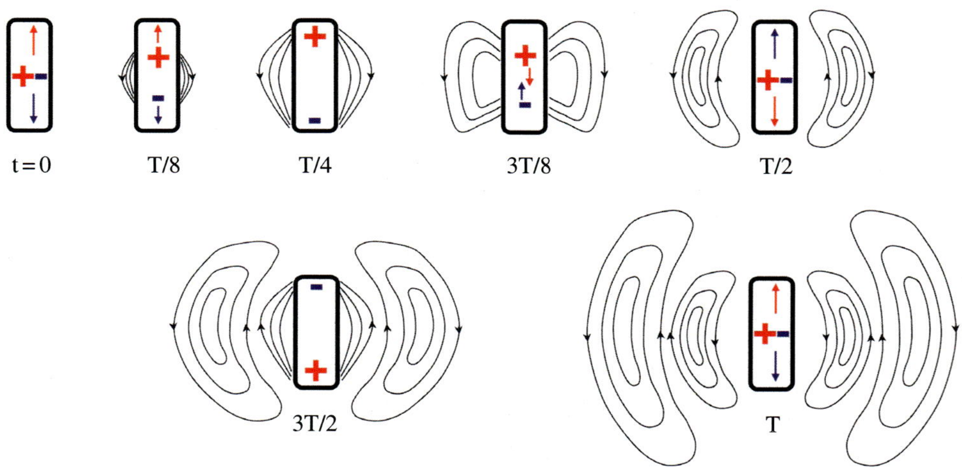

Figure 19 Schematic visualization of generation of RF EM waves by a dipole antenna. The magnetic fields are omitted here.

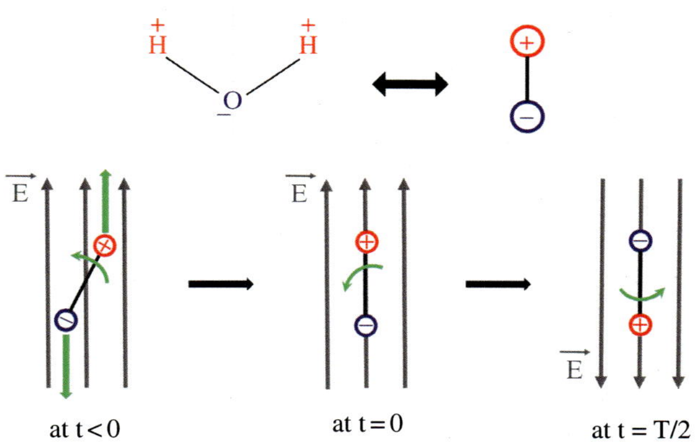

Figure 20 Water molecules are electric dipoles (top). When being in an electric field (bottom), they orient themselves parallel to the field.

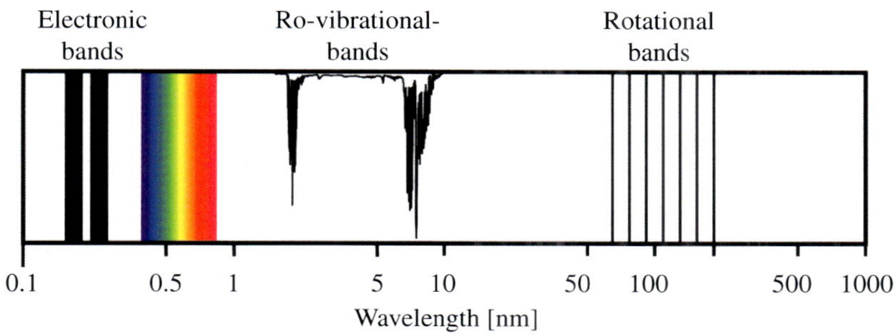

Figure 21 Schematic absorption spectrum of a hetero-nuclear diatomic molecule.

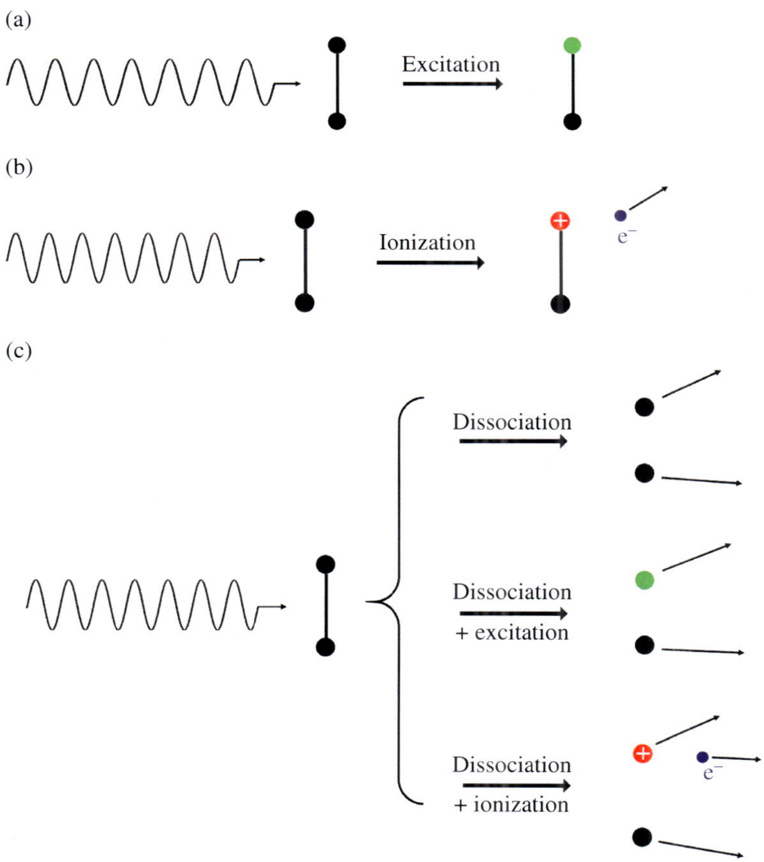

Figure 22 Three possible excitation processes of molecules by VIS or UV radiation.

Chapter 3

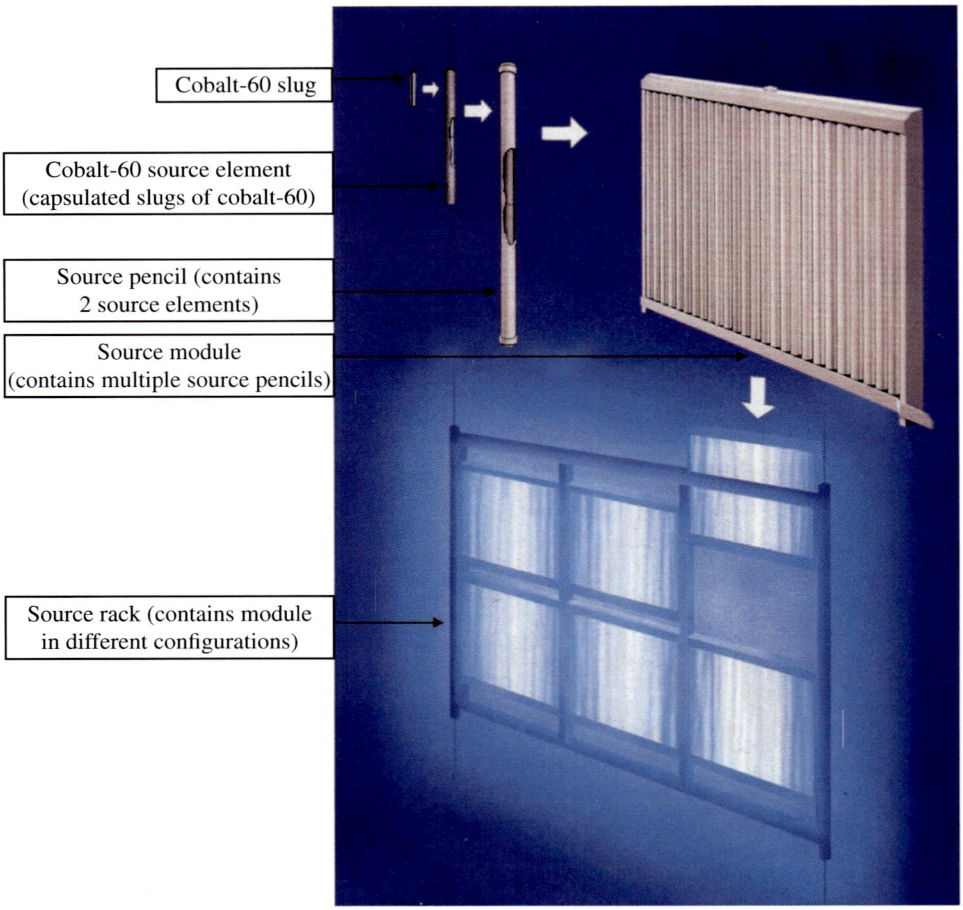

Figure 7 Illustration of a typical cobalt source rack built from slugs, pencils, and modules. *Source:* Courtesy of MDS Nordion.

Chapter 5

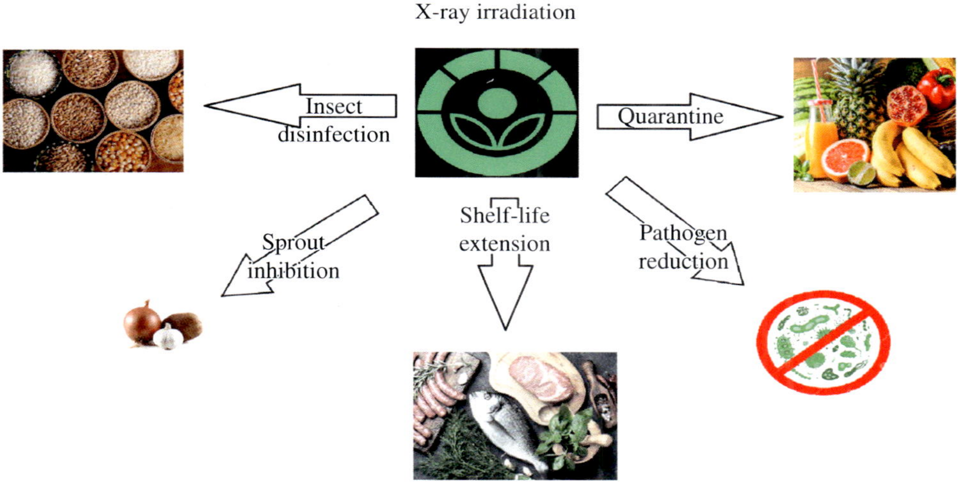

Figure 1 Applications of X-rays.

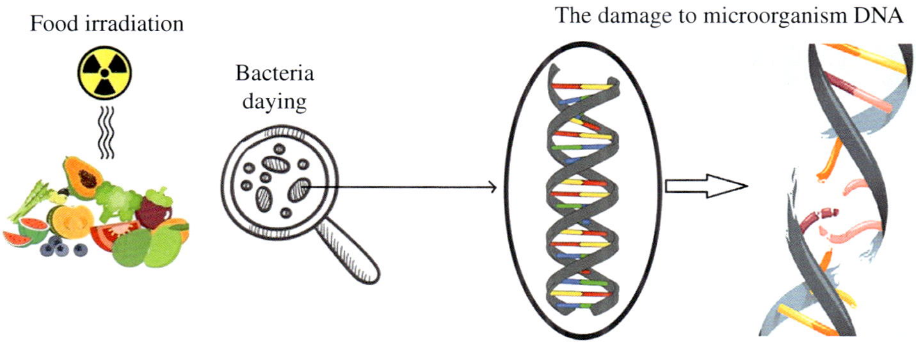

Figure 2 Mechanism of action of X-rays.

Chapter 6

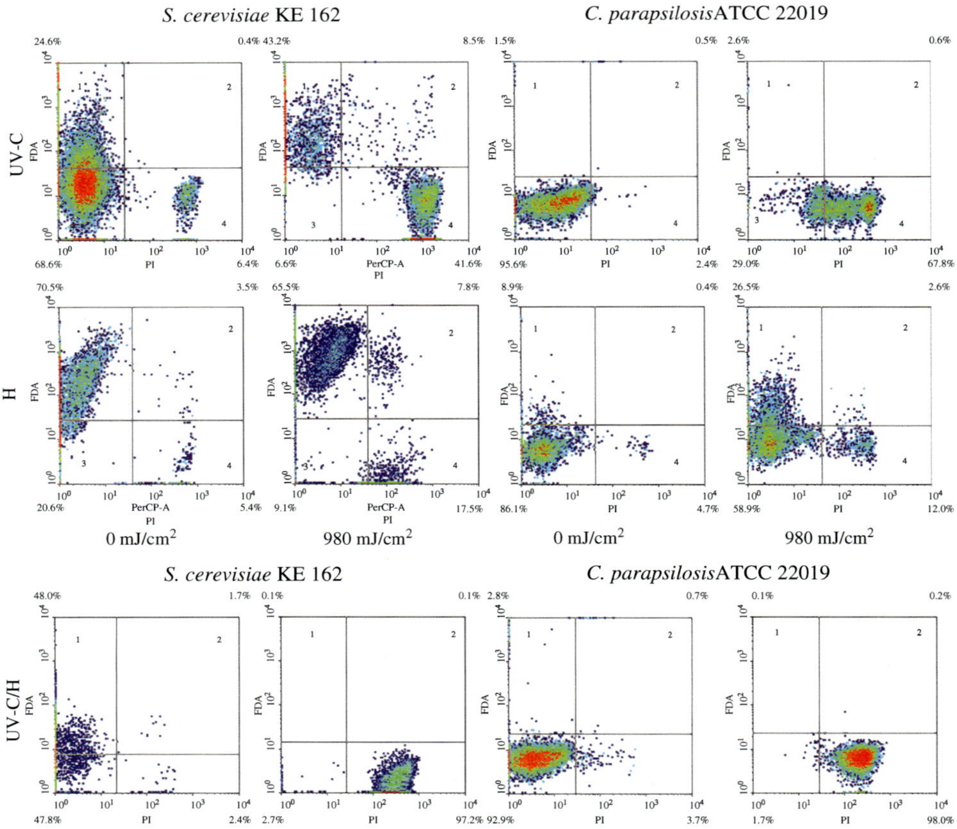

Figure 5 Fluorescence density plots of *S. cerevisiae* KE 162 and *C. parapsilosis* ATCC 22019 in a carrot–orange juice blend (pH: 4.2; 7.8°Brix; 7667 NTU) or not (control; 0 mJ cm^{-2}) by UV-C (985 mJ cm^{-2}; <25 °C); H (50 °C) and UV-C assisted by mild heat (UV-C/H). The percentages of microbial populations that fall in each gate are displayed in the four edges of each plot.

Chapter 7

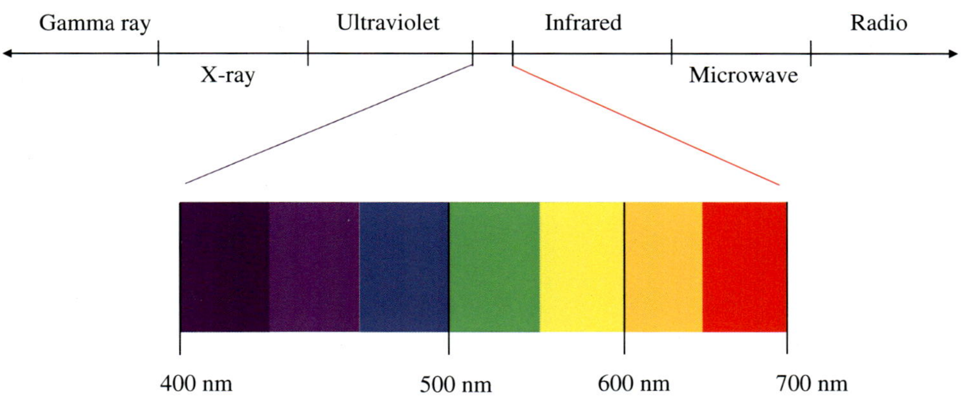

Figure 1 Visible light colors/wavelength within the electromagnetic spectrum.

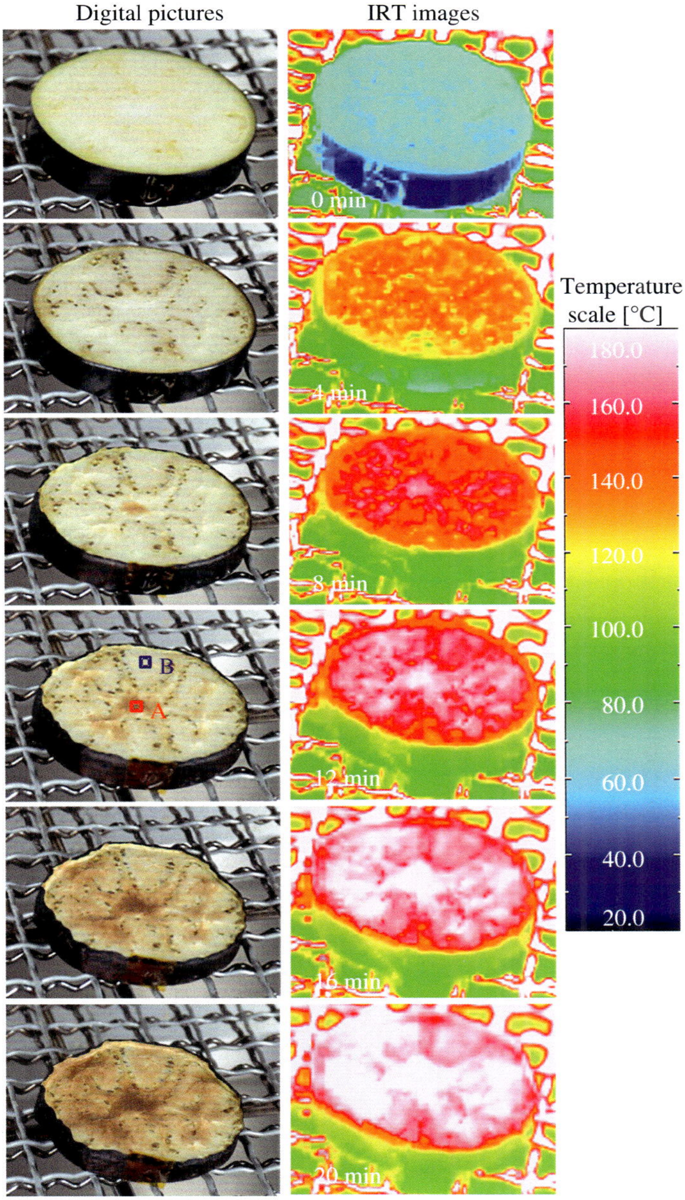

Figure 5 Temperature distribution on the upper surface of eggplant during FIR broiling, and digital pictures captured at several time intervals (infrared thermography images in °C). *Source:* Llave et al. (2017).

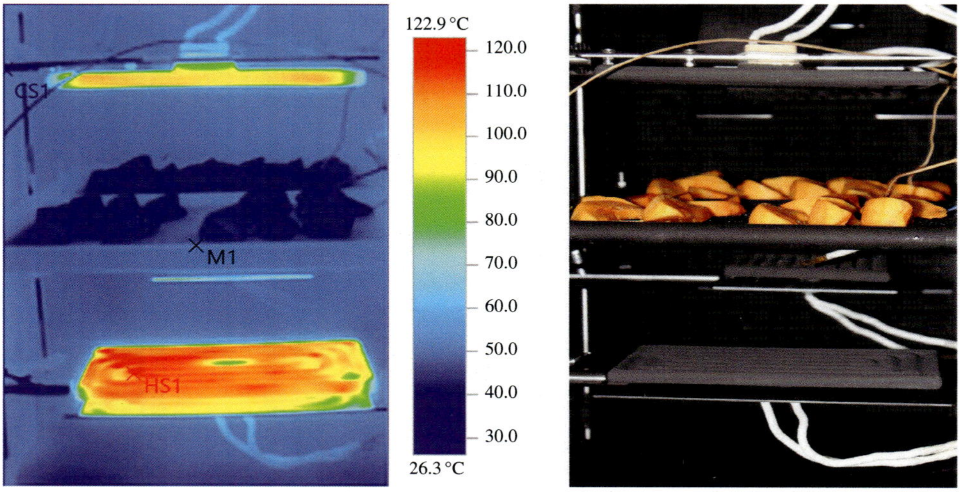

Figure 6 Infrared radiation temperature (IRT) image showing the final temperature distribution in the samples and heater surface (*left*), and an internal view of the oven with samples (*right*) of kaki chips drying in a hybrid convection oven with novel IR heaters.

Chapter 10

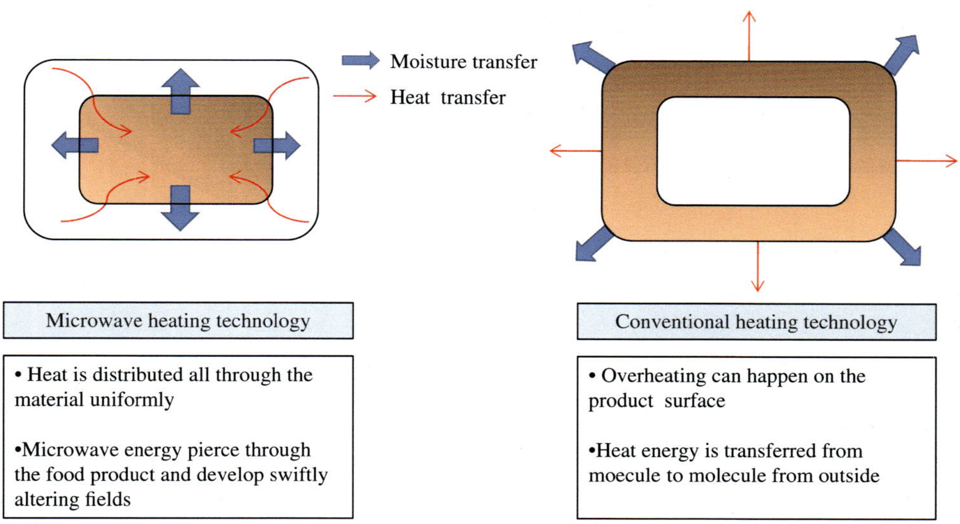

Figure 1 Microwave versus conventional heating technique.

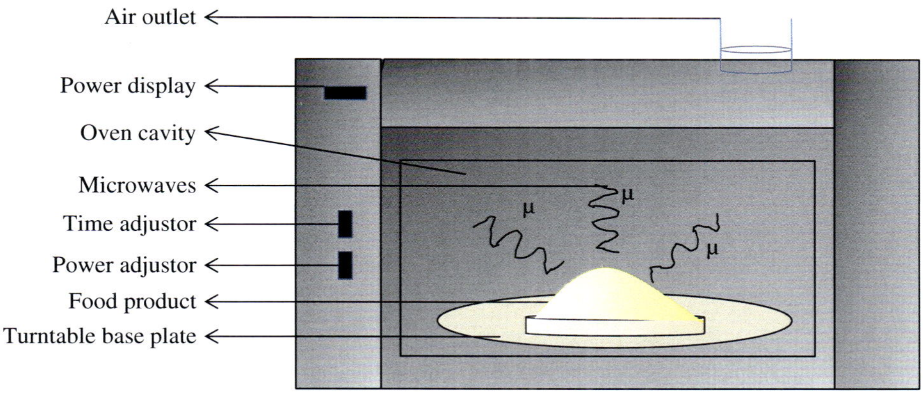

Figure 2 Domestic oven and important parts.

Chapter 11

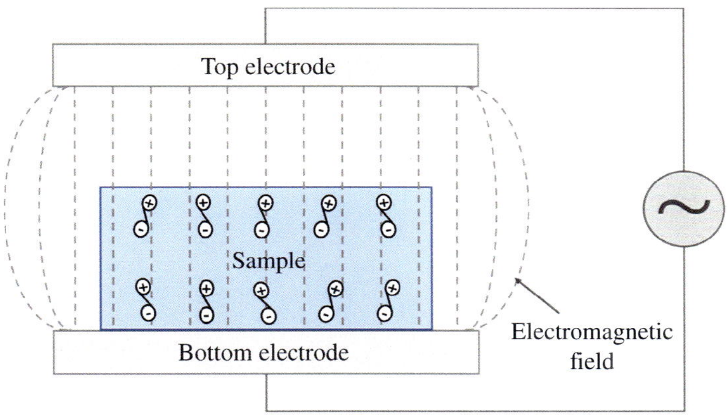

Figure 1 The simplified schematic diagram of and RF heating system. *Source:* Orfeuil (1987).

Chapter 13

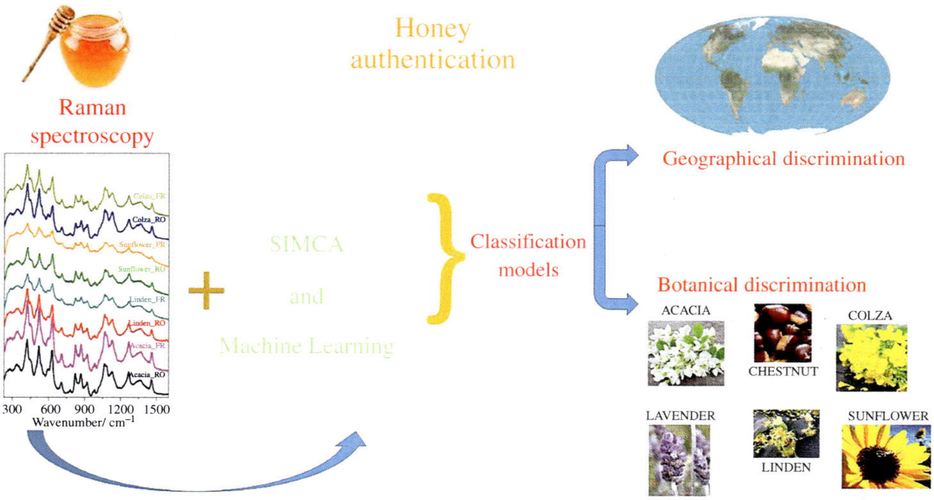

Figure 2 Schematic representation of the used approach for honey classification.

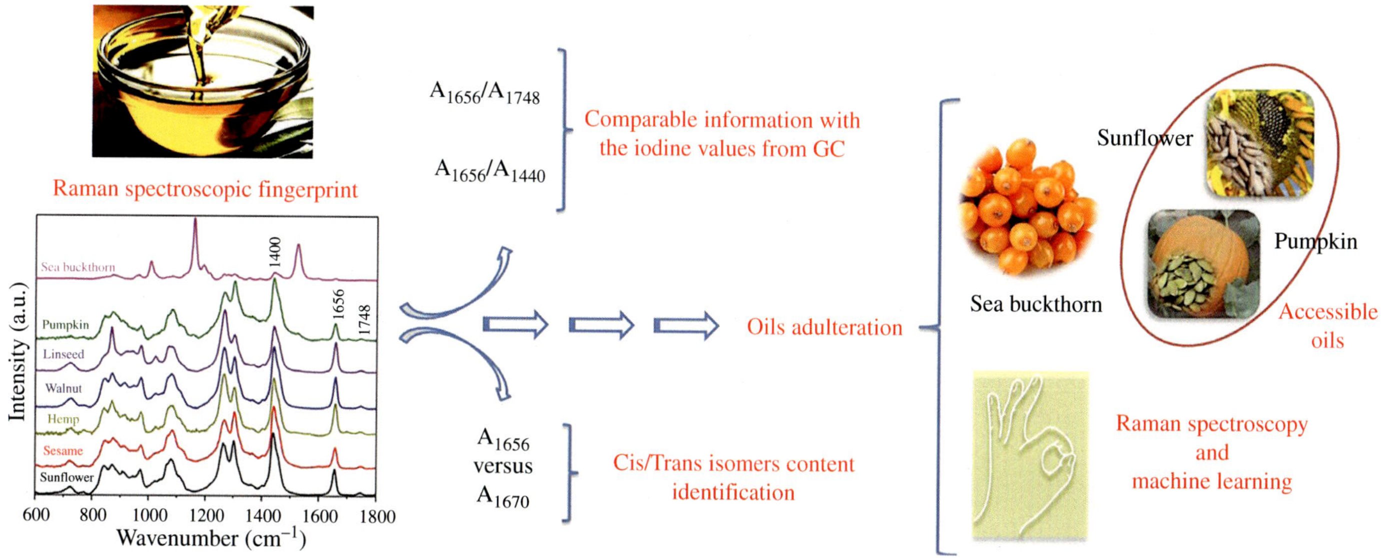

Figure 4 Schematic representation of some Raman approaches employed for the oils investigations.

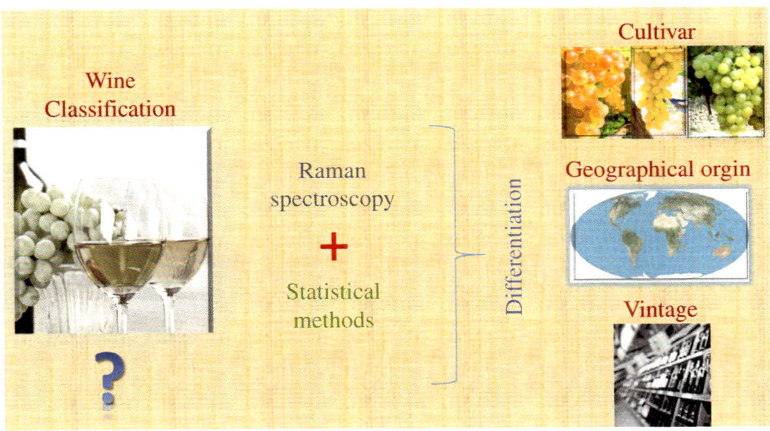

Figure 6 Schematic representation of the performed wine differentiation.

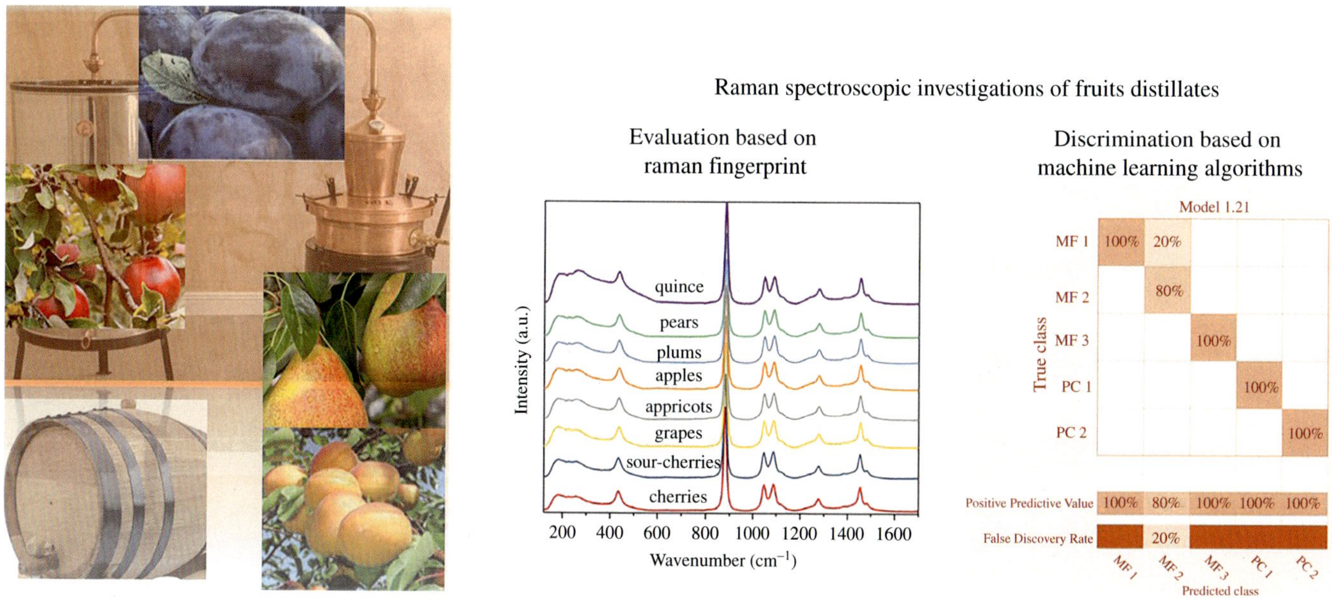

Figure 7 Schematic representation of fruit spirits Raman spectroscopic investigation. *Source:* Adapted from Quinlan (1986), Usoro (2015), Weinberger and Saul (2009), Dietterich (2000).

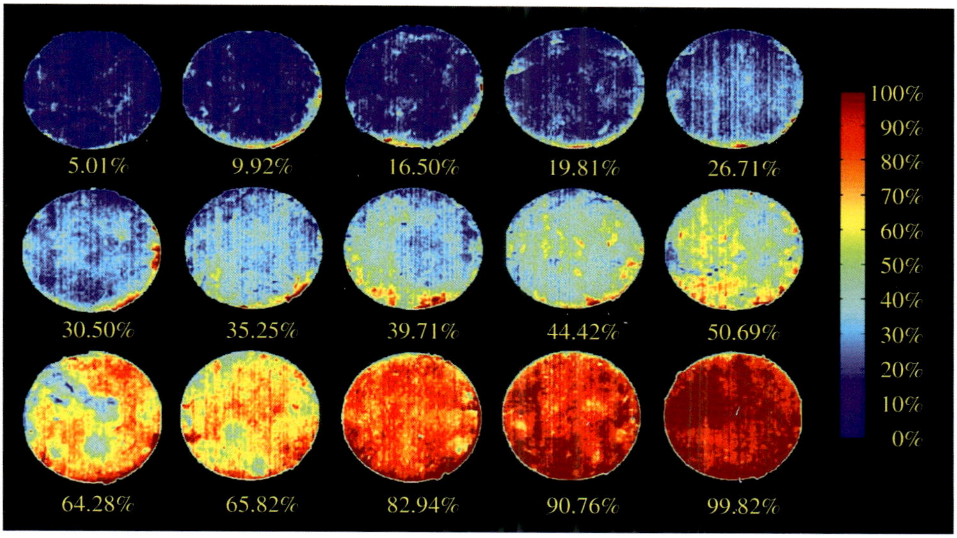

Chapter 14

Figure 3 Visualization of adulteration in fresh minced beef samples. *Source:* Zhao et al. (2019).

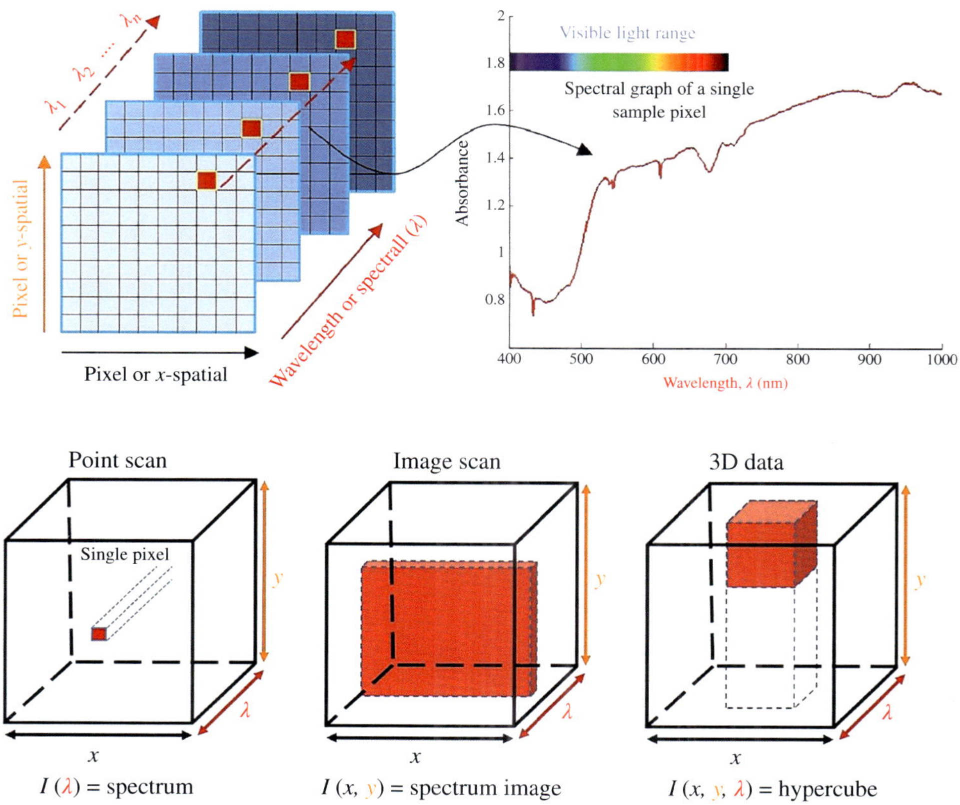

Figure 4 Visible images as a function of wavelength and data structures. *Source:* Munir et al. (2018).

Figure 6 Visible images of fried potato chips: (a) normal potato chip, (b) potato chip with mild acrylamide formation, (c) potato chip with moderate acrylamide formation, and (d) potato chip with severe acrylamide formation. *Source:* Yadav et al. (2018).

Chapter 15

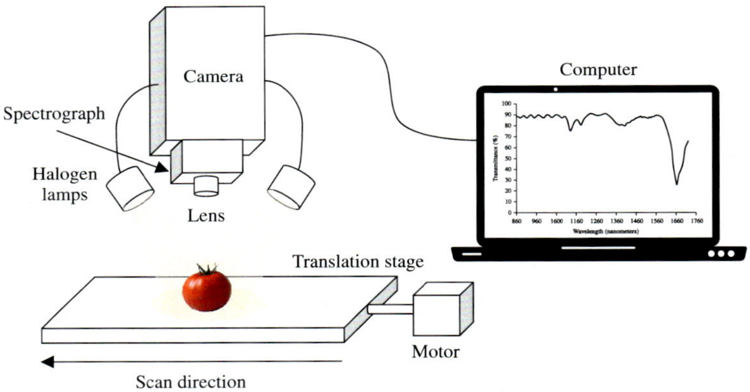

Figure 2 Line-scanning HSI equipment.

Chapter 16

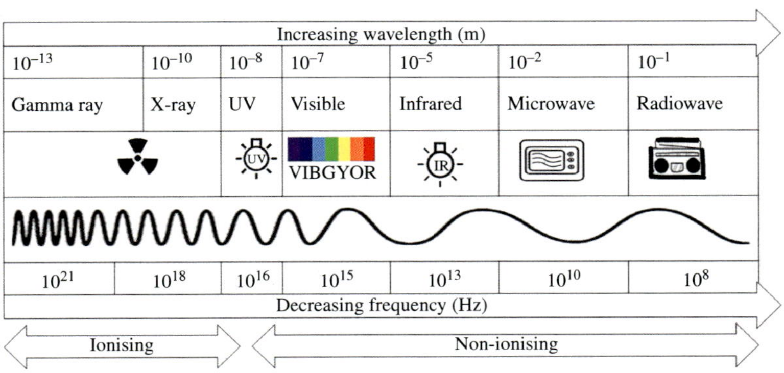

Figure 1 The electromagnetic spectrum composition with wavelength, frequency, and ionizing characteristics.